P9-AFH-740

The Herpesviruses

Volume 2

THE VIRUSES

Series Editors
HEINZ FRAENKEL-CONRAT, *University of California*
Berkeley, California
ROBERT R. WAGNER, *University of Virginia School of Medicine*
Charlottesville, Virginia

THE HERPESVIRUSES, Volumes 1, 2, 3, and 4
Edited by Bernard Roizman

THE REOVIRIDAE
Edited by Wolfgang K. Joklik

THE PARVOVIRUSES
Edited by Kenneth I. Berns

The Herpesviruses

Volume 2

Edited by
BERNARD ROIZMAN
University of Chicago
Chicago, Illinois

PLENUM PRESS • NEW YORK AND LONDON

Library of Congress Cataloging in Publication Data

Main entry under title:

The Herpesviruses.

Includes bibliographical references and index.
1. Herpesvirus diseases. 2. Herpesviridae. I. Roizman, Bernard, 1929–
[DNLM: 1. Herpesviridae. QW 165.5.H3 H5637]
RC147.H6H57 1982 616.9′25 82-15034
ISBN 0-306-40922-4 (v. 1)
ISBN 0-306-41083-4 (v. 2)

A Division of Plenum Publishing Corporation
233 Spring Street, New York, N.Y. 10013

Printed in the United States of America

Contributors

M. D. Daniel, New England Regional Primate Research Center, Harvard Medical School, Southborough, Massachusetts 01772

G. Darai, Institut für Medizinishe Virologie, Universität Heidelberg, Heidelberg, Germany

R. M. Flügel, Institut für Virusforschung, Heidelberg, Germany

H. Gelderblom, Robert Koch Institut des Bundesgesundheitsamtes, West Berlin, Germany

Glenn A. Gentry, Department of Microbiology, University of Mississippi Medical Center, Jackson, Mississippi 39216

Allan Granoff, Division of Virology, St. Jude Children's Research Hospital, Memphis, Tennessee 38101

Richard W. Hyman, Department of Microbiology and Cancer Research Center, The Pennsylvania State University College of Medicine, Hershey, Pennsylvania 17033

H.-G. Koch, Institut für Medizinische Virologie, Universität Heidelberg, Heidelberg, Germany

Hanns Ludwig, Institut für Virologie, Freie Universität Berlin, West Berlin, Germany

B. Norrild, Institute of Medical Microbiology, University of Copenhagen, Copenhagen, Denmark

Dennis J. O'Callaghan, Department of Microbiology, University of Mississippi Medical Center, Jackson, Mississippi 39216

G. Pauli, Institut für Virologie, Freie Universität Berlin, West Berlin, Germany

Charles C. Randall, Department of Microbiology, University of Mississippi Medical Center, Jackson, Mississippi 39216

Fred Rapp, Department of Microbiology and Cancer Research Center, The Pennsylvania State University College of Medicine, Hershey, Pennsylvania 17033

Mark F. Stinski, Department of Microbiology, College of Medicine, University of Iowa, Iowa City, Iowa 52242
Ken Wolf, National Fish Health Research Laboratory, U.S. Fish and Wildlife Service, Kearneysville, West Virginia 25430

Foreword

The first volume of the series entitled *Comprehensive Virology* was published in 1974 and the last is yet to appear. We noted in 1974 that virology as a discipline has passed through its descriptive and phenomenological phases and was joining the molecular biology revolution. The volumes published to date were meant to serve as an in-depth analysis and standard reference of the evolving field of virology. We felt that viruses as biological entities had to be considered in the context of the broader fields of molecular and cellular biology. In fact, we felt then, and feel even more strongly now, that viruses, being simpler biological models, could serve as valuable probes for investigating the biology of the far more complex host cell. During the decade-long compilation of a series of books like *Comprehensive Virology,* some of the coverage will obviously not remain up-to-date. The usual remedy to this aspect of science publishing is to produce a second edition. However, in view of the enormous increase in knowledge about viruses, we felt that a new approach was needed in covering virology in the 1980s and 1990s. Thus we decided to abandon the somewhat arbitrary subgrouping of the subject matter of *Comprehensive Virology* under the titles Reproduction, Structure and Assembly, Regulation and Genetics, Additional Topics, and Virus–Host Interactions. Instead we have organized a new series entitled *The Viruses.* This series will consist of individual volumes or groups of volumes, each to deal with a single virus family or group, each to be edited with full responsibility by an acknowledged authority on that topic, and each to cover all aspects of these viruses, ranging from physicochemistry to pathogenicity and ecology. Thus, over the next several years we plan to publish single volumes or multiple-volume sets devoted to each of the following virus families: Herpesviridae, Adenoviridae, Papovaviridae, Parvoviridae, Poxviridae, Reoviridae, Retroviridae, Picornaviridae, Togaviridae, Rhabdoviridae, Myxoviridae, and Paramyxoviridae, as well as hepatitis viruses, plant viruses, bacterial viruses, insect viruses, and perhaps other groups of viruses if and when they are deemed appropriate for comprehensive coverage and analysis.

This volume of THE VIRUSES is part of a set that will provide comprehensive coverage of herpesviruses. The editor of these books is Bernard Roizman. The Herpesviridae comprise a family of viruses widespread throughout the animal kingdom, many of which are extremely important pathogens. Diseases caused by herpesviruses are of ever-increasing significance as serious medical problems. In addition, research on the molecular biology, genetics, pathogenicity, and immunology of these complex viruses has in recent years undergone a veritable metamorphosis, which promises to continue for some years to come.

The first volume of this group dealt largely with viruses associated with malignancy in their natural or experimental hosts. The present volume covers the cytomegaloviruses, varicella-zoster virus, and bovine, equine, fish, reptilian, and amphibian herpes viruses.

Heinz Fraenkel-Conrat
Robert R. Wagner

Preface

Volume 1 of *Herpesviruses* appeared in December of 1982; it dealt with current nomenclature and classification of herpesviruses and with *Gamma herpesviridae*—the newest and most tantalizing of the various herpesvirus subfamilies. The viruses belonging to this subfamily that were discussed in Volume 1 included the Epstein–Barr virus, the virus associated with Marek's disease, and several herpesviruses isolated from New World monkeys. The organization of the topics by subfamilies is not, however, followed in the second volume. Rather, the editor (that is I) prevailed on Charles C. Randall and his disciples, Dennis J. O'Callaghan and Glenn Gentry who contributed much of what is known regarding the molecular biology of equine herpesviruses, and on Hanns Ludwig who, with his associates, has become the major contributor to our knowledge about bovine herpesviruses. Although the herpesviruses isolated from horses and bovines, respectively, belong to different subfamilies, it was not only convenient but also fortuitous to have each group discussed in one chapter. The same logic governed the inclusion of all amphibian herpesviruses in a single chapter written by Allan Granoff and of the fish and reptilian herpesviruses in another chapter by their main exponent—Ken Wolf. I am grateful to Fred Rapp and Mark F. Stinski for an extensive coverage of the structure, function and biology of human cytomegalovirus, to Richard W. Hyman for the chapter on varicella–zoster and again to Hanns Ludwig for bringing together the available knowledge on B virus.

Our knowledge concerning the more than 80 different herpesviruses is uneven. While one author could still cover authoritatively all that is known about the herpesviruses of fishes and reptiles, this is no longer possible for herpes simplex viruses and will no longer be possible for many others because of the exponentially increasing flow of information on just about every facet of the biology of these viruses. Whereas Volume 4 will be dedicated entirely to the Immunobiology of Human Herpesviruses, Volume 3 will consist of chapters exploring in depth specific aspects of the biology of individual and groups of herpesviruses, and, like Volume 2, will not have a unifying theme.

Editing, like writing, is a discipline one learns by making mistakes. The most profitable advice I can pass on to aspiring editors is to repeat Murphy's immortal words that if anything can go wrong, it will. Of the errors I attribute solely to myself, the most amusing is that Volume 1 credits me with a Ph.D. instead of an Sc.D.; the less amusing is the disappearance of my preface to Volume 1 between Chicago and the publisher's office in New York. As a consequence neither the intent of this series or acknowledgements appropriate to that volume appear in their proper place. The proverbial editorial license is most likely a myth; nevertheless it may not be too late to express at least some of the sentiments whose proper place would have been Volume 1.

Science is a systematic exploration not so much of the unknown as of the curious, the challenging, and of the observation that does not fit the conceptual rubric in which it has been placed. Among the virus families, few present the modern scholar with as much diversity, as many questions, puzzles, and ambiguities as the family *Herpesviridae.* The appreciation of both their significance in human and animal health and our ignorance concerning their biology is reflected in the ever increasing number of investigators focusing their research interests on these viruses.

It is characteristic of science that the facts of today become the foundation, the backdrop, and occasionally the ruins upon which the facts of tomorrow emerge. The *Herpesviruses* are intended to be a meaningful compilation of what we know, to help us interpret what we see today, and to assist us in designing the experiments of tomorrow. To dedicate the knowledge of the present to those who will contribute to our knowledge on herpesviruses in the future is superfluous; without them, all that is done today is of little value and of no consequences.

Bernard Roizman

Chicago
December 1982

Contents

Chapter 1

The Biology of Cytomegaloviruses

Fred Rapp

Chapter 2

Molecular Biology of Cytomegaloviruses

Mark F. Stinski

Chapter 3

Molecular Biology of Varicella–Zoster Virus

Richard W. Hyman

Chapter 4

Bovine Herpesviruses

Hanns Ludwig

Chapter 5

The Equine Herpesviruses

Dennis J. O'Callaghan, Glenn A. Gentry, and Charles C. Randall

Chapter 6

Biology and Properties of Fish and Reptilian Herpesviruses

Ken Wolf

Chapter 7

Amphibian Herpesviruses

Allan Granoff

Chapter 8

B Virus (*Herpesvirus simiae*)

Hanns Ludwig, G. Pauli, H. Gelderblom, G. Darai, H.-G. Koch, R.M. Flugel, B. Norrild, and M. D. Daniel

CHAPTER 1

The Biology of Cytomegaloviruses

FRED RAPP

I. PROPERTIES OF CYTOMEGALOVIRUSES

The initial isolations of human cytomegaloviruses (HCMVs) occurred in 1956 and 1957 (Rowe *et al.*, 1956; M. G. Smith, 1956; Weller *et al.*, 1957). Many investigators have since isolated CMVs or CMV-like agents from a variety of hosts including monkeys (Black *et al.*, 1963; Ablashi *et al.*, 1972; Asher *et al.*, 1974; Nigida *et al.*, 1979; Rangan and Chaiban, 1980), horses (Plummer and Waterson, 1963; Hsiung *et al.*, 1969), rats (Rabson *et al.*, 1969), squirrels (Diosi and Babusceac, 1970), pigs (Plowright *et al.*, 1976), mice (M. G. Smith, 1954; Kim *et al.*, 1975), guinea pigs (Jackson, 1920; Cole and Kuttner, 1926), and sheep (Hartley and Done, 1963).

As early as 1920, Leila Jackson reported the detection of "an intracellular protozoan parasite" in the salivary gland ducts of the guinea pig. This report was followed by a similar one (Cole and Kuttner, 1926) in which a filterable virus was detected in the submaxillary glands of guinea pigs. Cole and Kuttner recognized that the cells had characteristics of herpes simplex virus (HSV)-infected cells, although they were enlarged, which is not typical of HSV infection.

A. Physical Properties

The cytomegaloviruses (CMVs) are morphologically similar to other members of the herpesvirus group; they have an icosahedral capsid con-

FRED RAPP • Department of Microbiology and Cancer Research Center, The Pennsylvania State University College of Medicine, Hershey, Pennsylvania 17033.

figuration with 162 capsomeres, contain double-stranded linear DNA, are enveloped, and are therefore sensitive to ether and heat. The size of the virion varies among species, but ranges from 100 to 350 nm. The molecular weight, buoyant density, and guanine plus cytosine (G+C) content of the DNA differ among species of CMV (Table I). CMVs differ from HSV in that they are species-specific (usually grow only in cells from their natural host) and demonstrate slow *in vitro* growth. A few exceptions to species specificity include monkey CMV, which grows in human as well as in monkey cells; guinea pig, equine, and bovine CMVs, which replicate in rabbit cells; and rat CMV, which replicates in primary hamster kidney cells.

Krugman and Goodheart (1964) were the first to report thermal inactivation studies using HCMV and concluded that HCMV was a heat-labile virus. They did not observe a plateau in the inactivation curve during the first few hours, as had been observed by other investigators for HSV. However, Plummer and Lewis (1965) investigated thermoinactivation of HSV and HCMV and found a plateau in the inactivation curves at 36 and 22°C; extracellular virus was inactivated less rapidly than intracellular virus. In addition, they noted that there was a greater loss of HCMV infectivity at 4 and 10 than at 22°C.

Direct treatment of HCMV with phytohemagglutinin inactivated the virus (Ito *et al.*, 1978) and pretreatment of cells with phytohemagglutinin reduced HCMV infectivity. This work also confirmed an earlier study (Kim and Carp, 1973) of HCMV resistance to trypsin.

Most studies on the size of CMV DNA have utilized human and mouse CMVs. HCMV DNA was initially estimated to have a molecular weight of approximately 100×10^6 daltons by sedimentation velocity in neutral sucrose gradients and by contour length measurements (E.-S. Huang *et al.*, 1973; Sarov and Friedman, 1976). However, by 1977, Kilpatrick and Huang (1977) reported a second and less abundant size class of HCMV DNA with an approximate molecular weight of 150×10^6 daltons. This work was confirmed by Geelen *et al.* (1978). DeMarchi *et al.* (1978), by velocity sedimentation in sucrose gradients, contour meas-

TABLE I. Physical Properties of Cytomegaloviruses

Property	Human CMV	Mouse CMV
Number of capsomeres	162	162
Envelope	Present	Present
Configuration of DNA	Double-stranded linear	Double-stranded linear
Size of virion	100–150 nm	100–150 nm
Molecular weight of DNA	150×10^6 daltons	132×10^6 daltons
Bouyant density of DNA in cesium chloride	1.716 g/cm^3	1.718 g/cm^3
G + C content of DNA (%)	57	59

urements, and renaturation kinetics, detected only one size class of molecules with a molecular weight of 150×10^6 daltons.

The DNA of murine CMV (MCMV) was first analyzed by velocity sedimentation in neutral and alkaline sucrose gradients and by centrifugation in cesium chloride; an approximate molecular weight of 132×10^6 daltons and a G+C content of 59% were obtained (Mosmann and Hudson, 1973). It was noted, however, that after shearing, smaller fragments banded at two separate densities, approximately 57.5 and 61.5% G+C. The heterogeneity observed in the G+C content has been useful in studying transcriptional patterns (Mosmann and Hudson, 1974).

Antigenic relatedness of CMVs has been studied, but there is not much cross-reactivity among isolates from different species. Zablotney *et al.* (1978) studied 17 strains of HCMV for antigenic relatedness and found what appear to be four antigenic groups among the strains examined. On the basis of data from kinetic neutralization tests with hyperimmune guinea pig sera, it appears that HCMV strains are antigenically heterogeneous and that this heterogeneity is apparently not due to geography, age of host, or clinical disease. However, extensive cross-reactivity was observed among these 17 strains. Waner and Weller (1978) used high-titered rabbit antisera to examine the antigenic relatedness of six commonly used strains of HCMV: AD-169, Davis, Espaillat (Esp), C-87, Kerr, and Towne. They found that antisera to AD-169 or Davis strains distinguished three antigenic groups and that an antiserum to the Esp strain distinguished two groups. The Kerr and C-87 strains appear to have close antigenic similarity, and the Towne strain is neutralized least by each of the antisera. The Esp strain falls midway between the Davis and AD-169 strains in antigenic relatedness. Thus, the antigenic heterogeneity of CMV is complex. HCMV strain Colburn, isolated from the brain biopsy of a boy with encephalopathy (E.-S. Huang *et al.*, 1978a), was found to have more than 90% homology with simian CMV (SCMV) and to lack detectable homology with human strains AD-169 and TW-087. The significance of this observation is still obscure.

Hsiung *et al.* (1976) have observed no antigenic relationship between guinea pig CMV (GPCMV) and a guinea pig herpeslike virus (GPHLV) using antisera specific for each virus: rabbit antisera for GPHLV and guinea pig antisera for GPCMV. Cross *et al.* (1979) reported that a herpesvirus of mice, mouse thymic virus, does not cross-react serologically with MCMV in immunofluorescence (IF), complement-fixation (CF), or virus neutralization tests. However, all antigen studies will be more definitive with the advent of monoclonal antibodies against specific antigens of CMVs.

B. Biological Properties

Several biological properties are common to all CMVs including: high species specificity, tight association with the cell, low to intermediate virus yields, slow cytopathic effect (CPE), and sensitivity to ether (Table II). K. O. Smith and Rasmussen (1963) compared the morphology of HCMV with that of HSV and observed that HCMV is a DNA virus (by staining), is close to HSV in size, is enveloped, and exhibits protein membrane subunits that are very similar to those of HSV in number, size, and arrangement. However, they also observed that the particle/infectivity ratio with HCMV is very high and may be the result of incomplete nucleic acid cores. The morphology of HCMV was later confirmed to be that of a herpesvirus by negative staining (Wright *et al.*, 1964).

In 1963, McAllister *et al.* (1963) described the evolution of HCMV-induced intracytoplasmic and intranuclear inclusions in human embryo cells. This study provided a detailed description of the development of cytopathology induced by HCMV. Virus was not detectable until 72 hr postinfection (p.i.), and at that time, infectious HCMV was detected intracellularly and extracellularly. With the use of electron microscopy, virus particles were not detected until 48 hr p.i. and then only in the nucleus. After 72 hr, particles were observed in the nucleus and in the cytoplasm. DNA and virus antigens were detected in nuclear lesions 48 hr p.i. In cytoplasmic lesions, an RNA halo was observed 24 hr p.i. By 48 hr p.i., RNA and virus antigens were observed. However, by 72 hr p.i., DNA and virus antigens were observed in cytoplasmic lesions.

McGavran and Smith (1965) extended the studies of McAllister *et al.* (1963) and demonstrated that cytoplasmic inclusions are formed as virus enters the cytoplasm from the nucleus and are composed of aggregates of lysosomes. They also observed that the outer of the two coats of the virus originates from the inner nuclear membrane as the virus leaves the nucleus. A later study (McAllister *et al.*, 1967) described the effects of various inhibitors such as 5-iodo-2′-deoxyuridine (IUdR), mitomycin C, cytosine arabinoside (ARA-C), actinomycin D, and puromycin on HCMV-infected cells. They found that IUdR, mitomycin C, ARA-C (all inhibitors of DNA synthesis), and puromycin (an inhibitor of protein synthesis), if added when cells were infected, interfered with the production of infectious HCMV but not with the development of cytopathology and early cytoplasmic lesions. Actinomycin D, an inhibitor of RNA synthesis, if added at time of infection, deterred the development

TABLE II. Selected Biological Properties of Cytomegaloviruses

Epithelial target cell *in vivo*	No virus-coded thymidine kinase
Replication cycle in cell culture (diploid fibroblasts) 36–48 hr	Site of latency unknown
Low to intermediate virus yield (10^4 to 10^7) in cell culture	Sensitive to ether
Slow cytopathology	Labile at 37°C
Highly species-specific	Inactivated at 56°C
Highly cell-associated	Resistant to trypsin
	Inactivated by phytohemagglutinin

of all expected morphological changes and precluded the production of infectious virus. Another study by Wright *et al.* (1968) used time-lapse cinematography to observe cell changes resulting from HCMV infection. The earliest effects of HCMV infection, contraction and rounding of the cells, were not seen until 18 hr p.i. This was followed by increased cytoplasmic activity along one pole of the nucleus, movement of the area of cytoplasmic activity to a paranuclear position, and then invagination of the nucleus. A very slow rate of nuclear rotation occurred, which was followed by the development of intranuclear inclusions, their subsequent maturation, and death of the cells.

Additional morphological studies were conducted by Sarov and Abady (1975) on cytomegalovirions and dense bodies. They observed that HCMV has at least 23 polypeptides and that the dense bodies contain 22 of these. These polypeptides range in molecular weight from 24,500 to 171,000, with the major component (VP9) having a molecular weight of approximately 67,000. Other investigators (Severi *et al.*, 1979) studied the movement of the nucleocapsids from the nucleus to the cytoplasm and acquisition of the envelope. They suggested that the virus (after leaving the nucleus) acquires and then loses a temporary envelope and subsequently acquires the final envelope in the cytoplasm. Microvilli have also been detected in HCMV-infected human fibroblasts; their structure changes and their numbers increase as infection progresses (Garnett, 1979b). Albrecht and Weller (1980) demonstrated that different strains of HCMV produce different morphological features in infected cells and that these morphological characteristics are reproducible for each strain investigated. However, these characteristics were observed in HCMV-infected fibroblast cells from human embryo thyroid and skin muscle, but not in fibroblast cultures from human embryo lung (HEL) cells.

Two types of chromatin in HCMV-infected cells have been observed by electron microscopy: (1) a spheroid beaded pattern termed nucleosomes and (2) bipartite and oblate ellipsoid structures (Kierszenbaum and Huang, 1978). Both types can exist together, with the bipartite and oblate structures being longer than conventional nucleosomes but clearly less abundant. Kamata *et al.* (1978) demonstrated that HCMV infection causes rapid changes in conformation of chromatin and induces chromatin template activity as a result of the induction of at least two chromatin-associated factors. For induction of these factors to occur, virus-coded functions must be expressed early after infection. In a later study (Kamata *et al.*, 1979), this same group found that one chromatin-associated factor induced in HEL cells during early HCMV infection stimulates template activity of cell chromatin. This chromatin-associated factor corresponds to a major component of a nuclear antigen that can be detected 1 hr p.i. by anticomplement immunofluorescence (ACIF) staining of infected cells. A recent study (Radsak *et al.*, 1980) supports the idea that HCMV infection of human fibroblasts produces a complex between virus DNA and cellular chromatin that survives extraction procedures. Chromatin preparations from infected cells can continue virus and host-cell DNA synthesis under cell-free conditions. Virus DNA and virus DNA polymerase are involved in the *in vitro* activity of chromatin from HCMV-infected cells, and it is clear that viral or cellular factors or both are involved in alterations of chromatin in HCMV-infected cells.

Morphological studies have also been conducted with nonhuman types of CMV. Ruebner *et al.* (1964) investigated MCMV and found it ultrastructurally very similar to other CMVs, but distinct immunologically. The Smith strain of MCMV produced focal hepatic necrosis with degenerative alterations in Kupffer cells and intranuclear virus particles and cytoplasmic inclusions. A comparison was made of the features of MCMV particles in the salivary gland and in the liver (Ruebner *et al.*, 1966). In both organs, more intranuclear virus particles than protein coat material were produced in the early stages of infection, but this reversed in the later stages of infection. Cytoplasmic MCMV particles remained unchanged in the salivary gland up to 28 days, although in liver parenchymal cells, the MCMV particles were soon surrounded by dense material suggestive of lysosomes. After the 5th day, MCMV particles were seldom seen in liver cells, and the possibility exists that the persistence of virus in the salivary gland is due to low lysosomal activity.

Electron-microscopic studies (Duncan *et al.*, 1965; Valícek *et al.*, 1970) suggested that CMV is the etiological agent of porcine cytomegalic inclusion disease. Duncan and colleagues observed that the cytomegalic cells with inclusion bodies typical of rhinitis in swine are the result of a virus, most probably a herpesvirus. A study by Valícek *et al.* (1970) demonstrated the presence of enveloped virus particles within "membrane-enclosed clusters" in the nucleus and the duplication of nuclear membranes. Enveloped virus particles were found in vacuoles in the cytoplasm, but unenveloped virus particles were free in the cytoplasm and often exhibited a crystalline pattern. CMV particles in a crystal pattern were not detected in the nucleus.

In 1965, it was reported that an unknown virus had been isolated from submaxillary glands of rats (Ashe *et al.*, 1965). Investigation of the virus ruled out the possibility that it was a CMV of rats. It did not produce CPE in cell cultures of homologous species but did produce CPE in cell cultures of heterologous species, did not induce intranuclear inclusions, and was hemagglutination-positive and resistant to ether. A later study by Rabson *et al.* (1969) detailed the isolation of a CMV from Panamanian rats in rat kidney-cell cultures. This rat CMV was further studied (Berezesky *et al.*, 1971): the virion capsid structure was similar to that of HCMV and other herpesviruses, the nucleocapsid measured 138 nm, the virus grew in rat and hamster kidney cells, and virus particles accumulated in very large cytoplasmic aggregates.

Morphological studies have also been conducted on a CMV isolated from an owl monkey (Chopra *et al.*, 1972). These researchers observed that the owl monkey cytomegalovirus produced characteristic CMV cytopathology when inoculated into owl monkey kidney cells and observed intranuclear and cytoplasmic inclusions typical of other CMV-infected cells in virus-infected owl monkey kidney cells.

GPCMVs have been studied by many investigators. Middlekamp *et al.* (1967) found, by light and electron microscopy, that GPCMV produces nuclear and cytoplasmic inclusions *in vivo* but only nuclear inclusions *in vitro*. In addition, they observed three distinct virus particles in the nucleus and two distinct virus particles in the cytoplasm. Tubular structures characteristic of GPCMV-infected cells were observed by Fong *et al.* (1979), who suggested that these tubular structures are proteinaceous, since their synthesis does not require DNA synthesis but is dependent on *de novo* protein synthesis. The filamentous structures observed by Middlekamp *et al.* (1967) were not detected. However, Fong and her collaborators did observe the formation of enveloped dense virions and dense bodies in the cytoplasm of infected cells; virus capsids were always coreless on entrance into the dense matrix and acquired dense cores later during development into enveloped dense virions. Dense bodies without virus capsids were seen often in the cytoplasm of GPCMV-infected cells. In a recent study (Fong *et al.*, 1980), salivary glands of guinea pigs were infected with GPCMV and examined by electron microscopy. The investigators visualized several events including nucleocapsid assembly, envelopment of nucleocapsids at the inner nuclear membrane, and their enclosure by a thin vacuolar membrane. Enveloped virions within cytoplasmic vacuoles acquired surface spikes, and these vacuoles eventually released mature virions at the cell surface. The ultrastructural development of GPCMV therefore appears to be similar to that of HCMV.

C. Molecular Properties

1. Human CMV

a. Replication

Molecular characterization of HCMV DNA was first undertaken in 1973 by E.-S. Huang *et al.* (1973), who purified the AD-169 and C-87 strains in sucrose and cesium chloride gradients. They determined that the molecular weight of the DNA is 100×10^6 daltons and that the density is 1.716 g/cm^3. E.-S. Huang and Pagano (1974), using DNA–DNA renaturation kinetics, demonstrated that there was no relatedness of HCMV DNA to the DNAs of HSV type 1, type 2 (HSV-1, HSV-2), Epstein–Barr virus, or nonhuman CMV. Additional studies (E.-S. Huang, 1975a,b) demonstrated the presence of an HCMV-induced DNA polymerase found in the nuclei and cytoplasm of HCMV-infected cells and observed that this virus-induced DNA polymerase and HCMV DNA replication are inhibited by phosphonoacetic acid (PAA).

Restriction-endonuclease digestions and coelectrophoresis of HCMV DNA isolates were conducted to determine the structural and genetic differences among different HCMV genomes (Kilpatrick *et al.*, 1976). The results showed that no two isolates have the same *Hin*dIII or *Eco*R-I patterns, although there are more distinct similarities in the patterns of some isolates than in others. The electrophoretic profile of each isolate is distinctive.

Using contour-length measurements of HCMV DNA, Kilpatrick and Huang (1977) reported a second size class of HCMV DNA with a weight of approximately 150×10^6 daltons. This work was followed by the studies of Geelen *et al.* (1978), who found that the molecular weight of HCMV DNA is approximately $147 \pm 6.2 \times 10^6$ daltons. HCMV (strain AD-169) DNA purified by sucrose density-gradient centrifugation and studied by electron microscopy produced measurements corresponding to a molecular weight of approximately 150×10^6 daltons. Stinski *et al.* (1979a) also reported that most HCMV DNA from plaque-purified, low-multiplicity infections (m.o.i.) had an approximate molecular weight of 150×10^6 daltons. Serial high-m.o.i.-passaged HCMV resulted in production of defective HCMV. Most defective DNA had an approximate molecular weight of 100×10^6 daltons; however, there were some molecules with DNA molecular weights as low as 60×10^6. When the defective HCMV DNA was studied by restriction-enzyme analysis, unique DNA fragments were found. The presence of these unique fragments suggests that deletions, substitutions, or duplications may be responsible for the production of defective HCMV DNA. When the particle/plaque-forming unit (PFU) ratio is low, most of the HCMV DNA is nondefective; a high particle/PFU ratio will produce defective DNA. Studies by Ramirez *et al.* (1979) using contour-length measurements revealed several classes of HCMV DNA from defective virions ranging from 40×10^6 to 120×10^6 in molecular weight. They also observed production of defective HCMV particles when the virus was passaged at a high m.o.i. These defective HCMV virions banded at a lower buoyant density in CsCl than nondefective HCMV.

The replication of CMVs has now been studied for almost two decades. Early studies focused on the effects of drugs on virus synthesis and cytopathology as well as on methods to quantitate CMV research. Goodheart *et al.* (1963), using 5-fluorodeoxyuridine, found that CMV contains DNA as its nucleic acid and that the cell rounding and formation of intranuclear lesions occur even in the absence of detectable virus production. Goodheart and Jaross (1963) established an HCMV assay for more detailed investigation of this slow-growing virus. They counted infected cells and found that approximately 48 hr after inoculation, the majority of the cells originally infected were included in the count. Rapp *et al.* (1963) used an IF focus technique to establish a quantitative assay for CMV: approximately 6–7 days p.i., IF foci of antigen-containing cells were counted. This technique was used to study HCMV adsorption and replication in human embryo fibroblast (HEF) cells. Wentworth and French (1970) developed a plaque assay for CMV that used two sequential agarose overlays. This method, which is routinely used today, provides a reliable quantitation of CMV. Furukawa *et al.* (1973), in agreement with results obtained by Goodheart *et al.* (1963), noticed that when human fibroblasts were infected with a high m.o.i. of HCMV, early cell rounding occurred between 6 and 24 hr p.i. They also noticed that this CPE was not prevented by DNA inhibitors, but was prevented by protein inhibitors. As a result of these findings, they proposed that a protein is synthesized early (2 hr postinoculation) after infection of human fibroblasts.

When the sequence of HCMV replication was compared with that of HSV, it was found that HCMV required 3–4 days to replicate (Fig. 1) and release progeny virus, whereas HSV took only 8 hr. The differences between the two infectious cycles include the presence of cytoplasmic dense bodies in HCMV-infected cells only and the greater likelihood of condensation of chromatin and production of membrane changes in HSV-infected cells than in HCMV-infected cells (J.D. Smith and De Harven, 1973). J.D. Smith and De Harven (1974) studied early stages of virus penetration by electron microscopy and found that HCMV and HSV enter cells by fusion of the virus envelope with the cell membrane or by phagocytosis and that either event occurs within 3 min after penetration begins. Following fusion, the HCMV capsids are coated and visible up to 1.5 days p.i., in contrast to the HSV capsids, which are never coated and are visible only for 90–120 min. p.i. It appears that HCMV and HSV naked capsids migrate across the cytoplasm toward the nucleus and reach the nucleus within 5 min after penetration commences. Virus particles that enter the cells by phagocytosis usually remain within phagosomes.

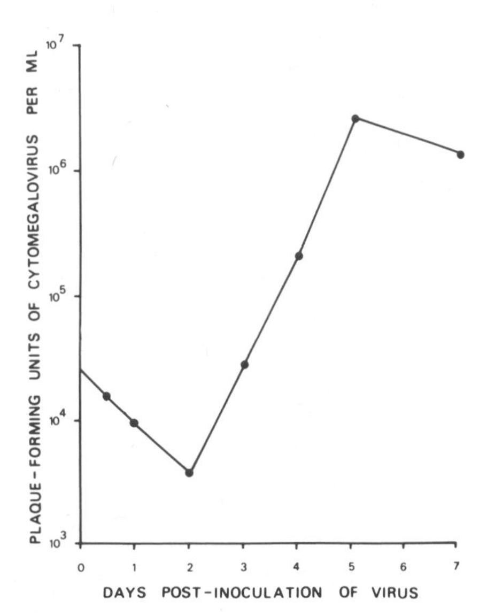

FIGURE 1. Replication of HCMV in human embryo fibroblasts.

J.D. Smith and De Harven (1978) investigated cytochemical localization of lysosomal enzymes in HCMV-infected WI-38 cells and observed dense bodies budding into acid-phosphatase- and arylsulfatase-positive vacuoles. They suggest that the dense bodies are not lysosomes but that they can bud into lysosomes like naked capsids and acquire an envelope. The budding may occur as a result of specific affinity between the lysosomal membrane and one or more of the HCMV structural proteins that are common to dense bodies and capsids. It appears that there is a close interaction between lysosomes and virus particles in HCMV-infected cells and that high titers of infectious virus persist within lysosomes. J.D. Smith and De Harven (1979) also observed that some dense-body proteins are rich in arginine and that HCMV may contain one or more arginine-rich structural proteins. They detected fluorescent cytoplasmic dense bodies in HCMV-infected cells by phenanthrene quinone fluorescence; this fluorescence was (not surprisingly) absent in HSV-infected cells in which dense bodies are not formed.

HCMV ultraviolet (UV) irradiated for 14 min and then inoculated onto HEF or mouse embryo fibroblast cells (Boldogh *et al.*, 1978) stimulates host-cell DNA synthesis, probably as the result of an early function of the HCMV genome. However, a very early complement-fixing nuclear antigen was detected only in the HEF cells, and an early cytoplasmic antigen was detected only in the mouse cells. The genes that code for the proteins detected by the ACIF and IF tests are probably not responsible for the induction of DNA synthesis (Boldogh *et al.*, 1978).

Heparin is known to inhibit the infectivity of several herpesviruses. Choi *et al.* (1978) have observed a similar effect using heparin in their work with GPCMV. When heparin is added to cultures after virus adsorption, the effect on virus infectivity is negligible; however, when it is present in cell-culture medium, infectivity is reduced and inhibition increases as the heparin concentration increases. Arginine deprivation has also been found to have an inhibitory effect on the replication of HCMV. When arginine-negative HCMV-infected cultures are re-fed medium containing arginine, HCMV maturation commences. However, if the cultures are re-fed arginine and a protein inhibitor simultaneously, then synthesis of infectious HCMV can be prevented (Gönczöl *et al.*, 1975). A second study on arginine deprivation (Garnett, 1975) reported that arginine is required for development of CPE in HCMV-infected human cells and for the production of infectious cytomegalovirions. The block in the HCMV replicative cycle due to arginine starvation occurs before formation of virus DNA, in contrast to HSV, adenovirus, polyoma, and SV40, which are blocked after the virus-specific DNA is produced. Thus, it appears that arginine is necessary in the initial stage of HCMV replication in HEF cells.

CMVs are highly species-specific and HCMV has demonstrated a preference for *in vitro* replication in human fibroblasts. However, some attempts to replicate HCMV in epithelial cells have been successful.

Human embryo kidney cells have been converted from a nonpermissive state to a permissive state by pretreatment with IUdR (St. Jeor and Rapp, 1973). Pretreatment with IUdR possibly reduces cell DNA synthesis, which in turn may permit the conversion of the cells from a nonpermissive to a permissive state. CMV has the ability to stimulate cell DNA synthesis in permissive and nonpermissive cells. St. Jeor *et al.* (1974) suggested that an HCMV-coded function expressed after infection is involved in the stimulation of cell DNA synthesis. Tanaka *et al.* (1975) investigated the effect of HCMV infection on cellular RNA synthesis and found that RNA synthesis is induced in HCMV-infected cells and precedes HCMV DNA synthesis and the production of progeny virus by 24 hr. Included in the RNA species synthesized are 28 S and 18 S ribosomal RNAs and 4 S transfer RNA. Stimulation of host-cell RNA synthesis is apparently dependent on a protein synthesized early in the infectious cycle.

Results from a similar study (Furukawa *et al.*, 1975b) demonstrated stimulation of cell RNA and DNA in guinea pig cells abortively infected with HCMV. Results reported for HCMV-infected human cells were similar to those in the HCMV-infected guinea pig cells: synthesis of the same classes of RNA species, 28 S and 18 S ribosomal RNA and 4 S transfer RNA. In addition, heat- and UV-inactivated CMV induce cellular RNA and DNA synthesis. This evidence was obtained from studies on the incorporation of thymidine and uridine, the use of sedimentation gradients to separate cell RNA and DNA, autoradiography, and mitotic activity. Furukawa *et al.* (1976) reported that the stimulation of cell DNA synthesis in HCMV-infected cells is due, in part, to an increase in the rate of mitochondrial DNA synthesis. This increase in the stimulation of mitochondrial DNA occurs in cells in which cell DNA synthesis is repressed by contact inhibition or by depletion of serum. In addition, stimulation of mitochondrial DNA precedes HCMV DNA replication. This work was followed shortly by the experiments of Albrecht *et al.* (1976) using abortively infected hamster embryo cells. They observed increases in the rate of cell DNA synthesis and mitotic activity following infection with HCMV; however, these effects could not be detected unless the cells were arrested with low serum concentrations before HCMV infection. When the cells were arrested before virus infection, susceptibility to HCMV infection increased. Unlike the findings with the HCMV abortively infected guinea pig cells described by Furukawa *et al.* (1975b), DNA replication and mitotic activity in hamster embryo cells abortively infected with HCMV could be increased by UV irradiating the HCMV before infection.

Hirai and Watanabe (1976) observed that α-type DNA polymerases are induced in HCMV-infected WI-38 cells, but β-type polymerase activity in the nuclei and cytoplasm is hardly affected. They suggest that α-type polymerases are involved in the synthesis of cell and HCMV DNA. DeMarchi and Kaplan (1976) treated HCMV-infected cells with 5-fluorouracil and thymidine and observed that HCMV replication continues in cells that do not synthesize cellular DNA from the time of infection and that HCMV DNA synthesis continues independently of host-cell DNA synthesis. Cells stimulated by HCMV infection to synthesize cell DNA had a delay or inhibition in productive infection. In addition, subconfluent, actively growing cells generally yield higher levels of infectious HCMV than confluent contact-inhibited cells. This higher yield is correlated to a large percentage of cells producing HCMV antigens within 48 hr p.i. (DeMarchi and Kaplan, 1977a). These same investigators (DeMarchi and Kaplan, 1977b) also studied the role of defective cytomegalovirions in the induction of host-cell DNA synthesis and observed that stocks from plaque-purified virus are poor inducers of cell DNA synthesis, but that undiluted, serially passaged virus stocks are highly effective in stimulating cell DNA synthesis. Following low doses of UV irradiation, the virus stocks have an increased capacity to stimulate host-cell DNA synthesis. Cells are abortively infected, as evidenced by the lack of detectable virus antigens as late as 96 hr p.i. DeMarchi and Kaplan conclude that stimulation of host-cell DNA synthesis occurs either from interactions between susceptible cells and virus particles that do not result in a productive infection (production of defective virus particles) or from interactions between virus particles and cells not susceptible to productive infection due to the physiological conditions of the cells at the time of infection.

Several herpesviruses, including HSV-1, HSV-2, varicella–zoster virus (VZV), marmoset herpesvirus, pseudorabies virus, turkey herpesvirus, and infectious laryngotracheitis virus induce their own virus-coded thymidine kinases (TKs). Závada *et al.* (1976), using the AD-169 strain of CMV, concluded that HCMV does not induce its own TK in productively infected cells. However, cellular TKs have been stimulated by infection with HCMV (Estes and Huang, 1977). Miller *et al.* (1977) also determined that HCMV does not induce TK based on studies exploring the effects of arabinofuranosylthymine (ARA-T) on HCMV replication.

Investigation of the effect of UV irradiation on expression of early CMV functions in HCMV-infected HEL cells (Hirai *et al.*, 1977) demonstrates the UV sensitivities of different virus functions. The induction of early antigen, cell DNA synthesis, and cell DNA polymerases is not as UV-sensitive as functions such as the synthesis of HCMV DNA, induction of CMV DNA polymerases, and plaque formation. Hirai and his colleagues have proposed that some mechanism must permit CMV to stimulate cell DNA synthesis when irradiated with high doses of UV light, even though some virus gene function (that is damaged by UV irradiation of the virus) is expressed after infection and is most probably involved in the stimulation of host-cell DNA synthesis.

Yamanishi *et al.* (1978) studied the effect of caffeine on nonirradiated and UV-irradiated HCMV and found that when HCMV was not irradiated, the addition of caffeine enhanced virus growth. However, when HCMV was irradiated with high doses of UV light, replication was inhibited. It is known that caffeine inhibits host-cell-mediated repair of UV-irradiated viruses. However, factors that determine cell sensitivity are not clearly understood, but the effect of caffeine on CMV replication is associated with the temporary inhibition of cell DNA synthesis, since virus synthesis occurs at a definite time after inhibition of cell DNA synthesis. Caffeine pretreatment also inhibited multiplicity reactivation of UV-irradiated CMV.

Plasminogen activator has also been examined in HCMV-infected human and hamster cells (Yamanishi and Rapp, 1979a). After exposure to virus, plasminogen activator is produced in permissive (human) and nonpermissive (hamster) cells. When human cells are exposed to UV-irradiated HCMV, the production of plasminogen activator increases. The level of plasminogen activator is significantly higher in cells transformed to malignancy.

Temperature-sensitive (ts) mutants of HCMV have been isolated and characterized by several groups of investigators to study early virus functions. Ihara *et al.* (1978) studied 15 HCMV ts mutants that could not form normal plaques at 39°C but could form plaques at 34°C. These mutants were isolated after mutagenesis by UV light; 5 were defective in virus DNA synthesis at 39°C. These five mutants fell into complementation groups A and B and represented mutants normal and defective in the induction of viral DNA polymerase. The A and B mutants were tested for ability to induce HCMV-specific DNA polymerase by dependency of polymerase activity on ammonium sulfate and by sensitivity to PAA. The only mutant belonging to group B (ts 256) was unable to induce ammonium sulfate polymerase activity at 39°C, but could induce it at 34°C. DNA polymerase activity induced by ts 256 at 39°C was sensitive to PAA. Questions remain as to whether HCMV actually induces two discrete species of DNA polymerases, one responding to PAA and the other to ammonium sulfate, or whether HCMV induces one species of DNA polymerase that is affected by PAA and ammonium sulfate.

Yamanishi and Rapp (1979b) used DNA-negative HCMV ts mutants to induce host DNA synthesis and DNA polymerase. Four DNA-negative ts mutants were studied in permissive (HEL) and nonpermissive [rabbit lung (RL)] cells. These four mutants stimulated host-cell DNA synthesis and DNA polymerase activity in HEL and RL cells at 33.5 and 39.5°C. Salt [$(NH_4)_2SO_4$] stimulation of DNA polymerase activity was used to distinguish between virus and cell DNA polymerase from HEL cells. DNA polymerase activity was stimulated by 100 mM $(NH_4)_2SO_4$ in cultures infected with three of the mutants, although not by the fourth mutant. It is possible that four, and maybe more, cistrons control the synthesis of HCMV DNA; HCMV DNA synthesis does not appear to be required for induction of cell DNA synthesis.

Using electron microscopy, Jean *et al.* (1978) examined intracellular forms of HCMV DNA before and after the initiation of HCMV DNA synthesis. Before HCMV DNA synthesis commenced, circular, concatemeric, and linear double-stranded molecules were observed. After HCMV DNA replication began, linear and concatemeric molecules were observed. The identification of circular molecules suggests the presence of reiterated sequences at or near the termini of the CMV DNA molecule and may be important in the ability of the virus to establish latent or persistent infections.

HCMV can induce endogenous RNA polymerase in human embryo cells (Tanaka *et al.*, 1978). However, when inhibitors of protein synthesis such as cycloheximide or actinomycin D are added to CMV-infected cells during the first 6 hr of infection, subsequent enhancement of endogenous RNA polymerase does not occur. This suggests that an HCMV-induced early protein may play an important role in the stimulation of this RNA polymerase. In HCMV-infected cells, a significant increase in the activity of three major classes of RNA polymerases is detected, with the activity of class II polymerase increasing most dramatically. In contrast to cells infected with vaccinia, HSV-1, and polioviruses (Landini *et al.*, 1979), when HEL cells are pretreated with actinomycin D and then infected with HCMV, HCMV infection does not progress past the early events of the HCMV replicative cycle. These results strongly suggest that HCMV needs a cellular function(s) for complete and productive virus replication and that this function(s) occurs between early transcription of the CMV genome and the initiation of CMV DNA synthesis. The early effects of CMV on protein and RNA synthesis in infected human cells are very similar to those seen with HSV infection: specifically, protein and RNA synthesis are inhibited within 3 hr of CMV infection (Garnett, 1979a). The differences between infection with HCMV and HSV begin to appear approximately 6–12 hours p.i.

Studying the patterns of transcription of HCMV in permissively infected cells, DeMarchi *et al.* (1980) observed that a small but significant amount of HCMV RNA is detectable in the first 3 hr p.i. and that the rate of transcript accumulation is static during the first 24 hr p.i. but increases thereafter. These virus transcripts can usually be divided into three classes: immediate–early RNA (the first 2–4 hr p.i.), early RNA (synthesized up to 24 hr p.i.), and late RNA (after 24 hr). Although rapid adsorption and penetration occur and the level of accumulation of transcripts is static during the first 24 hr, there is a steady increase in the number of cells that synthesize detectable amounts of CMV antigens.

When human diploid cells are infected with HCMV, there is a marked increase in the rate of putrescine uptake, with the highest levels attained after production of infectious progeny. HCMV replication is not affected by addition of methylglyoxalbis (guanylhydrazone), an inhibitor of spermidine and spermine synthesis, to the cells after completion of the eclipse phase of the virus growth cycle (Tyms and Williamson, 1980). Thus, polyamine metabolism is required only during the initial stages of CMV replication.

Albrecht *et al.* (1980b) studied the effects of temperature, virus strain, m.o.i., and cell type on the kinetics of replication of HCMV strains AD-169, C-87, Davis, Esp, and Kerr. These researchers found that almost no replication occurs at 40.5°C and that this block in replication occurs 12–16 hr p.i. with little synthesis of CMV DNA or late antigens. The extent of inhibition of the late functions at 40.5°C is virus-strain- and m.o.i.-dependent, but is not dependent on the type of fibroblastoid cells used. Supraoptimal temperatures appear to produce situations that favor the establishment of persistent CMV infections.

HCMV-infected human fibroblast cells infected with poliovirus exhibit enhanced poliovirus replication when compared to poliovirus replication in non-CMV-infected cells (Furukawa *et al.*, 1978). When HCMV is UV irradiated or treated with inhibitors of CMV replication, then enhancement of poliovirus replication is not seen, suggesting that a functional HCMV genome is required for the enhancement. However, further studies concerning this enhancement are necessary to establish the significance of the observation and the molecular events that regulate it.

b. Proteins

In 1975, Schmitz *et al.* (1975) published a report on a method to purify and concentrate CMV nucleocapsids and enveloped particles. Using the nucleocapsids and enveloped particles as antigens in the CF test, they found that the production of envelope-directed antibodies is delayed in patients with CMV-induced disease and cannot be detected before the 3rd week of disease. There is a stronger immunological reaction to nucleocapsid particles early in disease. The structural proteins of HCMV were studied in sucrose gradients by Edwards (1976), who detected three bands of densities: 1.175, 1.190, and 1.246 g/cm^3. The middle band, composed of mostly enveloped HCMV particles and some dense bodies, was subsequently used in polyacrylamide gel electrophoresis (PAGE) studies, which detected 26 HCMV proteins. The molecular weights of the 26 proteins range from 26,000 to 230,000 daltons. HCMV protein 14 appears to be a major protein with a molecular weight of 82,000 daltons.

The *et al.* (1974) detected an HCMV-induced early antigen (EA) in fibroblast cells infected in the presence of ARA-C. This EA was detected in only 0.6% of the cells at 8 hr p.i.; however, by 2 days p.i., antigen was detected in 65% of the cells. Sera from 45 patients acutely infected with HCMV, from healthy seropositive donors, and from patients with unrelated bacterial or viral infections were then examined. The and colleagues found that the presence of anti-EA antibodies in acute-phase serum (titer of $\geqq 1:80$) suggested a current primary infection. Furukawa *et al* (1975a) detected an immunoglobulin G (IgG) receptor induced in WI-38 diploid fibroblast cells by the Towne strain of HCMV. The IgG receptor was localized in the perinuclear region 48 hr p.i., and its synthesis was blocked by ARA-C and cycloheximide. It bound IgG of humans as well as of other species, including hamsters, rabbits, and guinea pigs. Geder (1976) reported that HCMV induces nuclear antigens similar to the Epstein–Barr nuclear antigen (EBNA) as early as 3 hr p.i. and that these EAs can be detected only by the ACIF technique. ARA-C did not influence the synthesis of these EAs. Immediate early antigens (IEAs) that can be detected within 1 hr p.i. do not require ACIF for detection. They appear sooner than EAs, are located in the nucleus, and appear in the absence of DNA synthesis (Michelson-Fiske *et al.*, 1977). Reynolds (1978) reported that these nuclear antigens can appear as early as 1 hr p.i., as detected by IF, and that the development of these early antigens occurs in the presence of inhibitors of DNA, RNA, and protein synthesis and is dependent on intact HCMV DNA in the inoculum. It is interesting that HCMV can induce a virus-specific antigen in the nuclei of permissive cells in the presence of inhibitors of RNA and protein synthesis. This antigen may be important in the initiation of early replicative events and later activation of the host-cell genome.

Musiani *et al.* (1979) studied the histochemical properties of IEAs and suggested that these antigens are proteins with DNA-binding properties. CMV-induced IEAs were analyzed by sodium dodecyl sulfate (SDS)–PAGE after immunoprecipitation with IEA-positive human sera (Michelson *et al.*, 1979). Within 90 min p.i., two polypeptides with molecular weights of 76,000 and 82,000 daltons were detected. Polypeptides of similar molecular weights were found in immunoprecipitates of nonpermissive cells infected with HCMV. More than 40% of normal blood donors react to IEA, and reactivity to IEA correlates highly with active virus excretion. HEL cells infected with HCMV can induce a pre-early nuclear antigen (PENA) within 1 hr p.i. (Tanaka *et al.*, 1979). Serum adsorption tests demonstrate that PENA differs immunologically from the EA and the nuclear antigens. The plaque-forming ability of HCMV corresponds to its PENA-forming ability, and plaque formation is inhibited by RNA and protein synthesis inhibitors and by UV irradiation. SDS-PAGE demonstrated that at least two species of CMV-specific polypeptides (molecular weights of 70,000 and 30,000 daltons) were synthesized *de novo* within 3 hr p.i. PENA is the only HCMV-specific antigen that has been detected in HCMV-transformed cells. Twenty antigens extracted from HCMV strain AD-169 by electroimmunodiffusion could be detected and were stable for up to 4 weeks (Sweet *et al.*, 1979). The C-87 and Davis strains and three recent isolates, VD14, 1694, and 1723, yielded 15, 15, 13, 11, and 11 antigens, respectively. As many as 8 antigens were common to all strains, which is suggestive of a higher degree of antigenic homology in HCMV than reported earlier.

CMV late antigens are inhibited by phosphonoformic acid (PFA), and the replication of HCMV is completely inhibited at concentrations of 500 mM and higher (Wahren and Öberg, 1979). At concentrations of 100 mM PFA, 50% inhibition was observed. Early HCMV nuclear antigens did not appear to be affected; however, formation of nuclear inclusion bodies, Fc receptors, and CPE was inhibited.

Ihara *et al.* (1980) described an early function of HCMV that is necessary for the structural integrity of the CMV-induced nuclear inclusion. Using an early HCMV ts mutant, it was shown that this mutant has a defect in CMV DNA synthesis and an inability to maintain the structural integrity of the nuclear inclusion. They proposed that an altered gene product is required for formation and maintenance of the nuclear inclusion and that the integrity of the inclusion is required for CMV DNA synthesis. An HCMV early nuclear antigen has been identified as a DNA-binding protein by Gergely *et al.* (1980). With ACIF, these investigators found that this antigen appears in the nucleus within 1 hr p.i. Experiments using polyacrylamide gel filtration established the molecular weight of this purified CMV nuclear antigen at approximately 80,000.

Blanton and Tevethia (1981) identified at least 20 HCMV-infected cell-specific polypeptides during lytic infections within 6 hr p.i. by immunoprecipitation. Four of these (78K, 77K, 73K, and 31K) have been classified immediate–early polypeptides on the basis of their synthesis in the presence of actinomycin D following removal from a cycloheximide-mediated protein synthesis block. The synthesis of most immediate–early polypeptides is enhanced after the removal of a cycloheximide block. The results suggest that several regulatory events occur during the early stage of the lytic cycle.

Fiala *et al.* (1976) purified HCMV (strain AD-169) virions and dense bodies by sucrose velocity and equilibrium centrifugation and detected 20 polypeptides with molecular weights ranging from 22,000 to more than 230,000 by SDS-PAGE. Six polypeptides were represented mainly in virions (2, 8, 14, 16, 17 and 18) and four were predominantly in dense bodies (5, 6, 7, and 9), while the remainder were shared by both types. Three of the four glycosylated polypeptides were shared by virions and dense bodies. These investigators propose that dense bodies are the products of abnormal assembly of the HCMV envelope and tegument polypeptides. Although the HCMV major capsid polypeptide (number 2) with a molecular weight of 150,000 is close in molecular weight (155,000) to the VP5 major polypeptide of the HSV capsid, comparison of other polypeptides shows major differences in the protein composition of these herpesviruses.

Stinski (1976) characterized the glycoproteins of HCMV associated with the membranes of virions and dense bodies, and routinely detected eight glycopolypeptides. Three other glycopolypeptides with higher molecular weights were occasionally detected and may be precursors of the lower-molecular-weight glycopolypeptides. It is thought that these glycoproteins are specified by the HCMV genome. A later study (Stinski, 1977) on the synthesis of proteins and glycoproteins in HCMV-infected cells detected two peaks of protein synthesis: the first in the early phase of infection and the second during the late phase after the initiation of HCMV DNA synthesis. Host and virus proteins were synthesized at the same time during both phases, although 70–90% of all proteins synthesized during the early phase were host proteins.

Two infected-cell-specific (ICS) glycoproteins were detected in the early phase of infection. In the late phase, 50–60% of the total protein synthesis was virus, with the ICS proteins and glycoproteins being virus structural proteins. No infectious HCMV was detected until 48–72 hr p.i. PAA prevented the appearance of late-phase ICS proteins and glycoproteins, but had almost no effect on early ICS glycoprotein synthesis. Thirty-five structural proteins were associated with purified HCMV virions and dense bodies; seven structural glycoproteins were identified as ICS glycoproteins.

Gupta *et al.* (1977), using SDS-PAGE, detected 32 polypeptides in HCMV with molecular weights of 13,500–235,000. Four strains of HCMV (AD-169, C-87, Towne, and Birch) studied were similar in polypeptide composition, which supports other studies that the strains are antigenically related. Talbot and Almeida (1977) purified enveloped HCMV virions and dense bodies with a combination of negative viscosity, positive density gradient centrifugation. This technique, to enhance the separation of enveloped virions and dense bodies, demonstrated minimal cross-contamination. Fiala *et al.* (1978) studied HCMV CF antigens and observed, after fractionation on agarose gel columns, that they contained five major polypeptides with molecular weights ranging from 60,000 daltons to 90,000 daltons. The 70,000-dalton polypeptide was not present in a soluble fraction from infected cells without CF activity. The polypeptides in the CF antigen did not elicit neutralizing antibody. Waner (1975) also isolated a soluble HCMV CF antigen with a molecular weight of 67,000–85,000 daltons.

Stinski (1978) detected at least ten early HCMV-induced polypeptides synthesized from 0–6 hr p.i. of permissive cells. These are synthesized before HCMV DNA replication, and many are also synthesized in nonpermissive cells. Their synthesis is sequential, since three proteins are synthesized before the others. Late virus proteins are synthesized at the onset of HCMV DNA replication (approximately 15 hr p.i.) and throughout the replicative cycle. The amount of HCMV DNA synthesis is directly correlated to the relative amount of late HCMV protein synthesis. When an inhibitor of HCMV DNA synthesis is added to permissive cells, late HCMV proteins are not detected because their synthesis depends on HCMV DNA replication. In nonpermissive cells, i.e., guinea pig embryo fibroblasts, early but not late HCMV ICS proteins are synthesized. The barrier for synthesis of HCMV DNA in nonpermissive cells may result from the absence of a required early protein(s), nonfunctional early proteins, or the lack of a host-cell factor that allows HCMV DNA replication.

Gibson (1981a) has investigated the structural and nonstructural proteins of HCMV strain Colburn. He was able to recover from infected cells four kinds of HCMV particles that are distinguished by intracellular compartmentalization, sedimentation properties, protein composition, and infectivity. The particles include A, B, and C capsids, and virions, A capsids being the simplest, with only three protein species (145K, 34K, and 28K daltons). In contrast, the virions contain at least 20 protein species and are structurally the most complex. The 145K protein is the major structural part of the capsid. The Colburn strain appears to have at least five nonstructural proteins in addition to structural proteins. The Colburn strain is more similar to SCMV strains than to HCMV strains.

Gupta and Rapp (1978), in work with HCMV-induced proteins, reported that hypertonic medium inhibited synthesis of host-cell proteins more readily than synthesis of HCMV-induced proteins. They were able to selectively suppress host-cell protein synthesis and observed that HCMV-induced proteins are synthesized in a cyclic manner; the time required for the cycle to repeat varied between 25 and 40 hr. This may be due to the synchronization of cells after HCMV infection, since it has been reported (St. Jeor and Hutt, 1977) that HCMV infection of HEL cells results in cyclic synthesis of HCMV and cell DNAs. After infection with HCMV, membrane glycoproteins and antigens are induced, with HCMV-specified antigens entering the plasma membrane 24 hr p.i. (Stinski *et al.*, 1979b). At 72 hr p.i., however, virus-specified antigens can be detected on the plasma, endoplasmic reticulum, and nuclear membranes. The appearance of HCMV-specified antigens on the internal membranes occurs at the same time as envelopment of cytomegalovirions and dense bodies.

Specific high-titered antisera to HCMV and SCMV can be produced in baboons and monkeys, but the availability of these primates is now very limited. Haines *et al.* (1971) successfully prepared specific, high-titered antisera to HCMV and SCMV in goats. Human strains C-87, AD-169, and Davis were used to induce complement-requiring neutralizing antibodies and complement-independent neutralizing antibodies. SCMV strain GR 2757 elicited only complement-independent neutralizing antibody. Crossing occurred between the neutralizing antigens of the human strains, but there was no interspecies crossing. Minamishima *et al.* (1971) detected neutralizing antibody to SCMV in sera of all of 12 African green monkeys, all of 7 baboons, and 8 of 10 rhesus monkeys captured in the wild. However, neutralizing antibody could not be detected in baboons and rhesus monkeys born in captivity. Y.-T. Huang *et al.* (1974) used CF and IF tests to study the relatedness of human and simian CMVs. They prepared specific, high-titered antisera in guinea pigs and found that human strains C-87 and AD-169 are closely related and have very little cross-reactivity with the simian strain GR 2757.

Waner (1975) found that soluble antigen preparations obtained from HCMV (strains Davis and AD-169)-infected cell cultures had a molecular weight range of 67,000–85,000. Antisera to this soluble antigen lacked neutralizing activity but elicited fluorescence limited to HCMV intranuclear inclusion material when the indirect fluorescent antibody test was used with Davis, AD-169, Kerr, or C-87 strains of HCMV. Antisera reacted only with specific HCMV antigens, which depend on HCMV DNA synthesis and are common to many strains. There was no reaction with HSV-1- and VZV-infected cells, EBV- and Cx-90-3B-transformed cells, or HCMV-infected cells treated with ARA-C.

E.-S. Huang *et al.* (1976) also used nucleic-acid hybridization, restriction endonuclease, and antigen analyses to study strain differences. They found that HCMV isolates share at least 80% homology and that several HCMV isolates share common comigrating fragments, although no two isolates had the same migration fragment patterns. By comparing the electrophoretic mobilities of CMV immediate–early proteins, Gibson (1981b) has demonstrated that the proteins from different HCMV strains are distinguishable and that the differences in size and broad banding pattern are not the result of slow posttranslational modification. His results also suggest that the differences in mobility are due to natural variation and not to passage in cell culture.

There is much heterogeneity in the CF antigen of HCMV isolates. It is not easy to place the HCMV strains and isolates into groups or types, as with HSVs, although differences do exist. Pritchett (1980) recently reported that approximately 10% heterogeneity exists for the DNAs of HCMV strains AD-169 and Towne even though they share approximately 90% of their nucleotide sequences. The heterogeneity is due to the presence of unique nucleotide sequences, as demonstrated by the identification of restriction-enzyme fragments of the strains. Summers (1980) also reported the use of restriction endonuclease cleavage to analyze individual strains of viruses by producing fingerprints of the virus genome.

2. Replication of Murine CMV

The properties of MCMV have been studied extensively, since many investigators have used MCMV as a model for CMV latency and pathogenesis of disease. MCMV DNA has been analyzed by velocity sedimentation in neutral and alkaline sucrose gradients and by centrifugation in CsCl (Mosmann and Hudson, 1973). The molecular weight of MCMV DNA is 132×10^6 daltons, and the DNA has single-stranded regions that are alkali-sensitive. MCMV DNA contains 59% G+C; however, there are smaller fragments that band at two distinct densities with G+C contents of 57.5 and 61.5%. These same investigators (Mosmann and Hudson, 1974) conducted further studies on the two density fragments to test for MCMV-specific RNA synthesized in mouse embryo cells. On the basis of their results, they propose that there is specific control of the synthesis of RNA sequences that is altered during the MCMV growth cycle.

Electron-microscopic studies (Hudson *et al.*, 1976a) revealed that the main form of the MCMV virion is similar to that of other herpesviruses, i.e., a collection of capsids in a dense matrix surrounded by an envelope. The number of capsids in each virion ranges from 1 to more than 20, with the multicapsid virions seen at all times after infection. All size classes of MCMV are infectious and exhibit enhanced infectivity on centrifugation. No RNA has been detected in MCMV virions.

The overall rate of DNA synthesis in MCMV-infected mouse embryo fibroblasts appears to decrease shortly after infection, reaching its lowest rate approximately 10–12 hr p.i., and then increases slowly (Moon *et al.*, 1976). Analysis of host and MCMV DNA synthesis demonstrates that by 10–12 hr p.i., the synthesis of host-cell DNA is inhibited by approximately 95%. MCMV DNA synthesis begins 10–12 hr p.i., and by 24 hr p.i., the rate is slightly elevated when compared to host-cell DNA synthesis at 1–2 hr p.i. Finally, MCMV DNA synthesis occurs without simultaneous synthesis of host-cell DNA. When MCMV is UV irradiated or heat inactivated, host DNA synthesis is inhibited and MCMV-induced proteins are not detectable, which suggests that suppression of host DNA synthesis during the first 12 hr p.i. does not require early protein synthesis induced by MCMV. It is possible that this suppression is due to a structural protein(s) of the infecting virus.

MCMV was purified to analyze the structural protein(s) (Kim *et al.*, 1976). Thirty-three structural proteins were detected with a total molecular weight of 2,462,000 daltons and with a range from 11,500 to 255,000. Further analysis showed that at least six of these were glycoproteins; the two heaviest corresponded to two virus polypeptides. There appears to be ample genetic information present to code for the 73,000 amino acids (Spear and Roizman, 1972). The total molecular weight of the 33 proteins is equivalent to the sequencing of approximately 23,000 amino acids, which would require approximately 30% of the MCMV DNA.

Mouse 3T3 cells were synchronized to investigate whether MCMV replication is dependent on cell cycle (Muller and Hudson, 1977a). In asynchronous 3T3 cells, a 12-hr latent period is usual; however, when cells are synchronized and infected early in the G_1 phase, the latent period is extended to more than 24 hr. In addition, initiation of MCMV DNA synthesis in mouse fibroblasts requirès events associated with the host S phase, and MCMV DNA synthesis does not begin until after the commencement of synthesis of host DNA. Consistent with findings for HCMV, Muller and Hudson (1977b) were unable to detect cell or virus TK in MCMV-infected cells and found that TK is not essential for normal virus growth, since MCMV replicates in synchronized TK-negative (TK^-) 3T3-4E cells. MCMV is unable to induce a detectable increase in TK by 27 hr and does not stimulate 3T3-4E TK^- cells to incorporate exogenous thymidine. Bromodeoxyuridine has no inhibitory effect on MCMV replicated in a TK^- host. Eizuru *et al.* (1978) confirmed earlier reports that MCMV does not require TK and does not induce TK activity in MCMV-infected cells.

The synthesis of MCMV DNA in infected cells was measured by DNA–DNA reassociation kinetics with ^{125}I-labeled MCMV DNA as a probe (Misra *et al.*, 1977). Synthesis of MCMV DNA was detected as early as 8 hr p.i. in mouse embryo cells, and RNA synthesis was measured by reassociation kinetics (Misra *et al.*, 1978). Analysis of the data using a computer program similar to the one designed by Frenkel and Roizman (1972) showed that approximately 25% of the genome was transcribed before DNA replication, with less than half the transcripts from the cytoplasm. After MCMV DNA replication, approximately 38% of the genome was transcribed, also with less than half the sequences detected in the cytoplasm. Early and late RNA each comprised two classes that differed from 8- to 10-fold in amounts, and the early RNA was a subset of late RNA. When cells were infected late in the presence of ARA-C or cycloheximide, the RNA sequences were similar to those of early RNA.

PAGE has also been used to study the synthesis of MCMV proteins in MCMV-infected mouse embryo cells (Chantler and Hudson, 1978). Three immediate–early proteins were detected 4 hr p.i., but no other proteins were detected until after the initiation of MCMV DNA synthesis when major structural proteins are made. Between 24 and 30 hr p.i., 52 proteins were labeled in infected cells. Of these, 29 had electrophoretic mobilities similar to those of structural virus proteins, and 8 precipitated by antisera to infected-cell antigens may be nonstructural proteins. The remaining 15 are classified as host proteins. MCMV-infected S-phase and G_0-phase nuclei incorporated more [^{3}H]-dTMP than did uninfected nuclei (Muller and Hudson, 1978). The infected nuclei contained a DNA polymerase that is stimulated by ammonium sulfate and inhibited by antibodies to infected-cell proteins. The G_0-phase cells also contain this DNA polymerase; however, only the S-phase nuclei are able to synthesize MCMV DNA, which suggests that a cell-cycle factor is necessary for MCMV DNA synthesis.

Following up similar studies with HCMV (Gupta and Rapp, 1978), Chantler (1978) reported that an increase in salt concentration in the medium of MCMV-infected cells produces a selective inhibition of cellular relative to viral messenger RNA translation. By establishment of this hypertonic condition, eight early virus proteins have been identified in MCMV-infected cells. Studies of MCMV gene expression during nonproductive infections in G_0-phase 3T3 cells (Muller *et al.*, 1978) detected infectious centers but no MCMV DNA replication. Of MCMV DNAs, 18% (early) and 21% (late) were transcribed during nonproductive infection (one abundance class), whereas 26% (early) and 38% (late) (comprising two abundance classes) were transcribed during productive infection. Five nonstructural MCMV polypeptides were detected in the G_0-phase cells. Virus DNA synthesis began and progeny virus appeared when serum and fresh medium were added to the G_0-phase cells.

Gönczöl *et al.* (1978) observed that heat-inactivated MCMV stimulates cell DNA synthesis, but that active or UV-irradiated MCMV does not. Heat-inactivated MCMV penetrates the cells, stimulates host DNA synthesis, and induces synthesis of early MCMV antigens. Another group

(Moon *et al.*, 1979), using SDS-PAGE and autoradiography, studied MCMV-induced protein synthesis. They detected synthesis of at least 14 MCMV-induced proteins in a lytic cycle, and these were distributed among three groups on the basis of rate of protein synthesis. They classified the proteins as immediate–early, early, and late virus-induced proteins according to the time sequences of the initiation of protein synthesis. Lakeman and Osborn (1979), using equilibrium centrifugation in CsCl gradients, reported that the infectious DNA from MCMV had an approximate molecular weight of 136×10^6 daltons in contrast to HCMV DNA that was detected in two peaks with molecular weights of 130×10^6 and 150×10^6 daltons.

Two strains of MCMV, Smith and K181, were studied because of their differences in virulence in mice (Misra and Hudson, 1980). Although the strains reacted the same in cell culture and the virus DNAs exhibited the same reassociation kinetics, minor differences in their base sequences were revealed by restriction-endonuclease techniques. Strain K181 was, in fact, more virulent and gave higher yields of infectious MCMV in salivary glands. The very subtle differences in base sequences may be related to MCMV pathogenesis. Hudson (1980) observed that MCMV infection of mouse fibroblasts resulted in an early inhibition of thymidine incorporation into cell DNA; however, the uptake of other nucleosides was normal or enhanced. In addition, nuclear monolayers from MCMV-infected cells incorporated thymidine triphosphate into cell DNA as fast as uninfected cell nuclei. There was no detectable difference between the total DNA content of infected and control cultures. Therefore, it appears as though MCMV does not inhibit or stimulate cellular DNA synthesis.

3. Simian CMVs

Waner *et al.* (1978) experimented with two SCMVs, rhesus CMV (RCMV) and vervet CMV (VCMV), to investigate the effect infection might have on cell DNA synthesis. RCMV or VCMV infection of permissive cells resulted in an inhibition of cell DNA synthesis. However, when cells were arrested with IUdR and then infected, cell DNA synthesis was induced. When the virus was UV-light-inactivated, there was no cell DNA stimulation in the IUdR-treated cells.

II. *IN VITRO* INFECTION OF HOST CELLS

A. Human Cytomegaloviruses

Cytomegaloviruses (CMVs) are found in man and lower mammalian species and are a very species-specific group that can cause asymptomatic as well as severe and fatal diseases (for a review, see Wright, 1973). *In vitro* and *in vivo* infections with CMV are characterized by enlarged cells that contain intranuclear (Figs. 2–4) and cytoplasmic (Figs. 5 and 6) inclusions. Waner and Weller (1974) were the first to report that human CMV (HCMV) is able to replicate in Vero (monkey) cells and to produce infectious, cell-associated virus. Cytopathic changes were also seen in HCMV-infected bovine cells, and although HCMV replication was not detected, synthesis of HCMV antigens was detected by immunofluorescence (IF) and complement fixation. Furukawa *et al.* (1975c) reported that HCMV replication was dramatically inhibited when human fibroblast cells were UV irradiated prior to virus infection. *In vitro* HCMV infection of human lung cells is possible with infectious virus released up to 8 weeks postinfection (p.i.) (Michelson-Fiske *et al.*, 1975). This report demonstrates that human epithelial cells can be productively infected with HCMV *in vitro*. Human diploid epithelial cells derived from floating human amniotic fluid are able to support the replication of HCMV (Vonka *et al.*, 1976). A later report (Figueroa *et al.*, 1978b) supports these findings and shows that HCMV can infect human amnion cells, producing slow virus growth, low virus titer, and characteristic cytopathic effects (CPE).

Waner and Budnick (1977) developed a method for measuring the blastogenic response of human lymphocytes to HCMV. Using a preparation of HCMV antigens, they were able to transform lymphocytes from asymptomatic seropositive individuals. Attempts to replicate HCMV in human leukocytes, lymphocyte subpopulations, and hemic cell lines were unsuccessful, although herpes simplex virus was shown to replicate in mitogen-stimulated B and T lymphocytes and in B-, T-, and myeloid-cell lines (Rinaldo *et al.*, 1978). Another study (Sarov *et al.*, 1978) demonstrated that HCMV and HCMV dense bodies are capable of stimulating human peripheral lymphocytes from CMV-positive patients; however, there is little lymphocyte response in seronegative donors. Booth *et al.* (1978) reported that large syncytia are produced routinely after prolonged *in vitro* infection of human embryo fibroblast (HEF) cells by HCMV. They propose that syncytium formation is a result of defective HCMV abortively infecting the cells. Figueroa *et al.* (1978a) studied HCMV replication in three human embryo cell lines transformed *in vitro* by HCMV. They found the cells completely resistant to infection by HCMV strains AD-169 and Mj, but detected the synthesis of HCMV DNA and HCMV proteins in CMV AD-169-transformed cells. They suggest that the resistance to HCMV infection demonstrated by HCMV-transformed human cells may indicate the presence of the HCMV genome in the cells.

Furukawa (1979a) investigated HCMV infection in human osteogenic sarcoma cells transformed by murine sarcoma virus and found that there is limited growth of virus in these cells and that infection might be dependent on the cell cycle. Results from experiments using synchronized cells suggest that establishment of infection may be more efficient in the S phase because host factors responsible for virus growth change during the cell cycle. A brief report (Färber *et al.*, 1979b) noted that HCMV infection of rabbit lung fibroblasts induces cytopathic changes and

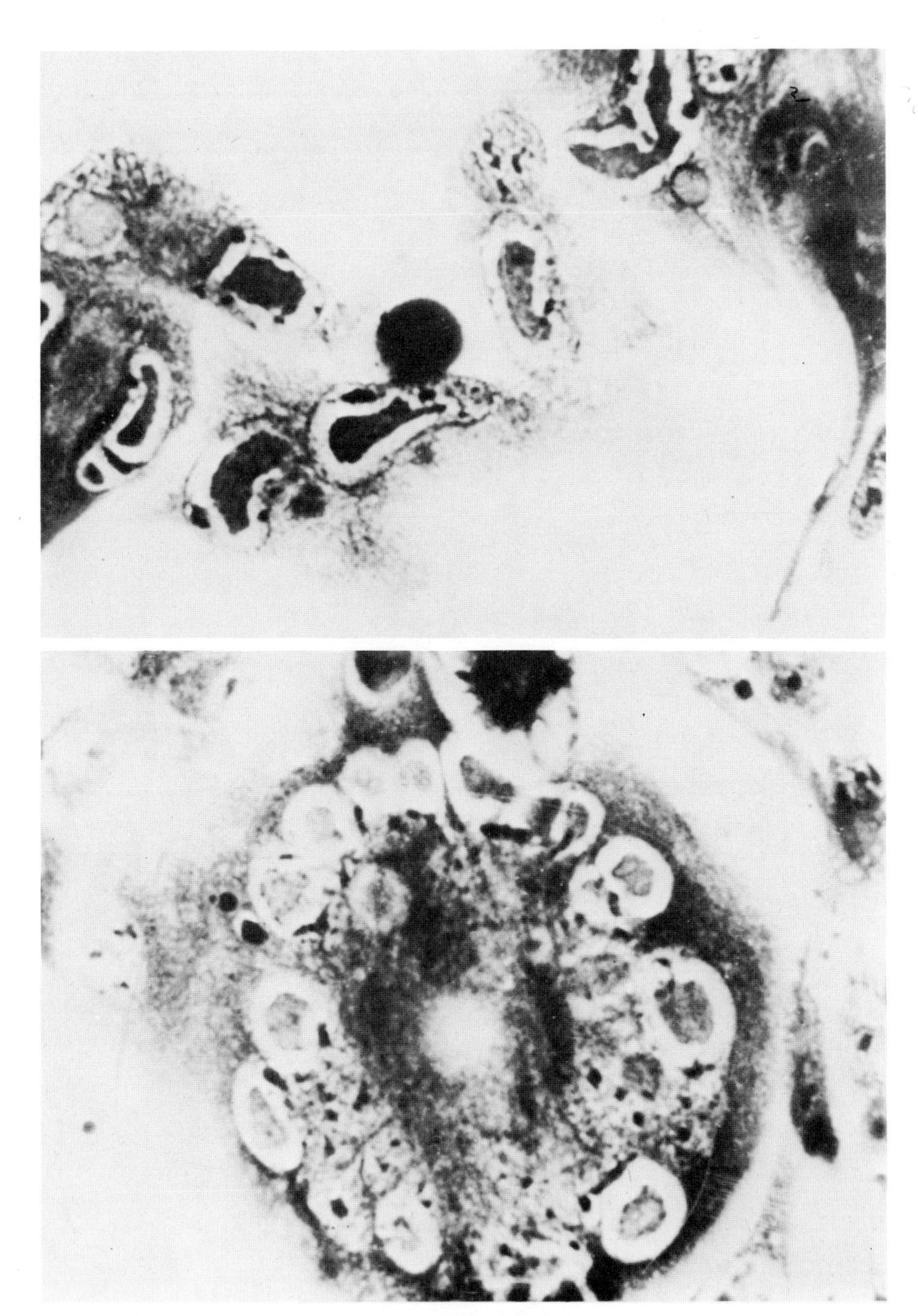

FIGURE 2. HCMV nuclear inclusions in human cells. Stained with hematoxylin and eosin. Kindly supplied by Dr. Mukta Webber.

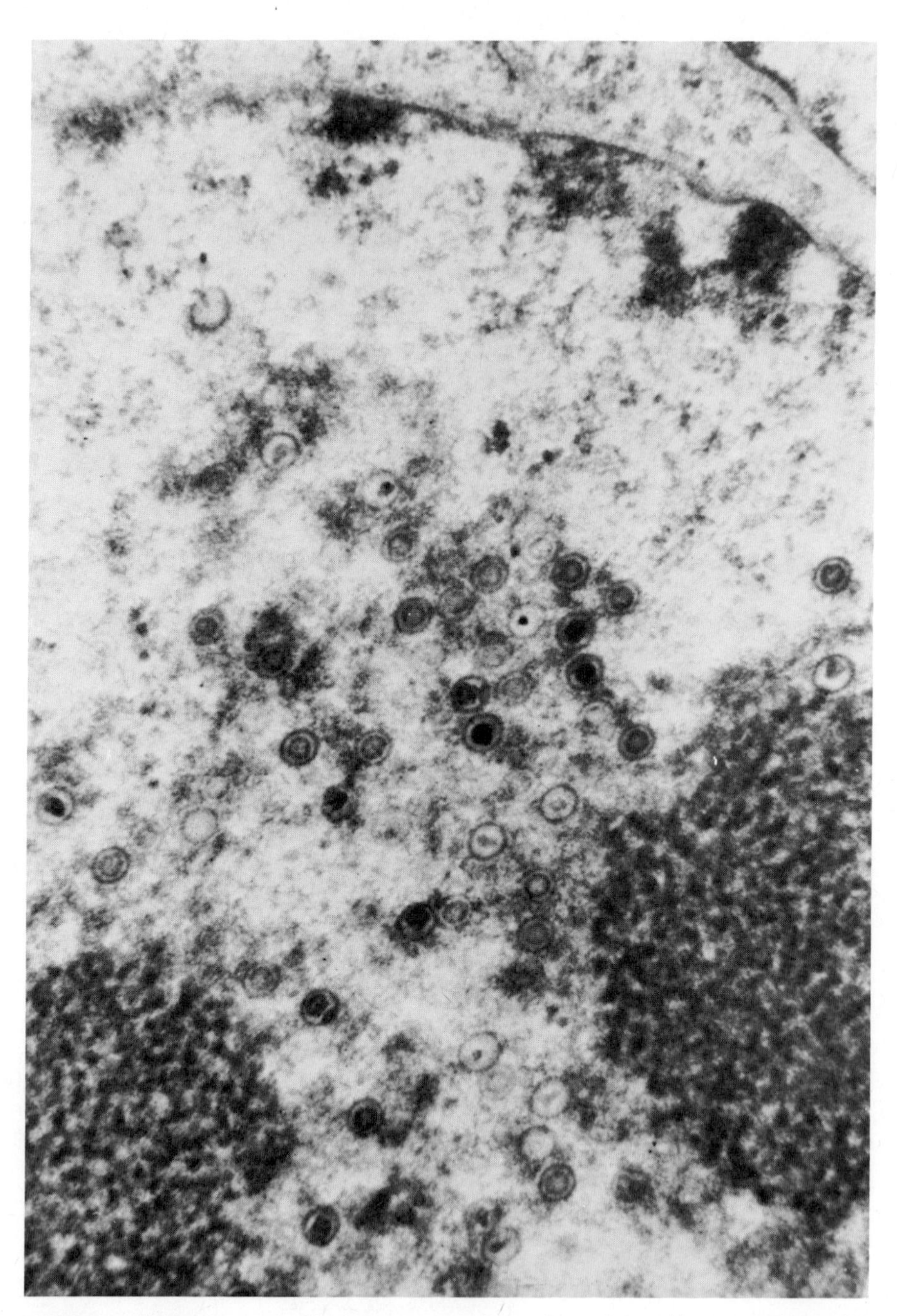

FIGURE 3. Electron micrograph of HCMV particles in the nucleus of human cells. Kindly supplied by Dr. Mukta Webber.

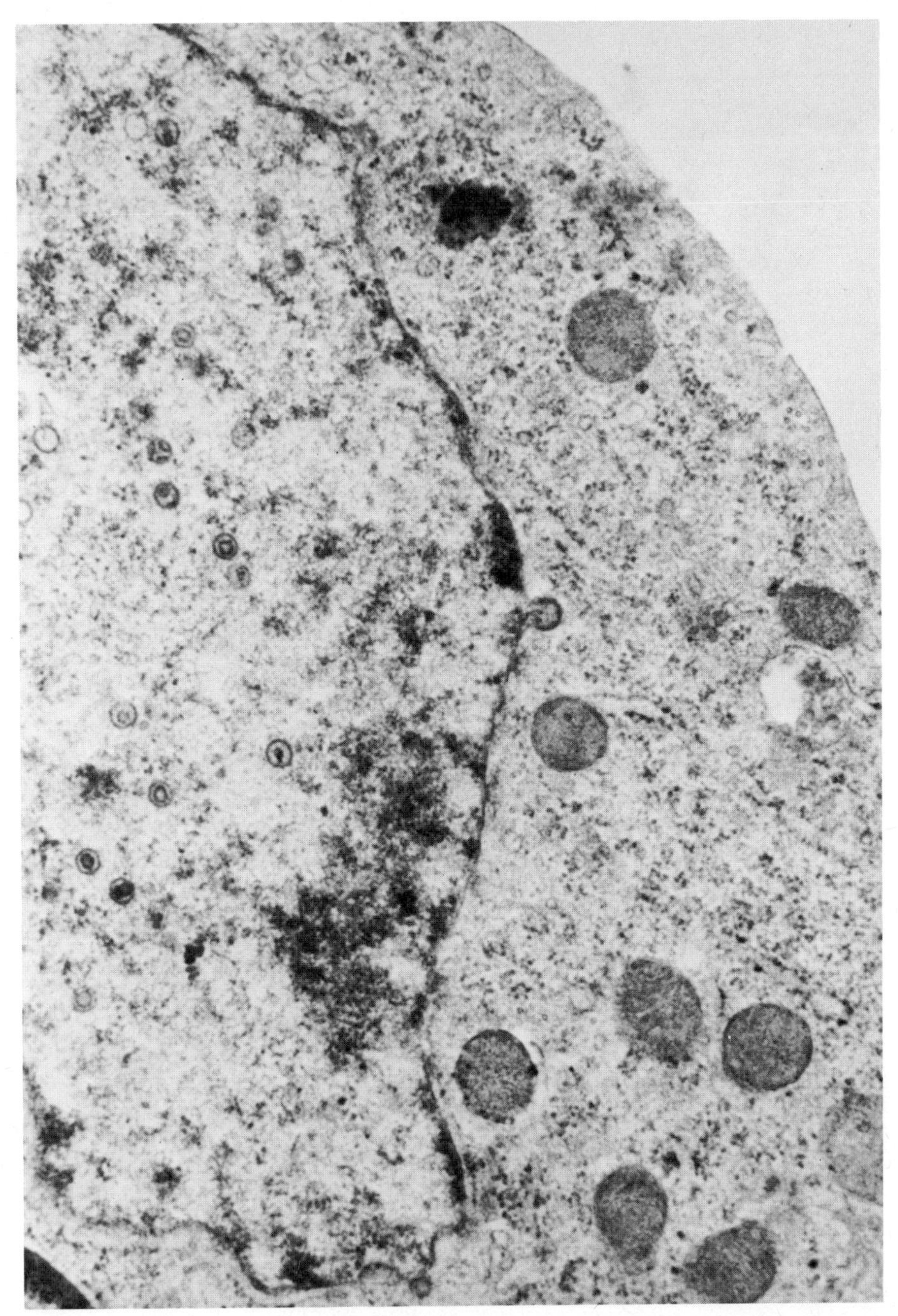

FIGURE 4. Electron micrograph of HCMV particles budding from the nuclear membrane of human cells. Kindly supplied by Dr. Mukta Webber.

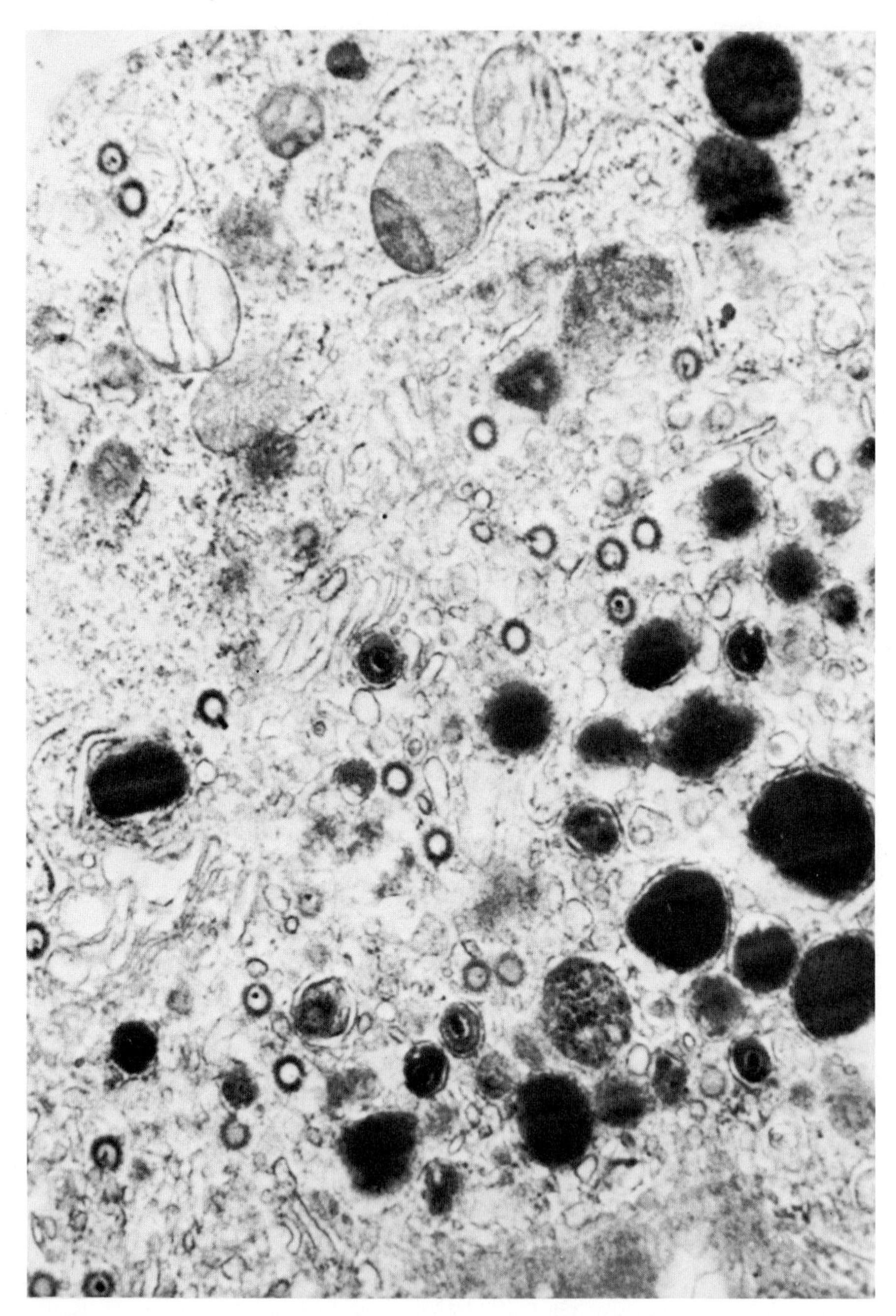

FIGURE 5. Electron micrograph of HCMV particles in the cytoplasm of human cells. Kindly supplied by Dr. Mukta Webber.

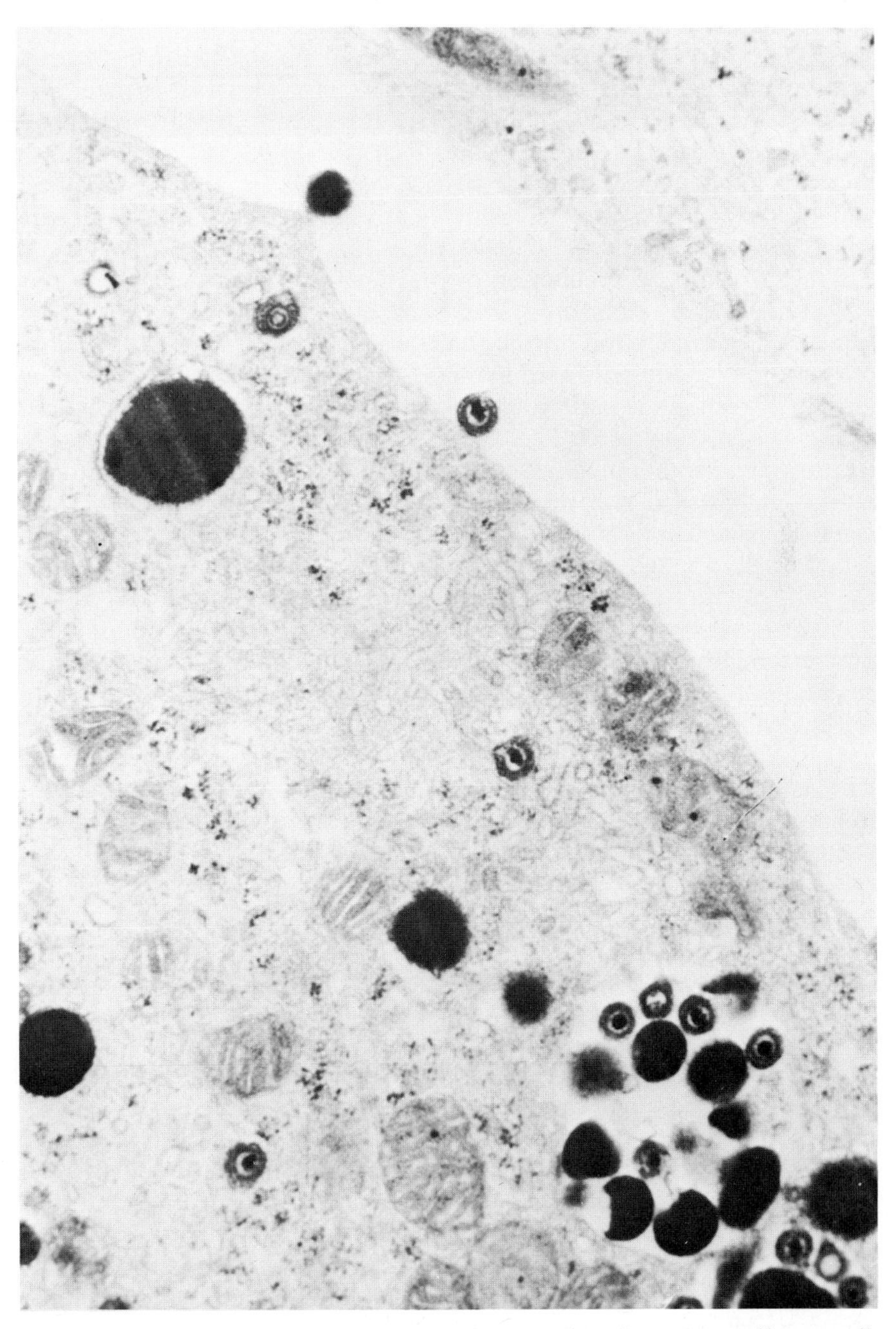

FIGURE 6. Electron micrograph of HCMV budding from the cell membrane of human cells. Kindly supplied by Dr. Mukta Webber.

HCMV-specific antigens. When these infected cells are cocultivated with human lung fibroblast cells, infectious virus can be recovered and HCMV-specific antigens (like those in HCMV-infected Vero cells) can be detected for as long as 28 days. Rabbit lung fibroblasts have been used to determine antibodies against HCMV-induced early antigens in patients with acute HCMV infection and in healthy controls (Färber *et al.*, 1979a). On HCMV infection of rabbit cells, only early HCMV-specific antigens are produced, which allows preparation of HCMV-early antigens without inhibiting DNA synthesis. Garnett (1979c) reported that fusion of HCMV-infected HEF cells results in production of multinucleate giant cells, an observation made earlier but only with phagocytosis (Diosi *et al.*, 1972). In addition, the production of giant cells was rapid in contrast to the slow production of giant cells seen by Booth *et al.* (1978). Lymphocyte transformation tests were performed to study *in vitro* stimulation of human colostral lymphocytes by HCMV (Meggs and Beer, 1979). These investigators found that lymphocytes from seropositive donors are stimulated by the HCMV antigen, whereas those from seronegative donors do not react. Tocci and St. Jeor (1979a) report that lymphoblastoid cells of B- and T-cell origin are susceptible to infection with HCMV, although few of the cells produce infectious virus.

The *in vitro* proliferation of lymphocytes induced by HCMV-infected human fetal fibroblasts and cell-free HCMV has been compared by Schirm *et al.* (1980). Lymphocytes from seropositive donors are stimulated by infected fibroblasts, whereas lymphocytes from seronegative donors are not. The lymphocytes react to cell-free HCMV in most seropositive donors, but not in most seronegative donors. Overall, lymphocyte reactivity to HCMV-infected human fetal fibroblasts is significantly higher than to cell-free CMV. These researchers suggest that HCMV-infected cells and cell-free HCMV are recognized by CMV-specific lymphocytes.

Albrecht *et al.* (1980a) examined the development and progression of CPE of five strains of HCMV in human cell cultures infected at a multiplicity of infection (m.o.i.) of 5 plaque-forming units (PFU)/cell at different times. By 5 hr p.i., cell rounding and early cytoplasmic inclusions were observed; however, nuclear inclusions were not observed until 24 hr p.i. and then as a homogeneous eosinophilic bead. Between 96 and 120 hr p.i., both cytoplasmic and intranuclear inclusions underwent extensive morphogenesis. Discrete beadlike subunits of nuclear inclusions (Fig. 7) were observed at 48–72 hr p.i., after 96 hr p.i., they began to break up. Strain-related distinctions in cytopathology that were independent of m.o.i. and the type of fibroblast cells used were also noted. One study reported that HCMV infection of human fibroblasts does not impair histone synthesis until late after infection (Radsak and Schmitz, 1980); therefore, HCMV does not induce host-cell DNA replication. This work has not yet been extended by these researchers, and there is no other work to support this view. Nishiyama and Rapp (1980) observed that HCMV infection of primary rabbit kidney and Vero cells enhances the

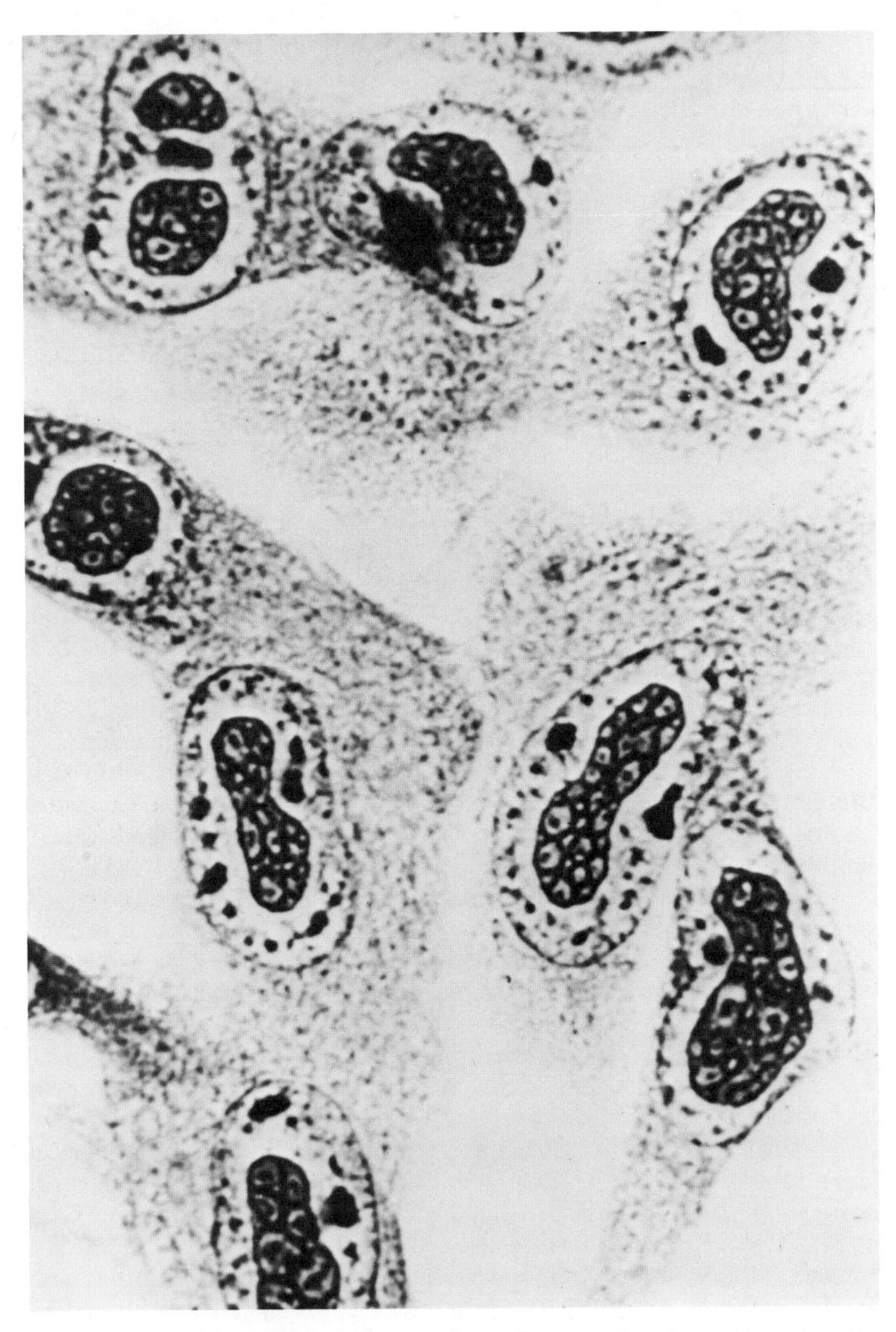

FIGURE 7. Photomicrograph of inclusion bodies in human embryo lung cells produced by HCMV strain AD-169 after 120 hr. Kindly supplied by Dr. Thomas Albrecht.

survival of UV-irradiated HSV. This HCMV reactivation is sensitive to UV irradiation, although HCMV reactivation is more resistant to irradiation than is infectivity. The enhancement induced by HCMV infection is further amplified when the cells are pretreated with cycloheximide. It is possible that a functional HCMV genome may be required for enhancement of reactivation and that HCMV infection of cells may trigger repair mechanisms in host cells.

B. Murine Cytomegalovirus

Plummer and Goodheart (1974) reported the replication of murine CMV (MCMV) in rabbit kidney cultures, but virus yields were lower than those in MCMV-infected mouse cells. MCMV growth was significantly enhanced when the cells were grown in the presence of 5-iodo-2′-deoxyuridine. Mäntyjärvi *et al.* (1977) developed a differentiated tissue model to study MCMV infection *in vitro*. When tracheal rings from mice were infected with MCMV, infectious virus at titers of $10^{5.5}$ PFU/ml was obtained. In addition, MCMV antigens were present in MCMV-infected epithelial cells. Tracheal organ cultures were infected with MCMV to study the susceptibility of different cells to infection (Nedrud *et al.*, 1979). In tracheal organ cultures from MCMV-susceptible strains of mice, CPE developed and virus titers of 10^5 PFU/ml were obtained within 2–3 hr p.i. Cultures from MCMV-resistant strains of mice failed to develop CPE, and the titers of virus were 10- to 100-fold lower. MCMV-infected fibroblast cells demonstrated no differences in susceptible and resistant strains of mice. Resistance may be due to a genetic predisposition of epithelial cells to MCMV infection.

Centrifugal enhancement of infectivity has been detected in mouse embryo 3T3 cells, rat kidney cells (Hudson *et al.*, 1976b), and primary mouse embryo cells (Osborn and Walker, 1968). Three different strains of MCMV, 20 plaque-purified strains of MCMV, and at least one HCMV strain (AD-169) exhibit this property. Ten- to 100-fold more infectious virus can be detected by this method than by standard inoculation methods. However, the reason for this characteristic is unknown and may be an intrinsic property of CMV.

MCMV infects and replicates in primary and continuous African green monkey kidney, primary rabbit kidney, baby hamster kidney, primary fetal sheep brain, L cells, and rabbit kidney cells (Kim and Carp, 1971). The CPE observed in all these cell lines is similar to that in MCMV-infected mouse embryo fibroblast cells. However, human embryo brain, human embryo kidney, WI-38, HeLa, and Hep-2 cells are not susceptible to infection with MCMV. Kim and Carp (1972) reported the production of an abortive infection of human diploid cells (WI-38) after inoculation with MCMV. MCMV CPE was detected within 12 hr p.i., and MCMV adsorbed to the WI-38 cells with the same efficiency that it adsorbed to mouse embryo fibroblast cells; however, infection did not result in the synthesis of infectious MCMV.

Tegtmeyer and Craighead (1968), using the Smith strain of MCMV to infect adult mouse peritoneal macrophages, found that virus production and the appearance of typical intranuclear inclusions are slower than in mouse fibroblast cultures, although the level of virus production is approximately the same for both. MCMV-infected spleen cells from different strains of mice were studied to determine their extent of permissiveness to MCMV (Hudson *et al.*, 1977). It was found that infectious centers were established in 1% or less of the cells, but that virus could be rescued by cocultivation. A small amount of the cell population replicated MCMV, but DNA synthesis was inhibited. Loh and Hudson (1979) continued studies with MCMV-infected spleen cells and observed, with autoradiography, that macrophages were infected and formed infectious centers, but supported only limited MCMV replication. B lymphocytes contained a small cell fraction that allowed replication. T lymphocytes did not produce infectious centers and were nonpermissive for replication. Wu and Ho (1979) also studied MCMV infection of B and T lymphocytes and found that viremia persisted in lymphocytes for as long as 4 weeks. Sonicated lymphocytes had no detectable free virus 8–10 days p.i.; however, when these lymphocytes were enclosed in a Millipore chamber on a fibroblast feeder layer, the T cells produced free virus, but the B cells did not. Hyperimmune mouse serum reduced B-cell-associated infectious centers by 74% but T-cell-associated infectious centers by only 38%.

Three strains of mice were latently infected at birth with MCMV, and by adulthood, the testes of these mice contained 4–6 genome equivalents per 100 cells when tested by hybridization with mouse DNA and MCMV DNA (Dutko and Oldstone, 1979). Thus, it appears that MCMV can replicate in spermatogenic cells and is harbored in the testes during acute and latent infections.

Adult, female C_3H mice were infected with MCMV, and the macrophage and reticuloendothelial system functions were studied (Schleupner *et al.*, 1979). Macrophages were tumoricidal *in vitro* and also showed antiviral activity by inhibiting the replication of vaccinia virus. In addition, there was more protection against *Listeria monocytogenes*, which suggests that MCMV infection had stimulated the reticuloendothelial system. Infection of spleen cells with MCMV inhibited their ability to respond to mitogens *in vitro* (Loh and Hudson, 1980), and this inhibition was proportional to the m.o.i. Infected macrophages were found to have a defective capacity to mediate the response of T lymphocytes to concanavalin A, and it is possible that immunosuppression due to MCMV occurs because of an interference with macrophage function. Barabas *et al.* (1980) observed that MCMV produces CPE in mouse brain, guinea pig embryo brain, human brain, and human fibroblast cells and replicates in mouse embryo brain and guinea pig embryo brain cells; they also detected MCMV-specific antigens by IF. MCMV is able to replicate in mouse lung and peritoneal macrophages, but cytopathic changes develop and kill the macrophages (Shanley and Pesanti, 1980). Typical CMV particles were seen in the nuclei and cytoplasm of the infected macrophages, and MCMV antigens could be detected by indirect IF. At 24 hr p.i., MCMV rapidly disrupts phagocytosis by these MCMV-infected lung macrophages.

Manischewitz and Quinnan (1980) demonstrated that MCMV-infected mouse cells can by lysed by antibody-dependent cell-mediated cytotoxicity and that this mechanism might be necessary for the control of chronic and latent infections. When splenic lymphocytes from MCMV-infected mice are treated *in vivo* with antilymphocyte globulin and prednisolone, *in vitro* proliferative responses to CMV antigen are lost (Mattsson *et al.*, 1980). This loss of specific cell-mediated immunity and depression of humoral immunity suggest that immunosuppression creates an inability to respond to an active MCMV infection.

C. Guinea Pig Cytomegalovirus

At least two herpesviruses are harbored by guinea pigs: guinea pig herpeslike virus (GPHLV) and guinea pig CMV (GPCMV) (for a review, see Hsiung *et al.*, 1980). GPCMV infection of guinea pig kidney cells does not produce CPE or significant amounts of intranuclear inclusions; however, the virus persists in kidney cells for 8–10 days with low levels of infectivity. Other cells of the guinea pig can also be involved (Fig. 8). GPHLV replicates faster than GPCMV in guinea pig embryo cells. Tubular structures are seen in nuclear inclusions of GPCMV but not in GPHLV, and enveloped virus particles enclosed in large vacuoles that are seen in GPHLV-infected cells are not usually observed in GPCMV-infected cells. Dense bodies and virus particles are often observed in GPCMV-infected embryo cells. Patrizi *et al.* (1967) observed that the nuclei of guinea pig embryonic skin muscle cells infected with GPCMV contain central inclusions, tubular structures, and virus particles in varying states of development. Multilamellated structures between the nuclear envelope and the intranuclear inclusion are present only in infected cells. These structures are located close to or near nuclear membranes and appear in two forms: needle-shaped and spherical. They are similar in that they are found in infected cell nuclei, seem to arise from the nuclear envelope, and have inner lamellae that appear thicker than outer lamellae. Virus particles within the needle-shaped structures have acquired their second coat and are morphologically mature. Obviously, studies with this virus need to be extended for a better understanding of its properties.

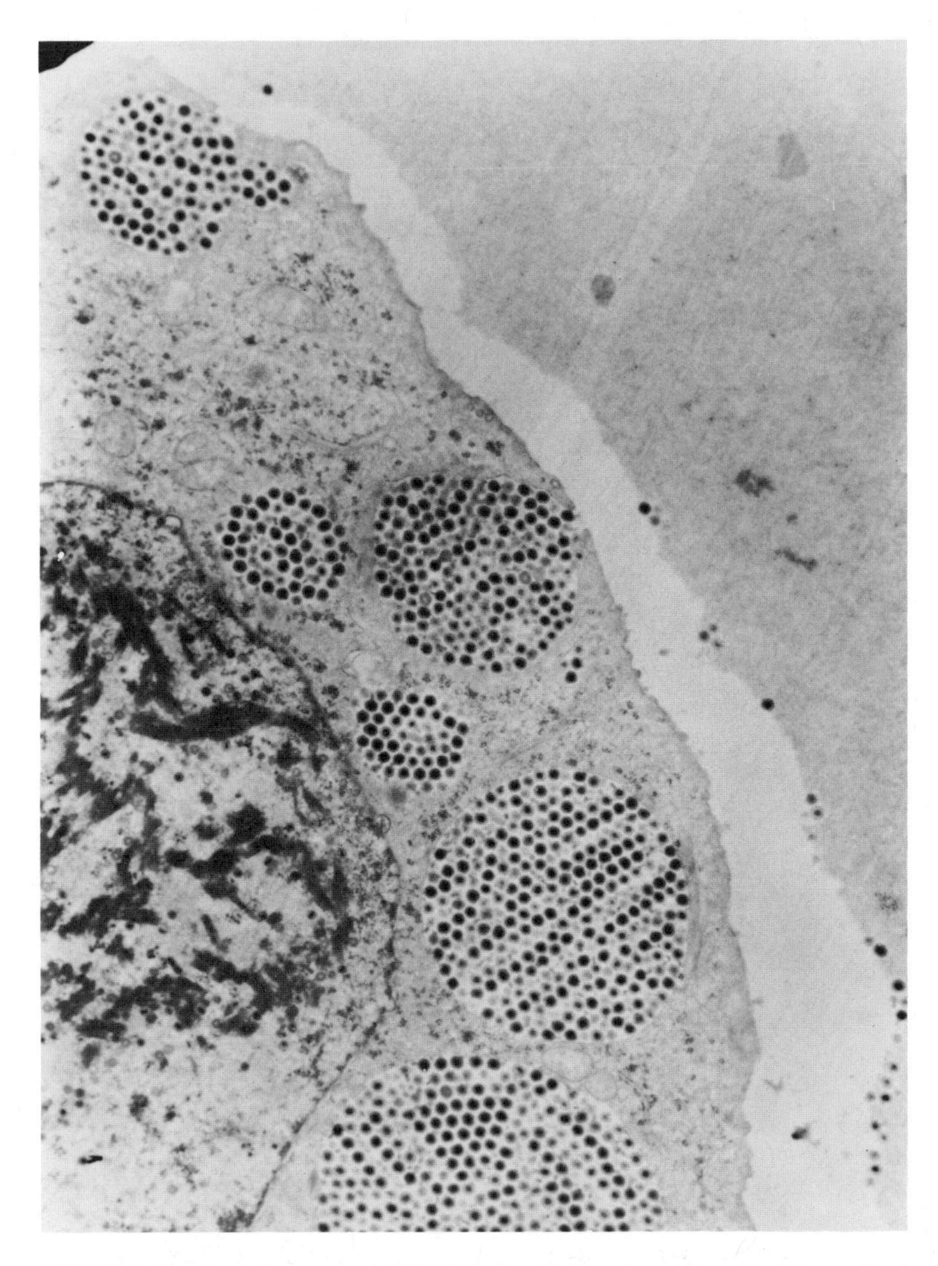

FIGURE 8. Electron micrograph of GPCMV isolated from submaxillary gland duct cells of a guinea pig. Kindly supplied by Dr. Caroline K. Y. Fong (Fong *et al.*, 1980).

D. Other Cytomegaloviruses

Primary African green monkey kidney cells have been found to harbor simian CMVs (SCMVs) (K.O. Smith *et al.*, 1969). Of 52 lots tested, 27 contained CMVs. The isolation of these viruses and serological data demonstrate that SCMV is prevalent as a latent infection among healthy African green monkeys. These SCMVs are hard to detect and difficult to study, but are worth the further effort as possible analogues to HCMV.

III. PATHOGENESIS IN NATURAL HOSTS

In 1956, M.G. Smith (1956) and Rowe *et al.* (1956) separately isolated cytopathic viruses from human tissues. M.G. Smith (1956) isolated the virus from human salivary gland tissue and also from the kidney of an infant with cytomegalic inclusion disease (CID). Rowe *et al.* (1956) isolated similar viruses from the adenoids of children. These isolations were quickly followed by the isolation of cytopathic agents from three infants with CID by Weller *et al.* (1957). These agents produced intranuclear inclusions and were obtained from liver biopsy material and urine of children with microcephaly, jaundice, hepatosplenomegaly, cerebral calcification, and chorioretinitis. E.-S. Huang *et al.* (1980) studied cytomegaloviruses (CMVs) from mothers and their children to detect whether recurrent infections are a result of reinfection or reactivation of endogenous virus. Restriction endonuclease analysis of purified human CMV (HCMV) DNA was used to detect genetic relatedness among 21 strains of HCMV. The strains from five of six congenitally infected infants were the same as or very similar to strains of their mothers; strains from congenitally infected siblings were similar. Endogenous HCMV appears to be the source of recurrent infection that can be transmitted intrauterinally in immune women. Reinfection occurs, but less commonly than anticipated.

A. Pathogenicity in Rodents

The pathogenicity of murine CMV (MCMV) infection has been studied extensively over the past 25 years (for reviews, see Wright, 1973; Hudson, 1979). M.G. Smith (1954) was the first to propagate MCMV in explant cultures of mouse embryo tissue. MCMV intranuclear inclusions were detected in sections from the salivary glands of the mice. The virus was serially passed in MCMV-free mice and then serially passed in culture. When mice were inoculated intraperitoneally (i.p.) with culture fluids, salivary gland virus disease, with characteristic large intranuclear inclusions, was produced. Medearis (1964a) was able to induce long-lasting infections (>6 months) in mice inoculated with MCMV subcutaneously (s.c.) or i.p. and to recover virus from eye swabs, throat swabs, and urine specimens from these mice. He found a positive correlation between the amount of virus recovered from urine and the kidneys and also between the amount recovered from the throat swabs and the salivary glands. Serum neutralizing antibody and complement-fixing antibody were demonstrated in the sera. Medearis (1964b) also found that MCMV could not be recovered from fetuses or suckling mice conceived during prolonged MCMV infection of the mother or during maternal viremia, or from mothers inoculated intravenously (i.v.) and exposed to stresses. When young adult mice were infected i.p. with sublethal doses of MCMV, an acute and nonfatal hepatitis resulted (Henson *et al.*, 1966). This infection was accompanied by MCMV replication, interferon production in the liver, and the presence of intranuclear inclusions in hepatic cells. The severity of the hepatitis was dependent on the amount of MCMV inoculated: the mice recovered although they were chronically infected.

Osborn *et al.* (1968) studied immunosuppression during acute MCMV infection. They reported that MCMV suppressed the immune response when 4-week-old mice were acutely infected with a sublethal dose of MCMV. Even as late as 14 days postinfection (p.i.), neutralizing antibody titers against MCMV were minimal.

When the pathogenesis of MCMV infection in pregnant mice was investigated (Johnson, 1969), it was found that fetal loss was dramatically increased, although there was no virological or histological evidence of MCMV infection in fetal tissue at any time during pregnancy. If fetal or suckling mice were inoculated, then active infection was initiated. Placental infection occurred with the production of virus and inclusion-bearing cells.

The pathogenesis of chronic MCMV infection in the submaxillary glands of two types of mice was examined (Henson and Neapolitan, 1970). In mice inoculated i.p., virus titers reached maximum levels in the submaxillary glands within 3 weeks, but decreased more quickly in C_3H mice than in ICR mice. ICR mice had inclusions and infectious virus that persisted for a longer period of time. These investigators concluded that chronic MCMV infection in the submaxillary glands is dependent on persistence of intact inclusion-bearing cells and that chronic infection is terminated by chronic inflammatory cells. Henson *et al.* (1972) also isolated MCMV from the spleen and abdominal lymph nodes of chronically infected mice and established that MCMV could be recovered from lymphoid tissue after prolonged cocultivation with mouse embryo fibroblast cells. They could not detect intranuclear inclusions in the spleens, nor were herpesviruses detected by electron microscopy, but virus could be isolated from the mice up to 55 days p.i. When spleen cells from chronically infected mice were injected into uninfected mice, MCMV was transmitted.

Henson and Strano (1972) examined MCMV chronically infected submaxillary glands by electron microscopy and observed that lymphocytes are responsible for the termination of chronic MCMV infection, that MCMV infection of acinar epithelium is not cytolytic, and, finally, that normal cells undergo necrosis during termination of chronic MCMV infection. They suggest that the lymphocytes release a substance that produces indiscriminate necrosis of acinar cells. Howard and Najarian (1974) also studied immunosuppression due to MCMV and found that maximal suppression of the immune function occurs in the 1st week p.i., with the degree of suppression dependent on the virus inoculum. Howard *et al.* (1974) reported that MCMV infection depresses cell-mediated immunity and found that in mice previously infected with MCMV, skin grafts survived for a longer period across the H-2 histocompatibility barrier.

Another study (Gardner *et al.*, 1974) established that wild mice chronically infected with MCMV in nature have a high incidence of disseminated and virulent MCMV infection if immunosuppressed with antithymocytic sera. Olding *et al.* (1975) reported that MCMV could be recovered from spleen lymphocytes of MCMV-infected animals up to 5 months p.i. These spleen lymphocytes had been cocultivated with histoincompatible mouse embryo cells from uninfected animals; however, cocultivation of spleen lymphocytes with mouse embryo cells from histocompatible animals did not activate MCMV. The lymphoid cells from the MCMV-infected mice were purified, and it was observed that MCMV resides in the B lymphocyte population.

MCMV infection was enhanced by host-graft reactions (Wu *et al.*, 1975). These investigators reported that MCMV chronically infected mice received skin allografts from histoincompatible donors with a subsequent increase in MCMV titers in the spleens and kidneys within 3 days after the graft was made. This report may have bearing on what happens in human recipients of renal transplants, since the host-vs.-graft reaction alone may amplify the virus in a chronically infected host. A later study (Dowling *et al.*, 1977) on MCMV infection during graft-vs.-host reaction reported that MCMV was recovered more often from mice undergoing graft-vs.-host reaction than from control mice and that the reaction can enhance MCMV infection in a chronically infected host.

In 1975, Schwartz *et al.* (1975) observed the effect of MCMV on the developing retina and noted significant lymphoid cell necrosis, thymic atrophy, growth retardation, bacteremia, and death following intraocular inoculation of newborn mice. They proposed that thymic injury was responsible for the subsequent abnormal physical development and that thymic atrophy was clearly a result of virus infection.

Margolis and Kilham (1976) experimentally induced a severe meningoencephalitis in neonatal and suckling mice by intracerebral (i.c.) inoculation of MCMV. Neuronal involvement and giant-cell formation were prominent features, and they suggested that cell division is not necessary for activation of MCMV. Olding *et al.* (1976) found three distinct pathological groups in MCMV-infected mice: (1) mice that died within 4 weeks p.i. with tissue injury as a result of MCMV infection; (2)

mice that survived initial MCMV infection and were chronically infected; and (3) mice in which shedding of virus could not be detected. These studies demonstrated that MCMV infection of mice is an excellent model for studying latency, tissue tropism, and genetic susceptibility.

Young *et al.* (1977) artificially inseminated female mice with a mixture of sperm and MCMV. On fertilization, maternal infection was induced. Tissue cultures were prepared after 14 days, and in one case, MCMV was recovered from embryonic fibroblasts. Neighbour (1978) reported that mouse preimplantation embryos infected with MCMV *in vitro* developed normally to the blastocyst stage with no evidence of embryonic infection. However, when female mice were inoculated i.p. with MCMV, a generalized infection developed and embryonic development was retarded *in vivo*. When the embryos were transferred to noninfected mice, they developed normally.

Hamilton *et al.* (1978) reported that inoculation of EL_4, a mouse ascites homograft, had no effect on (1) the host response to MCMV, (2) duration of MCMV persistence, or (3) anti-MCMV complement-fixing antibody titer. However, prior MCMV infection resulted in reduced numbers of splenocytes. These investigators suggest that in MCMV-infected animals, there is a meaningful reduction in the number and function of B- and T cells directed against alloantigens in EL_4.

Macrophages in CD1 mice can be productively infected by MCMV from salivary glands (Mims and Gould, 1978), and the susceptibility of the macrophages to MCMV infection increases 5-fold after one passage in mouse embryo fibroblasts and 15-fold by passage eight. This increase in the susceptibility of macrophages to virus infections parallels a decrease in pathogenicity. Serum obtained from MCMV-infected mice 3–5 days p.i. contained complement-requiring neutralizing antibody that, when passively transferred, protected the mice from lethal infection with homologous virus (Araullo-Cruz *et al.*, 1978). This active antibody is a heat-stable 7 S immunoglobulin.

An *in vitro* lymphocyte-proliferation assay has been used to test cell-mediated immunity to MCMV (Howard *et al.*, 1978). Splenic lymphocytes from MCMV-infected mice undergo proliferation when incubated with MCMV antigen; however, lymphocytes from uninfected mice do not proliferate. Lymphocyte proliferation is specific and directly related to the number of lymphocytes in each culture and to the amount of MCMV antigen.

Mims and Gould (1979) compared MCMV infection in salivary glands, kidneys, adrenal glands, ovaries, and epithelial surfaces by fluorescent-antibody staining, histology, and virus titration. The submaxillary gland was most affected by MCMV infection, and virus could be recovered from the saliva for up to 58 days. Ovaries were also infected, especially after i.v. inoculation of pregnant mice, but the corpora lutea and ova were not infected. The medulla was affected more in suckling mice, while the cortex was more affected in old mice. MCMV could not be recovered from mammary glands or milk. Only the dermis became infected; respiratory epithelial cells did not. Almost all surviving mice had chronic glomerulonephritis and hyaline acellular deposits in the adrenal and salivary glands.

Howard *et al.* (1979) reported that mice immunosuppressed at approximately the same time of MCMV infection or from 18 to 30 days p.i. developed humoral immunity but not cell-mediated immunity to MCMV. When they were immunosuppressed from 18–30 days p.i., cell-mediated immunity that was present disappeared. Eizuru and Minamishima (1979) reported that MCMV virulence and antigenicity depend not only on whether the virus is propagated *in vitro* or *in vivo* but also on the organ in which it is replicated.

Davis *et al.* (1979) studied the role of MCMV in nerve tissue by explanting trigeminal ganglia from newborn, suckling, and weanling mice 3–36 days after i.c. inoculation of virus. The MCMV replicated in fibroblasts, Schwann cells, satellite cells, and neurons from newborn and suckling, but not weanling mice. Neurons are infected, but appear to be resistant to MCMV replication. Mice with nonlethal MCMV infections demonstrate a depressed anti-sheep erythrocyte plaque-forming cell response followed by an enhanced plaque-forming cell response (Tinghitella and Booss, 1979). Thus, MCMV may induce an impairment in immune regulation.

Brautigam *et al.* (1979) found in continued studies that macrophages are permissive for MCMV infection and that 20–50% of the MCMV-infected macrophages formed infectious centers. More than 82% of the MCMV-infected macrophages contained MCMV DNA using *in situ* hybridization. Macrophages from latently MCMV-infected mice harbored between 4 and 7 virus genome DNA copies per 100 cells. This work indicates that macrophages from MCMV-susceptible or -resistant mice replicate MCMV in the same amounts and that they are a reservoir of MCMV during latent and chronic infections. Activation of the macrophages may be required to reactivate *in vivo* latent infections.

Mayo and Rapp (1980) established another mouse model and observed that Friend-leukemia-virus-infected mice subsequently infected with MCMV developed a more severe MCMV infection than normal mice. These two viruses appear to interact by suppressing the ability of infected spleen cells to respond to mitogen-induced stimulation. *In vitro* phenotypic mixing occurs between mouse retroviruses and MCMV, suggesting that *in vivo* interactions between these two viruses are possible and may be important in the initiation of disease (Schnitzer and Gönczöl, 1979). An experimental mouse model was also developed by R.D. Smith and Wehner (1980) to study acute MCMV glomerulonephritis. They inoculated mice i.p. with MCMV and established a transient acute glomerulonephritis characterized by cytomegalic intranuclear inclusions in the mesangial cells. They then used electron microscopy and immunofluorescence (IF) to study the course of infection and reported that at least one route results in viruria and involves the glomerular mesangium, the juxtaglomerular zone, and the distal convoluted tubule.

Bixler and Booss (1980) reported that during MCMV-induced immunosuppression, the ability to be primed for a secondary direct plaque-forming cell (DPFC) response remains intact. When MCMV-infected mice were immunized 5 days p.i. with sheep erythrocytes, the DPFC response was significantly suppressed. However, when immunization occurred 10 days p.i., the primary response was similar to that of controls. If mice were primed with sheep erythrocytes at 5 days p.i. and then rechallenged at 10 days p.i., a secondary immune response similar to that in the uninfected controls was observed. The mice maintained the ability to be primed, which may be due to an alteration in immune regulation during acute MCMV infection. Booss (1980) reported that MCMV can infect a subpopulation of macrophage-enriched cells following i.p. injection and can promote the establishment of persistent infection even in the presence of minimal MCMV replication.

Brautigam and Oldstone (1980) also demonstrated MCMV in the reproductive tissues of MCMV-infected newborn mice. MCMV could be detected in ovarian stromal cells and in squamous epithelial cells of the testes using a ^{3}H-labeled complementary RNA probe made from MCMV DNA. The presence of MCMV DNA in reproductive tissues during acute infections suggests that these tissues might be a reservoir for the virus and might be responsible for latent virus in adult mice. Hybridization kinetic analysis of DNA isolated from ovaries and testes of 5- and 6-month-old latently infected mice yielded evidence that the MCMV genome was present.

In 1975, Kilham and Margolis (1975) reported induction of encephalitis in suckling rats inoculated i.c. with a rat CMV that had been passaged 5 times through rats. Only crude preparations of the virus induced sickness and death. The animals exhibited typical signs of infection with CMV: cytomegaly, nuclear and cytoplasmic inclusions, and polykaryocytosis.

The pathogenesis of GPCMV has been reviewed in detail (Hsiung *et al.*, 1980). GPCMV was repeatedly isolated from the salivary glands of guinea pigs 2 weeks after inoculation, but was not isolated as frequently from blood, spleen, and kidneys (Tenser and Hsiung, 1976). When GPCMV-infected tissue suspensions were inoculated into salivary glands, intranuclear inclusions and virus particles were observed, but were seldom seen when tissue-culture virus was used as inoculum. GPCMV-infected guinea pigs produced high titers of GPCMV-specific neutralizing antibody to homologous virus. Weanling guinea pigs were inoculated s.c. and i.c. with GPCMV (Connor and Johnson, 1976). No illness was detected by the 24th day after s.c. inoculation, although persistent infection was present in many tissues and virus was obtained in high titer from salivary gland and thymus tissue homogenates. In i.c. inoculated guinea pigs, encephalitis developed and virus spread to other organs.

Two distinct phases of infection without evidence of clinical disease have been associated with the pathogenesis of GPCMV infection of guinea pigs (Hsiung *et al.*, 1978). In acute primary infection, viremia can be demonstrated and infectious virus is recoverable from lungs, spleen, and kidneys 2–10 days p.i. This stage is followed by establishment of chronic persistent infection. When animals in this phase have high levels of circulating antibody, infectious GPCMV can be recovered from the pancreas and salivary glands. GPCMV is seldom detected in the blood; however, it can be detected when a large inoculum of virus is used, and it is likely that the virus persists in the blood of healthy animals for long periods of time.

Kumar and Nankervis (1978) developed a guinea pig animal model to study congenital infections with GPCMV. Guinea pigs in the second half of pregnancy were inoculated i.p. with GPCMV; 40% of them delivered at least one infected newborn. Of mothers without complement-fixing antibody to GPCMV, 100% delivered infected litters compared to 25% of immune mothers. Neonatal infection occurred as frequently whether low-passaged tissue-culture-adapted virus or salivary-gland-passaged virus was used. Infected newborns did not appear to have congenital abnormalities.

Griffith and Hsiung (1980) followed the effects of GPCMV infection at different stages of gestation. GPCMV was excreted from the cervix at a higher level in pregnant animals 1–30 days postinoculation. All fetuses were infected no matter when maternal infection occurred; however, there was a higher number of stillbirths and virus infections in newborns from mothers infected late in gestation than from mothers infected during the first trimester of pregnancy. GPCMV could be detected more frequently in fetuses and newborns within 2 weeks after inoculation of the mothers. After that, GPCMV gradually disappeared in the offspring, although virus could be isolated from the salivary glands up to 14 weeks after delivery. Thus, GPCMV infection of the guinea pig may be a suitable model for congenital transmission of HCMV.

Persistent GPCMV infections were investigated in random-bred and inbred strains of guinea pigs (Bia *et al.*, 1979). Infectious GPCMV was isolated from urine, kidneys, spleen, pancreas, salivary glands, and cervix, but not buffy coats. Strain 2 guinea pigs developed a high rate of chronic viruria that was twice as prevalent in females as in males. When buffy coats from persistently infected inbred guinea pigs were transfused into isogenic and allogenic animals, CMV infection was produced; however, GPCMV was not activated in these same animals when the buffy coats were transfused from uninfected donors.

To conclude with a nonrodent system, Corner *et al.* (1964) reported two outbreaks of generalized disease in swine that was characterized by cytomegalic inclusions. This disease occurred in suckling piglets, accompanied by high mortality, stunting, and an encephalomyelitis in many cases. It appears that the piglets in the two herds had a disseminated form of CID.

B. Pathogenesis in Monkeys

Ablashi *et al.* (1969) prepared antiserum in rabbits with neutralizing activity against a simian CMV (SCMV) isolated from a vervet monkey. Neutralization studies using this rabbit antiserum suggest the existence of antigenic heterogeneity in SCMV strains. African green monkeys have been used extensively in the preparation of kidney cell cultures for virus research. It was found that CMVs could be recovered from 36% of 22 culture lots of primary vervet monkey kidney cultures (Yamashiroya *et al.*, 1971). Swack *et al.* (1971) also examined kidney cell cultures from African green monkeys and studied the serum from these animals for the presence of neutralizing antibody to two SCMVs. They found that more than 95% of the monkeys had antibody titers to African green monkey CMV, but that the same virus was detected in only 4 of 119 lots of African green monkey cell cultures examined. Muchmore (1971) reported the possible transmission of CMV to a human who was bitten by a chimpanzee.

It was found that other nonhuman primates are susceptible to African green monkey CMV and that animals held in captivity for 1 year or more have a high incidence of antibody against the virus. Persistent shedding of rhesus CMV (RCMV) in the urine of healthy adult rhesus monkeys has been detected 4 years after it was originally isolated from the same young monkeys (Asher *et al.*, 1974). RCMV is similar to African green monkey CMV in that it grows in human, rhesus, and African green monkey fibroblast cultures. It is antigenically different from African green monkey CMV, although the two SCMVs and HCMV apparently share some antigenic determinants.

Marmosets experimentally infected with the Colburn strain of HCMV (Nigida *et al.*, 1975) did not develop visible clinical disease during an 8-month observation period but did develop antibodies, as detected by IF and neutralization tests. The virus could not be isolated from blood, peripheral lymphocytes, vaginal, or oropharyngeal swabs, but was recovered by cocultivating kidney tissues from an adult marmoset that died.

Waner and Biano (1979) identified common early antigens of HCMV and vervet CMV (VCMV) and found that early VCMV antigens were more useful in distinguishing active and latent infections in humans than early HCMV antigens used to detect the corresponding antibodies.

Two strains of CMV were isolated from the same chimpanzee: one from the left frontal lobe, the other from a lymph node (Wroblewska *et al.*, 1979). This chimpanzee had been inoculated i.c. 3 years earlier at birth with brain cell cultures from a patient with multiple sclerosis. Both isolates produced cytopathic effects characteristic of CMV when inoculated into brain, ganglion, or fibroblast cultures of human or simian origin. The virus was identified as CMV by electron microscopy, by the presence of intranuclear and intracytoplasmic inclusion bodies, and by the use of convalescent human anti-HCMV serum. The two isolates were identical by restriction endonuclease analysis and were closely related to chimpanzee CMV. There was no similarity in restriction endonuclease patterns when the chimpanzeelike CMV was compared with two strains of HCMV.

IV. TRANSMISSION

Mouse models have been used to study transmission of murine cytomegalovirus (MCMV) in the hope of obtaining information to explain transmission of human CMV in humans. Cheung and Lang (1977a) followed the transmission of MCMV after transfusing blood from latently infected mice into uninfected allogeneic and isogeneic hosts. MCMV was detected in allogeneic hosts, but seldom in inbred hosts. When blood from uninfected donors was transfused into latently infected mice, latent MCMV was activated. Another investigation (Jordan, 1978) attempted to determine whether MCMV infection could be transmitted by a respiratory route. To do this, virus was inoculated intranasally (i.n.). Subclinical infection was established in 50% of the mice after inoculation with 100 plaque-forming units (PFU) of MCMV. Virus replicated in the lungs and, within 7 days, spread to other organs through the blood. The remaining mice, given 10^4 PFU or more MCMV i.n., developed a diffuse interstitial pneumonia with a 20% mortality rate. Thus, the respiratory route is another way in which MCMV can be spread.

The possibility existed that MCMV could be sexually transmitted, since it is known that the virus can replicate in spermatogenic cells *in vitro*. Ten-week-old male mice inoculated intraperitoneally with MCMV presented an acute and generalized infection (Neighbour and Fraser, 1978). The virus was isolated from sperm and seminal vesicles and from sperm collected from mated females. There was no evidence that MCMV affected embryonic development, since development was normal. Chantler *et al.* (1979), who studied vertical transmission of MCMV from mother to fetus, frequently detected latent MCMV in fetal tissues by immunofluorescence, and *in situ* hybridization. It is not clear whether the transmission is the result of transplacental passage or direct vertical transmission though the germ line.

Choi and Hsiung (1978) reported that transplacental transmission of guinea pig CMV (GPCMV) occurred, but only during acute primary maternal infection of guinea pigs. Fetuses were infected from 28–60 days of gestation. However, GPCMV could not be detected in any fetuses of pregnant guinea pigs that had been infected for more than 30 days. Significant levels of circulating neutralizing antibody were detected in the salivary glands of sacrificed pregnant guinea pigs. Horizontal transmission also occurred, but was most efficient by sexual contact. When uninfected males were housed with infected females, all males demonstrated increases in antibody titer, and 5 males had GPCMV in the salivary glands. When same-sex guinea pigs were housed together, only 1 of 13 uninfected had virus in the salivary gland and 4 of 13 had rises in antibody titer. Johnson and Connor (1979) also observed transplacental transmission of GPCMV and noted that infection occurred during the second trimester of pregnancy. GPCMV could be isolated from fetal tissues without histological evidence of infection. Thus, it can be stated that GPCMV crosses the placenta and infects the fetus *in vitro* with no apparent pathological changes in the fetus.

V. LATENCY, PERSISTENCE, AND REACTIVATION

Cytomegaloviruses (CMVs) have been studied seriously in recent years since they have the ability to latently infect cells, to establish persistent infections, and to reactivate under unknown conditions (for a review, see Rapp and Geder, 1978). For obvious reasons, many of these investigations have been carried out in cultured cells.

Gönczöl and Váczi (1973) reported that human CMV (HCMV) was present in at least 1 of 70 cells infected in the presence of cytosine arabinoside (ARA-C) and that 40–50% of the cells contained HCMV-specific antigens as detected by immunofluorescence (IF). Infectious virus could also be detected 4–5 days after the removal of ARA-C. In the presence of ARA-C, latent HCMV infections could be maintained indefinitely; these cultures could be superinfected with HCMV. Rapp *et al.* (1975) were able to establish a human cell line (CMV-Mj-P) from prostatic tissue cells of a child donor in which foci could not be detected and virus could not be rescued, although CMV-specific antigens were detected by indirect IF. Approximately 10–15 genome equivalents of HCMV DNA were detected by nucleic-acid hybridization tests. The long-term persistence of these cells appears to be due to cell transformation by HCMV or to the establishment of a chronic HCMV infection. In the same year, Joncas *et al.* (1975) reported that the HCMV genome was unexpressed in an Epstein–Barr virus-positive lymphoblastoid cell line persistently infected with the virus. The line contained 8–10 HCMV genome equivalents per cell. A later study by E.-S. Huang *et al.* (1978b) continued and extended the earlier study of Joncas *et al.* (1975) in two lymphoblastoid-cell lines, D4 and M1. Huang and co-workers were unable to detect HCMV antigens and could not rescue virus even after cocultivation with permissive fibroblasts. However, using DNA–DNA reassociation kinetic analysis, they could detect less than one genome equivalent per cell of HCMV

DNA. HCMV has the ability to persist in lymphoid cells with B-cell characteristics, and it may be that the HCMV genome persists in a defective form in the cells.

HCMV was also used to persistently infect a line of human lymphocytes (Furukawa, 1979b; Furukawa *et al.*, 1979). Approximately 1–10% of the cells were infected, on the basis of results from infectious-center assays and detection of antigens by fluorescent antibodies. When these persistently infected cells were treated with anti-HCMV serum, the virus infection was cured and the cured cells became resistant to HCMV infection.

St. Jeor and Weisser (1977) reported that persistent infections could be established in human peripheral lymphocytes and lymphoblastoid cells infected *in vitro.* If infection of the lymphocytes was accompanied by a blastogenic response, HCMV persisted longer and at a higher titer than in cells in which a blastogenic response did not occur. Only a small number of the cells that appeared to be lymphocytes seemed to be involved in virus persistence, as determined by infectious-center assays and autoradiography. Another report (Tocci and St. Jeor, 1979b) demonstrated that HCMV can replicate and persist in B- and T-lymphoblastoid cell lines. These cell lines that support the replication of HCMV DNA also harbor multiple copies of the entire HCMV genome.

Mocarski and Stinski (1979) isolated and characterized HCMV-infected human fibroblasts carrying the HCMV genome. Cells in the nonproductive phase of infection (as evidenced by the lack of HCMV structural antigens and HCMV particles) contained approximately 120 genome equivalents of HCMV DNA per cell as determined by reassociation kinetics. Most of the HCMV DNA persisted with limited gene expression. Growth of nonproductive cells was continual, and some progressed to replicate virus productively. The production of infectious virus could be curtailed by growth in anti-HCMV serum, but approximately 45 genome equivalents of HCMV DNA persisted per cell. This curtailment of virus production resumed when antiserum was removed.

Colberg-Poley *et al.* (1979) developed a model for *in vitro* latency of herpes simplex virus (HSV). In this human fibroblast system, they were able to stimulate the replication of latent HSV-2 using HCMV. Reactivation of HSV-2 occurred when HCMV was used to superinfect latently infected cultures. The interaction between HSV-2 and HCMV may serve an important role in herpesvirus latency *in vivo.*

Williams *et al.* (1980) have investigated whether HCMV genomes are present in human skin fibroblasts. Using DNA–DNA reassociation kinetics, they examined six cultured human skin lines and detected HCMV genomes in three of them. The three skin-fibroblast lines with detectable HCMV genomes were from patients with Charcot–Marie–Tooth disease, although no HCMV could be detected in a fourth line from a patient with the disease. Two of the three lines that were negative for HCMV DNA were control lines that had been derived from individuals with other diseases. It is possible that human skin may be a site of HCMV latency in some circumstances.

Murine CMV (MCMV) has also been studied for its ability to establish persistent infections. Cheung and Lang (1977b) observed that MCMV could be detected in salivary gland suspensions for as long as 3 months. After virus was no longer detectable, latent MCMV could be detected in explants of salivary and prostate glands and in cell lines from these explants. Oldstone *et al.* (1977) demonstrated that MCMV could be activated from spleen lymphocytes of latently infected mice and that the activation depended on appropriate antigenic stimulation: activating cells that had the genetic capacity to express MCMV information, were susceptible to infection, and allowed the virus to replicate. They demonstrated that bone marrow-derived lymphocytes harbor MCMV, but that thymus-derived lymphocytes do not.

MCMV can also be activated *in vivo* after allogeneic stimulation. After MCMV inoculation of mice, virus becomes undetectable after 4 months, but can be reactivated after immunosuppression with antilymphocyte serum and corticosteroids for a 2-week period (Jordan *et al.*, 1977). Mayo *et al.* (1978) also studied the activation of latent MCMV infection. No virus was detectable in any organ 16 weeks postinfection (p.i.); however, spleen cells were shown to be latently infected using cocultivation, cell transfer, and immunosuppression. Latent MCMV infection of mouse spleens was investigated using an explant tissue culture technique as well as cocultivation, and nucleic acid hybridization (Wise *et al.*, 1979). The explant tissue culture technique was superior to other tests for detecting latent MCMV infections of lymphoid tissues. These investigators found that the frequency of latent infection in BALB/c mouse spleen cells is much less than that reported in other studies.

Gould and Mims (1980) examined MCMV reactivation in outbred mice. They observed replicating virus in the kidneys of neonatal mice infected intraperitoneally (i.p.) with the Smith strain of MCMV for up to 90 weeks p.i.; the results were similar in pregnant mice. However, when neonatal mice were injected i.p. with the attenuated strain of MCMV, the kidneys yielded fewer virus isolations. When pregnant mice received the attenuated strain, the rate of virus isolation increased 2-fold and there was an 8-fold increase in mean virus titer. These workers are now investigating whether the reactivation is a result of natural immunosuppression during pregnancy or of hormonal changes that occur during pregnancy.

Hudson *et al.* (1979) have reported two model systems currently available to study latent MCMV infections: (1) MCMV in G_0-phase 3T3 cells, as opposed to the required S phase of the cell cycle for MCMV replication; and (2) reactivation of MCMV from mouse spleen cells that survive an acute infection. The pathogenesis of MCMV reactivated from a latent state was investigated by Shanley *et al.* (1979). MCMV was detected only in the hepatocytes of immunosuppressed mice for 1 week (17 weeks p.i. with MCMV). This was followed by a viremia with infection spreading to the other organs. MCMV titers reached their highest levels in salivary glands.

Cheung *et al.* (1980) were able to detect MCMV-specific DNA in cell cultures and salivary gland tissues from latently infected animals using nucleic acid hybridization tests. They found approximately 4.5 and 0.2 genome equivalents per cell of MCMV DNA in cultures of salivary and prostate gland cells, respectively. The cells were negative for MCMV-specific antigens by IF, and no MCMV was detectable by electron microscopy. CMV pathogen-free mice were tested for MCMV-specific DNA. MCMV-specific DNA was detected in pooled salivary gland extracts of these uninoculated "pathogen-free" mice; however, the animals were seronegative and virus-negative when tested by routine infectivity assays. This work confirms results by others that nucleic acid hybridization techniques are useful in detecting the presence of latent MCMV. It is also important to know that mice that are supposedly CMV pathogen-free may, in fact, be harboring latent MCMV.

VI. CELL TRANSFORMATION BY CYTOMEGALOVIRUSES (TABLE III)

The morphological transformation of human and nonhuman cells is a property of some DNA viruses that has been demonstrated for several members of the herpesvirus group. Duff and Rapp (1971, 1973) were the first investigators to demonstrate the oncogenic transformation of cells using a human herpesvirus, herpes simplex virus. In 1973, Albrecht and Rapp (1973) reported that human cytomegalovirus (HCMV) can transform mammalian cells and demonstrated the malignant transformation of hamster cells after exposure to UV-irradiated HCMV. They isolated clones that were not contact-inhibited and were able to establish a continuous cell line, Cx-90-3B, from one of these clones. Cells from this line were oncogenic when inoculated into weanling golden Syrian hamsters, and HCMV antigens were detected on the surface and in the cytoplasm of the tumor cells. Serum from animals with tumors contained antibody to HCMV-specific antigens. A later report (Rapp *et al.*, 1975) described the establishment of a long-term persistent infection of human cells. HCMV was used to oncogenically transform human embryo lung (HEL) cells (Geder *et al.*, 1976), and HCMV-specific antigens were detected by immunofluorescence (IF) techniques. The transformed cells were able to induce nondifferentiated tumors when injected into weanling athymic nude mice (Geder *et al.*, 1977). Geder and Rapp (1977) reported intranuclear CMV-related antigens in CMV-transformed human cells by the anticomplement IF (ACIF) technique. The nuclear reactivity appears mainly with immune sera from prostatic cancer patients or patients recovering from an acute CMV infection. HCMV-transformed HEL cells were inoculated into athymic nude mice, with approximately 62% of the

TABLE III. Properties of Cells Transformed by Cytomegaloviruses

Morphological changes
Immortalization in culture
Retention of virus DNA fragment(s)
Synthesis of virus antigens in nucleus, cytoplasm, and cellular membrane
Resistant to superinfection
Malignant and metastatic

mice developing tumors within an average latent period of 19 days (Geder *et al.*, 1977). The tumor cells were poorly differentiated and may have been of epithelioid origin. There was also evidence of CMV-related intracellular and membrane antigens, as detected by indirect IF and ACIF in cells that had been cultured from the tumors.

The chemical carcinogen 4-nitroquinoline-1-oxide (4-NQO) has been shown to inactivate HCMV, and the rate of inactivation is directly dependent on the concentration of the chemical (Albrecht *et al.*, 1979). This study resulted from earlier unpublished work of Li and Albrecht in which they demonstrated that HEL cells persistently infected with HCMV developed a morphologically transformed phenotype following exposure to 4-NQO.

Glaser *et al.* (1977) studied a herpesvirus isolated from a cell culture that had been established from a tumor biopsy specimen from a patient with Kaposi's sarcoma. They characterized this virus (K9V) and observed that its host range was limited to human cells and that it exhibited typical CMV-induced cytopathic effects. The presence of a CMV was suggested by CMV-specific antigens that were detected by IF and complement-fixation tests and also because the density of K9V DNA was consistent with that of CMV DNA. K9V was more cell-associated in human fibroblasts than standard laboratory strains of HCMV. The association of the K9V strain of CMV with Kaposi's sarcoma is still not clear.

Chromosomal anomalies in HCMV-infected cells were reported by Lüleci *et al.* (1980), who observed that chromosomes 2, 3, 4, and 21 showed more anomalies than other chromosomes and that the number of anomalies was not proportional to the length of the chromosomes. It is not known whether the selective chromosomal damage induced by HCMV is important in the oncogenicity of this virus. The role of HCMV in human neoplasia is unknown, but the virus has many properties associated with oncogenesis (Table IV) and has been linked to tumors of

TABLE IV. Properties of Human Cytomegalovirus Associated with Oncogenesis

Production of many naturally defective particles
Persistence in various tissues in host for extended period
Stimulation of host DNA and RNA synthesis
Transformation of cells to malignancy *in vitro*
Induction of tumors in nude mice by transformed cells

the genitourinary tract and to Kaposi's sarcoma (for a review, see Rapp and McCarthy, 1979).

VII. CONTROL OF CYTOMEGALOVIRUS INFECTIONS

A. Antiviral Therapy

Antiviral therapy, which includes the use of drugs, natural substances, and vaccines to treat or prevent viral infections, has had more success in the development of efficacious regimens recently than in previous years. Ch'ien *et al.* (1974) described a clinical study on the effect of adenine arabinoside (ARA-A) on human cytomegalovirus (HCMV) infections. They treated 12 patients with HCMV infections with 5–20 mg ARA-A intravenously for 6–18 days. Of 5 immunosuppressed patients treated, none improved. However, 2 congenitally infected infants and 2 patients with CMV mononucleosis improved. Urinary excretion of HCMV in the congenitally infected and CMV mononucleosis patients was suppressed by the drug. Viremia and urinary excretion of CMV persisted in the immunosuppressed patients who received the drug.

Overall *et al.* (1976) demonstrated that murine CMV (MCMV) and HCMV were sensitive to the antiviral effects of ARA-A and phosphonoacetic acid (PAA) *in vitro*. However, *in vivo* studies of mice with lethal MCMV infections showed that ARA-A had no effect on the mortality rate, whereas PAA treatment resulted in a significant reduction in mortality with complete inhibition of MCMV replication in the liver. When nonlethal MCMV infections were treated with PAA, clinical signs of illness were eliminated. Another drug, arabinofuranosylthymine, a thymidine analogue, did not inhibit the replication of HCMV (Miller *et al.*, 1977), although it was effective in inhibiting the replication of herpes simplex virus and varicella–zoster virus. This is probably because HCMV does not induce a virus thymidine kinase in infected cells. At this time, antiviral therapy has not been used extensively against HCMV.

B. Interferon

Glasgow *et al.* (1967) investigated the ability of HCMV to induce production of interferon in human cells *in vitro* and also studied the sensitivity of HCMV to inhibition by interferon. They found that HCMV is relatively insensitive to inhibition by interferon *in vitro*. In cultured human cells, HCMV was unable to induce production of interferon and was also able to suppress the interferon response in human tissue cultures that were infected with other viruses.

Human exogenous leukocyte interferon has been used in the treatment of HCMV infections in humans (Emödi *et al.*, 1976). A complete antiviral effect on HCMV excretion was observed in three of nine patients and there was transient inhibition of HCMV excretion in the urine of five patients. Another report (Arvin *et al.*, 1976) demonstrated that human leukocyte interferon may suppress viruria in infants with HCMV infection if it is administered in large doses. The dosage and toxicity of the interferon and the state of the host's immune response may be important in potentiating the effect of human leukocyte interferon on HCMV infections in infants.

Holmes *et al.* (1978) reported that cell-free virus is much more sensitive to the effects of interferon than virus-infected cells and that interferon sensitivity decreases with an increase in the dose of infectious units. MCMV was able to induce interferon as early as 4 hr postinfection, which was approximately 8–12 hr before infectious virus could be detected (Oie *et al.*, 1975). Small doses of MCMV are very sensitive to interferon, but higher doses of MCMV are resistant. Rytel and Hooks (1977) developed a mouse model to study cell-mediated immunity and observed that immune interferon was produced in response to MCMV antigens in spleen lymphocyte cultures from MCMV-infected mice. Lymphocyte transformation did not occur when measured by antigen-specific enhancement of thymidine uptake. Cruz *et al.* (1981) injected CD-1 mice intraperitoneally (i.p.) with interferon within 24 hr of birth and inoculated them with MCMV 24 hr later. They found that the administration of interferon was prophylactic and prevented growth retardation. In addition, the interferon reduced the harmful effects of the virus and effected lower virus titers in tissues, although lymph node necrosis and marked splenomegaly were present. Again, these studies represent only the beginning of work with natural substances and HCMV.

C. Vaccines

The desire and the need to control HCMV infections have prompted the study and development of HCMV vaccines (for reviews, see Hanshaw, 1974; Betts, 1979; Phillips, 1979; Plotkin, 1979). Elek and Stern (1974) were the first to describe an HCMV vaccine for use in humans. They developed a live, tissue-culture-adapted strain (from AD-169) of HCMV and inoculated it subcutaneously (s.c.) into volunteers with successful stimulation of neutralizing and complement-fixing antibodies. Intradermal inoculation of the virus was not successful because too large a dose was necessary to produce conversion and induced severe reactions. When vaccine was administered by the s.c. route, there was a 96% conversion rate with only rare minor side effects. HCMV was not detected in the throat or urine of vaccinated volunteers.

Plotkin *et al.* (1975) developed an HCMV vaccine from a fresh isolate (Towne) that they passaged 125 times in WI-38 cells; during this period, the HCMV was cloned three times. At passage 128, it was decided that pathogenicity of the virus had been attenuated and that it could be administered to humans. The Towne live attenuated vaccine was administered intranasally and s.c. to adult volunteers; neither clinical infection nor side effects resulted, except for local reaction and lymphocytosis in some volunteers. Plotkin and colleagues also tested the Towne strain for oncogenicity and found it to be negative.

Glazer *et al.* (1978) reported on a volunteer who had received the live Towne CMV vaccine and had subsequently undergone renal transplantation. The transplanted kidney was rejected, but there was no histological evidence of HCMV infection. A second HCMV vaccinee received a kidney transplant and was discharged from the hospital with normal kidney function. She did not show excretion of HCMV during immunosuppressive therapy. Glazer *et al.* (1979) reported a clinical study in which they vaccinated 12 seronegative patients with the live Towne 125 strain of HCMV. These patients were renal transplant candidates, and all seroconverted after vaccination. Of the 12 vaccinees, 10 received transplants; 8 of these had functioning transplants up to 1 year later. This vaccine virus could not be isolated from the 8 patients who were vaccinated and then were immunosuppressed following transplantation. Thus, this vaccine produced cellular and humoral immunity and does not appear to reactivate following immunosuppression.

Neff *et al.* (1979) reported on a vaccine that was prepared from the AD-169 vaccine virus of Elek and Stern (1974). Twenty-four adult male priests and seminarians were inoculated with the vaccine. All seronegative volunteers who were vaccinated developed antibodies, and these neutralizing antibodies remained at a high level for a year. The side effects were minor: a small number of individuals experienced soreness, induration, and erythema at the site of inoculation as well as headache, chills, fever, fatigue, and myalgia. HCMV could not be recovered from the urine, throat, or peripheral leukocytes of seronegative vaccine recipients. Individuals who were susceptible to HCMV did not excrete virus or develop antibody following exposure to the vaccinated individuals.

Medearis and Prokay (1978) have studied MCMV to investigate the effect of MCMV immunization on MCMV infection in suckling mice. They examined route of inoculation, age of host at inoculation, effect of time of immunization on pregnant mice, and the immune response. They confirmed that i.p. inoculation can produce infection of the central nervous system and that immunization of female weanling mice protects offspring from central nervous system infection. Suckling mice born to nonimmunized mothers but nursed by immune foster mothers were as protected as sucklings born to immune mothers. Immunization reduced mortality and the amount of virus that could be recovered from the brain, liver, spleen, and salivary gland.

A model to examine immunoprophylaxis of MCMV infection in mice has been established by Minamishima *et al.* (1978). They reported that MCMV serially propagated in mouse embryo fibroblast cells lost pathogenicity in weanling mice. The cell-culture-adapted MCMV was as effective as live, attenuated virus vaccine when challenged by mouse-passaged MCMV. When the mice were immunized against MCMV infection i.p., they were protected against MCMV replication, clinical manifestations, histopathology, and mortality. MCMV infection can be prevented by immunization. Jordan (1980) reported, however, that attenuation of MCMV is unstable and that reversion of the vaccine virus to a virulent state occurs during the acute stage of infection. Virus from an attenuated MCMV vaccine was able to establish latent infection, could be reactivated after immunosuppression, and could revert to a virulent state in the salivary glands.

Hsiung and her colleagues (Bia *et al.*, 1980) investigated the effect of vaccination on guinea pig CMV (GPCMV) infection using a live GPCMV vaccine and a noninfectious envelope antigen vaccine. Immunized Hartley strain guinea pigs were compared with guinea pigs receiving virulent salivary-gland-passaged GPCMV, passively immunized animals, and controls. These workers found that guinea pigs inoculated with the cell-culture-passaged GPCMV or virulent virus were protected against acute viremia and death after challenge with virulent GPCMV. Pregnant animals that received the live vaccine had a lower incidence of viremia and generalized maternal and fetal infection. Pregnant animals that received the envelope antigen vaccine or that were passively immunized displayed acute viremia when challenged with virulent virus, although their infections were not as generalized as those of control animals. GPCMV could be cocultivated from the internal organs of 27% of the fetuses from unvaccinated control guinea pigs and from fewer than 1% of the fetuses from any of the vaccinated guinea pigs. This work suggests that in the guinea pig, maternal immunity to virus before pregnancy affects the course of maternal and fetal infection by ameliorating clinical illness.

The possibility of reversion and lack of a suitable experimental model to test for attenuation mandates caution in the widespread use of HCMV vaccines. For the time being, and until more data are available, such vaccines should probably be restricted to high-risk populations (i.e., seronegative individuals facing organ transplants).

REFERENCES

Ablashi, D.V., Martos, L.M., Gilden, R.V., and Hampar, B., 1969, Preparation of rabbit immune serum with neutralizing activity against a simian cytomegalovirus (SA6), *J. Immunol.* **102:**263.

Ablashi, D.V., Chopra, H.C., and Armstrong, G.R., 1972, A cytomegalovirus isolated from an owl monkey, *Lab. Anim. Sci.* **22:**190.

Albrecht, T., and Rapp, F., 1973, Malignant transformation of hamster embryo fibroblasts following exposure to ultraviolet-irradiated human cytomegalovirus, *Virology* **55:**53.

Albrecht, T., and Weller, T.H., 1980, Heterogeneous morphologic features of plaques induced by five strains of human cytomegalovirus, *Am. J. Clin. Pathol.* **73:**648.

Albrecht, T., Nachtigal, M., St. Jeor, S.C., and Rapp, F., 1976, Induction of cellular DNA synthesis and increased mitotic activity in Syrian hamster embryo cells abortively infected with human cytomegalovirus, *J. Gen. Virol.* **30:**167.

Albrecht, T., Speelman, D.J., and Li, J.L.H., 1979, Inactivation of human cytomegalovirus by the chemical carcinogen 4-nitroquinoline 1-oxide, *J. Gen. Virol.* **45:**231.

Albrecht, T., Cavallo, T., Cole N.L., and Graves, K., 1980a, Cytomegalovirus: Development and progression of cytopathic effects in human cell culture, *Lab. Invest.* **42:**1.

Albrecht, T., Li, M.-L., Cole, N., Downing, E., and Funk, F.D., 1980b, Replication of human cytomegalovirus at supra-optimal temperatures is dependent on the virus strain, multiplicity of infection, and phase of virus replication, *J. Gen. Virol.* **51:**83.

Araullo-Cruz, T.P., Ho, M., and Armstrong, J.A., 1978, Protective effect of early serum from mice after cytomegalovirus infection, *Infect. Immun.* **21:**840.

Arvin, A.M., Yeager, A.S., and Merigan, T.C., 1976, Effect of leukocyte interferon on urinary excretion of cytomegalovirus by infants, *J. Infect. Dis.* **133**(Suppl.):A205.

Ashe, W.K., Scherp, H.W., and Fitzgerald, R.J., 1965, Previously unrecognized virus from submaxillary glands of gnotobiotic and conventional rats, *J. Bacteriol.* **90:**1719.

Asher, D.M., Gibbs, C.J., Lang, D.J., and Gajdusek, D.C., 1974, Persistent shedding of cytomegalovirus in the urine of healthy rhesus monkeys, *Proc. Soc. Exp. Biol. Med.* **145:**794.

Barabas, G., Wroblewska, Z., and Gilden, D.H., 1980, Growth of murine cytomegalovirus in murine and heterologous brain cell cultures, *Arch. Virol.* **65:**193.

Berezesky, I.K., Grimley, P.M., Tyrrell, S.A., and Rabson, A.S., 1971, Ultrastructure of a rat cytomegalovirus, *Exp. Mol. Pathol.* **14:**337.

Betts, R.F., 1979, Cytomegalovirus vaccine in renal transplants, *Ann. Intern. Med.* **91:**780.

Bia, F.J., Hastings, K., and Hsiung, G.D., 1979, Cytomegalovirus infection in guinea pigs. III. Persistent viruria, blood transmission, and viral interference, *J. Infect. Dis.* **140:**914.

Bia, F.J., Griffith, B.P., Tarsio, M., and Hsiung, G.D., 1980, Vaccination for the prevention of maternal and fetal infection with guinea pig cytomegalovirus, *J. Infect. Dis.* **142:**732.

Bixler, G.S., Jr., and Booss, J., 1980, Establishment of immunologic memory concurrent with suppression of the primary immune response during acute cytomegalovirus infection of mice, *J. Immunol.* **125:**893.

Black, P.H., Hartley, J.W., and Rowe, W.P., 1963, Isolation of a cytomegalovirus from African green monkey, *Proc. Soc. Exp. Biol. Med.* **112:**601.

Blanton, R.A., and Tevethia, M.J., 1981, Immunoprecipitation of virus-specific immediate–early and early polypeptides from cells lytically infected with human cytomegalovirus strain AD169, *Virology* **112:**262.

Boldogh, I., Gönczöl, É., Gärtner, L., and Váczi, G., 1978, Stimulation of host DNA synthesis and induction of early antigens by ultraviolet light irradiated human cytomegalovirus, *Arch. Virol.* **58:**289.

Booss, J., 1980, Establishment of cytomegaloviral infection in mice: Role of a macrophage-enriched subpopulation, *J. Infect. Dis.* **141:**466.

Booth, J.C., Beesley, J.E., and Stern, H., 1978, Syncytium formation caused by human cytomegalovirus in human embryonic lung fibroblasts, *Arch. Virol.* **57:**143.

Brautigam, A.R., and Oldstone, M.B.A., 1980, Replication of murine cytomegalovirus in reproductive tissues, *Am. J. Pathol.* **98:**213.

Brautigam, A.R., Dutko, F.J., Olding, L.B., and Oldstone, M.B.A., 1979, Pathogenesis of murine cytomegalovirus infection: The macrophage as a permissive cell for cytomegalovirus infection, replication and latency, *J. Gen. Virol.* **44:**349.

Chantler, J.K., 1978, The use of hypertonicity to selectively inhibit host translation in murine cytomegalovirus-infected cells, *Virology* **90:**166.

Chantler, J.K., and Hudson, J.B., 1978, Proteins of murine cytomegalovirus: Identification of structural and nonstructural antigens in infected cells, *Virology* **86:**22.

Chantler, J.K., Misra, V., Hudson, J.B., 1979, Vertical transmission of murine cytomegalovirus, *J. Gen. Virol.* **42:**621.

Cheung, K.-S., and Lang, D.J., 1977a, Transmission and activation of cytomegalovirus with blood transfusion: A mouse model, *J. Infect. Dis.* **135:**841.

Cheung, K.-S., and Lang, D.J., 1977b, Detection of latent cytomegalovirus in murine salivary and prostate explant cultures and cells, *Infect. Immun.* **15**:568.

Cheung, K.-S., Huang, E.-S., and Lang, D., 1980, Murine cytomegalovirus: Detection of latent infection by nucleic acid hybridization technique, *Infect. Immun.* **27**:851.

Ch'ien, L.T., Cannon, N.J., Whitley, R.J., Diethelm, A.G., Dismukes, W.E., Scott, C.W., Buchanan, R.A., and Alford, C.A., Jr., 1974, Effect of adenine arabinoside on cytomegalovirus infections, *J. Infect. Dis.* **130**:32.

Choi, Y.C., and Hsiung, G.D., 1978, Cytomegalovirus infection in guinea pigs. II. Transplacental and horizontal transmission, *J. Infect. Dis.* **138**:197.

Choi, Y.C., Swack, N.S., and Hsiung, G.D., 1978, Effect of heparin on cytomegalovirus replication, *Proc. Soc. Exp. Biol. Med.* **157**:569.

Chopra, H.C., Lloyd, B.J., Jr., Ablashi, D.V., and Armstrong, G.R., 1972, Morphologic studies of a cytomegalovirus isolated from an owl monkey, *J. Natl. Cancer Inst.* **48**:1333.

Colberg-Poley, A.M., Isom, H.C., and Rapp, F., 1979, Reactivation of herpes simplex virus type 2 from a quiescent state by human cytomegalovirus, *Proc. Natl. Acad. Sci. U.S.A.* **76**:5948.

Cole, R., and Kuttner, A.G., 1926, A filterable virus present in the submaxillary glands of guinea pigs, *J. Exp. Med.* **44**:855.

Connor, W.S., and Johnson, K.P., 1976, Cytomegalovirus infection in weanling guinea pigs, *J. Infect. Dis.* **134**:442.

Corner, A.H., Mitchell, D., Julian, R.J., and Meads, E.B., 1964, A generalized disease in piglets associated with the presence of cytomegalic inclusions, *J. Comp. Pathol. Ther.* **74**:192.

Cross, S.S., Parker, J.C., Rowe, W.P., and Robbins, M.L., 1979, Biology of mouse thymic virus, a herpesvirus of mice, and the antigenic relationship to mouse cytomegalovirus, *Infect. Immun.* **26**:1186.

Cruz, J.R., Dammin, G.J., and Waner, J.L., 1981, Protective effect of low-dose interferon against neonatal murine cytomegalovirus infection, *Infect. Immun.* **32**:332.

Davis, G.L., Krawczyk, K.W., and Hawrisiak, M.M., 1979, Age-related neurocytotropism of mouse cytomegalovirus in explanted trigeminal ganglions, *Am. J. Pathol.* **97**:261.

DeMarchi, J.M., and Kaplan, A.S., 1976, Replication of human cytomegalovirus DNA: Lack of dependence on cell DNA synthesis, *J. Virol.* **18**:1063.

DeMarchi, J.M., and Kaplan, A.S., 1977a, Physiological state of human embryonic lung cells affects their response to human cytomegalovirus, *J. Virol.* **23**:126.

DeMarchi, J.M., and Kaplan, A.S., 1977b, The role of defective cytomegalovirus particles in the induction of host cell DNA synthesis, *Virology* **82**:93.

DeMarchi, J.M., Blankenship, M.L., Brown, G.D., and Kaplan, A.S., 1978, Size and complexity of human cytomegalovirus DNA, *Virology* **89**:643.

DeMarchi, J.M., Schmidt, C.A., and Kaplan, A.S., 1980, Patterns of transcription of human cytomegalovirus in permissively infected cells, *J. Virol.* **35**:277.

Diosi, P., and Babusceac, L., 1970, Biological and physicochemical properties of ground squirrel cytomegalovirus, *Am. J. Vet. Res.* **31**:157.

Diosi, P., Babusceac, L., and Gherman, D., 1972, Cytophagia in cell cultures infected with cytomegalovirus, *J. Infect. Dis.* **125**:669.

Dowling, J.N., Wu, B.C., Armstrong, J.A., and Ho, M., 1977, Enhancement of murine cytomegalovirus infection during graft-vs.-host reaction, *J. Infect. Dis.* **135**:990.

Duff, R., and Rapp, F., 1971, Properties of hamster embryo fibroblasts transformed *in vitro* after exposure to ultraviolet-irradiated herpes simplex virus type 2, *J. Virol.* **8**:469.

Duff, R., and Rapp, F., 1973, Oncogenic transformation of hamster embryo cells after exposure to inactivated herpes simplex virus type 1, *J. Virol.* **12**:209.

Duncan, J.R., Ramsey, F.K., and Switzer, W.P., 1965, Electron microscopy of cytomegalic inclusion disease of swine (inclusion body rhinitis), *Am. J. Vet. Res.* **26**:939.

Dutko, F.J., and Oldstone, M.B.A., 1979, Murine cytomegalovirus infects spermatogenic cells, *Proc. Natl. Acad. Sci. U.S.A.* **76**:2988.

Edwards, R.L., 1976, Structural proteins of human cytomegalovirus, *Yale J. Biol. Med.* **49**:65.

Eizuru, Y., and Minamishima, Y., 1979, Co-variation of pathogenicity and antigenicity in murine cytomegalovirus, *Microbiol. Immunol.* **23**:559.

Eizuru, Y., Minamishima, Y., Hirano, A., and Kurimura, T., 1978, Replication of mouse cytomegalovirus in thymidine kinase-deficient mouse cells, *Microbiol. Immunol.* **22:**755.

Elek, S.D., and Stern, H., 1974, Development of a vaccine against mental retardation caused by cytomegalovirus infection *in utero*, *Lancet* **1:**1.

Emödi, G., O'Reilly, R., Müller, A., Everson, L.K., Binswanger, U., and Just, M., 1976, Effect of human exogenous leukocyte interferon in cytomegalovirus infections, *J. Infect. Dis.* **133:**(Suppl.):A199.

Estes, J.E., and Huang, E.-S., 1977, Stimulation of cellular thymidine kinases by human cytomegalovirus, *J. Virol.* **24:**13.

Färber, I., Wutzler, P., Schweizer, H., and Sprössig, M., 1979a, Human cytomegalovirus induced changes in rabbit cells, *Arch. Virol.* **59:**257.

Färber, I., Wutzler, P., Sprössig, M., and Schweizer, H., 1979b, Determination of antibodies against cytomegalovirus-induced early antigens by using rabbit lung fibroblasts: Brief report, *Arch. Virol.* **62:**273.

Fiala, M., Honess, R.W., Heiner, D.C., Heine, J.W., Murnane, J., Wallace, R., and Guze, L.B., 1976, Cytomegalovirus proteins. I. Polypeptides of virions and dense bodies, *J. Virol.* **19:**243.

Fiala, M., Heiner, D.C., Murnane, J., and Guze, L.B., 1978, Cytomegalovirus proteins. II. Polypeptide composition of cytomegalovirus complement-fixing antigen, *J. Med. Virol.* **2:**39.

Figueroa, M.E., Geder, L., and Rapp, F., 1978a, Replication of herpesviruses in human cells transformed by cytomegalovirus, *J. Gen. Virol.* **40:**391.

Figueroa, M.E., Geder, L., and Rapp, F., 1978b, Infection of human amnion cells with cytomegalovirus, *J. Med. Virol.* **2:**369.

Fong, C.K.Y., Bia, F., Hsiung, G.-D., Madore, P., and Chang, P.-W., 1979, Ultrastructural development of guinea pig cytomegalovirus in cultured guinea pig embryo cells, *J. Gen. Virol.* **42:**127.

Fong, C.K.Y., Bia, F., and Hsiung, G.D., 1980, Ultrastructural development and persistence of guinea pig cytomegalovirus in duct cells of guinea pig submaxillary gland, *Arch. Virol.* **64:**97.

Frenkel, N., and Roizman, B., 1972, Ribonucleic acid synthesis in cells infected with herpes simplex virus: Control of transcription and of RNA abundance, *Proc. Natl. Acad. Sci. U.S.A.* **69:**2654.

Furukawa, T., 1979a, Cell cycle-dependent chronic infection of human cytomegalovirus in human osteogenic sarcoma cells, *J. Gen. Virol.* **45:**81.

Furukawa, T., 1979b, Persistent infection with human cytomegalovirus in a lymphoblastoid cell line, *Virology* **94:**214.

Furukawa, T., Fioretti, A., and Plotkin, S., 1973, Growth characteristics of cytomegalovirus in human fibroblasts with demonstration of protein synthesis early in viral replication, *J. Virol.* **11:**991.

Furukawa, T., Hornberger, E., Sakuma, S., and Plotkin, S.A., 1975a, Demonstration of immunoglobulin G receptors by human cytomegalovirus, *J. Clin. Microbiol.* **2:**332.

Furukawa, T., Tanaka, S., and Plotkin, S.A., 1975b, Stimulation of macromolecular synthesis in guinea pig cells by human CMV, *Proc. Soc. Exp. Biol. Med.* **148:**211.

Furukawa, T., Tanaka, S., and Plotkin, S.A., 1975c, Restricted growth of human cytomegalovirus in UV-irradiated WI-38 human fibroblasts, *Proc. Soc. Exp. Biol. Med.* **148:**1249.

Furukawa, T., Sakuma, S., and Plotkin, S. A., 1976, Human cytomegalovirus infection of WI-38 cells stimulates mitochondrial DNA synthesis, *Nature (London)* **262:**414.

Furukawa, T., Jean, J.-H., and Plotkin, S.A., 1978, Enhanced poliovirus replication in cytomegalovirus-infected human fibroblasts, *Virology* **85:**622.

Furukawa, T., Yoshimura, N., Jean, J.-H., and Plotkin, S.A., 1979, Chronically persistent infection with human cytomegalovirus in human lymphoblasts, *J. Infect. Dis.* **139:**211.

Gardner, M.B., Officer, J.E., Parker, J., Estes, J., and Rongey, R.W., 1974, Induction of disseminated virulent cytomegalovirus infection by immunosuppression of naturally chronically infected wild mice, *Infect. Immun.* **10:**966.

Garnett, H.M., 1975, The effect of arginine deprivation on the cytopathogenic effect and replication of human cytomegalovirus, *Arch. Virol.* **48:**131.

Garnett, H.M., 1979a, The early effects of human cytomegalovirus infection on macromolecular synthesis in human embryonic fibroblasts, *Arch. Virol.* **60:**147.

Garnett, H.M., 1979b, Early alterations in the morphology of cytomegalovirus-infected human fibroblasts, *Intervirology* **11:**359.

Garnett, H.M., 1979c, Fusion of cytomegalovirus infected fibroblasts to form multinucleate giant cells, *J. Med. Virol.* **3:**271.

Geder, L., 1976, Evidence for early nuclear antigens in cytomegalovirus-infected cells, *J. Gen. Virol.* **32:**315.

Geder, L., and Rapp, F., 1977, Evidence for nuclear antigens in cytomegalovirus-transformed human cells, *Nature (London)* **265:**184.

Geder, L., Lausch, R., O'Neill, F., and Rapp, F., 1976, Oncogenic transformation of human embryo lung cells by human cytomegalovirus, *Science* **192:**1134.

Geder, L., Kreider, J., and Rapp, F., 1977, Human cells transformed *in vitro* by human cytomegalovirus: Tumorigenicity in athymic nude mice, *J. Natl. Cancer Inst.* **58:**1003.

Geelen, J.L.M.C., Walig, C., Wertheim, P., and van der Noordaa, J., 1978, Human cytomegalovirus DNA. I. Molecular weight and infectivity, *J. Virol.* **26:**813.

Gergely, L., Czeglédy, J., and Váczi, L., 1980, Early nuclear antigen as DNA-binding protein in cytomegalovirus-infected cells, *Intervirology* **13:**352.

Gibson, W., 1981a, Structural and nonstructural proteins of strain Colburn cytomegalovirus, *Virology* **111:**516.

Gibson, W., 1981b, Immediate–early proteins of human cytomegalovirus strains AD169, Davis, and Towne differ in electrophoretic mobility, *Virology* **112:**350.

Glaser, R., Geder, L., St. Jeor, S., Michelson-Fiske, S., and Haguenau, F., 1977, Partial characterization of a herpes-type virus (K9V) derived from Kaposi's sarcoma, *J. Natl. Cancer Inst.* **59:**55.

Glasgow, L.A., Hanshaw, J.B., Merigan, T.C., and Petralli, J.K., 1967, Interferon and cytomegalovirus *in vivo* and *in vitro, Proc. Soc. Exp. Biol. Med.* **125:**843.

Glazer, J.P., Friedman, H.M., Grossman, R.A., Barker, C.F., Starr, S.E., and Plotkin, S.A., 1978, Cytomegalovirus vaccination and renal transplantation, *Lancet* **1:**90.

Glazer, J.P., Friedman, H.M., Grossman, R.A., Starr, S.E., Barker, C.F., Perloff, L.J., Huang, E.-S., and Plotkin, S.A., 1979, Live cytomegalovirus vaccination of renal transplant candidates: A preliminary trial, *Ann. Intern. Med.* **91:**676.

Gönczöl, É., and Váczi, L., 1973, Cytomegalovirus latency in cultured human cells, *J. Gen. Virol.* **18:**143.

Gönczöl, É., Boldogh, I., and Váczi, L., 1975, Effect of arginine deficiency on the reproduction of human cytomegalovirus, *Acta Microbiol. Acad. Sci. Hung.* **22:**263.

Gönczöl, É., Stone, J., and Melero, J.M., 1978, The effect of heat-inactivated murine cytomegalovirus on host DNA synthesis of different cells, *J. Gen. Virol.* **39:**415.

Goodheart, C.R., and Jaross, L.B., 1963, Human cytomegalovirus: Assay by counting infected cells, *Virology* **19:**532.

Goodheart, C.R., Filbert, J.E., and McAllister, R.M., 1963, Human cytomegalovirus: Effects of 5-fluorodeoxyuridine on viral synthesis and cytopathology, *Virology* **21:**530.

Gould, J.J., and Mims, C.A., 1980, Murine cytomegalovirus: Reactivation in pregnancy, *J. Gen. Virol.* **51:**397.

Griffith, B.P., and Hsiung, G.D., 1980, Cytomegalovirus infection in guinea pigs. IV. Maternal infection at different stages of gestation, *J. Infect. Dis.* **141:**787.

Gupta, P., and Rapp, F., 1978, Cyclic synthesis of human cytomegalovirus-induced proteins in infected cells, *Virology* **84:**199.

Gupta, P., St. Jeor, S., and Rapp, F., 1977, Comparison of the polypeptides of several strains of human cytomegalovirus, *J. Gen. Virol.* **34:**447.

Haines, H.G., Von Essen, R., and Benyesh-Melnick, M., 1971, Preparation of specific antisera to cytomegaloviruses in goats, *Proc. Soc. Exp. Biol. Med.* **138:**846.

Hamilton, J.D., Fitzwilliam, J.F., Cheung, K.S., Shelburne, J., Lang, D.J., and Amos, D.B., 1978, Viral infection–homograft interactions in a murine model, *J. Clin. Invest.* **62:**1303.

Hanshaw, J.B., 1974, A cytomegalovirus vaccine?, *Am. J. Dis. Child.* **128:**141.
Hartley, W.J., and Done, J.T., 1963, Cytomegalic inclusion-body disease in sheep: A report of two cases, *J. Comp. Pathol. Ther.* **73:**84.
Henson, D., and Neapolitan, C., 1970, Pathogenesis of chronic mouse cytomegalovirus infection in submaxillary glands of C3H mice, *Am. J. Pathol.* **58:**255.
Henson, D., and Strano, A.J., 1972, Mouse cytomegalovirus: Necrosis of infected and morphologically normal submaxillary gland acinar cells during termination of chronic infection, *Am. J. Pathol.* **68:**183.
Henson, D., Smith, R.D., and Gehrke, J., 1966, Non-fatal mouse cytomegalovirus hepatitis: Combined morphologic, virologic and immunologic observations, *Am. J. Pathol.* **49:**871.
Henson, D., Strano, A.J., Slotnik, M., and Goodheart, C., 1972, Mouse cytomegalovirus: Isolation from spleen and lymph nodes of chronically infected mice, *Proc. Soc. Exp. Biol. Med.* **140:**802.
Hirai, K., and Watanabe, Y., 1976, Induction of α-type DNA polymerases in human cytomegalovirus-infected WI-38 cells, *Biochim. Biophys. Acta* **447:**328.
Hirai, K., Maeda , F., and Watanabe, Y., 1977, Expression of early virus functions in human cytomegalovirus infected HEL cells: Effect of ultraviolet light-irradiation of the virus, *J. Gen. Virol.* **38:**121.
Holmes, A.R., Rasmussen, L., and Merigan, T.C., 1978, Factors affecting the interferon sensitivity of human cytomegalovirus, *Intervirology* **9:**48.
Howard, R.J., and Najarian, J.S., 1974, Cytomegalovirus-induced immune suppression. I. Humoral immunity, *Clin. Exp. Immunol.* **18:**109.
Howard, R.J., Miller, J., and Najarian, J.S., 1974, Cytomegalovirus-induced immune suppression. II. Cell-mediated immunity, *Clin. Exp. Immunol.* **18:**119.
Howard, R.J., Mattsson, D.M., Seidel, M.V., and Balfour, H.H., Jr., 1978, Cell-mediated immunity to murine cytomegalovirus, *J. Infect. Dis.* **138:**597.
Howard, R.J., Mattsson, D.M., and Balfour, H.H., Jr., 1979, Effect of immunosuppression on humoral and cell-mediated immunity to murine cytomegalovirus, *Proc. Soc. Exp. Biol. Med.* **161:**341.
Hsiung, G.D., Fischman, H.R., Fong, C.K.Y., and Green, R.H., 1969, Characterization of a cytomegalo-like virus isolated from spontaneously degenerated equine kidney cell culture, *Proc. Soc. Exp. Biol. Med.* **130:**80.
Hsiung, G.D., Tenser, R.B., and Fong, C.K.Y., 1976, Comparison of guinea pig cytomegalovirus and guinea pig herpes-like virus: Growth characteristics and antigenic relationship, *Infect. Immun.* **13:**926.
Hsiung, G.D., Choi, Y.C., and Bia, F., 1978, Cytomegalovirus infection in guinea pigs. I. Viremia during acute primary and chronic persistent infection, *J. Infect. Dis.* **138:**191.
Hsiung, G.D., Bia, F.J., and Fong, C.K.Y., 1980, Viruses of guinea pigs: Considerations for biomedical research, *Microbiol. Rev.* **44:**468.
Huang, E.-S., 1975a, Human cytomegalovirus. III. Virus-induced DNA polymerase, *J. Virol.* **16:**298.
Huang, E.-S., 1975b, Human cytomegalovirus. IV. Specific inhibition of virus-induced DNA polymerase activity and viral DNA replication by phosphonoacetic acid, *J. Virol.* **16:**1560.
Huang, E.-S., and Pagano, J.S., 1974, Human cytomegalovirus. II. Lack of relatedness to DNA of herpes simplex I and II, Epstein–Barr virus, and nonhuman strains of cytomegalovirus, *J. Virol.* **13:**642.
Huang, E.-S., Chen, S.-T., and Pagano, J.S., 1973, Human cytomegalovirus. I. Purification and characterization of viral DNA, *J. Virol.* **12:**1473.
Huang, E.-S., Kilpatrick, B.A., Huang, Y.-T., and Pagano, J.S., 1976, Detection of human cytomegalovirus and analysis of strain variation, *Yale J. Biol. Med.* **49:**29.
Huang, E.-S., Kilpatrick, B., Lakeman, A., and Alford, C.A., 1978a, Genetic analysis of a cytomegalovirus-like agent isolated from human brain, *J. Virol.* **26:**718.
Huang, E.-S., Leyritz, M., Menezes, J., and Joncas, J.H., 1978b, Persistence of both human cytomegalovirus and Epstein–Barr virus genomes in two human lymphoblastoid cell lines, *J. Gen. Virol.* **40:**519.

Huang, E.-S., Alford, C.A., Reynolds, D.W., Stagno, S., and Pass, R.F., 1980, Molecular epidemiology of cytomegalovirus infections in women and their infants, *N. Engl. J. Med.* **303:**958.

Huang, Y.-T., Huang, E.-S., and Pagano, J.S., 1974, Antisera to human cytomegaloviruses prepared in the guinea pig: Specific immunofluorescence and complement fixation tests, *J. Immunol.* **112:**528.

Hudson, J.B., 1979, The murine cytomegalovirus as a model for the study of viral pathogenesis and persistent infections, *Arch. Virol.* **62:**1.

Hudson, J.B., 1980, The problem of host DNA synthesis in murine cytomegalovirus-infected cells, *Virology* **101:**545.

Hudson, J.B., Misra, V., and Mosmann, T.R., 1976a, Properties of the multicapsid virions of murine cytomegalovirus, *Virology* **72:**224.

Hudson, J.B., Misra, V., and Mosmann, T.R., 1976b, Cytomegalovirus infectivity: Analysis of the phenomenon of centrifugal enhancement of infectivity, *Virology* **72:**235.

Hudson, J.B., Loh, L., Misra, V., Judd, B., and Suzuki, J., 1977, Multiple interactions between murine cytomegalovirus and lymphoid cells *in vitro*, *J. Gen. Virol.* **38:**149.

Hudson, J.B., Chantler, J.K., Loh, L., Misra, V., and Muller, M.T., 1979, Model systems for analysis of latent cytomegalovirus infections, *Can. J. Microbiol.* **25:**245.

Ihara, S., Hirai, K., and Watanabe, Y., 1978, Temperature-sensitive mutants of human cytomegalovirus: Isolation and partial characterization of DNA-minus mutants, *Virology* **84:**218.

Ihara, S., Saito, S., and Watanabe, Y., 1980, An early function of human cytomegalovirus required for structural integrity of the virus-induced nuclear inclusion, *Microbiol. Immunol.* **24:**891.

Ito, M., Girvin, L., and Barron, A.L., 1978, Inactivation of human cytomegalovirus by phytohemagglutinin, *Arch. Virol.* **57:**97.

Jackson, L., 1920, An intracellular protozoan parasite of the ducts of the salivary glands of the guinea pig, *J. Infect. Dis.* **26:**347.

Jean, J.-H., Yoshimura, N., Furukawa, T., and Plotkin, S.A., 1978, Intracellular forms of the parental human cytomegalovirus genome at early stages of the infective process, *Virology* **86:**281.

Johnson, K.P., 1969, Mouse cytomegalovirus: Placental infection, *J. Infect. Dis.* **120:**445.

Johnson, K.P., and Connor, W.S., 1979, Guinea pig cytomegalovirus: Transplacental transmission, *Arch. Virol.* **59:**263.

Joncas, J.H., Menezes, J., and Huang, E.-S., 1975, Persistence of CMV genome in lymphoid cells after congenital infection, *Nature (London)* **258:**432.

Jordan, M.C., 1978, Interstitial pneumonia and subclinical infection after intranasal inoculation of murine cytomegalovirus, *Infect. Immun.* **21:**275.

Jordan, M.C., 1980, Adverse effects of cytomegalovirus vaccine in mice, *J. Clin. Invest.* **65:**798.

Jordan, M.C., Shanley, J.C., and Stevens, J.G., 1977, Immunosuppression reactivates and disseminates latent murine cytomegalovirus, *J. Gen. Virol.* **37:**419.

Kamata, T., Tanaka, S., and Watanabe, Y., 1978, Human cytomegalovirus-induced chromatin factors responsible for changes in template activity and structure of infected cell chromatin, *Virology* **90:**197.

Kamata, T., Tanaka, S., and Watanabe, Y., 1979, Characterization of the human cytomegalovirus-induced chromatin factor responsible for activation of host cell chromatin template, *Virology* **97:**224.

Kierszenbaum, A.L., and Huang, E.-S., 1978, Chromatin pattern consisting of repeating bipartite structures in WI-38 cells infected with human cytomegalovirus, *J. Virol.* **28:**661.

Kilham, L., and Margolis, G., 1975, Encephalitis in suckling rats induced with rat cytomegalovirus, *Lab. Invest.* **33:**200.

Kilpatrick, B.A., and Huang, E.-S., 1977, Human cytomegalovirus genome: Partial denaturation map and organization of genome sequences, *J. Virol.* **24:**261.

Kilpatrick, B.A., Huang, E.-S., and Pagano, J.S., 1976, Analysis of cytomegalovirus genomes with restriction endonucleases *Hin*D III and *Eco*R-1, *J. Virol.* **18:**1095.

Kim, K.S., and Carp, R.I., 1971, Growth of murine cytomegalovirus in various cell lines, *J. Virol.* **7**:720.
Kim, K.S., and Carp, R.I., 1972, Abortive infection of human diploid cells by murine cytomegalovirus, *Infect. Immun.* **6**:793.
Kim, K.S., and Carp, R.I., 1973, Effect of proteolytic enzymes on the infectivity of a number of herpesviruses, *J. Infect. Dis.* **128**:788.
Kim, K.S., Sapienza, V., and Carp, R.I., 1975, Comparative characteristics of three cytomegaloviruses from rodents, *Am. J. Vet. Res.* **36**:1495.
Kim, K.S., Sapienza, V.J., Carp, R.I., and Moon, H.M., 1976, Analysis of structural proteins of purified murine cytomegalovirus, *J. Virol.* **17**:906.
Krugman, R.D., and Goodheart, C.R., 1964, Human cytomegalovirus: Thermal inactivation, *Virology* **23**:290.
Kumar, M.L., and Nankervis, G.A., 1978, Experimental congenital infection with cytomegalovirus: A guinea pig model, *J. Infect. Dis.* **138**:650.
Lakeman, A.D., and Osborn, J.E., 1979, Size of infectious DNA from human and murine cytomegaloviruses, *J. Virol.* **30**:414.
Landini, M.P., Musiani, M., Zerbini, M., Falcieri, E., La Placa, M., 1979, Inhibition of a complete replication cycle of human cytomegalovirus in actinomycin pre-treated cells, *J. Gen. Virol.* **42**:423.
Loh, L., and Hudson, J.B., 1979, Interaction of murine cytomegalovirus with separated populations of spleen cells, *Infect. Immun.* **26**:853.
Loh, L., and Hudson, J.B., 1980, Immunosuppressive effect of murine cytomegalovirus, *Infect. Immun.* **27**:54.
Lüleci, G., Sakízlí, M., and Günalp, A., 1980, Selective chromosomal damage caused by human cytomegalovirus, *Acta Virol.* **24**:341.
Manischewitz, J.E., and Quinnan, G.V., Jr., 1980, Antivirus antibody-dependent cell-mediated cytotoxicity during murine cytomegalovirus infection, *Infect. Immun.* **29**:1050.
Mäntyjärvi, R.A., Selgrade, M.K., Collier, A.M., Hu, S.-C., Pagano, J.S., 1977, Murine cytomegalovirus infection of epithelial cells in mouse tracheal ring organ culture, *J. Infect. Dis.* **136**:444.
Margolis, G., and Kilham, L., 1976, Neuronal parasitism and cell fusion in mouse cytomegalovirus encephalitis, *Exp. Mol. Pathol.* **25**:20.
Mattsson, D.M., Howard, R.J., and Balfour, H.H., Jr., 1980, Immediate loss of cell-mediated immunity to murine cytomegalovirus upon treatment with immunosuppressive agents, *Infect. Immun.* **30**:700.
Mayo, D.R., and Rapp, F., 1980, Acute cytomegalovirus infections in leukemic mice, *Infect. Immun.* **29**:311.
Mayo, D., Armstrong, J.A., and Ho, M., 1978, Activation of latent murine cytomegalovirus infection: Cocultivation, cell transfer, and the effect of immunosuppression, *J. Infect. Dis.* **138**:890.
McAllister, R.M., Straw, R.M., Filbert, J.E., and Goodheart, C.R., 1963, Human cytomegalovirus: Cytochemical observations of intracellular lesion development correlated with viral synthesis and release, *Virology* **19**:521.
McAllister, R.M., Filbert, J.E., and Goodheart, C.R., 1967, Human cytomegalovirus: Studies on the mechanism of viral cytopathology and inclusion body formation, *Proc. Soc. Exp. Biol. Med.* **124**:932.
McGavran, M.H., and Smith, M.G., 1965, Ultrastructural, cytochemical and microchemical observations on cytomegalovirus (salivary gland virus) infection of human cells in tissue culture, *Exp. Mol. Pathol.* **4**:1.
Medearis, D.N., Jr., 1964a, Mouse cytomegalovirus infection. II. Observations during prolonged infections, *Am. J. Hyg.* **80**:103.
Medearis, D.N., Jr., 1964b, Mouse cytomegalovirus infection. III. Attempts to produce intrauterine infections, *Am. J. Hyg.* **80**:113.
Medearis, D.N., Jr., and Prokay, S.L., 1978, Effect of immunization of mothers on cytomegalovirus infection in suckling mice, *Proc. Soc. Exp. Biol. Med.* **157**:523.

Meggs, P.D., and Beer, A.D., 1979, *In vitro* stimulation of human colostral lymphocytes by cytomegalovirus, *Am. J. Obstet. Gynecol.* **133:**703.
Michelson, S., Horodniceanu, F., Kress, M., and Tardy-Panit, M., 1979, Human cytomegalovirus-induced immediate early antigens: Analysis in sodium dodecyl sulfate–polyacrylamide gel electrophoresis after immunoprecipitation, *J. Virol.* **32:**259.
Michelson-Fiske, S., Arnoult, J., and Febvre, H., 1975, Cytomegalovirus infection of human lung epithelial cells *in vitro, Intervirology* **5:**354.
Michelson-Fiske, S., Horodniceanu, F., and Guillon, J.-C., 1977, Immediate early antigens in human cytomegalovirus infected cells, *Nature* (*London*) **270:**615.
Middlekamp, J.N., Patrizi, G., and Reed, C.A., 1967, Light and electron microscopic studies of the guinea pig cytomegalovirus, *J. Ultrastruct. Res.* **18:**85.
Miller, R.L., Iltis, J.P., and Rapp, F., 1977, Differential effect of arabinofuranosylthymine on the replication of human herpesviruses, *J. Virol.* **23:**679.
Mims, C.A., and Gould, J., 1978, The role of macrophages in mice infected with murine cytomegalovirus, *J. Gen. Virol.* **41:**143.
Mims, C.A., and Gould, J., 1979, Infection of salivary glands, kidneys, adrenals, ovaries and epithelia by murine cytomegalovirus, *J. Med. Microbiol.* **12:**113.
Minamishima, Y., Graham, B.J., and Benyesh-Melnick, M., 1971, Neutralizing antibodies to cytomegaloviruses in normal simian and human sera, *Infect. Immun.* **4:**368.
Minamishima, Y., Eizuru, Y., Yoshida, A., and Fukunishi, R., 1978, Murine model for immunoprophylaxis of cytomegalovirus infection. I. Efficacy of immunization, *Microbiol. Immunol.* **22:**693.
Misra, V., and Hudson, J.B., 1980, Minor base sequence differences between the genomes of two strains of murine cytomegalovirus differing in virulence, *Arch. Virol.* **64:**1.
Misra, V., Muller, M.T., and Hudson, J.B., 1977, The enumeration of viral genomes in murine cytomegalovirus-infected cells, *Virology* **83:**458.
Misra, V., Muller, M.T., Chantler, J.K., and Hudson, J.B., 1978, Regulation of murine cytomegalovirus gene expression. I. Transcription during productive infection, *J. Virol.* **27:**263.
Mocarski, E.S., and Stinski, M.F., 1979, Persistence of the cytomegalovirus genome in human cells, *J. Virol.* **31:**761.
Moon, H.M., Sapienza, V.J., Carp, R.I., and Kim, K.S., 1976, DNA synthesis in mouse embryo fibroblast cells infected with murine cytomegalovirus, *Virology* **75:**376.
Moon, H.M., Sapienza, V.J., Carp, R.I., and Kim, K.S., 1979, Murine cytomegalovirus-induced protein synthesis, *J. Gen. Virol.* **42:**159.
Mosmann, T.R., and Hudson, J.B., 1973, Some properties of the genome of murine cytomegalovirus (MCV), *Virology* **54:**135.
Mosmann, T.R., and Hudson, J.B., 1974, Structural and functional heterogeneity of the murine cytomegalovirus genome, *Virology* **62:**175.
Muchmore, E., 1971, Possible cytomegalovirus infection in man following chimpanzee bite, *Lab. Anim. Sci.* **21:**1080.
Muller, M.T., and Hudson, J.B., 1977a, Cell cycle dependency of murine cytomegalovirus replication in synchronized 3T3 cells, *J. Virol.* **22:**267.
Muller, M.T., and Hudson, J.B., 1977b, Thymidine kinase activity in mouse 3T3 cells infected by murine cytomegalovirus (MCV), *Virology* **80:**430.
Muller, M.T., and Hudson, J.B., 1978, Murine cytomegalovirus DNA synthesis in nuclear monolayers, *Virology* **88:**371.
Muller, M., Misra, V., Chantler, J.K., and Hudson, J.B., 1978, Murine cytomegalovirus gene expression during nonproductive infection in G_0-phase 3T3 cells, *Virology* **90:**279.
Musiani, M., Zerbini, M., Landini, M.P., and La Placa, M., 1979, Preliminary evidence that cytomegalovirus-induced immediate early antigens are DNA-binding proteins, *Microbiologica* **2:**281.
Nedrud, J.G., Collier, A.M., and Pagano, J.S., 1979, Cellular basis for susceptibility to mouse cytomegalovirus: Evidence from tracheal organ culture, *J. Gen. Virol.* **45:**737.
Neff, B.J., Weibel, R.E., Buynak, E.B., McLean, A.A., and Hilleman, M.R., 1979, Clinical

and laboratory studies of live cytomegalovirus vaccine Ad-169, *Proc. Soc. Exp. Biol. Med.* **160:**32.

Neighbour, P.A., 1978, Studies on the susceptibility of the mouse preimplantation embryo to infection with cytomegalovirus, *J. Reprod. Fertil.* **54:**15.

Neighbour, P.A., and Fraser, L.R., 1978, Murine cytomegalovirus and fertility: Potential sexual transmission and the effect of this virus on fertilization *in vitro, Fertil. Steril.* **30:**216.

Nigida, S.M., Falk, L.A., Wolfe, L.G., Deinhardt, F., Lakeman, A., and Alford, C.A., 1975, Experimental infection of marmosets with a cytomegalovirus of human origin, *J. Infect. Dis.* **132:**582.

Nigida, S.M., Falk, L.A., Wolfe, L.G., and Deinhardt, F.:, 1979, Isolation of a cytomegalovirus from salivary glands of white-lipped marmosets (*Saguinus fuscicollis*), *Lab. Anim. Sci.* **29:**53.

Nishiyama, Y., and Rapp, F., 1980, Enhanced survival of ultraviolet-irradiated herpes simplex virus in human cytomegalovirus-infected cells, *Virology* **100:**189.

Oie, H.K., Easton, J.M., Ablashi, D.V., and Baron, S., 1975, Murine cytomegalovirus: Induction of and sensitivity to interferon *in vitro, Infect. Immun.* **12:**1012.

Olding, L.B., Jensen, F.C., and Oldstone, M.B.A., 1975, Pathogenesis of cytomegalovirus infection. I. Activation of virus from bone marrow-derived lymphocytes by *in vitro* allogenic reaction, *J. Exp. Med.* **141:**561.

Olding, L.B., Kingsbury, D.T., and Oldstone, M.B.A., 1976, Pathogenesis of cytomegalovirus infection: Distribution of viral products, immune complexes and autoimmunity during latent murine infection, *J. Gen. Virol.* **33:**267.

Oldstone, M.B.A., Olding, L.B., and Brautigam, A.R., 1977, Adventures in latency and reactivation of cytomegalovirus: Facts and fantasies, in: Excerpta Medica International Congress Series No. 423, Transplantation and Clinical Immunology, Volume IX, Proceedings of the Ninth International Course, Lyon, France (June 6–8, 1977), p. 11.

Osborn, J.E., and Walker, D.L., 1968, Enhancement of infectivity of murine cytomegalovirus *in vitro* by centrifugal inoculation, *J. Virol.* **2:**853.

Osborn, J.E., Blazkovec, A.H., and Walker, D.L., 1968, Immunosuppression during acute murine cytomegalovirus infection, *J. Immunol.* **100:**835.

Overall, J.C., Jr., Kern, E.R., and Glasgow, L.A., 1976, Effective antiviral chemotherapy in cytomegalovirus infection of mice, *J. Infect. Dis.* **133:**(Suppl.):A237.

Patrizi, G., Middlekamp, J.N., and Reed, C.A., 1967, Reduplication of nuclear membranes in tissue-culture cells infected with guinea-pig cytomegalovirus, *Am. J. Pathol.* **50:**779.

Phillips, C.F., 1979, Cytomegalovirus vaccine: A realistic appraisal, *Hosp. Pract.*, p. 75.

Plotkin, S.A., 1979, Vaccination against herpes group viruses, in particular, cytomegalovirus, *Monogr. Paediatr.* **11:**58.

Plotkin, S.A., Furukawa, T., Zygraich, N., and Huygelen, C., 1975, Candidate cytomegalovirus strain for human vaccination, *Infect. Immun.* **12:**521.

Plowright, W., Edington, N., and Watt, R.G., 1976, The behaviour of porcine cytomegalovirus in commercial pig herds, *J. Hyg.* **76:**125.

Plummer, G., and Goodheart, C.R., 1974, Growth of murine cytomegalovirus in a heterologous cell system and its enhancement by 5-iodo-2′-deoxyuridine, *Infect. Immun.* **10:**251.

Plummer, G., and Lewis, B., 1965, Thermoinactivation of herpes simplex virus and cytomegalovirus, *J. Bacteriol.* **89:**671.

Plummer, G., and Waterson, A.P., 1963, Equine herpes viruses, *Virology* **19:**412.

Pritchett, R.F., 1980, DNA nucleotide sequence heterogeneity between the Towne and AD169 strains of cytomegalovirus, *J. Virol.* **36:**152.

Rabson, A.S., Edgcomb, J.H., Legallais, F.Y., and Tyrrell, S.A., 1969, Isolation and growth of rat cytomegalovirus *in vitro, Proc. Soc. Exp. Biol. Med.* **131:**923.

Radsak, K., and Schmitz, B., 1980, Unimpaired histone synthesis in human fibroblasts infected by human cytomegalovirus, *Med. Microbiol. Immunol.* **168:**63.

Radsak, K., Furukawa, T., and Plotkin, St., 1980, DNA synthesis in chromatin preparations from human fibroblasts infected by cytomegalovirus, *Arch. Virol.* **65:**45.

Ramirez, M.L., Virmani, M., Garon, C., and Rosenthal, L.J., 1979, Defective virions of human cytomegalovirus, *Virology* **96:**311.

Rangan, S.R.S., and Chaiban, J., 1980, Isolation and characterization of a cytomegalovirus from the salivary gland of a squirrel monkey (*Saimiri sciureus*), *Lab. Anim. Sci.* **30:**532.

Rapp, F., and Geder, L., 1978, Persistence and transformation by human cytomegalovirus, in: *Persistent Viruses*, ICN–UCLA Symposia on Molecular and Cellular Biology (J.G. Stevens, G.J. Todaro, and C.F. Fox, eds.), pp. 767–785, Academic Press, New York.

Rapp, F., and McCarthy, B.A., 1979, The oncogenic properties of human cytomegalovirus, in: *Antiviral Mechanisms in the Control of Neoplasia* Vol. 20 (P. Chandra, ed.), pp. 263–283, Plenum Press, New York.

Rapp, F., Rasmussen, L.E., and Benyesh-Melnick, M., 1963, The immunofluorescent focus technique in studying the replication of cytomegalovirus, *J. Immunol.* **91:**709.

Rapp, F., Geder, L., Murasko, D., Lausch, R., Ladda, R., Huang, E.-S., and Webber, M.M., 1975, Long-term persistence of cytomegalovirus genome in cultured human cells of prostatic origin, *J. Virol.* **16:**982.

Reynolds, D.W., 1978, Development of early nuclear antigen in cytomegalovirus infected cells in the presence of RNA and protein synthesis inhibitors, *J. Gen. Virol.* **40:**475.

Rinaldo, C.R., Jr., Richter, B.S., Black, P.H., Callery, R., Chess, L., and Hirsch, M.S., 1978, Replication of herpes simplex virus and cytomegalovirus in human leukocytes, *J. Immunol.* **120:**130.

Rowe, W.P., Hartley, J.W., Waterman, S., Turner, H.C., and Huebner, R.J., 1956, Cytopathic agent resembling human salivary gland virus recovered from tissue cultures of human adenoids, *Proc. Soc. Exp. Biol. Med.* **92:**418.

Ruebner, B.H., Miyai, K., Slusser, R.J., Wedemeyer, P., and Medearis, D.N., Jr., 1964, Mouse cytomegalovirus infection: An electron microscopic study of hepatic parenchymal cells, *Am. J. Pathol.* **44:**799.

Ruebner, B.H., Hirano, T., Slusser, R., Osborn, J., and Medearis, D.N., Jr., 1966, Cytomegalovirus infection: Viral ultrastructure with particular reference to the relationship of lysosomes to cytoplasmic inclusions, *Am. J. Pathol.* **48:**971.

Rytel, M.W., and Hooks, J.J., 1977, Induction of immune interferon by murine cytomegalovirus, *Proc. Soc. Exp. Biol. Med.* **155:**611.

St. Jeor, S.C., and Hutt, R., 1977, Cell DNA replication as a function in the synthesis of human cytomegalovirus, *J. Gen. Virol.* **37:**65.

St. Jeor, S., and Rapp, F., 1973, Cytomegalovirus: Conversion of nonpermissive cells to a permissive state for virus replication, *Science* **181:**1060.

St. Jeor, S., and Weisser, A., 1977, Persistence of cytomegalovirus in human lymphoblasts and peripheral leukocyte cultures, *Infect. Immun.* **15:**402.

St. Jeor, S.C., Albrecht, T.B., Funk, F.D., and Rapp, F., 1974, Stimulation of cellular DNA synthesis by human cytomegalovirus, *J. Virol.* **13:**353.

Sarov, I., and Abady, I., 1975, The morphogenesis of human cytomegalovirus: Isolation and polypeptide characterization of cytomegalovirions and dense bodies, *Virology* **66:**464.

Sarov, I., and Friedman, A., 1976, Electron microscopy of human cytomegalovirus DNA, *Arch. Virol.* **50:**343.

Sarov, I., Larson, A.M., Heron, I., and Anderson, H.K., 1978, Stimulation of human lymphocytes by cytomegalovirions and dense bodies, *Med. Microbiol. Immunol.* **166:**81.

Schirm, J., Roenhorst, H.W., and The, T.H., 1980, Comparison of *in vitro* lymphocyte proliferations induced by cytomegalovirus-infected human fibroblasts and cell-free cytomegalovirus, *Infect. Immun.* **30:**621.

Schleupner, C.J., Olsen, G.A., and Glasgow, L.A., 1979, Activation of reticuloendothelial cells following infection with murine cytomegalovirus, *J. Infect. Dis.* **139:**641.

Schmitz, H., Doerr, H.W., and Orbrig, M., 1975, Envelope and nucleocapsid antigens of cytomegalovirus (CMV), *Med. Microbiol. Immunol.* **161:**155.

Schnitzer, T.J., and Gönczöl, É., 1979, Phenotypic mixing between murine oncoviruses and murine cytomegalovirus, *J. Gen. Virol.* **43:**691.

Schwartz, J.N., Daniels, C.A., and Klintworth, G.K., 1975, Lymphoid cell necrosis, thymic atrophy, and growth retardation in newborn mice inoculated with murine cytomegalovirus, *Am. J. Pathol.* **79:**509.

Severi, B., Landini, M.P., Musiani, M., and Zerbini, M., 1979, A study of the passage of human cytomegalovirus from the nucleus to the cytoplasm, *Microbiologica* **2:**265.

Shanley, J.D., and Pesanti, E.L., 1980, Replication of murine cytomegalovirus in lung macrophages: Effect on phagocytosis of bacteria, *Infect. Immun.* **29:**1152.

Shanley, J.D., Jordan, M.C., Cook, M.L., and Stevens, J.G., 1979, Pathogenesis of reactivated latent murine cytomegalovirus infection, *Am. J. Pathol.* **95:**67.

Smith, J.D., and De Harven, E., 1973, Herpes simplex virus and human cytomegalovirus replication in WI-38 cells. I. Sequence of viral replication, *J. Virol.* **12:**919.

Smith, J.D., and De Harven, E., 1974, Herpes simplex virus and human cytomegalovirus replication in WI-38 cells. II. An ultrastructural study of viral penetration, *J. Virol.* **14:**945.

Smith, J.D., and De Harven, E., 1978, Herpes simplex virus and human cytomegalovirus replication in WI-38 cells. III. Cytochemical localization of lysosomal enzymes in infected cells, *J. Virol.* **26:**102.

Smith, J.D., and De Harven, E., 1979, Localization of arginine-rich protein in cytomegalovirus dense bodies using phenanthrenequinone fluorescence, Virology **96:** 335.

Smith, K.O., and Rasmussen, L., 1963, Morphology of cytomegalovirus (salivary gland virus), *J. Bacteriol.* **85:**1319.

Smith, K.O., Thiel, J.F., Newman, J.T., Harvey, E., Trousdale, M.D., Gehle, W.D., and Clark, G., 1969, Cytomegaloviruses as common adventitious contaminants in primary African green monkey kidney cell cultures, *J. Natl. Cancer Inst.* **42:**489.

Smith, M.G., 1954, Propagation of salivary gland virus of the mouse in tissue cultures, *Proc. Soc. Exp. Biol. Med.* **86:**435.

Smith, M.G., 1956, Propagation in tissue cultures of a cytopathogenic virus from human salivary gland virus (SGV) disease, *Proc. Soc. Exp. Biol. Med.* **92:**424.

Smith, R.D., and Wehner, R.W., 1980, Acute cytomegalovirus glomerulonephritis, *Lab. Invest.* **43:**278.

Spear, P.G., and Roizman, B., 1972, Proteins specified by herpes simplex virus. V. Purification and structural proteins of the herpesvirion, *J. Virol.* **9:**143.

Stinski, M.F., 1976, Human cytomegalovirus: Glycoproteins associated with virions and dense bodies, *J. Virol.* **19:**594.

Stinski, M.F., 1977, Synthesis of proteins and glycoproteins in cells infected with human cytomegalovirus, *J. Virol.* **23:**751.

Stinski, M.F., 1978, Sequence of protein synthesis in cells infected by human cytomegalovirus: Early and late virus-induced polypeptides, *J. Virol.* **26:**686.

Stinski, M.F., Mocarski, E.S., and Thomsen, D.R., 1979a, DNA of human cytomegalovirus: Size heterogeneity and defectiveness resulting from serial undiluted passage, *J. Virol.* **31:**231.

Stinski, M.F., Mocarski, E.S., Thomsen, D.R., and Urbanowski, M.L., 1979b, Membrane glycoproteins and antigens induced by human cytomegalovirus, *J. Gen. Virol.* **43:** 119.

Summers, W.C., 1980, Molecular epidemiology of DNA viruses: Applications of restriction endonuclease cleavage site analysis, *Yale J. Biol. Med.* **53:**55.

Swack, N.S., Liu, O.C., and Hsiung, G.D., 1971, Cytomegalovirus infections of monkeys and baboons, *Am. J. Epidemiol.* **94:**397.

Sweet, G.H., Tegtmeier, G.E., and Bayer, W.L., 1979, Antigens of human cytomegalovirus: Electroimmunodiffusion assay and comparison among strains, *J. Gen. Virol.* **43:**707.

Talbot, P., and Almeida, J.D., 1977, Human cytomegalovirus: Purification of enveloped virions and dense bodies, *J. Gen. Virol.* **36:**345.

Tanaka, S., Furukawa, T., and Plotkin, S.A., 1975, Human cytomegalovirus stimulates host cell RNA synthesis, *J. Virol.* **15**:297.

Tanaka, S., Ihara, S., and Watanabe, Y., 1978, Human cytomegalovirus induces DNA-dependent RNA polymerases in human diploid cells, *Virology* **89**:179.

Tanaka, S., Otsuka, M., Ihara, S., Maeda, F., and Watanabe, Y., 1979, Induction of pre-early nuclear antigen(s) in HEL cells infected with human cytomegalovirus, *Microbiol. Immunol.* **23**:263.

Tegtmeyer, P.J., and Craighead, J.E., 1968, Infection of adult mouse macrophages *in vitro* with cytomegalovirus, *Proc. Soc. Exp. Biol. Med.* **129**:690.

Tenser, R.B., and Hsiung, G.D., 1976, Comparison of guinea pig cytomegalovirus and guinea pig herpes-like virus: Pathogenesis and persistence in experimentally infected animals, *Infect. Immun.* **13**:934.

The, T.H., Klein, G., and Langenhuysen, M.M.A.C., 1974, Antibody reactions to virus-specific early antigens (EA) in patients with cytomegalovirus (CMV) infection, *Clin. Exp. Immunol.* **16**:1.

Tinghitella, T.J., and Booss, J., 1979, Enhanced immune response late in primary cytomegalovirus infection of mice, *J. Immunol.* **122**:2442.

Tocci, M.J., and St. Jeor, S.C., 1979a, Susceptibility of lymphoblastoid cells to infection with human cytomegalovirus, *Infect. Immun.* **23**:418.

Tocci, M.J., and St. Jeor, S.C., 1979b, Persistence and replication of the human cytomegalovirus genome in lymphoblastoid cells of B and T origin, *Virology* **96**:664.

Tyms, A.S., and Williamson, J.D., 1980, Polyamine metabolism in MRC-5 cells infected with human cytomegalovirus, *J. Gen. Virol.* **48**:183.

Valícek, L., Smíd, B., Pleva, V., and Mensík, J., 1970, Porcine cytomegalic inclusion disease virus, *Arch. Gesamte Virusforsch.* **32**:19.

Vonka, V., Anisimová, E., and Macek, M., 1976, Replication of cytomegalovirus in human epithelioid diploid cell line, *Arch. Virol.* **52**:283.

Wahren, B., and Öberg, B., 1979, Inhibition of cytomegalovirus late antigens by phosphonoformate, *Intervirology* **12**:335.

Waner, J.L., 1975, Partial characterization of a soluble antigen preparation from cells infected with human cytomegalovirus: Properties of antisera prepared to the antigen, *J. Immunol.* **114**:1454.

Waner, J.L., and Biano, S.A., 1979, Antibody reactivity to human and vervet cytomegalovirus early antigen(s) in sera from patients with active cytomegalovirus infections and from asymptomatic donors, *J. Clin. Microbiol.* **9**:134.

Waner, J.L., and Budnick, J.E., 1977, Blastogenic response of human lymphocytes to human cytomegalovirus, *Clin. Exp. Immunol.* **30**:44.

Waner, J.L., and Weller, T.H., 1974, Behavior of human cytomegaloviruses in cell cultures of bovine and simian origin, *Proc. Soc. Exp. Biol. Med.* **145**:379.

Waner, J.L., and Weller, T.H., 1978, Analysis of antigenic diversity among human cytomegaloviruses by kinetic neutralization tests with high-titered rabbit antisera, *Infect. Immun.* **21**:151.

Waner, J.L., Budnick, J.E., and Albrecht, T.B., 1978, Effect of rhesus or vervet cytomegalovirus infection on DNA synthesis in untreated and 5-iodo-2′-deoxyuridine-arrested cells, *J. Virol.* **25**:465.

Weller, T.H., Macauley, J.C., Craig, J.M., and Wirth, P., 1957, Isolation of intranuclear inclusion producing agents from infants with illnesses resembling cytomegalic inclusion disease, *Proc. Soc. Exp. Biol. Med.* **94**:4.

Wentworth, B.B., and French, L., 1970, Plaque assay of cytomegalovirus strains of human origin, *Proc. Soc. Exp. Biol. Med.* **135**:253.

Williams, L.L., Blakeslee, J.R., Jr., Boldogh, I., and Huang, E.S., 1980, Detection of cytomegalovirus genomes in human skin fibroblasts by DNA hybridization, *J. Gen. Virol.* **51**:435.

Wise, T.G., Manischewitz, J.E., Quinnan, G.V., Aulakh, G.S., and Ennis, F.A., 1979, Latent cytomegalovirus infection of BALB/c mouse spleens detected by an explant culture technique, *J. Gen. Virol.* **44:**551.

Wright, H.T., Jr., 1973, Cytomegaloviruses, in: *The Herpesviruses* (A.S. Kaplan, ed.), pp. 353–388, Academic Press, New York.

Wright, H.T., Jr., Goodheart, C.R., and Lielausis, A., 1964, Human cytomegalovirus: Morphology by negative staining, *Virology* **23:**419.

Wright, H.T., Kasten, F.H., and McAllister, R.M., 1968, Human cytomegalovirus: Observations of intracellular lesion development as revealed by phase contrast, time-lapse cinematography, *Proc. Soc. Exp. Biol. Med.* **127:**1032.

Wroblewska, Z., Gilden, D., Devlin, M., Huang, E.-S., Rorke, L.B., Hamada, T., Furukawa, T., Cummins, L., Kalter, S., and Koprowski, H., 1979, Cytomegalovirus isolation from a chimpanzee with acute demyelinating disease after inoculation of multiple sclerosis brain cells, *Infect. Immun.* **25:**1008.

Wu, B., and Ho, M., 1979, Characteristics of infection of B and T lymphocytes from mice after inoculation with cytomegalovirus, *Infect. Immun.* **24:**856.

Wu, B.C., Dowling, J.N., Armstrong, J.A., and Ho, M., 1975, Enhancement of mouse cytomegalovirus infection during host-versus-graft reaction, *Science* **190:**56.

Yamanishi, K., and Rapp, F., 1979a, Production of plasminogen activator by human and hamster cells infected with human cytomegalovirus, *J. Virol.* **31:**415.

Yamanishi, K., and Rapp, F., 1979b, Induction of host DNA synthesis and DNA polymerase by DNA-negative temperature-sensitive mutants of human cytomegalovirus, *Virology* **94:**237.

Yamanishi, K., Fogel, M., and Rapp, F., 1978, Effect of caffeine on the replication of non-irradiated and ultraviolet-irradiated cytomegalovirus, *Intervirology* **10:**241.

Yamashiroya, H.M., Reed, J.M., Blair, W.H., and Schneider, M.D., 1971, Some clinical and microbiological findings in vervet monkeys (*Cercopithecus aethiops pygerythrus*), *Lab. Anim. Sci.* **21:**873.

Young, J.A., Cheung, K.-S., and Lang, D.J., 1977, Infection and fertilization of mice after artificial insemination with a mixture of sperm and murine cytomegalovirus, *J. Infect. Dis.* **135:**837.

Zablotney, S.L., Wentworth, B.B., and Alexander, E.R., 1978, Antigenic relatedness of 17 strains of human cytomegalovirus, *Am. J. Epidemiol.* **107:**336.

Závada, V., Erban, V., Rezácová, D., and Vonka, V., 1976, Thymidine-kinase in cytomegalovirus infected cells, *Arch. Virol.* **52:**333.

CHAPTER 2

Molecular Biology of Cytomegaloviruses

MARK F. STINSKI

I. INTRODUCTION

In the human host, infection by cytomegaloviruses (CMVs) induces a variety of syndromes ranging from the classic cytomegalic inclusion disease to intrauterine death, prematurity, congenital defects, infectious mononucleosis, postperfusion syndrome, and interstitial pneumonia in transplantation patients (for reviews, see Weller, 1971; Rapp, 1980; Ho, 1982). These diseases caused by human CMV (HCMV) frequently represent infections after reactivation of latent virus and are usually associated with immuno-suppression due to a variety of situations such as malignant disease and chemotherapy (Weller, 1971; Plummer, 1973). Therefore, the HCMV genome has evolved to remain in human cells in a quiescent state from the time of primary infection. Although the majority of humans carry this virus, overt disease is a rare event. Unfortunately, little is known about the molecular biology of the latent CMV genome; consequently, this chapter will emphasize the events after reactivation of the viral genome. Reactivation is the event that allows for replication of the viral DNA genome and eventual release of infectious virus to cause overt disease. Latency and reactivation of HCMV replication relate directly to the regulation of the viral genome by host- and/or virus-specified genes in the eukaryotic cell.

In the human host, CMV can infect a variety of cell types and pre-

MARK F. STINSKI • Department of Microbiology, College of Medicine, University of Iowa, Iowa City, Iowa 52242.

sumably enters both nonproductive (nonpermissive) and productive (permissive) cells. It is proposed from recent studies (Stinski, 1978; Mocarski and Stinski, 1979; Wathen *et al.*, 1981; Tanaka *et al.*, 1981) that CMV early gene expression but not late gene expression occurs in nonproductive cells. It is hypothesized here that the nonproductive cell may favor latency of the viral genome. Physiological change may cause the conversion of a nonproductive cell to a productive cell; consequently, the latent viral genome is reactivated. The possible pathways of the CMV genome in an infected cell are diagrammed in Fig. 1. This chapter will attempt to review the data that suggest these virus–cell interactions.

Our understanding of viral induced diseases and their relationship to the eukaryotic cell have frequently emerged from comparisons between different virus–cell interactions. In this regard, recent evidence suggests that herpes simplex virus (HSV) and HCMV may have evolved from a common stem, since their genome structure is similar. This suggests that the two viruses employ a similar mechanism for viral DNA replication. Nevertheless, the interaction with the host cell is very different in terms of biochemical and biological effects on the host cell, and this suggests that HCMV carries several genes that function differently from those of HSV. Throughout this chapter, comparisons will be made between HSV and CMV in an attempt to better understand these virus–cell interactions.

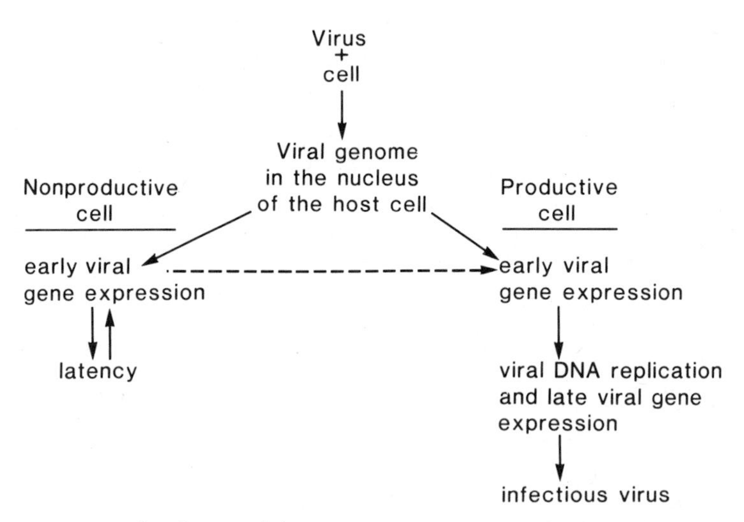

FIGURE 1. Proposed pathways of the CMV genome in nonproductive and productive cells. It is implied that CMV encounters both nonproductive and productive cells in the human. The type of cell may be influenced by the physiology of the cell at the time of infection. It is proposed that early viral gene expression occurs in both types of cells. Latency occurs in the nonproductive cell when the genome does not proceed to replication. It is also proposed that physiological changes or the eventual expression of a required early viral gene can convert the nonproductive cell to a productive cell; consequently, the viral genome can go from a state of latency to replication and production of infectious progeny.

CMVs have three properties that are uniquely characteristic of the herpesviruses: (1) The genomes of human (Kilpatrick and Huang, 1977; Geelen *et al.*, 1978; DeMarchi *et al.*, 1978; Lakeman and Osborn, 1979), simian (Fleckenstein, personal communication; Hayward, personal communication), and murine (Mosmann and Hudson, 1973; Lakeman and Osborn, 1979) CMVs have a considerably higher order of sequence complexity than do those of the other herpesviruses. (2) The viruses replicate preferentially only in the cells of the species of origin; consequently, these viruses frequently display species specificity for induction of disease. (3) Although there is considerable variability in the rate of replication, most CMVs replicate slowly when compared to HSV. Therefore, HSV is representative of a rapidly replicating herpesvirus that can infect cells of both human and animal origin, and HCMV is representative of a slowly replicating herpesvirus that has a limited host range. Nevertheless, both these viruses share the characteristic herpesvirus features of a double-stranded DNA enveloped virus with a nuclear site for capsid assembly and similar modes for maturation and release. The observed differences presumably reflect differences in the genetic makeup of the viruses; therefore, a comparison will be made between the genetic arrangement, genome transcription, and viral gene products.

II. RATE OF INFECTIOUS VIRUS PRODUCTION: COMPARISONS BETWEEN HERPES SIMPLEX VIRUS AND CYTOMEGALOVIRUS

After infection of cells with herpes simplex virus (HSV), there is an eclipse period of at least 4 hr, and plaques are detectable in monolayers of cells within 24–48 hr (Ben-Porat and Kaplan, 1973; Darlington and Granoff, 1973). In contrast, human cytomegalovirus (HCMV) generally has an eclipse period of approximately 2 days, and plaques are not detectable until 10–14 days after infection. In addition, the yield of infectious virus for CMV is frequently as much as 10-fold lower than that of HSV. Slow replication and lower levels of infectious virus production are two major reasons that studies on the molecular biology of CMVs have lagged behind studies on that of HSV.

Smith and DeHarven (1973) compared the replication of HSV and CMV in human diploid fibroblast cells at the same multiplicity of infection (m.o.i.) of 10 plaque-forming units (PFU) per cell. These data are summarized in Fig. 2. Infectious progeny were released from the HSV-infected cells within 8 hr, but not from the CMV-infected cell until 4 days. In addition, Smith and DeHarven (1973) compared the timing of major events related to productive infection in the HSV- and CMV-infected cells by electron microscopy. The majority of the HSV-induced events that caused significant changes in the ultrastructural properties of the host cell occurred within hours after infection (Table I). In contrast, the CMV-induced events required days, even though the same cell strain was infected at the same multiplicity (Table I). The limitations associated with CMV infection do not appear to involve viral adsorption or penetration, since capsids of both viruses reach the periphery of the nucleus in approximately the same time (Smith and DeHarven, 1973).

The long eclipse period that is characteristic of HCMV infection may be related to the effect of the early virus-specified gene products on host-cell macromolecular synthesis and to the biochemical events associated with viral DNA replication. In addition, these events may be related to the phenomenon of species specificity for CMV infectious virus production. These relationships will be discussed in more detail, after the genome structure and its transcription and translation are described.

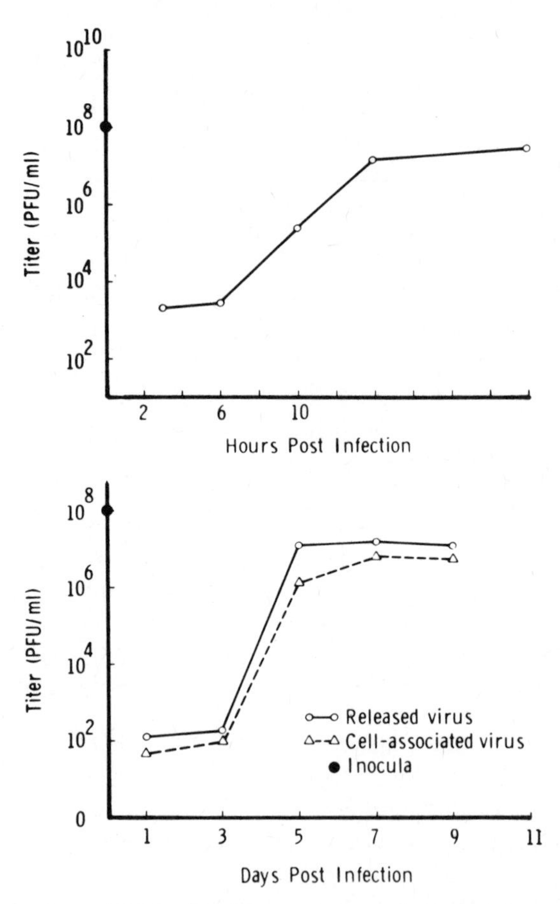

FIGURE 2. Comparison of HSV and CMV one-step growth curves in human fibroblast cells. (A) Titers of released infectious HSV (○). (B) Titers of released (○) and cell-associated (△) infectious CMV. The input m.o.i for both virus infections was 10 PFU/cell, which required an inoculum (●) of 10^8 PFU/ml. Data from Smith and DeHarven (1973).

TABLE I. Comparison of Timing of Major Events in CMV- and HSV-Infected Cells as Determined by Electron Microscopy[a]

Event	Onset	
	CMV (days)	HSV (hr)
Cell fusion and rounding	0.5	6
Golgi alteration	1	8
Viral eclipse	2	2
Appearance of increased number of perichromatin like granules	2	3
Condensation of the chromatin	Not observed	3
Assembly of the first capsids in the nucleus	3	4
Appearance of the first capsids with dense cores	3.5	5
Envelopment at the nuclear membrane	3.5	6
Appearance of the naked capsids in the cytoplasm	4	6
Appearance of dense cytoplasmic aggregates	4	Not observed
First released particles	4	8
Nuclear membrane reduplication	4	8
Envelopment of cytoplasmic membranes	4	8
Cell lysis	7–8	24–48

[a] Data from Smith and DeHarven (1973).

III. VIRAL GENOME

A. Physical Properties of the Cytomegalovirus Genome

Huang *et al.* (1973) demonstrated that human cytomegalovirus (HCMV) DNA could be isolated from purified virions. This viral DNA was found to be relatively pure because as little as 1 μg would significantly accelerate the reassociation of CMV DNA under renaturation conditions, whereas 2 mg of host-cell DNA had no effect. When isolated from purified virions, the CMV genome is a linear double-stranded molecule. Like other herpesviruses, HCMV has nicks and gaps that are randomly distributed over the genome; consequently, careful manipulations are required to isolate the intact genome.

The molecular weight of the viral genome associated with purified virions has been determined by electron microscopy (Kilpatrick and Huang, 1977; Geelen *et al.*, 1978; DeMarchi *et al.*, 1978) and velocity centrifugation (DeMarchi *et al.*, 1978; Lakeman and Osborn, 1979) to be approximately 155 million daltons (240 kilobases). The majority of this viral genome represents unique sequences, while about 11% is repeated twice. The HCMV genome is about 50% greater in size and molecular complexity than that of the herpes simplex virus (HSV) genome. Likewise, the murine CMV genome is approximately 30% larger than the HSV genome (Mosmann and Hudson, 1973; Lakeman and Osborn, 1979).

The density of the HCMV genome is slightly higher than that of host-cell DNA; consequently, the two molecules can be separated by equilibrium centrifugation in gradients of heavy salts such as cesium chloride. Although there may be slight variations among different strains of HCMV, the density of the viral DNA is approximately 1.716 g/cm^3 (Crawford and Lee, 1964; Huang *et al.*, 1973), which corresponds to a base composition of 58% guanine plus cytosine (G + C). The HCMV DNA molecules have characteristically a lower G + C composition than that of approximately 68% G + C for the HSV DNA molecule.

Thermodenaturation studies of the HCMV DNA indicated a biphasic denaturation (Kilpatrick and Huang, 1977). The lower T_m occurred at 86°C and resulted from 21% of the DNA with an average of 43% G + C content. The higher T_m occurred at 94°C and was composed of the remaining 79% of the DNA with an average of 60% G + C content. Whether this nonhomogeneous distribution reflects the presence of both defective and nondefective DNA molecules or a unique feature of the CMV genome is at present not known. The CMV DNA molecule is presumably a normal double-stranded DNA maintained in a duplex by hydrogen bonding and probably does not have any unusual linkages such as cross-links or RNA linkers.

B. Viral Genome Arrangement

While analyzing partially denatured HCMV DNA molecules by electron microscopy, Kilpatrick and Huang (1977) noted that the HCMV DNA molecule contain six unique adenine plus thymidine (A + T)- and G + C-rich regions. At temperatures below total denaturation, the hydrogen bonding would break between the A + T-rich regions, which were recognizable by electron microscopy as separated single-stranded DNA, whereas the G + C-rich regions remained double-stranded. Five of these regions were found always in a specific arrangement; i.e., three unique A + T-rich zones were separated by two unique G + C-rich zones. There was an additional unique region that was located either to the outer right or outer left of an A + T zone (Kilpatrick and Huang, 1977). This was the first indication that the HCMV DNA molecule contained a small transposable region, i.e., a region that could be in more than one location in the molecule. The alignment of these A + T- and G + C-rich regions indicated that the HCMV DNA molecule of approximately 155 million daltons could be arranged into four populations that differed from each other in the relative positions of these A + T- and G + C-rich regions (Kilpatrick and Huang, 1977). These data suggested that the CMV DNA molecule could be present in virions in four possible isomeric arrangements similar to the genome arrangement of HSV (Sheldrick and Bertholot,

1974; Hayward *et al.*, 1975; Delius and Clements, 1976). After reannealing of single-stranded CMV DNA (AD169) molecules and subsequent analysis by electron microscopy, LaFemina and Hayward (1980) and Sheldrick *et al.* (to be published) noted two types of foldback structures that consisted of single-stranded circles joined by a duplex. It was proposed that the HCMV DNA molecule (Towne strain) was composed of a long (L) unique sequence of about 197 kilobases (kb) and a short (S) unique sequence of about 42 kb. At each end of the long unique (U_L) sequence, there were repeat sequences of about 11 kb that were inverted relative to each other. Likewise, at each end of the short unique (U_S) sequence, there were repeat sequences of about 2–2.5 kb that were inverted relative to each other. Repeat sequences at the end of the molecule were referred to as terminal repeats (TR), and those located internally were referred to as internal repeats (IR). Figure 3 summarizes the proposed genome arrangement of the HCMV genome. Even though the genome arrangement of HCMV is similar to that of HSV, the two viruses share less than 5% sequence homology (Huang and Pagano, 1974). The striking difference between the two viral genomes is that HCMV is approximately 50% greater in size and sequence complexity. This difference is primarily due to the unique sequences in the L and S section of the viral genome. (LaFemina and Hayward, 1980).

The presence of four populations of DNA molecules is the result of leftward and rightward orientations of the L and S sections of the viral genome. The four populations of CMV DNA molecules would consist of a prototype (P), an inversion of the S component (I_S), an inversion of the L component (I_L), and inversions of both the S and L components (I_{SL}) similar to that proposed for HSV (Hayward *et al.*, 1975; Wilkie, 1976; Share and Summers, 1977).

Restriction endonucleases that cleave outside the repeat sequences would generate four fragments in 0.5 M concentrations relative to the molarity of intact viral DNA and four fragments in 0.25 M concentrations relative to the molarity of intact DNA. All other fragments situated between the terminal and junction fragments would be present in concentrations of 1.0 M relative to the concentrations of intact DNA. Weststrate

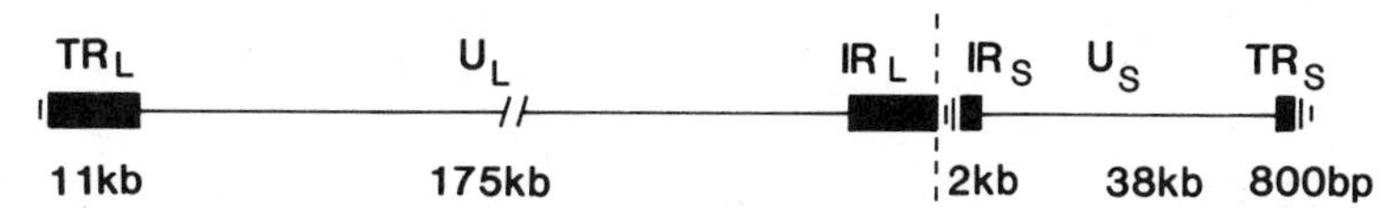

FIGURE 3. Structural arrangement of the CMV (Towne) genome. The genome consists of unique (U) sequences in a long (L) and short (S) section that are bounded by terminal repeats (TR) and inverted internal repeats (IR). The size in kilobases (kb) of each section of the genome is designated. Approximately 10% of the L-end molecules have an additional 800 base pairs (bp) at the terminus. Some 60% of the S-end molecules have 800 bp added on, and an additional 10% have 2 × 800 bp added to the terminus. This heterogeneity is also observed in the population of joint molecules formed by the inversion of L and S. Data from LaFemina and Hayward (to be published).

et al. (1980), LaFemina and Hayward (1980), and DeMarchi *et al.* (1980), using the AD169, Towne, and Davis strains of HCMV, respectively, were able to demonstrate the presence of 0.5 and 0.25 M fragments, which confirmed the presence of four equimolar populations differing in the relative orientations of the L and S components.

The terminal 0.5 M fragments were proven to be located at the ends of the DNA molecule because brief treatment with lambda exonuclease preferentially degraded these submolar fragments. When the molecular weights of the restriction-enzyme fragments as determined by electron microscopy and mobility in agarose gels was totaled, the molecular weight of the viral genome was calculated to be approximately 155 million daltons. This figure was in agreement with molecular weights determined by electron microscopy and velocity centrifugation and therefore supported the validity of the restriction-enzyme analysis.

The HSV genome can be circularized after brief digestion with a processive exonuclease. The cohesive ends of the molecule were designated the "a" sequence(s). Since the long and short terminal repeats are present internally, the same "a" sequence(s) would be present internally. The HCMV DNA presumably has the same kind of "a" sequence(s), since after brief processive exonuclease treatment followed by denaturation and reannealing, Geelen and Weststrate (1982) found unit length circles.

The similarities in the genome arrangements between CMV and HSV suggest that these two human viruses may have evolved from a common stem. There was presumably a strong selective pressure to retain the inverted repeats and the "a" sequence(s) in both the large and the small repeats. The repeat sequences could play an important role in the orientation of the sequential viral gene expression and in the mechanism of viral DNA replication.

C. Comparisons among Different Strains of Human Cytomegaloviruses

Commonly shared gene sequences or molecular arrangements among various strains of DNA viruses can be resolved by nucleic-acid hybridization and DNA restriction-endonuclease analysis.

The commonly used laboratory strains of HCMV (AD169, Towne, C87) have approximately 80% genome homology when analyzed by nucleic-acid reassociation kinetics (Huang *et al.*, 1976). Pritchett (1980) demonstrated by reciprocal DNA–DNA reassociation kinetics that CMV strains Towne and AD169 share approximately 90% of their nucleotide sequences. In addition, restriction-endonuclease analysis (Kilpatrick *et al.*, 1976) and cross-blot hybridization of restriction-endonuclease digests (Geelen and Weststrate, 1982) detected extensive comigration of viral DNA fragments with a few unique regions. These observations suggest an extensive conservation of DNA sequence order among the various strains. At present, there is no clear-cut subgrouping of isolates based on characteristic restriction-endonuclease cleavage patterns. However, recent mapping data have detected a small amount of heterogeneity among different strains (Pritchett, 1980). LaFemina and Hayward (1980) have demonstrated that this variability is associated primarily with the large repeat sequence. The data suggests that the Davis strain of HCMV lacks approximately 8000 base pairs (bp) in both large repeats relative to the Towne strain. The significance of this region of variability remains unclear.

DNA molecules of the same strains also have a region of sequence variability. This is due to a "stepwise" heterogeneity in the repeat sequences (LaFemina and Hayward, 1980; Spector *et al.*, 1982). For example, the Towne strain of CMV has a short repeat that is in three subspecies differing in size by 800 bp and the long repeat that is in two subspecies differing in size by 800 bp (Fig. 3). It is possible that the smallest of the S ends contains these same 800 bases. In addition, the same 800 bp could be at both the L and S ends. This 800-bp sequence may be analogous to the HSV "a" sequence. This sequence variability may result from insertions and/or deletions of sequences within the repeated regions, as has been reported for HSV by Wagner and Summers (1978). The significance of sequence variability in this region of the genome is not clear.

Last, human, simian, and murine CMVs do not share any significant sequence homology (Huang and Pagano, 1974).

D. Defective Cytomegalovirus DNA

Early reports on HCMV DNA, analyzed by sedimentation in sucrose gradients (Huang *et al.*, 1973) or by electron microscopy (Sarov and Friedman, 1976), indicated that the molecular weight of HCMV DNA was approximately 100 million daltons, like that of HSV. It was subsequently demonstrated that the DNA molecule of approximately 100 million daltons was characteristic of defective DNA and could be induced by high-multiplicity serial passage (Stinski *et al.*, 1979a; Ramirez *et al.*, 1979). Defective virions of HCMV are detectable even after one high-multiplicity infection. The development of defective virions is associated with a decrease in infectivity (PFU/ml) and in the percentage of DNA molecules having the standard molecular weight of 155 million daltons. Simultaneously, there is an increase in the particles/PFU ratio and in DNA molecules having molecular weights of approximately 60–100 million daltons for the Towne strain (Stinski *et al.*, 1979a) and 40–120 million daltons for the AD169 strain (Ramirez *et al.*, 1979). In addition, two viral particles of different density can be isolated by equilibrium centrifugation. The particles containing defective viral DNA have a density that is lighter than those particles containing the standard viral DNA. Particles containing the unit length genome have a higher percentage of DNA and, consequently, have a higher density (Ramirez *et al.*, 1979).

Restriction-enzyme analysis with *Eco*RI and *Xba*I of both nondefective and defective viral DNA detected the presence of specific DNA fragments associated with defective DNA. These "defective-specific" DNA fragments were detected in several separate isolations of defective viral DNA (Stinski *et al.*, 1979a).

Alterations in the restriction-enzyme fragment profile suggest that defective CMV DNA is generated by either deletions, substitutions, or duplications. After high-multiplicity infection, it is possible that some of the CMV DNA becomes rearranged during replication. This DNA may then be cut from defective concatemeric DNA; consequently, defective molecules would be packaged.

Defective herpesviruses, such as pseudorabies (Rubenstein and Kaplan, 1975; Ben-Porat and Kaplan, 1976), HSV (Bronson *et al.*, 1973; Frenkel *et al.*, 1975, 1976), and *Herpesvirus saimiri* (Fleckenstein *et al.*, 1975) also develop as a result of serial high-multiplicity passage in cell culture. The DNA associated with these defective viruses is generally highly deleted, containing multiple tandem duplications of a specific region of the viral genome (Ben-Porat and Kaplan, 1976; Frenkel *et al.*, 1976). In none of these systems is defectiveness associated with a change in the size of the virion DNA. Even though HCMV DNA has the same genome arrangement as HSV DNA, it appears that the generation of defective DNA molecules is different from that of HSV. This difference may be related to the size and rate of replication of the CMV DNA molecules. HSV DNA is 50% smaller and replicates within a short period of time (Roizman and Furlong, 1974). In contrast, replication of HCMV DNA is a relatively slow process (Stinski, 1978).

IV. TRANSCRIPTION OF THE VIRAL GENOME

A. Sequential Viral Genome Expression

Since human cytomegaloviruses (HCMVs) contain a complicated genome of approximately 155 million daltons and since this viral genome can remain in cells in a quiescent state (Weller, 1971; Lang, 1972; Plummer, 1973; Rapp *et al.*, 1975; Ho, 1977; Mocarski and Stinski, 1979), a partially expressed state (Mocarski and Stinski, 1979), or a productive state (Stinski, 1977), presumably a number of control features regulate its gene expression.

Chua *et al.*, (1981) have analyzed the percentage of transcription of the HCMV genome at various times after infection. At early times after infection, 4 hr, approximately 20% of the asymmetric coding capacity of the genome is transcribed, whereas at late times, 40 hr, approximately

40% is transcribed. With murine CMV (MCMV), approximately 23% is transcribed early and 37% late (Misra *et al.*, 1978). At both early and late times after infection, the relative abundance of RNA was higher in the nucleus than in the cytoplasm for both HCMV and MCMV. These observations suggest that there is a sequential expression of the HCMV and MCMV genomes in that one set of genes is expressed before another set. In addition, there may be precursor–product relationships between the nuclear and cytoplasmic viral RNA.

To study sequential gene expression of HCMV, virus-specified RNA in an infected cell was analyzed before and after *de novo* protein synthesis. Viral RNA synthesized in the absence of protein synthesis was referred to as immediate early viral RNA, and RNA synthesized after 2 hr or more of protein synthesis and in the absence of viral DNA replication was referred to as early RNA. Two approaches were employed to study HCMV transcription. In the first approach (Geelen and Weststrate, 1982; Stinski *et al.*, 1980; DeMarchi *et al.*, 1980; DeMarchi, 1981; Wathen *et al.*, 1981; Wathen and Stinski, 1982), viral DNA was digested by treatment with a restriction endonuclease, the DNA fragments were fractionated by agarose gel electrophoresis, and the DNA fragments were transferred for immobilization onto a nitrocellulose filter. This is referred to as a Southern blot (Southern, 1975). With this type of filter, radioactively labeled viral RNA could be hybridized to the immobilized single-stranded viral DNA to identify the regions of viral genome expression. In the second approach (Stinski *et al.*, 1980; Wathen *et al.*, 1981; Wathen and Stinski, 1982), polyadenylated [poly(A)] viral RNA was fractionated by electrophoresis in an agarose gel containing the denaturant methylmercury hydroxide and the RNA was transferred for immobilization onto diazobenzyloxymethyl cellulose paper. This is referred to as a Northern blot (Alwine *et al.*, 1977). With this type of filter, radioactively labeled whole viral DNA or cloned sections of the viral DNA could be hybridized to the immobilized viral RNA to identify the size classes of virus-specified RNA and determine the origin of the viral RNA on the genetic map.

Southern blot analysis of whole-cell RNA demonstrated that in the absence of protein synthesis, the immediate early transcripts of HCMV originated from restricted regions of the viral genome (Fig. 4). After infection and 2 or 4 hr of protein synthesis, whole-cell virus-specified RNA that accumulated in the presence of an inhibitor of protein synthesis, cycloheximide, or in the presence of an inhibitor of viral DNA replication, phosphonoacetic acid (PAA), was complementary to all restriction-enzyme DNA fragments fractionated by agarose gel electrophoresis (Fig. 4). Therefore, protein synthesis, either host- or virus-specified, permitted a switch from restricted to extensive transcription, whereas transcription at 24 hr after infection in the presence or absence of an inhibitor of viral DNA replication in productive human fibroblast (HF) cells appeared qualitatively similar. In addition, after 2 hr or more of protein synthesis in nonproductive guinea pig embryo fibroblast (GPEF) cells, extensive tran-

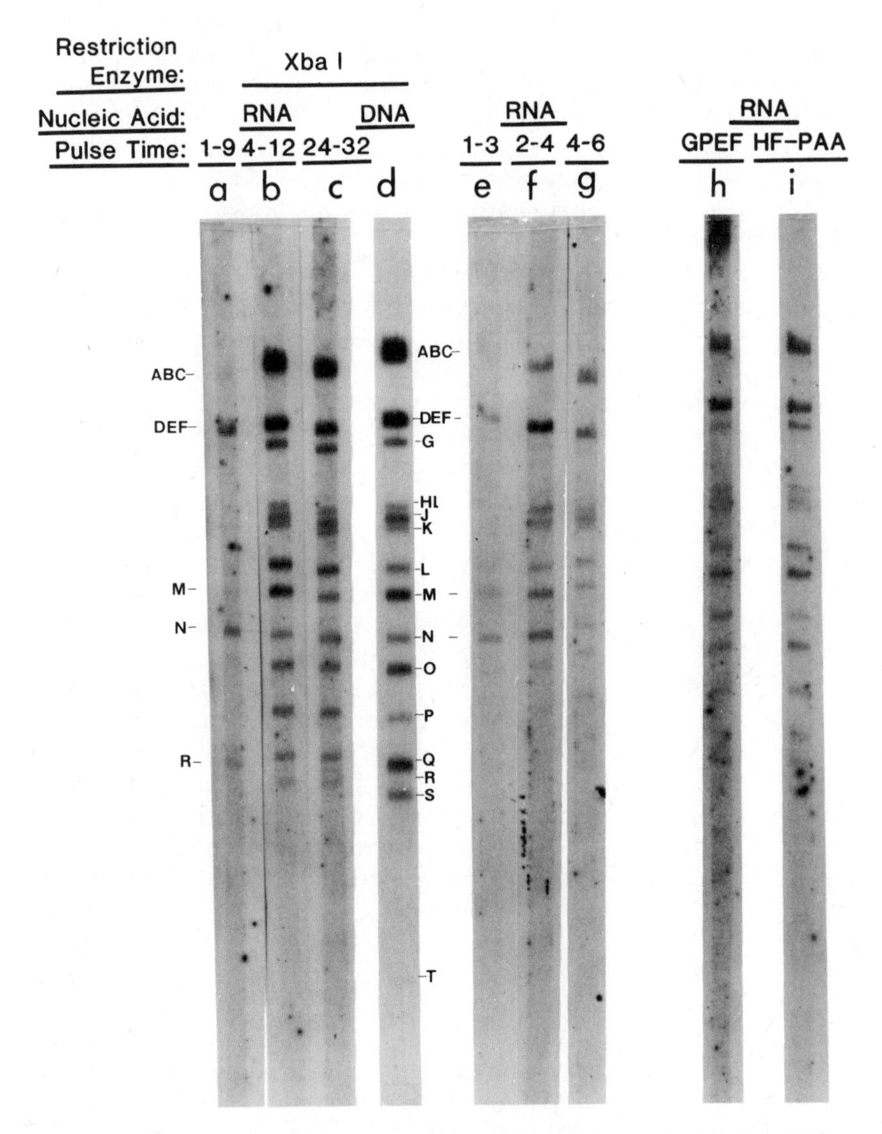

FIGURE 4. Autoradiographs of viral RNA hybridized to restriction-enzyme digest of CMV DNA. To label virus-specified immediate early RNA, the cells were treated with 100 μg/ml of cycloheximide before infection, during the viral adsorption period, and during pulse-labeling with ^{32}P from 1 to 9 (a) or 1 to 3 hr postinfection (e). For early RNA, infected cells were maintained in the absence of cycloheximide for various periods of time and then pulse-labeled for 8 hr (b, c) or 2 hr (f, g) in the presence of cycloheximide, or nonproductive GPEF cells (h) and productive HF cells (i) treated with 200 μg/ml of PAA were pulse-labeled with ^{32}P for 8-hr periods. RNA was isolated from the whole cell and hybridized to *XbaI* restriction-enzyme digest of CMV DNA immobilized onto nitrocellulose filters by the method of Southern (1975). The immobilized viral DNA was detected by hybridization of ^{32}P-labeled viral DNA (d). All major immobilized viral DNA fragments obtained with the restriction enzymes are alphabetically labeled according to size. Data from Wathen *et al.* (1981).

scription of the CMV genome occurred (Fig. 4). It is possible that one or more of the virus-specified immediate early proteins may be regulatory proteins that bind to the viral chromatin and regulate transcription. Alternatively, these virus-specified proteins could bind to host-cell chromatin and have an influence on cellular metabolism as well as viral replication.

Northern blot analysis of polysome-associated poly(A) RNA demonstrated that three size classes, 4.8–4.6, 2.2, and 1.9 kilobases (kb), of immediate early RNA were present on the polyribosomes in intermediate to high abundance (Fig. 5). Also present in extremely low abundance were a few size classes that were detectable only when specific probes representing defined regions of the viral genome were employed (Wathen and Stinski, 1982). After 2 hr of protein synthesis, there were changes in the size classes of RNA found to be associated with the polyribosomes (Wathen *et al.*, 1981). RNA size classes that were different from the immediate early RNA size classes appeared in high abundance at 24 hr after infection in the presence or absence of an inhibitor of viral DNA replication in productive cells (Fig. 5). These same classes of viral RNA were also detected in the nonproductive cell along with a large size class of 5.4 kb (Fig. 5). This large size class of RNA was not prominent in the productive cell. Once the immediate early viral proteins were synthesized, early viral messenger RNA (mRNA) size classes were present on the polyribosomes of productive or nonproductive cells. Therefore, expression of the HCMV genome is regulated, and the primary factor(s) that influences regulation is *de novo* protein synthesis. Even though extensive transcription occurs at early times after infection as determined by RNA–DNA hybridization of nuclear RNA (Chua *et al.*, 1981) and by Southern blot analysis of whole-cell RNA (Stinski *et al.*, 1980; DeMarchi *et al.*, 1980; Wathen *et al.*, 1981), a definable number of viral RNA size classes is present on the polyribosomes as determined by Northern blot analysis (Wathen *et al.*, 1981; Wathen and Stinski, 1982). It is possible that posttranscriptional control mechanisms cause only a few regions of the viral genome to be represented on the polyribosomes.

Wathen *et al.* (1981) translated the virus-specified mRNAs *in vitro* and subsequently identified the virus-specified gene products. Figure 6A illustrates a transition from immediate early gene expression to early gene expression. The majority of the immediate early virus-specified proteins had electrophoretic mobilities different from those of the early proteins. A similar transition in HCMV gene expression was observed when *in vitro* translation products were compared with infected cells pulse-labeled with [^{35}S]methionine (Fig. 6B). Although many of the virus-specified proteins have similar apparent molecular weights after *in vitro* translation or *in vivo* pulse-labeling, differences *in vivo* are expected due to posttranslational modifications of the viral proteins. Even though there appears to be a relatively transient expression of immediate early proteins, the virus-specified early proteins are expressed for a prolonged period of

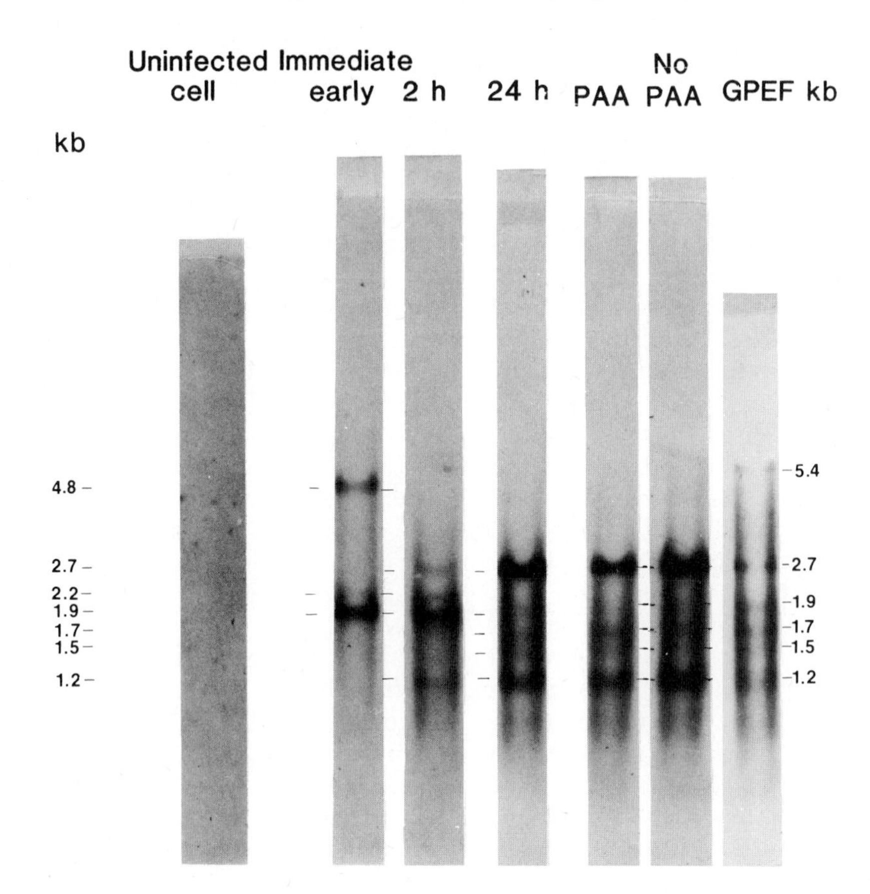

FIGURE 5. Autoradiographs of virus-specified poly(A) RNA associated with the polyribosomes at various times after infection or at 25 hr postinfection in nonproductive GPEF cells or in the presence or absence of an inhibitor of viral DNA replication in productive cells. Quantities of 1 μg (productive cells) or 2 μg (nonproductive cells) of polysome-associated poly(A) RNA for each sample were fractionated according to size in 1.0% agarose gels containing 10 mM methylmercury hydroxide and transferred to diazobenzyloxymethyl cellulose paper. Viral RNA was detected by hybridization of ^{32}P-labeled CMV DNA followed by autoradiography. Virus-specified RNA is designated according to kilobase (kb) size determined by using 3.3 and 1.7 kb *Escherichia coli* ribosomal RNA and 5.3 and 2.0 kb human cell ribosomal RNA. Data summarized from Wathen *et al.* (1981).

time (Wathen *et al.*, 1981). The biochemical event of prolonged early viral gene expression may be related to the relatively slow rate of HCMV DNA replication.

The predominant factor(s) that influences transcription of the CMV genome is apparently synthesized within the first 2 hr after infection. It

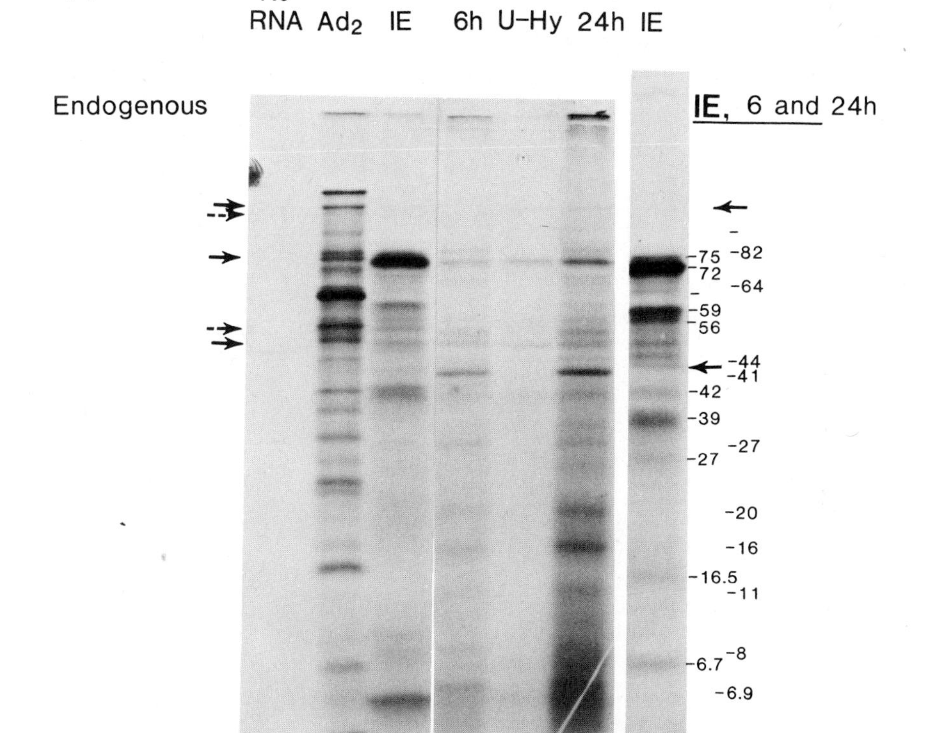

FIGURE 6. Autoradiography of the *in vitro* translation products of virus-specified immediate early and early mRNA. (A) Virus-specified mRNA was isolated by hybridization to CMV DNA bound to cellulose and the poly(A)-containing RNA was selected by oligodeoxythymidylic acid cellulose chromatography. After *in vitro* translation in a rabbit reticulocyte lysate system, the polypeptides were fractionated by electrophoresis in a denaturing discontinuous sodium dodecyl sulfate (SDS)-containing polyacrylamide gel. *In vitro* translation was with no added RNA (No RNA), adenovirus type 2 cytoplasmic RNA (Ad_2), CMV-specified immediate early RNA (IE), CMV-specified early RNA associated with polyribosomes after 6 or 24 hr of protein synthesis (6h, 24h), and uninfected-cell RNA that was incubated under hybridization conditions with CMV DNA bound to cellulose and subjected to oligodeoxythymidylic acid cellulose chromatography for a control (U-Hy). (→) Major endogenous proteins; (→) minor endogenous proteins. The immediate early mRNA translation profile at far right represents a separate analysis of immediate early mRNA by *in vitro* translation and SDS–polyacrylamide gel electrophoresis (PAGE). (B) Comparison between viral mRNA *in vitro* translation with CMV-infected cells pulse-labeled with [^{35}S]methionine for 2-hr periods at various times after infection. In addition to infected-cell-specific polypeptides, cells pulse-labeled at 6 hr after infection have some host-cell-specific polypeptides. Apparent molecular weights ($\times 10^3$) of virus-specified polypeptides coded for by immediate early (IE) or early viral mRNA after 6 or 24 hr of protein synthesis in the infected cell are designated. Data summarized from Wathen *et al.* (1981).

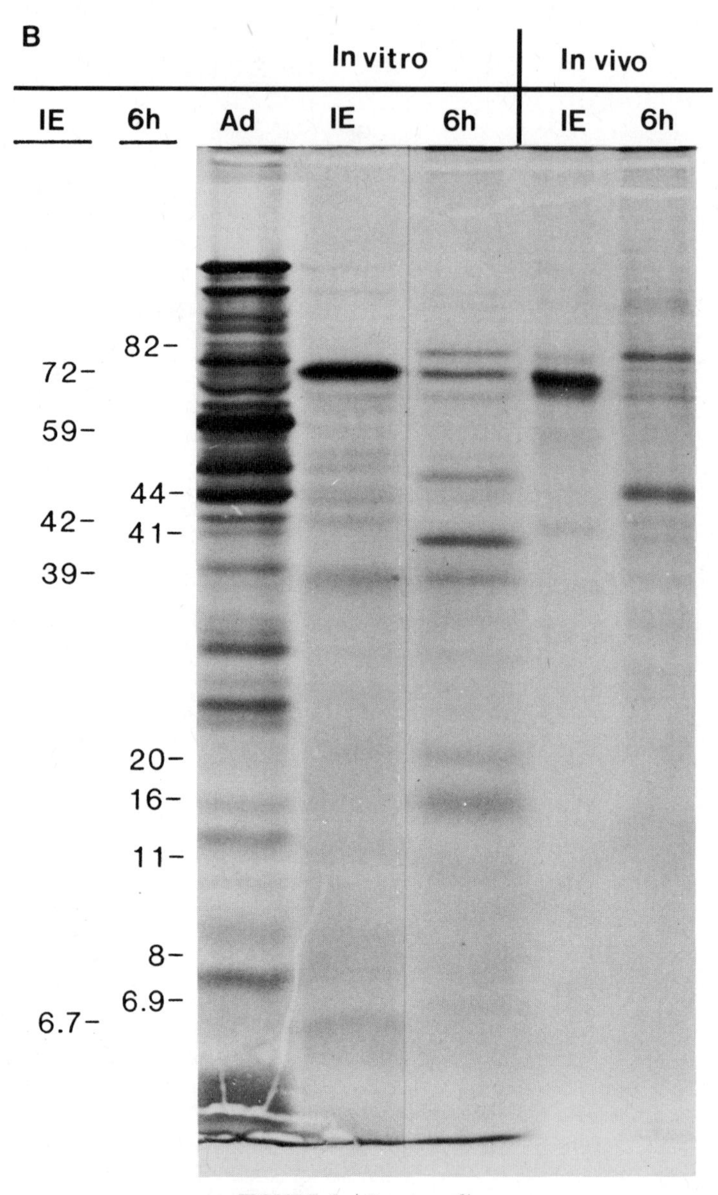

FIGURE 6. (*Continued*)

is proposed that the immediate early proteins of CMV are virus-specified proteins that control the temporal transcription of the HCMV genome.

B. Mapping the Immediate Early Viral Genes

HCMV, like herpes simplex virus (HSV) (Jones *et al.*, 1977; Clements *et al.*, 1977, 1979), has restricted transcription in the presence of an inhibitor of protein synthesis. However, recent data, obtained by Southern blot analysis, indicates that CMV, unlike HSV, has a prominent region of immediate early transcription in the large unique region of the viral genome (Stinski *et al.*, 1980; DeMarchi, 1981; Wathen *et al.*, 1981; Wathen and Stinski, 1982).

Northern blot analysis demonstrated that the abundant immediate early RNA size classes described above for the Towne strain originate from the large unique region (Wathen and Stinski, 1982). These results are summarized for the Towne strain of human CMV in Fig. 7A. The thickness of the bar depicting the regions of the immediate early transcription illustrates diagrammatically the relative amount of transcription. There was a direct correlation between the regions of abundant viral gene transcription and those regions of the viral genome represented in relatively high concentrations on the polyribosomes as poly(A) RNA. For example, the region between 0.660 and 0.770 map units hybridized with approximately 88% of the immediate early whole-cell RNA, and the abundant polyribosome-associated poly(A) RNAs originated from this region of the viral genome (Fig. 7A). Immediate early RNAs originating from the long repeat sequences represented only 6.1% of the whole-cell RNA and were associated with the polyribosomes in relatively low abundance.

C. Mapping the Early and Late Viral Genes

At early times after infection, there was not a direct correlation between the regions of abundant viral gene transcription and those regions of the viral genome represented in relatively high concentrations on the polyribosomes. Viral RNA from extensive regions of the long unique section of the viral genome was preferentially retained in the nucleus (Wathen and Stinski, 1982). In contrast, the long repeat sequences hybridized with 20.1% of the whole-cell RNA, and Fig. 7B demonstrates that five size classes of RNA were abundantly associated with the polyribosomes as poly(A) RNA (Wathen and Stinski, 1982).

At late times after infection, intermediate to abundant RNA size classes originating from various regions of the long unique section of the viral genome were detected on the polyribosomes; however, most of the abundant polyribosome-associated RNA size classes at late times originated from the long repeat sequences. These RNAs had sizes similar to the RNA size classes originating in the long repeat sequences at early times after infection. The high abundance of these RNA size classes late in infection could be due to a continuous buildup from early times. Transcription at late times after infection was detected in all *Xba*I fragments, and RNA size classes associated with the polyribosomes originated from all *Xba*I fragments (Fig. 7C). As expected at late times, the number of different RNA size classes detected in the various regions of the genome was considerable, with sizes ranging from 8.0 to 0.6 kb (Wathen and Stinski, 1982).

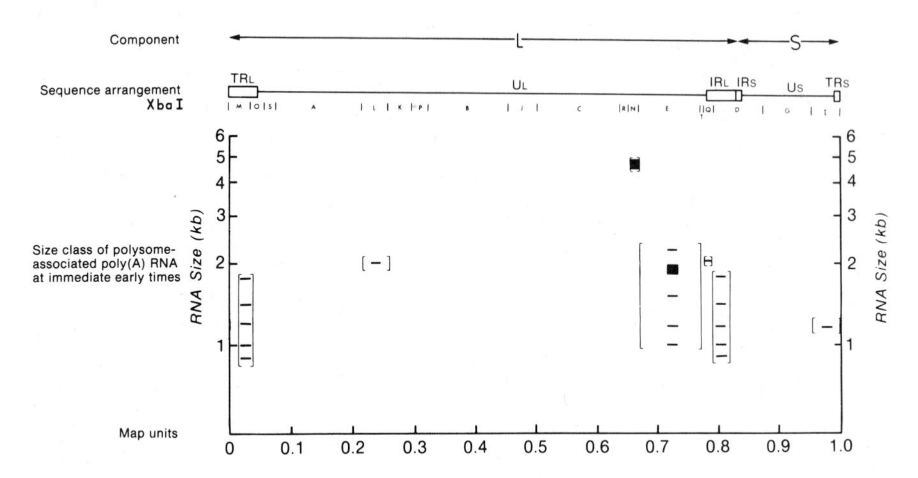
Component
L
S
Sequence arrangement
Xba I
TRL
UL
IRL
IRS
US
TRS
Size class of polysome-associated poly(A) RNA at immediate early times
RNA Size (kb)
Map units
0
0.1
0.2
0.3
0.4
0.5
0.6
0.7
0.8
0.9
1.0

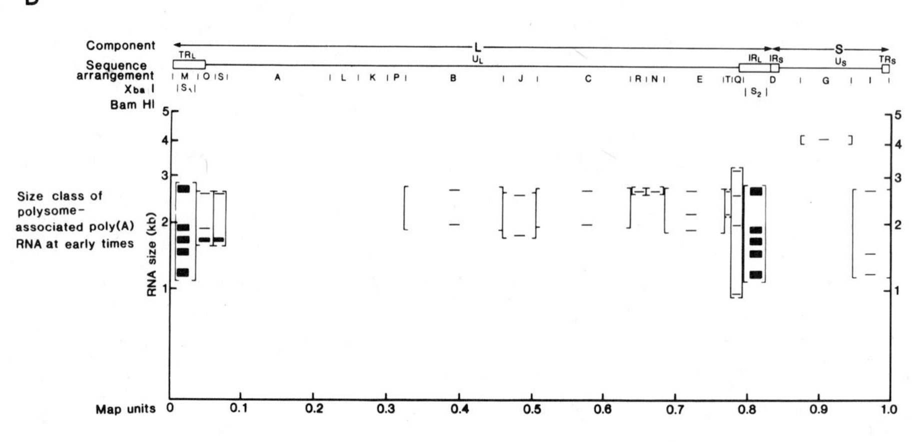
Component
Sequence arrangement
Xba I
Bam HI
L
S
TRL
UL
IRL
IRS
US
TRS
M
O
S
A
L
K
P
B
J
C
R
N
E
T
Q
D
G
I
S1
S2
Size class of polysome-associated poly(A) RNA at early times
RNA size (kb)
Map units
0
0.1
0.2
0.3
0.4
0.5
0.6
0.7
0.8
0.9
1.0

C

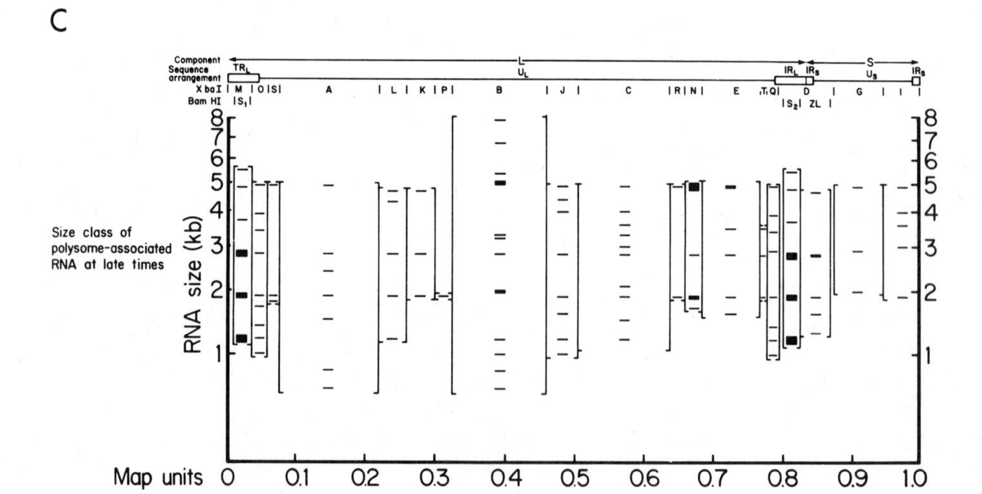
Component
Sequence arrangement
Xba I
Bam HI
L
S
TRL
UL
IRL
IRS
US
M
O
S
A
L
K
P
B
J
C
R
N
E
T
Q
D
G
I
S1
S2
ZL
Size class of polysome-associated RNA at late times
RNA size (kb)
Map units
0
0.1
0.2
0.3
0.4
0.5
0.6
0.7
0.8
0.9
1.0

D. Comparisons between Cytomegalovirus and Herpes Simplex Virus Transcription

CMV RNA from the region of 0.685–0.770 map unit (*Xba*I E), which as transcribed abundantly at immediate early and early times, is present on the polyribosomes in relatively low concentrations at early times after infection. Therefore, preferential transcription from this region of the viral genome occurs at immediate early and early times, but expression of the viral genome as mRNA associated with polyribosomes is higher at immediate early times. Even though the region of 0.325–0.460 map unit (*Xba*I B) is abundantly transcribed at early times, little or no viral RNA is associated with the polyribosomes as poly(A) RNA or as non-polyadenylated RNA (Wathen and Stinski, 1982). The region of 0.075–0.325 map unit which represents an extensive section of the long unique region of the viral genome, is transcribed at early times as evidenced by hybridization of whole-cell RNA, but there was little or no RNA found associated with the polyribosomes as poly(A) RNA.

Transcription from the long repeat sequences of CMV is very limited immediate early, but replete at early times. Some of the same size classes of RNA found in low abundance from these sequences at immediate early times are in high abundance at early times. It is possible that a promoter(s) in or near the long repeat sequences initiates transcription inefficiently under the conditions used to inhibit protein synthesis. Following *de novo* protein synthesis, this promoter(s) may operate efficiently. In the presence of cycloheximide, protein synthesis is inhibited approximately 97%. The residual protein synthesis might be enough to allow a limited amount of transcription from an early promoter(s) that cannot function in the absence of *de novo* protein synthesis, or the promoter(s) may be minimally active in the absence of protein synthesis and require *de novo* protein synthesis for increased activity. Alternatively, different promoters in the repeat sequences could function at immediate early and early times. Last, it is possible that at immediate early times, viral RNA from the long repeat sequences has a rapid rate of turnover, whereas viral RNA from the region of 0.660–0.770 map unit (*Xba*I N and E) is stable.

FIGURE 7. Origin, relative abundance, and size class of CMV-specified RNA at various times after infection. The locations of the *Xba*I and certain *Bam*HI restriction-enzyme cleavage sites are shown for the prototype arrangement of the viral genome. The brackets depict the limits in map units of the probes used to detect the viral RNAs. The thickness of the bar represents the relative abundance of the various RNA size classes determined by the relative amount of hybridization and the incubation of the Northern blot autoradiograms. The size class (kb) of the polysome-associated viral RNA was determined by electrophoresis in denaturing methylmercury hydroxide gels using 5.3, 3.3, 2.0, and 1.7 Kb RNA as standard size markers. The RNA was transferred to diazobenzyloxymethyl cellulose paper by the Northern blot method of Alwine *et al.* (1977). The size classes of the viral RNAs are indicated in kilobases (kb). The viral RNAs were isolated at immediate early (A), early (B), and late (C) times after infection as described in the text. The physical map of the *Xba*I fragments was determined by LaFemina and Hayward (to be published). Data summarized from Stinski *et al.* (1980) and Wathen and Stinski (1982).

CMV differs from HSV in several aspects of its transcription program. Unlike HSV, CMV appears to require certain gene products from the long unique region of the genome before it begins abundant transcription from the long repeat sequences. Abundant transcription from the long repeat sequences then continues for an extended period of time. Also, transcription of CMV early mRNA destined for the polyribosomes as poly(A) RNA is clustered in the long repeat sequences and adjoining regions, whereas most regions in the long and short unique regions are either not expressed or expressed in relatively low concentrations on the polyribosomes. In contrast to CMV, HSV early polyribosome-associated poly(A) RNA maps throughout its genome. This suggests that important early genes of CMV are located in the long repeat sequences. These genes continue to be abundantly expressed on the polyribosomes even at late times after infection. Since early viral gene products are presumably involved in viral DNA replication, the continuous expression of the gene products from the long repeat sequences could be related to the extended phase of viral DNA replication, which does not reach its peak until 72–96 hr postinfection.

V. VIRUS-INDUCED PROTEIN AND GLYCOPROTEIN SYNTHESIS IN INFECTED CELLS

A. Protein Synthesis

Infection of cells by herpes simplex virus (HSV) (Honess and Roizman, 1973, 1974; Powell and Courtney, 1975; Strnad and Aurelian, 1976) causes inhibition of host-cell protein synthesis and, subsequently, a specialization in viral protein synthesis. In contrast, infection of cells by human cytomegalovirus (HCMV), even at high multiplicities of infection, causes an increase in host-cell DNA, RNA, and protein synthesis above the level of that in the mock-infected cell (St. Jeor *et al.*, 1974; Furukawa *et al.*, 1975b; Tanaka *et al.*, 1975; DeMarchi and Kaplan, 1976; Stinski, 1977). It is assumed that the early viral proteins of CMV have an effect on the host cell that is different from that of HSV. In cell cultures infected with CMV, there is an initial decrease in the rate of protein synthesis, followed by an increase at two different periods after infection (Stinski, 1977). Selective suppression of host-cell protein synthesis by hypertonic conditions indicated that virus-induced proteins are also synthesized in a cyclic manner (Gupta and Rapp, 1978). Virus inactivated by ultraviolet irradiation or neutralized with specific antiserum causes the initial decline in the rate of protein synthesis, but the phasic sequence in the rate of protein synthesis is not detectable. The failure of inactivated virus to induce an increase in the rate of protein synthesis suggests that a virus-coded protein expressed early after infection is necessary and that the expression of this protein was inactivated.

To differentiate between viral and host-cell protein synthesis at various times after infection, Stinski (1977) employed a double-isotopic-label difference analysis according to the methods of Hightower and Bratt (1974). Figure 8 illustrates that the relative amount of virus-specific protein synthesized in infected cells was very low during early times after infection with 10–20 plaque-forming units (PFU)/cell. Even at 10–12 hr postinfection, virus-specific protein represented only about 10% of the total protein synthesis. Hence, protein synthesis during the early phase of HCMV infection represents primarily host-cell protein synthesis. A linear increase in the rate of virus-specific protein synthesis occurs after the initiation of viral DNA replication, which is estimated to occur at approximately 12 hr postinfection when cells are infected with a high multiplicity of the Towne strain of HCMV (Stinski, 1978). At 48 hr postinfection, virus-specific protein synthesis represents approximately 60% of the total and remains at this level for as long as 5 days. Conversely, host-cell protein synthesis remains at approximately 40% during the late period of infection even though 10^7 PFU/ml was produced by the infected culture (Fig. 8).

B. Criteria for the Identification of Virus-Induced Proteins and Glycoproteins

Several criteria have been employed to identify viral gene products in the HCMV-infected cell. Unique proteins induced by viral infection,

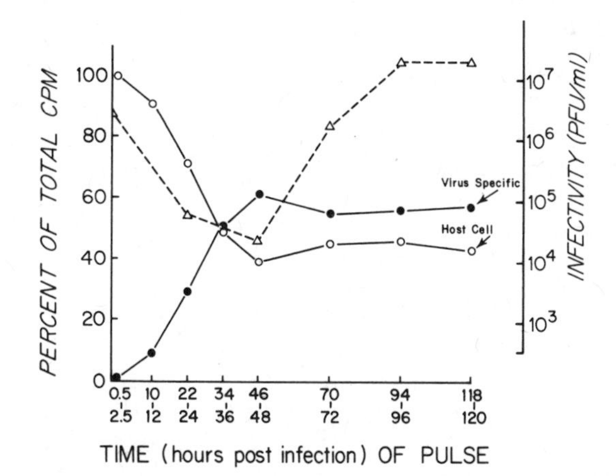

FIGURE 8. Relationship between protein synthesis in the CMV-infected cell and infectious virus production. The percentage of total radioactivity in virus-specified (●) or host-cell-specified (○) proteins was determined by double-isotopic-label difference analysis according to the methods of Hightower and Bratt (1974). Cells were infected with a multiplicity of 10 PFU/cell, and at various times after infection, the amount of infectious virus (△) associated with the cell and in the extracellular fluid was determined. Data summarized from Stinski (1977).

referred to as infected-cell specific proteins (ICSPs) or glycoproteins (ICSGPs), were identified by their relative rates of synthesis, by their different electrophoretic mobilities compared with those of host-cell proteins and glycoproteins, and by their similar electrophoretic mobilities compared to those proteins and glycoproteins associated with cytomegalovirions and dense bodies, an aberrant particle of the virus. Virus-induced proteins are not detected in cells inoculated with virus inactivated by ultraviolet irradiation.

Many of the virus-induced proteins were identified as virus-specified proteins by immunoprecipitation using HCMV-specific antibodies and by *in vitro* translation of virus-specified messenger RNA (mRNA).

The virus-induced proteins were classified as either immediate early, early, or late. Immediate early proteins were defined as those proteins synthesized in relatively high concentration in the presence of actinomycin D and immediately after the removal of cycloheximide when the inhibitor of protein synthesis was added prior to infection. Early proteins were present in the infected cell after 4 hr or more of protein synthesis. Late proteins were present after the replication of a significant amount of viral DNA. The sequence of ICSP synthesis in HCMV-infected cells was analyzed by doing 2-hr pulses with [^{35}S]methionine and separating the proteins by electrophoresis in denaturing SDS-containing polyacrylamide gels (Fig. 9).

C. Immediate Early Proteins

To identify the immediate early proteins, it was necessary to enhance CMV-induced protein synthesis relative to ongoing host-cell protein synthesis. This was accomplished by treatment with cycloheximide to allow for an accumulation of viral mRNA followed by radioisotope pulse-labeling in hypertonic medium for preferential translation of viral mRNAs (Saboris *et al.*, 1974). The medium also contained actinomycin D to inhibit further transcription of the viral genome. *In vitro* translation of virus-specific mRNA detected nine proteins with apparent molecular weights ($\times 10^3$) of 75, 72, 59, 56, 42, 39, 27, 16.5, and 6.7 (see Fig. 6A). When cells were treated with cycloheximide 1 hr before infection, during the 1-hr viral adsorption period, and from 1 to 9 hr postinfection and then pulsed for 2 hr with [^{35}S]methionine, at least three ICSPs, 75, 72, and 68, were synthesized at relatively high concentrations and several other ICSPs were present in relatively low concentrations (Fig. 9) (Stinski, 1978). At 2 hr postinfection, three additional ICSPs were detected at relatively high concentrations and were designated as ICSPs 59, 56, and 53. Four proteins at relatively lower concentrations were also detectable and were designated as ICSPs 39, 27, and 19 (Fig. 9) (Stinski, 1978). In addition, one protein, ICSP 135, which was previously not identified (Stinski, 1978), was detected when an inhibitor of proteolytic enzyme activity was used.

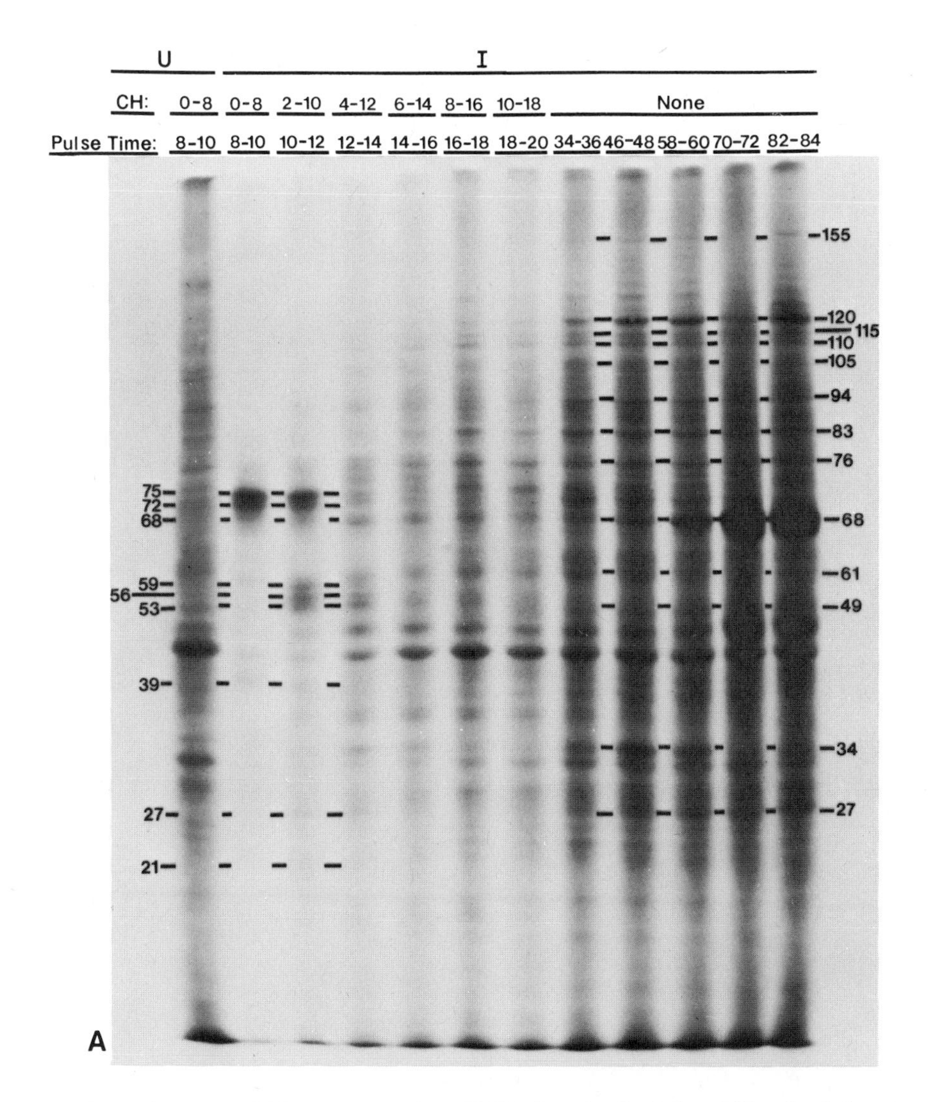

FIGURE 9. Sequence of ICSP synthesis in HCMV-infected cells. Infected (I) and uninfected (U) cells were treated with cycloheximide (CH) for 8-hr intervals or left untreated and then pulse-labeled for 2 hr with [^{35}S]methionine as indicated in the text. (A) Equal volumes of cell lysates were analyzed by slab SDS-PAGE followed by autoradiography. (B) Levels of representative proteins synthesized at each point were determined by scintillation counting of ^{35}S radioactivity contained in individual protein bands of the SDS–polyacrylamide gels. Radioactivity present in the same regions of the corresponding uninfected cell sample was subtracted for each value, and the net amount was plotted. Viral DNA synthesis was measured by pulse-labeling with [^{3}H]thymidine for 2-hr periods and then hybridizing DNA extracted from the whole cell to viral DNA immobilized onto nitrocellulose filters. ICSPs are denoted by their apparent molecular weights ($\times 10^3$). Host-cell proteins are either labeled with the letter "h" or not labeled at all. Data from Stinski (1978).

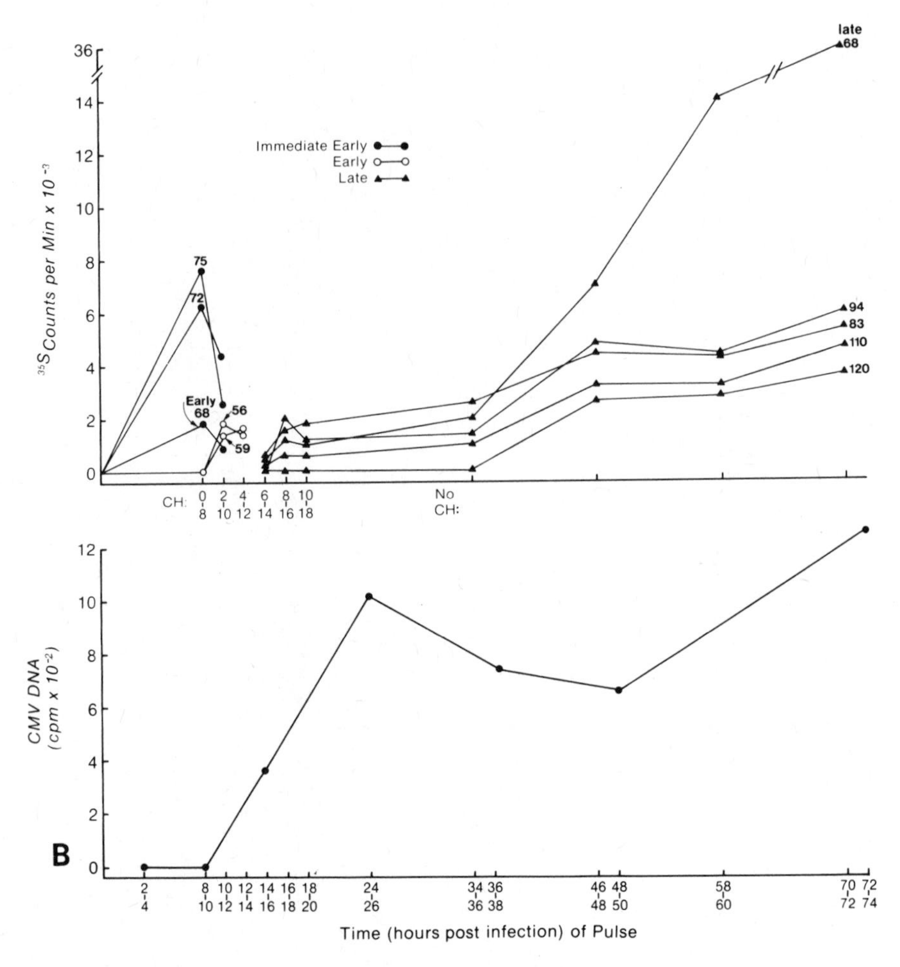

FIGURE 9. (*Continued*)

Additional proteins were detected in the infected cell that were not detected by *in vitro* translation. These proteins could be products of unstable mRNAs or mRNAs present in relatively low concentrations. Alternatively, these proteins could represent posttranslational modifications that are not detectable by the *in vitro* translation assay. In general, the *in vivo* protein studies are in good agreement with the *in vitro* translation studies, indicating that many of the ICSPs are virus-specified.

Michelson *et al.* (1979) and Tanaka *et al.* (1979) have identified immediate early proteins as virus-specific, and Blanton and Tevethia (1981) have identified immediate early and early proteins as virus-specific by immunoprecipitation with human convalescent serum specific for CMV antigens. Analysis of these immunoprecipitates by SDS-PAGE identified proteins with apparent molecular weights that approximate those detected by *in vitro* translation or by pulse-labeling with [^{35}S]methionine.

There is a possibility that many of the immediate early proteins specified by CMV are regulatory proteins that interact with viral or host-cell DNA or chromatin. HCMV infection induces at least two unique species of proteins at very early stages of infection (Tanaka *et al.*, 1979). One is responsible for the stimulation of chromatin template activity as measured by the incorporation of [^{3}H]-UMP in the presence of *E. coli* RNA polymerase, and the other is responsible for configuration changes of chromatin as measured by circular dichroism spectra Kamata *et al.*, 1978).

The immediate early proteins of CMV play a very important role in productive infection, since sequential expression of the viral genome requires their synthesis. Further investigation is necessary to determine the role of these virus-specified proteins.

D. Early Proteins

In vitro translation of virus-specified mRNA associated with polyribosomes after 6 or 24 hr of protein synthesis detected approximately ten proteins (see Fig. 6). The majority of these early virus-specified proteins had electrophoretic mobilities different from those of the immediate early proteins and the major late structural proteins of CMV described below. Many of these virus-specified proteins were synthesized in infected cells after 4 hr or more of protein synthesis at relatively low concentrations; consequently, they were not identified as ICSPs in previous studies (Stinski, 1978). The virus-specified early proteins are expressed in the cell for a prolonged period of time. The phenomenon of prolonged early viral gene expression may be related to the relatively slow rate of viral DNA replication. Like the immediate early proteins, little is known about the function of the virus-specified early proteins. Presumably, one or more of the early proteins constitutes the CMV-induced DNA polymerase (Huang, 1975a).

E. Late Proteins

By pulse-labeling with [^{3}H]thymidine for 2-hr periods after a high-multiplicity infection, extracting DNA from the cells, and hybridizing the ^{3}H-labeled DNA to CMV DNA immobilized on nitrocellulose filters, the initiation of HCMV DNA (Towne strain) replication was estimated to occur at 12 hr postinfection (Stinski, 1978). Because viral DNA replication proceeds at a slow rate, late mRNAs would be present in relatively low concentrations; consequently, late ICSPs would not be de-

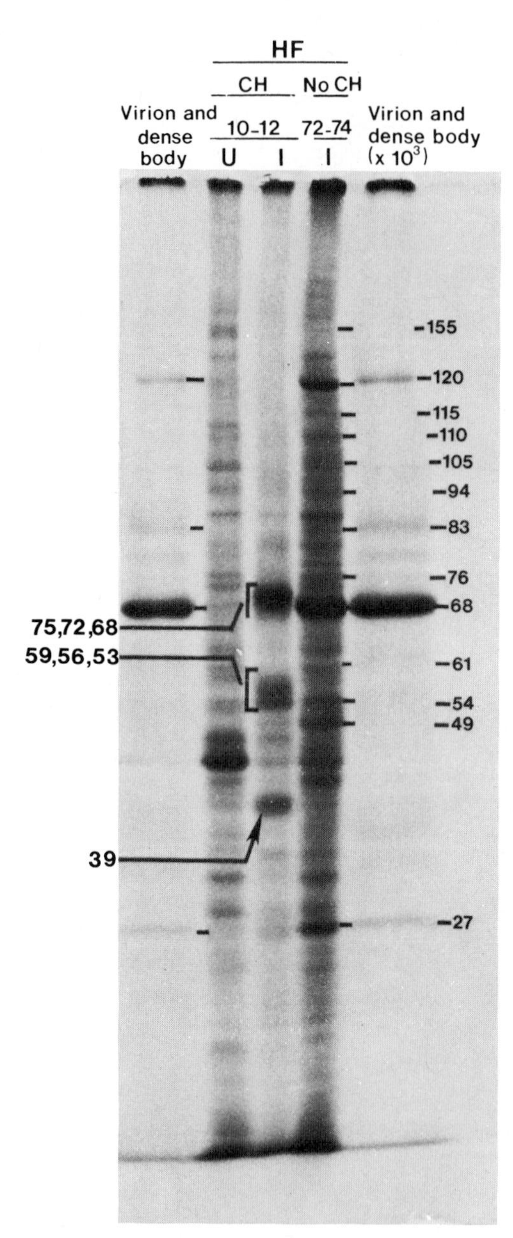

FIGURE 10. Autoradiogram of electrophoretically separated polypeptides from CMV-infected cells at early and late times after infection. Cells were either treated with cycloheximide at 2 hr post infection or left untreated. The cycloheximide was removed and the cells were pulse-labeled for 2 hr with [^{35}S]methionine in hypertonic medium as indicated in the text. Polypeptides associated with purified virions plus dense bodies were subjected to electrophoresis in parallel. Structural viral polypeptides and ICSPs are indicated according to their apparent molecular weights ($\times 10^3$). Data from Stinski (1978).

tectable until after a significant amount of viral DNA replication. Many of the virus-induced late ICSPs were just within the limits of detectability by pulse-labeling for 2 hr with [^{35}S]methionine at approximately 36 hr postinfection (Fig. 9). The relative rate of synthesis of these ICSPs increased linearly with time (Fig. 9A and B) and, simultaneously, with the accumulation of newly synthesized DNA (Fig. 9B). Many of these late

ICSPs were considered virus-specific because they were preferentially immunoprecipitated with an antiserum prepared against purified virions (Stinski, 1977). In addition, many of the late ICSPs had the same electrophoretic mobilities as the proteins associated with purified virions (Fig. 10).

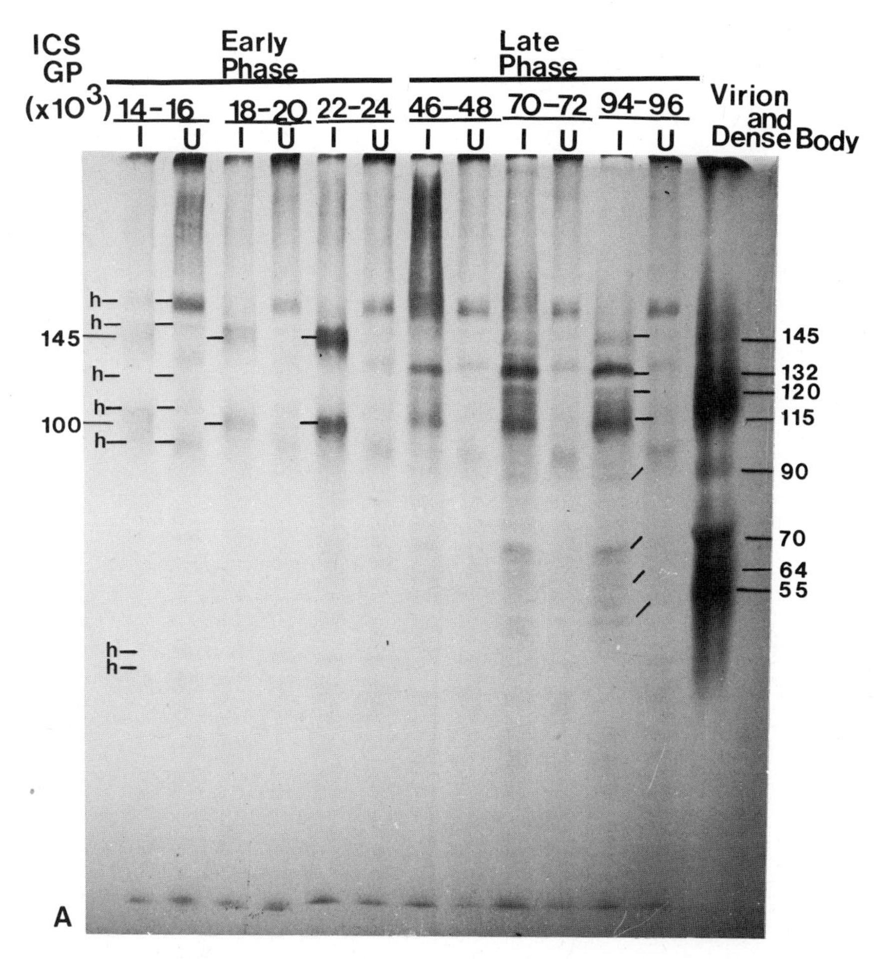

FIGURE 11. Synthesis of glycoproteins in CMV-infected cells. (A) Infected (I) and uninfected (U) cells were labeled with [^{3}H]glucosamine for 2 hr at various times after infection as indicated above. [^{3}H]Glycoproteins associated with purified virions plus dense bodies were subjected to electrophoresis in parallel with the 94–96 hr sample. ICSGPs and structural glycoproteins are designated according to their apparent molecular weights ($\times 10^3$). Host-cell glycoproteins are either labeled with the letter "h" or not labeled at all. (B) Densitometer scans of fluorograms of glycoproteins synthesized in uninfected and infected cells. Uninfected cells, early-phase-infected cells (22–24 hr postinfection), and late-phase-infected cells (94–96 hr postinfection) were labeled with [^{3}H]glucosamine and analyzed as described above. The molecular weights ($\times 10^3$) of ICSGPs are indicated. Data from Stinski (1977).

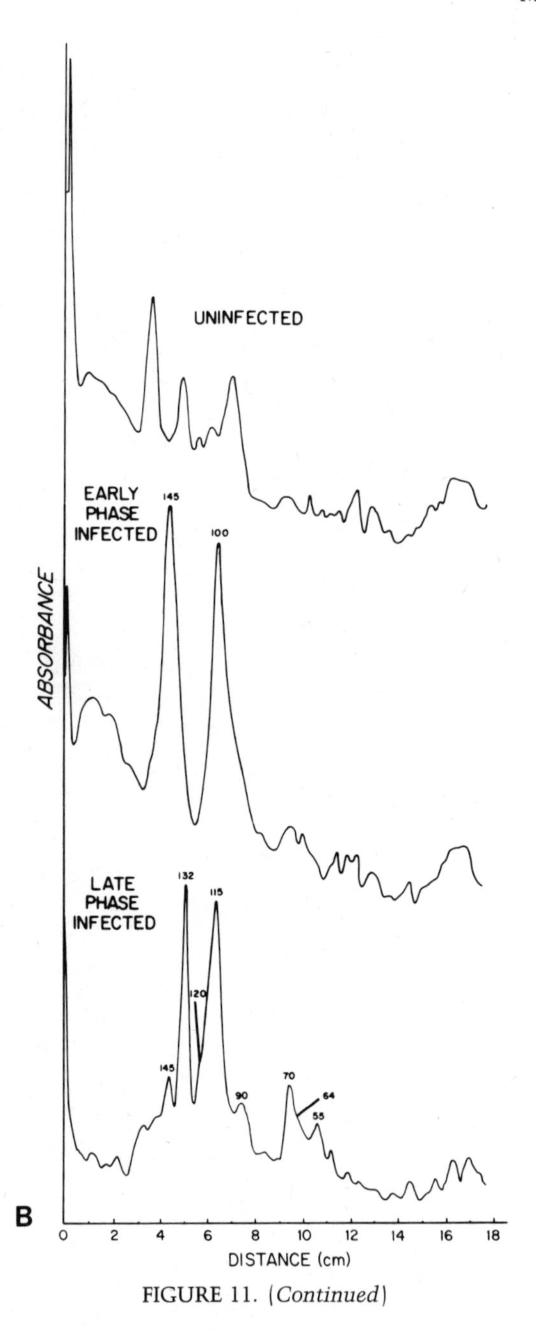

FIGURE 11. (*Continued*)

F. Early and Late Virus-Induced Glycoproteins

Virus-induced glycoproteins were not detectable in the infected cell until 14–16 hr postinfection (Fig. 11A). At least two glycoproteins, ICSGPs 145 and 100, were synthesized preferentially at early times (Fig.

11B) and in the absence of viral DNA replication (Stinski, 1977). At late times, 70–72 and 94–96 hr postinfection, eight to nine ICSGPs were detected (Fig. 11A and B). Many of these ICSGPs were considered to be virus-specific because they were immunoprecipitated from infected-cell extracts with an antiserum prepared against the glycoproteins associated with purified virions of CMV (Stinski, 1977). The synthesis of ICSGPs at early times was easier to detect because the virus suppresses host-cell glycoprotein synthesis.

The early ICSGPs modified the immunological specificity of the plasma membrane at approximately 24 hr postinfection of productive (Stinski *et al.*, 1979b) or nonproductive (Tanaka *et al.*, 1981) cells, which is as much as 2 days before virion maturation. Although these early ICSGPs were found bound to both plasma and microsomal membranes as determined by pulse-labeling with [^{3}H]glucosamine, virus-specific antigen associated with these early ICSGPs accumulated primarily on the plasma membrane during the early stage of infection as determined by electron microscopy of specific immunoperoxidase reaction product (Stinski *et al.*, 1979b). In contrast, at late times (72 hr) after infection when virus nucleocapsids can be detected in the nucleus, virus-specified antigen associated with the early and late ICSGPs of CMV was prominent on the plasma, endoplasmic reticulum, and nuclear membranes (Stinski *et al.*, 1979b).

Immunity to virus antigens on the cell surface is probably the primary immunological mechanism of host resistance to CMV infection. Antibody against membrane antigens of CMV-infected cells can be found in the sera of patients with CMV infection (The and Largenhuysen, 1972). In addition, lymphocytes from CMV-seropositive individuals are stimulated by purified virus or infected cells, whereas lymphocytes from seronegative individuals do not respond (Moller-Larsen *et al.*, 1976). The appearance of virus antigens on the plasma membrane during the early phase of virus replication may be recognized by host immunological defense mechanisms and lead to the elimination of the virus-infected cell.

VI. VIRUS-INDUCED ENZYMES

A. Host-Cell Enzymes

Mammalian cells contain two species of thymidine kinase that are separable by disk PAGE (Kit *et al.*, 1973). There is a rapidly migrating enzyme with low specific activity associated with the mitochondria and a slowly migrating enzyme with high specific activity associated with the cytoplasm. The latter enzyme is cell-cycle-dependent and demonstrates maximal activity in fetal tissue and after cells are stimulated to divide *in vitro* (Taylor *et al.*, 1972; Adler and McAuslan, 1974). Infection by human cytomegalovirus (HCMV) stimulates the activity of both species of thymidine kinase, with maximum stimulation associated with the slow-migrating species (Zavada *et al.*, 1976; Estes and Huang, 1977). Estes and Huang (1977) separated these enzymes according to net charge or molecular size and were unable to detect a unique thymidine kinase different from the uninfected-cell thymidine kinases. The thymidine kinase activity stimulated in infected cells has the same phosphate donor specificity, pH optima, thermostability, salt sensitivity, and electrophoretic mobilities as the uninfected-cell thymidine kinases (Zavada *et al.*, 1976; Estes and Huang, 1977). Therefore, HCMV, unlike herpes simplex virus (HSV), does not code for the synthesis of a virus-specific thymidine kinase. Since the rate of CMV DNA replication is slow, compared to that of HSV, CMV presumably does not require a mechanism to ensure nucleotide triphosphate precursors for the viral DNA polymerase. In contrast, HSV codes for thymidine kinase, deoxycytosine kinase (Kit *et al.*, 1967; Jamieson *et al.*, 1974), ribonucleotide reductase (Roller and Cohen, 1976), and DNase (Morrison and Keir, 1976), all of which directly or indirectly could provide nucleotide triphosphate precursors for the viral DNA polymerase.

Infection of cells by HCMV also stimulates the synthesis of host-cell DNA polymerases as well as a new virus-specific DNA polymerase (Huang, 1975a). In addition, it stimulates the activity of ornithine decarboxylase (Isom, 1979), which is an enzyme prominent in the G_1 phase of the cell cycle and is a regulatory enzyme in the synthesis of polyamines.

B. Virus-Specified Enzymes

Virus-specified DNA polymerase can be distinguished from host-cell DNA polymerase by its different behavior in phosphocellulose chromatography, template specificity, salt sensitivity, and sedimentation property (Huang, 1975a). The virus-specified DNA polymerase can easily be distinguished from host-cell DNA polymerase by its enhanced activity in high salt such as NaCl or $(NH_4)_2SO_4$ (Huang, 1975a; Miller and Rapp, 1976; Hirai *et al.*, 1976). In addition, the virus-specified DNA polymerase is very sensitive to phosphonoacetic acid (PAA) (Huang, 1975b). As little as 50 μg PAA/ml can completely inhibit viral DNA replication, whereas this concentration has little or no effect on the host-cell DNA polymerase (Huang, 1975b). Failure to overcome the inhibition of the viral enzyme activity by adding increasing concentrations of template suggests that the PAA interacts directly with the viral enzyme and not with the template primer (Huang, 1975b). When the PAA is removed, viral DNA replication resumes.

The presence of a gene for virus-specified DNA polymerase is characteristic of many herpesviruses (Mar and Huang, 1979) including HSV. Even though HCMV stimulates host-cell DNA polymerase activity, it has evolved to ensure the synthesis of its DNA genome by coding for a viral DNA polymerase.

VII. VIRAL DNA REPLICATION

On the basis of analysis of the amount of viral DNA and cellular DNA synthesized at various times after infection and the cyclic pattern of this DNA synthesis, St. Jeor and Hutt (1977) have proposed that human cytomegalovirus (HCMV) DNA replication is dependent on cell DNA synthesis and that, consequently, viral DNA is not synthesized until the cell enters the synthetic or S phase of the cell cycle. In contrast, DeMarchi and Kaplan (1976) proposed that viral DNA synthesis proceeds independently of host-cell DNA synthesis because late viral antigen synthesis was less likely to occur in the cells in which cellular DNA was synthesized. The effect of CMV was to delay or in some cases abort productive infection; consequently, it was proposed that viral DNA synthesis is not dependent on the S phase of the cell cycle. Although cellular DNA synthesis is activated after infection, this activation is apparently not essential for viral DNA or viral antigen synthesis.

The possibility remains that early virus-specified proteins, which interact with viral DNA to regulate viral gene expression, also interact with host-cell chromosomal DNA. One effect of this interaction is the stimulation of cellular DNA synthesis. This stimulation would actually be a process that suppresses or even inhibits viral DNA synthesis. The rate of CMV DNA replication is extremely slow compared to that of herpes simplex virus (HSV) (Roizman and Furlong, 1974). Even though viral DNA replication is initiated at approximately 12 hr after high-multiplicity infection (Towne strain), maximum viral DNA replication does not occur until 72–96 hr postinfection (Huang *et al.*, 1973; Stinski, 1978). One explanation for the slow rate of CMV DNA synthesis is that this synthetic process does not compete well with host-cell DNA synthesis. Low concentrations of viral DNA relative to cellular DNA and the competitive process between the two synthetic events result in a relatively slow accumulation of viral DNA and a long eclipse period prior to the production of detectable amounts of infectious virus.

Although the CMV genome is approximately 50% larger than the HSV genome, this physical difference does not solely explain the long eclipse period of 2–3 days when the same type of cells is infected at the same multiplicity (see Fig. 2).

Roizman (1979a,b) has proposed a model for the replication of the HSV genome that may be applicable to the replication of the CMV genome. The HSV DNA can circularize by uncovering a complementary "a" sequence that is a redundancy in one or more copies in the terminal repeat as well as the internal repeat sequences. Replication could be initiated in the L or the S section of the viral genome, or both, and proceed in both directions to a rolling circle yielding a head-to-tail concatemer. The excision of unit length viral DNA would then occur at the modified junctions. It has been proposed that the missing "a" sequence would be recovered by using internal inverted repeats as template (Roizman, 1979a,b).

HCMV presumably employs a similar mechanism of replication because it has a viral genome with an L and S section containing terminal and internal inverted repeats similar to HSV. Although few studies have been done on CMV DNA replication, the following observations suggest that CMV DNA may replicate according to the model proposed by Roizman (1979a,b). Jean *et al.*, (1978) found unit length circular DNA molecules in addition to linear double-stranded molecules after infection of cells with HCMV. In addition, Geelen and Weststrate (1982) treated purified CMV DNA with exonuclease III, incubated the DNA under reannealing conditions, and found unit length circles of the viral genome. These observations suggest that CMV DNA has a repetitive sequence that is located at or near the termini, and after exonuclease digestion, the single-stranded regions can complement each other to form circles. Although further research is necessary, studies by Jean *et al.* (1978) suggest the presence of concatemers of CMV DNA with replicative "eye" loops or forks that suggest that the viral DNA may be replicating according to the rolling-circle model as proposed for HSV. Last, purified virions of CMV contain DNA molecules in four isometric arrangements as discussed in Section III.B. Inversions of the L and S components of the CMV DNA genome may occur by the mechanisms proposed for HSV DNA by Roizman (1979a,b).

VIII. VIRIONS AND DENSE BODIES OF CYTOMEGALOVIRUS

The typical inoculum of cytomegalovirus (CMV), but not that of herpes simplex virus (HSV), contains membrane-bounded bodies composed of a finely granular, homogeneous, electron-dense material (Fig. 12). These particles adsorb and fuse with the plasma membrane in the same manner as virus (Smith and DeHarven, 1974). The inherent problem in characterizing dense bodies is that it is difficult to obtain quantitative separation of virions from dense bodies (Sarov and Abady, 1975; Stinski, 1976; Fiala *et al.*, 1976). The best separation has been obtained by Talbot and Almeida (1977), who employed negative viscosity/positive density gradients. At present, a characterization of the polypeptide and glycopolypeptide composition of virion and dense bodies purified by their procedure has not been reported. Therefore, most studies on the structural analysis of the cytomegalovirions is complicated by the contamination with dense bodies. An analysis of the polypeptide composition of preparations that contain high or low ratios of virions to dense bodies suggests that dense bodies share most of the polypeptides contained in virions (Sarov and Abady, 1975; Fiala *et al.*, 1976). However, dense bodies do not contain the viral DNA genome (Sarov and Abady, 1975). It has been proposed that dense bodies represent an aberrant assembly of virion structural proteins (Sarov and Abady, 1975), while others have proposed that dense bodies represent an aberrant assembly of just the viral envelope and tegument polypeptides (Fiala *et al.*, 1976). Unfortunately, the nature of these dense bodies and their importance, if any, in the infectious cycle of CMV remain unclear. What is clear is that virions and dense bodies share antigenic determinants in common as demonstrated by immune electron microscopy (Craighead *et al.*, 1972).

Earlier analyses of purified virions and dense bodies underestimated the number of polypeptides associated with these particles (Sarov and Abady, 1975; Fiala *et al.*, 1976). Purified virions and dense bodies of human CMV (HCMV) are found to contain 33–35 polypeptides with molecular weights ranging from approximately 200,000 to 10,000 daltons (Kim *et al.*, 1976; Stinski, 1977; Gupta *et al.*, 1977). Figure 13 illustrates the polypeptide profile of purified virions and dense bodies of the Towne strain (Stinski, 1977). None of these polypeptides was found to be of host-cell origin, acquired during the morphogenesis of virions and dense bodies (Stinski, 1977). The greatest relative amounts of viral protein were associated with virus polypeptide (VP) 155 (4.4%), VP 120 (7.1%), VP 83 (8.4%), and VP 68 (15.4%). Note that there are many structural polypeptides incorporated into virions and dense bodies at relatively low concentrations, based on [^{35}S]methionine incorporation.

A comparison of the CMV polypeptide composition with that of HSV demonstrated a profound difference in protein composition of these two herpesviruses (Fiala *et al.*, 1976). On the other hand, the polypeptide composition of four different strains of HCMV (Birch, AD169, C87, and Towne) demonstrated remarkable similarities with respect to relative amounts of a polypeptide and electrophoretic mobilities (Gupta *et al.*, 1977). These findings are in agreement with the observation of Kilpatrick *et al.* (1976) that strains AD169, C87, and Towne share approximately 80% DNA sequence homology and similar profiles of restriction-enzyme digests with the viral DNAs.

The glycopolypeptides associated with purified virions and dense bodies have also been characterized by their relative mobility, percentage of glucosamine incorporation, and molecular weight. The glycoproteins associated with the membranes were on the surface of the virions and dense bodies, as determined by iodination with ^{125}I in the presence of lactoperoxidase and H_2O_2. The membranes were stabilized by purification in gradients of D-sorbitol. Figure 14 compares the electrophoretic mobilities of the [^{3}H]glycopolypeptides of virions and dense bodies with the viral surface polypeptides. Eight glycopolypeptides were repeatedly detectable, and three glycopolypeptides of higher molecular weight with

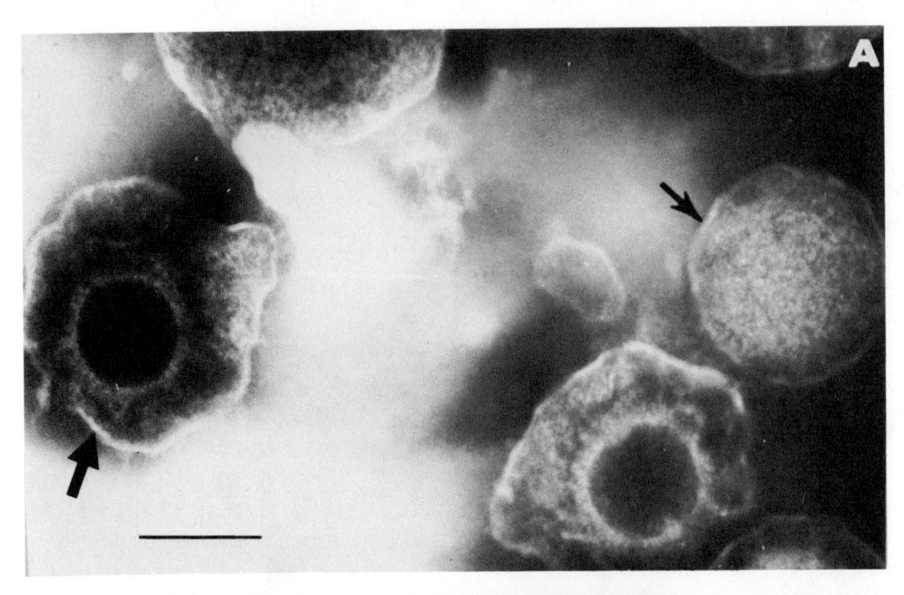

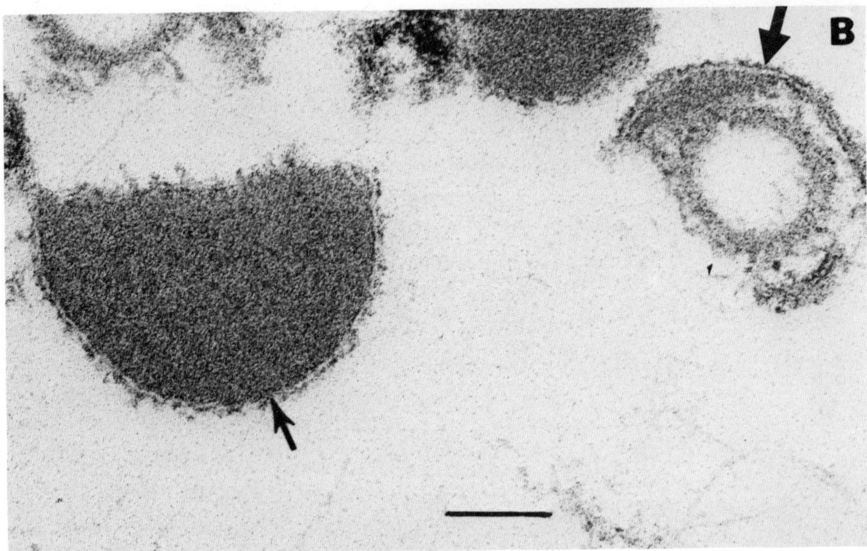

FIGURE 12. Electron microscopy of purified CMV (large arrows) and dense bodies (small arrows). (A) Stained with 2% sodium phosphotungstic acid containing 100 μg bacitracin/ml. (B) Fixed with 1% osmium tetroxide and 2% glutaraldehyde in 0.2 M sodium cacodylate–hydrochloride buffer, pH 7.0. Fixed samples were dehydrated in ethanol, embedded in Epon, and thin-sectioned. Sections were stained with uranyl acetate and lead citrate. (C) Dense body prepared as described in (B). Scale bars: 100 nm. Data from Stinski (1976).

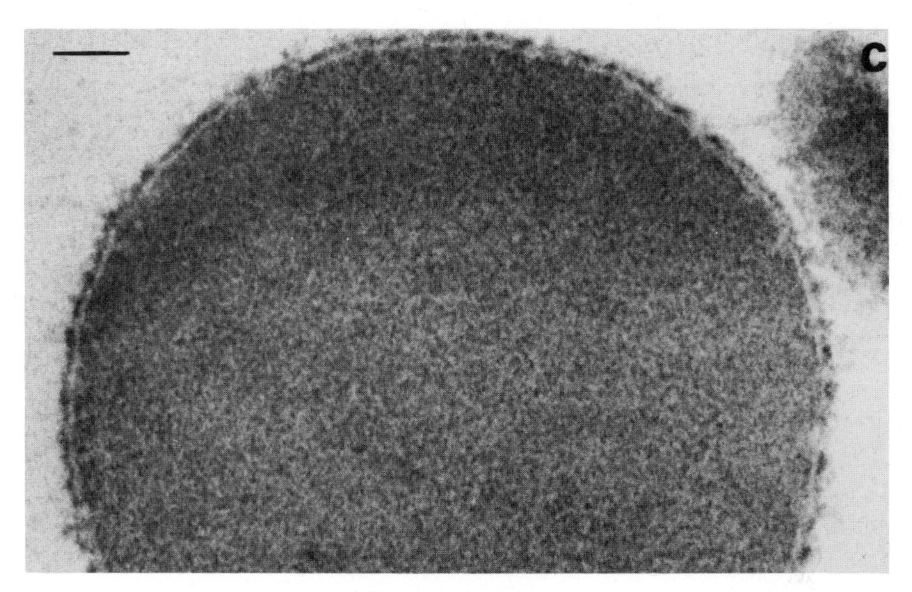

FIGURE 12. (*Continued*)

low levels of glucosamine incorportion were occasionally detectable. These latter glycopolypeptides may be precursors or aggregates of the glycopolypeptides with lower molecular weights. The glycopolypeptides associated with the membrane of purified virions and dense bodies have been also extracted by using Triton X-100 and characterized according to electrophoretic mobility (Stinski, 1976). Antisera to these glycopolypeptides contain antibodies that neutralize viral infectivity and react with antigens in cells infected with CMV (Stinski, 1976). These antisera did not contain antibodies that reacted with uninfected cells. Therefore, the glycoproteins associated with the membranes of cytomegalovirions and dense bodies were considered to be specified by the CMV genome.

IX. NONPRODUCTIVE INFECTION

Most cytomegaloviruses (CMVs) have the property of replicating only in cells of the species from which they were originally isolated. Human CMV (HCMV) does not replicate in cells of animal origin. *In vitro*, the virus replicates more efficiently in human fibroblasts than in human epithelioid cells (Knowles, 1976; Vonka *et al.*, 1976). Hence, the type and physiology of the cell that the virus encounters may determine whether infectious virus is produced. It is proposed that the HCMV genome encounters both nonproductive and productive cells in the human host. Viral regulatory mechanisms or the physiology of the cell, or both, may influence whether the infection is nonproductive or productive. It is hy-

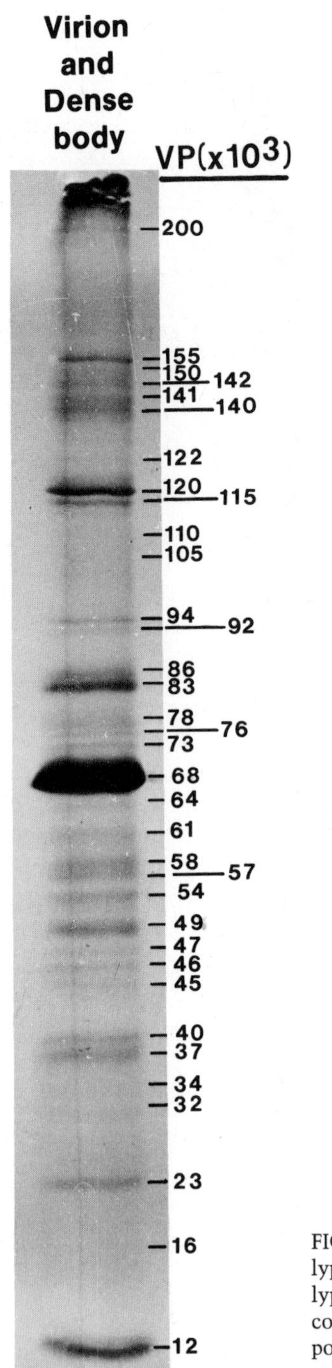

FIGURE 13. Autoradiogram of [^{35}S]methionine-labeled polypeptides associated with CMV plus dense bodies. The polypeptides were fractionated by electrophoresis in a discontinuous SDS-containing polyacrylamide gel. Viral polypeptides (VP) are indicated by their apparent molecular weights ($\times 10^3$). Data summarized from Stinski (1977).

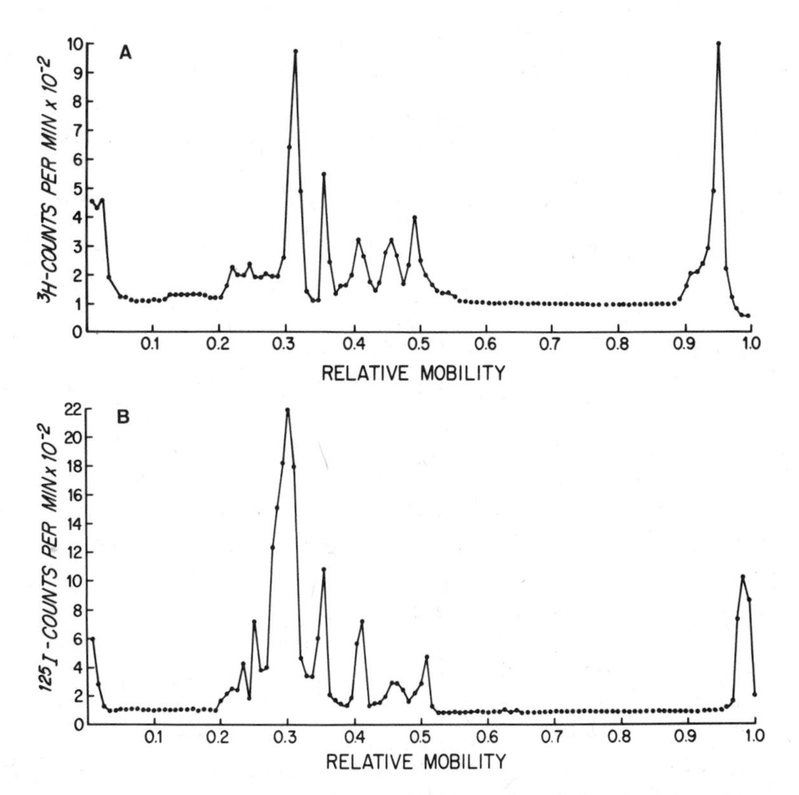

FIGURE 14. Discontinuous SDS-PAGE of [^{3}H]glycopolypeptides or surface polypeptides associated with cytomegalovirions and dense bodies. (A) The cytomegalovirions and dense bodies were pulse-labeled with [^{3}H]glucosamine from 36 hr postinfection until harvest and then purified. (B) The surface polypeptides of virions and dense bodies were iodinated with ^{125}I in the presence of lactoperoxidase and H_2O_2 after partial purification in a sorbitol gradient designed to stabilize the viral membranes. The particles were further purified by velocity and equilibrium centrifugation in CsCl. Gels were sliced into 1-mm fractions and analyzed. The mobility of the polypeptides was relative to a bromophenol blue dye marker. The apparent molecular weights of the viral glycopolypeptides and surface polypeptides were determined using a series of standard-molecular-weight proteins and are described in Stinski (1976). Data summarized from Stinski (1976).

pothesized here that the nonproductive cell may favor persistent infection or latency of the viral genome or both (see Fig. 1).

A. Nonproductive Cells of Animal Origin

HCMV enters cells of animal origin and some viral gene expression occurs, but infectious virus is not produced. Transcription of the HCMV genome in nonproductive cells is restricted in the absence of *de novo* protein synthesis, and a switch from restricted to extensive transcription occurs after 2 hr or more of protein synthesis as described above for the productive infection (Wathen *et al.*, 1981; Stinski, unpublished data). Michelson *et al.* (1979) demonstrated the synthesis of CMV-specified immediate early antigens in nonproductive cells that had the same molecular-weight properties as the virus-specified proteins synthesized in the productive infection. Figure 15 compares the synthesis of immediate early and early infected-cell-specific proteins (ICSPs) in productive human fibroblast (HF) cells and non-productive guinea pig embryo fibroblast (GPEF) cells. The majority of the immediate early ICSPs appear to be synthesized at early times in the productive and nonproductive cells. Several ICSPs were detected in the productive cells that are not detectable under the same conditions in the nonproductive cells (Fig. 15). Hiari *et al.* (1976) were unable to detect the virus-induced, salt-dependent, DNA polymerase after infection of nonproductive GPEF with HCMV. However, the activity of host-cell DNA polymerase was stimulated.

Jean *et al.* (to be published) demonstrated that HCMV DNA enters the nuclei of the nonproductive GPEF cells, but there was no evidence for the replication of the viral DNA molecule. In contrast, density-shift experiments in CsCl gradients with bromodeoxyuridine substituted for thymidine demonstrated semiconservative replication in the productive cells. In addition, electron-microscopic observations demonstrated various replicative structures in the viral DNA molecule. Therefore, cells of animal origin are nonproductive for HCMV replication as a result of a block that occurs prior to the initiation of viral DNA replication. Without viral DNA replication, late viral gene expressions, such as the late ICSPs, are not detectable in the nonproductive cells (Stinski, 1978).

B. Nonproductive Cells of Human Origin

Furukawa *et al.* (1975a) reported that HCMV would not replicate in HF cells exposed to ultraviolet light prior to infection, whereas herpes simplex virus (HSV) would replicate in these cells. Landini *et al.* (1979) found that HSV replicated in HF cells pretreated with actinomycin D, but HCMV did not. In these cells, CMV-specific immediate early and early antigens were detectable, but late antigens and viral nucleocapsids were not detectable. These reports suggest that HCMV replication depends on some cellular function(s) that is required between early transcription of the viral genome and viral DNA replication. Ultraviolet light or actinomycin D pretreatment impaired the cell such that a function(s) is missing for HCMV replication, but not HSV replication.

St. Jeor and Rapp (1973) were able to induce replication of CMV in nonpermissive human epithelioid cells by treatment with iododeoxyuridine. These results also suggest a dependence on some host-cell function for HCMV replication that must be induced. A host-cell protein(s) may be required to regulate the viral genome for complete early gene expression to provide a crucial protein(s) for the initiation of viral DNA repli-

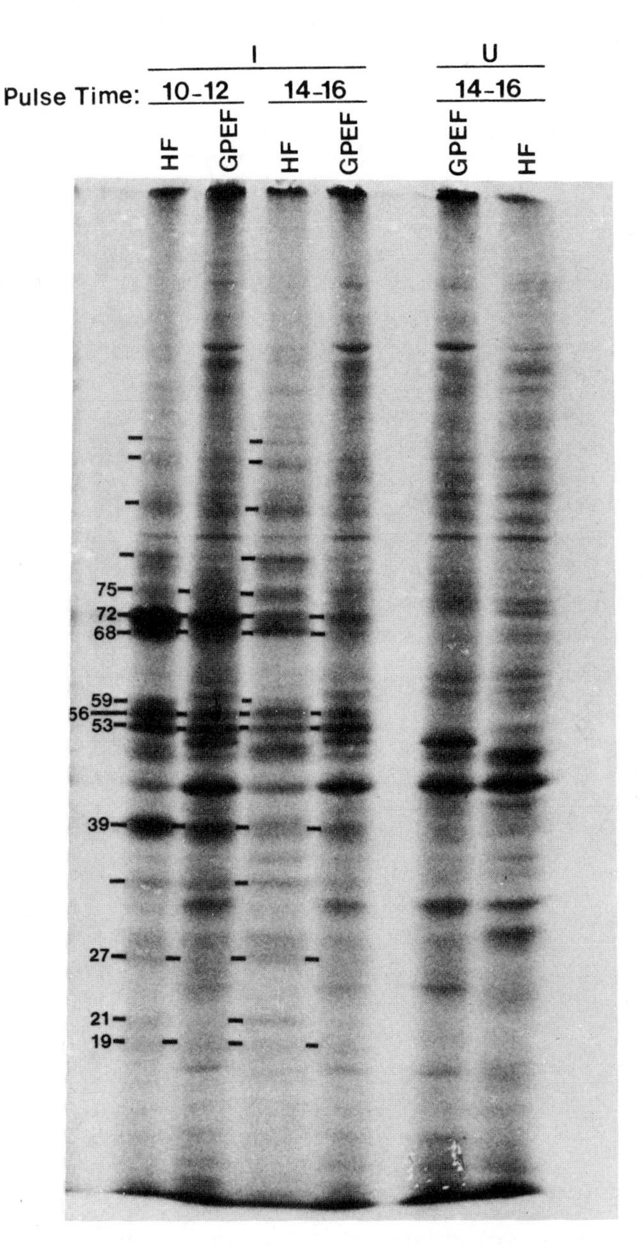

FIGURE 15. Autoradiogram of electrophoretically separated polypeptides synthesized in HCMV-infected productive HF cells and nonproductive GPEF cells at immediate early and early times after infection. Infected (I) and uninfected (U) cells were treated with cycloheximide at 2 hr postinfection. After removal of the cycloheximide, the cells were pulse-labeled for 2 hr with [^{35}S]methionine in hypertonic medium containing actinomycin D to inhibit further transcription. The ICSPs are denoted by their apparent molecular weights ($\times 10^3$). The dash markers indicate several ICSPs detectable in the HF cells that were not detectable under the same conditions in the GPEF cells. Data from Stinski (1978).

cation. Limited production of this crucial protein(s) may influence persistent infection or latency. For example, CMV infection of human lymphoblastoid cells does not produce a normal one-step virus growth curve (Tocci and St. Jeor, 1979a). Instead, a slow persistent infection occurs in which fewer than 1% of the cells produce infectious virus (St. Jeor and Weisser, 1977; Tocci and St. Jeor, 1979a). The remaining cells in the population may be either abortively or latently infected with CMV.

This type of infection can also be induced in productive HF cells by infecting with CMV at a high multiplicity. A small percentage of cells survive this type of infection and can be isolated as persistently infected cultures. These cells have been referred to as HF cells persistently infected with CMV (Mocarski and Stinski, 1979). Approximately 30% of the cells were in the productive phase of infection, since virus-specific structural antigens and virions were associated with these cells. The remaining cells contained neither viral structural antigens nor viral particles. Nevertheless, the nuclear DNA of these cells contained a significant number of viral genome equivalents (Mocarski and Stinski, 1979). These nonproductive cells continued to grow, and there was a slow, spontaneous transition of some of these cells to productive viral replication. The continued presence of the CMV genome in nonproductive cells with restricted genome expression and the slow conversion of a nonproductive cell to a productive cell may be representative of persistent infection in the host. The relationship between persistent infection and restricted genome expression in these cells is at present not understood. However, there may be a relationship to defective CMV genomes. The DNA of virions produced by the persistently infected cultures contains the "defective-specific" restriction-enzyme DNA fragments described in Section III.D. (Mocarski and Stinski, unpublished data). These data suggest a link between persistent infection and the presence of defective viral DNA. If a deletion were to occur in the area of a gene required for viral DNA replication, then the infected cells would express immediate early and some early genes but have difficulty in entering the late phase of infection. When the nonproductive cells of the persistently infected culture were analyzed for either late ICSPs or late viral antigens, these late viral gene products were not detectable (Mocarski and Stinski, 1979). However, the cells had early antigens and ICSPs (Mocarski and Stinski, 1979). The restriction of CMV gene expression in the nonproductive human cells of a persistently infected culture may not only reflect the physiology of the cell but also indicate the deletion or blockage of a certain gene(s) necessary for the efficient replication of the viral genome.

Unlike nonproductive cells of animal origin, nonproductive human cells retain the capacity for transition to virus production. This transition may be due to (1) a change in the physiology of the host cell, (2) the eventual expression of a required viral gene, or (3) the recombination of defective genomes to form a nondefective genome. These explanations are admittedly vague; consequently, further investigation is necessary to understand nonproductive CMV infection. It is postulated here that immediate early and some early CMV genome expression occurs in all cells of animal or human origin, but viral DNA replication is blocked in some cells and the viral genome consequently remains latent. The mechanism by which the CMV genome enters latency and the physical arrangement of the genome during latency remain unknown. Since the CMV genome can persist in cells for extended periods of time without viral DNA replication or late gene expression, it is proposed that latency is associated with a nonproductive cell that permits only limited viral gene expression (see Fig. 1). This event represents one of two possible pathways for the viral genome. In the productive cell, the temporal synthesis of virus-specified RNA and protein and the continued synthesis of host-cell protein provide the essential regulatory and structural elements for the production of infectious virus. Persistent infection, which is frequently characteristic of CMV infections, may represent the dynamic balance between the two possible pathways of the CMV genome.

X. CONCLUSION

After entry into a productive or nonproductive cell, the human cytomegalovirus (HCMV) genome is presumably transcribed by host-cell RNA polymerase II in restricted regions of the genome that contain the coding sequences of the virus-specified immediate early genes. Discrete size classes of viral messenger RNAs (mRNAs) are found associated with the polyribosomes in high abundance at immediate early times. These viral RNAs originate from a region in the large unique section of the viral genome. *In vitro* translation studies indicate that these mRNAs code for the majority of the immediate early viral proteins. Since some of these proteins are considered to be chromatin-binding proteins, it is hypothesized that these virus-specified proteins bind to either viral or cellular genomes. Hence, immediate early viral proteins may have a regulatory function that influences either transcription of the viral genome or host-cell metabolism. After the synthesis of the immediate early proteins, extensive transcription of the viral genome occurs in both the productive and the nonproductive cell. Even though extensive transcription occurs, a definable number of mRNAs are found on the polyribosomes of the productive or nonproductive cell. Early viral gene expression continues to be the major viral gene expression for a prolonged period of 1 day or more. Although initiation of viral DNA replication occurs after the synthesis of early viral proteins in the productive cell, there is no initiation of viral DNA replication in the nonproductive animal cell and delayed or limited initiation in the nonproductive human cell. Therefore, either a viral or a host-cell factor(s) is necessary for viral DNA replication, and this factor(s) is absent in the nonproductive animal cell and present in limited amounts in the nonproductive human cell. It is hypothesized here that this stage in the viral replication cycle influences the species specificity of HCMV replication and the propensity for persistent infection in certain cell types of the human host. Viral DNA replication proceeds at an extremely slow rate compared to that of HSV. Once viral DNA replication occurs, the late genes are expressed and the cell is eventually killed due to productive infectious virus replication. Since the CMV genome can persist in cells for extensive periods of time without viral DNA replication or late gene expression, it is proposed that latency is associated with a nonproductive cell that permits only immediate early and some early viral gene expression.

HCMV infection differs from that of herpes simplex virus (HSV) in that HCMV stimulates host-cell macromolecular synthesis. The virus replicates as cellular macromolecular synthesis continues; consequently, the virus either depends on or prefers an actively metabolizing cell. In contrast, productive infection by HSV demonstrates independence from host-cell functions to the point of even providing enzymes to ensure the presence of deoxynucleotide triphosphates for viral DNA replication. Nevertheless, these two viruses presumably evolved from a common stem to have a similar viral genome arrangement and sequence of virion morphogenesis. The predominant differences between the two viruses are genetically determined and not influenced by the host cell. Knowledge of the organization and function of the regulatory genes associated with HCMV and HSV may provide new insights into virus infection and replication as well as an alternative approach to investigating the molecular biology of the eukaryotic cell.

ACKNOWLEDGMENTS. Studies conducted at the University of Iowa were aided by U.S. Public Health Service Grant 2RO1AI13562 from the National Institute of Allergy and Infectious Diseases and by Grant 1-697 from the National Foundation March of Dimes. M.F.S. is the recipient of U.S. Public Health Service Career Development Award 1K04AI100373 from the National Institute of Allergy and Infectious Diseases. Drs. J. DeMarchi, T. Furukawa, J. Geelen, G. Hayward, L. Rosenthal, and P. Sheldrick have kindly provided information on work about to be published. Drs. G. Hayward, E.-S. Huang, B. Kilpatrick, R. LeFemina, and J. Shanley contributed to a critical review of this chapter.

REFERENCES

Adler, R., and McAuslan, B.R., 1974, Expression of thymidine kinase variants is a function of the replicative state of cells, *Cell* **2**:113.

Alwine, J.C., Kemp, D.J., and Stark, G.R., 1977, Method for detection of specific RNAs in agarose gels by transfer to diazobenzyloxymethyl-paper and hybridization with DNA probes, *Proc. Natl. Acad. Sci. U.S.A.* **74**:5350.

Ben-Porat, T., and Kaplan, A.S., 1973, Replication biochemical aspects, in: *The Herpesviruses* (A.S. Kaplan, ed.), pp. 163–220, Academic Press, New York.

Ben-Porat, T., and Kaplan, A.S., 1976, A comparison of two populations of defective, interfering pseudorabies virus particles, *Virology* **72:**471.

Blanton, R.A., and Tevethia, M.J., 1981, Immunoprecipitation of virus-specific immediate-early and early polypeptides from cells lytically infected with human cytomegalovirus strain AD169, *Virology* **112:**262.

Bronson, D.L., Dressman, G.R., Biswal, N., and Benyesh-Melnick, M., 1973, Defective virions of herpes simplex virus, *Intervirology* **1:**141.

Chua, C.C., Carter, T.H., St. Jeor, S., 1981, Transcription of the human cytomegalovirus genome in productively infected cells, *J. Gen. Virol.* **56:**1.

Clements, J.B., Watson, R.J., and Wilkie, N.W., 1977, Temporal regulation of herpes simplex virus type 1 transcription: Location of transcripts on the viral genome, *Cell* **12:**275.

Clements, J.B., McLauchlan, J., and McGeoch, D.J., 1979, Orientation of herpes simplex virus type 1 immediate early mRNAs, *Nucleic Acid Res.* **7:**77.

Craighead, J.E., Kanich R., and Almeida, J.D., 1972, Nonviral microbodies with viral antigenicity produced in cytomegalovirus-infected cells, *J. Virol.* **10:**766.

Crawford, L.V., and Lee, A.J., 1964, The nucleic acid of human cytomegalovirus, *Virology* **23:**105.

Darlington, R.W., and Granoff, A., 1973, Replication—biological aspects, in: *The Herpesviruses* (A.S. Kaplan, ed.), pp. 93–132, Academic Press, New York.

Delius, H., and Clements, J. B., 1976, A partial denaturation map of herpes simplex virus type 1 DNA: Evidence for inversion of the unique DNA regions, *J. Gen. Virol.* **33:**125.

DeMarchi, J.M., 1981, Human cytomegalovirus DNA: Restriction enzyme cleavage maps and map locations for immediate-early, early, and late RNAs. *Virology* **114:**23.

DeMarchi, J.M., and Kaplan, A.S., 1976, Replication of human cytomegalovirus DNA: Lack of dependence on cell DNA synthesis, *J. Virol.* **18:**1063.

DeMarchi, J.M., Blankship, M.L., Brown, G. D., and Kaplan, A.S., 1978, Size and complexity of human cytomegalovirus DNA, *Virology* **89:**643.

DeMarchi, J.M., Schmidt, C.A., and Kaplan, A.S., 1980, Patterns of transcription of human cytomegalovirus in permissively infected cells, *J. Virol.* **35:**277.

Estes, J.E., and Huang, E.S., 1977, Stimulation of cellular thymidine kinase by human cytomegalovirus, *J. Virol.* **24:**13.

Fiala, M., Honess, R.W. Heiner, D.C., Heine, J.W., Murnane, J., Wallace, R., and Guze, L.B., 1976, Cytomegalovirus proteins. I. Polypeptides of virions and dense bodies, *J. Virol.* **19:**243.

Fleckenstein, B., Bornkamm, G.W., and Ludwig, H., 1975, Repetitive sequences in complete and defective genomes of *Herpesvirus saimiri*, *J. Virol.* **15:**318.

Frenkel, N., Jacob, R. J., Honess, R.W., Hayward, G. S., Locker, H., and Roizman, B., 1975, Anatomy of herpes simplex virus DNA. III. Characterization of defective DNA molecules and biological properties of virus population containing them, *J. Virol.* **16:**153.

Frenkel, N., Locker, H., Batterson, W., Hayward, G.S., and Roizman, B., 1976, Anatomy of herpes simplex virus DNA. VI. Defective DNA originates from the S component, *J. Virol.* **20:**527.

Furukawa, T., Tanaka, S., and Plotkin, S.A., 1975a, Restricted growth of human cytomegalovirus in UV-irradiated WI-38 human fibroblasts, *Proc. Soc. Exp. Biol. Med.* **148:**1249.

Furukawa, T., Tanaka, S., and Plotkin, S.A., 1975b, Stimulation of macromolecular synthesis in guinea pig cells by human CMV, *Proc. Soc. Exp. Biol. Med.* **148:**211.

Geelen, J.L.M.C., and Weststrate, M.W., 1982, Organization of the human cytomegalovirus genome, in: *Herpesvirus DNA* (Y. Becker, ed.) 1982, Martinus Nijhoff Medical Publishers.

Geelen, J.L.M.C., Walig, C., Wertheim, P., and Vander Noordaa, J., 1978, Human cytomegalovirus DNA. I. Molecular weight and infectivity, *J. Virol.* **26:**813.

Gupta, P., and Rapp, F., 1978, Cyclic synthesis of human cytomegalovirus induced proteins in infected cells, *Virology* **84:**199.

Gupta, P., St. Jeor, S., and Rapp, F., 1977, Comparison of the polypeptides of several strains of human cytomegalovirus, *J. Gen. Virol.* **34:**447.

Hayward, G.S., Jacob, R.J., Wadsworth, S.C., and Roizman, B., 1975, Anatomy of herpes simplex virus DNA: Evidence for four populations of molecules that differ in the relative orientations of their long and short components, *Proc. Natl. Acad. Sci. U.S.A.* **72:**4243.

Hightower, L.W., and Bratt, M.A., 1974, Protein synthesis in Newcastle disease virus-infected chicken embryo cells, *J. Virol.* **13:**788.

Hirai, K., Furukawa, T., and Plotkin, S.A., 1976, Induction of DNA polymerase in WI-38 and guinea pig cells infected with human cytomegalovirus (HCMV), *Virology* **70:**251.

Ho, M., 1977, Virus infections after transplantation in man, *Arch. Virol.* **55:**1.

Ho, M., 1982. Cytomegalovirus biology and infection (W.B. Greenough and T. C. Merigan, eds.), pp. 1–275. Plenum Publishing Corp., New York.

Honess, R.W., and Roizman, B., 1973, Proteins specified by herpes simplex virus. XI. Identification and relative molar rates of synthesis of structural and nonstructural herpes virus polypeptides in the infected cell, *J. Virol.* **12:**1347.

Honess, R.W., and Roizman, B., 1974, Regulation of herpesvirus macromolecular synthesis. I. Cascade regulation of the synthesis of three groups of viral proteins, *J. Virol.* **12:**1347.

Huang, E.-S.,1975a, Human cytomegalovirus. III. Virus-induced DNA polymerase, *J. Virol.* **16:**298.

Huang, E.-S.,1975b, Human cytomegalovirus. IV. Specific inhibition of virus-induced DNA polymerase activity and viral DNA replication by phosphonoacetic acid, *J. Virol.* **16:**1560.

Huang, E.-S.,and Pagano, J. S., 1974, Human cytomegalovirus. II. Lack of relatedness to DNA of herpes simplex I and II, Epstein–Barr virus, and nonhuman strains of cytomegalovirus, *J. Virol.* **13:**642.

Huang, E.-S., Chen, S.T., and Pagano, J. S., 1973, Human cytomegalovirus. I. Purification and characterization of viral DNA, *J. Virol.* **12:**1473.

Huang, E.-S., Kilpatrick, B.A., Huang, Y.-T., and Pagano, J.S., 1976, Detection of human cytomegalovirus and analysis of strain variation, *Yale J. Biol. Med.* **49:**29.

Isom, H.C., 1979, Stimulation of ornithine decarboxylase by human cytomegalovirus, *J. Gen. Virol.* **42:**265.

Jamieson, A.T., Gentry, G.A., and Subak-Sharpe, J.H., 1974, Induction of both thymidine and deoxycytidine kinase activity by herpes viruses, *J. Gen. Virol.* **24:**465.

Jean, J.H., Yoshimara, N., Furukawa, T., and Plotkin, S.A., 1978, Intracellular forms of the parental human cytomegalovirus genome at early stages of the infective process, *Virology* **86:**281.

Jones, P.C., Hayward, G.S., and Roizman, B., 1977, Anatomy of herpes simplex virus DNA. VII. αRNA is homologous to noncontiguous sites in both the L and S components of viral DNA, *J. Virol.* **21:**268.

Kamata, T., Tanaka, S., and Watanabe, Y., 1978, Human cytomegalovirus induced chromatin factors responsible for changes in template activity and structure of infected cell chromatin, *Virology* **90:**292.

Kilpatrick, B.A., and Huang, E.-S., 1977, Human cytomegalovirus genome: Partial denaturation map and organization of genome sequences, *J. Virol.* **24:**261.

Kilpatrick, B.A., Huang, E.-S., and Pagano, J.S., 1976, Analysis of cytomegalovirus genomes with restriction endonuclease HindIII and EcoRI, *J. Virol.* **18:**1095.

Kim, K.S., Sapienza, V.J., Carp, R.I., and Moon, H.M., 1976, Analysis of structural polypeptides of purified human cytomegalovirus, *J. Virol.* **20:**604.

Kit, S., Dubbs, D.R., and Anken, M., 1967, Altered properties of thymidine kinase after infection of mouse fibroblast cells with herpes simplex virus, *J. Virol.* **1:**238.

Kit, S., Leung, W.C., and Trkula, D., 1973, Properties of mitochondrial thymidine kinase in parental and enzyme-deficient Hela cells, *Arch. Biochem. Biophys.* **158:**503.

Knowles, W.A., 1976, *In vitro* cultivation of human cytomegaloviruses in thyroid epithelial cells, *Arch. Virol.* **50:**119.

LaFemina, R.L., and Hayward, G.S., 1980, Structural organization of the DNA molecules from human cytomegalovirus, in: *Animal Virus Genetics* (B.N. Fields and R. Jaenisch, eds.), pp. 39–55, Academic Press, New York.

Lakeman, A.D., and Osborn, J.E., 1979, Size of infections DNA from human and murine cytomegalovirus, *J. Virol.* **30**:414.

Landini, M.P., Musiani, M., Zerbini, M., Falcieri, E., and LaPlaca, M., 1979, Inhibition of a complete replication cycle of human cytomegalovirus in actinomycin pre-treated cells, *J. Gen. Virol.* **42**:423.

Lang, D.L., 1972, Cytomegalovirus infections in organ transplantation and post perfusion: An hypothesis, *Arch. Gesamte Virusforsch.* **37**:365.

Mar, E.C., and Huang, E.S., 1979, Comparative study of herpes group virus-induced DNA polymerases, *Intervirology* **12**:73.

Michelson, S., Horodniceanu, F., Kress, M., and Tardy-Panit, M., 1979, Human cytomegalovirus-induced immediate early antigens: Analysis in sodium dodecyl sulfate–polyacrylamide gel electrophoresis after immunoprecipitation, *J. Virol.* **32**:259.

Miller, R.L., and Rapp, F., 1976, Distinguishing cytomegalovirus, mycoplasma, and cellular DNA polymerases, *J. Virol.* **20**:564.

Misra, V., Muller, M.T., Chantler, J.K., and Hudson, J.B., 1978, Regulation of murine cytomegalovirus gene expression. I. Transcription during productive infection, *J. Virol.* **27**:263.

Mocarski, E.S., and Stinski, M.F., 1979, Persistence of the cytomegalovirus genome in human cells, *J. Virol.* **31**:761.

Moller-Larsen, A., Andersen, H.K., Heron, I., and Sarov, I., 1976, *In vitro* stimulation of human lymphocytes by purified cytomegalovirus, *Intervirology* **6**:249.

Morrison, J. M., and Keir, H. M., 1967, Heat-sensitive deoxyribonuclease activity in cells infected with herpes simplex virus, *Biochem. J.* **98**:37.

Mosmann, T.R., and Hudson, J.B., 1973, Some properties of the genome of murine cytomegalovirus, *Virology* **54**:135.

Plummer, G., 1973, Cytomegalovirus of man and animals, *Prog. Med. Virol.* **15**:92.

Powell, K.L., and Courtney, R.L., 1975, Polypeptides synthesized in herpes simplex virus type 2-infected HEp-2 cells, *Virology* **66**:217.

Pritchett, R.F., 1980, DNA nucleotide sequence heterogeneity between the Towne and AD169 strains of cytomegalovirus, *J. Virol.* **36**:152.

Ramirez, M.L., Virmani, M., Garon, C., and Rosenthal, L.G., 1979, Defective virions of hyman cytomegalovirus, *Virology* **96**:311.

Rapp, F., 1980, Persistence and transmission of cytomegalovirus, in: *Virus-Host Interactions: Viral Invasion, Persistence, and Disease* (M. Fraenkel-Conrat and Wagner, eds.), Vol. 16, pps. 193–232, Plenum Publishing Corp., New York.

Rapp, F., Geder, L., Muraski, D., Lausch, R., Ladda, R., Huang, E-S., and Weber, M.W., 1975, Long-term persistence of cytomegalovirus genome in cultured human cells of prostatic origin, *J. Virol.* **16**:982.

Roller, B., and Cohen, G.H., 1976, Deoxyribonucleotide triphosphate pools in synchronized human cells infected with herpes simplex virus types 1 and 2, *J. Virol.* **18**:58.

Roizman, B., 1979a, The structure and isomerization of herpes simplex virus genomes, *Cell* **16**:481.

Roizman, B., 1979b, The organization of the herpes simplex virus genomes, *Annu. Rev. Genet.* **13**:25.

Roizman, B., and Furlong, D., 1974, The replication of herpesviruses, in: *Comprehensive Virology*, Vol. 3, (H. Fraenkel-Conrat and Wagner, eds.), pp. 229–403, Plenum Publishing Corp., New York.

Rubenstein, A.S., and Kaplan, A.S., 1975, Electron microscopy studies of the DNA of defective and standard pseudorabies virions, *Virology* **66**:385.

Saboris, J.L., Pong, S.S., and Koch, G., 1974, Selective and reversible inhibition and initiation of protein synthesis in mammalian cells, *J. Mol. Biol.* **85**:195.

St. Jeor, S., and Hutt, R., 1977, Cell DNA replication as a function in the synthesis of human cytomegalovirus, *J. Gen. Virol.* **37**:65.

St. Jeor, S., and Rapp, F., 1973, Cytomegalovirus conversion of nonpermissive cells to a permissive state for virus replication, *Science* **181**:1060.

St. Jeor, S.C., and Weisser, A., 1977, Persistence of cytomegalovirus in human lymphoblasts and peripheral leukocyte cultures, *Infect. Immun.* **15**:402.

St. Jeor, S.C., Albrecht, T.B., Funk, F.D., and Rapp, F., 1974, Stimulation of cellular DNA synthesis by human cytomegalovirus, *J. Virol.* **13**:353.

Sarov, I., and Abady, I., 1975, The morphogenesis of human cytomegalovirus: Isolation and polypeptide characterization of cytomegalovirus and dense bodies, *Virology* **66**:464.

Sarov, I., and Friedman, A., 1976, Electron microscopy of human cytomegalovirus DNA, *Arch. Virol.* **50**:343.

Share, J., and Summers, W.C., 1977, Structure and function of herpesvirus genomes. II. EcoRI, XbaI and HindIII endonuclease cleavage sites on herpes simplex virus type 1 DNA, *Virology* **76**:581.

Sheldrick, P., and Berthelot, N., 1974, Inverted repetitions in the chromosome of herpes simplex virus, *Cold Spring Harbor Symp. Quant. Biol.* **39**:677.

Smith, J.D., and DeHarven, E., 1973, Herpes simplex virus and human cytomegalovirus replication in WI-38 cells. I. Sequence of viral replication, *J. Virol.* **12**:919.

Smith, J.D., and DeHarven, E., 1974, Herpes simplex virus and human cytomegalovirus replication in WI-38 cells. II. An ultrastructural study of viral penetration, *J. Virol.* **14**:945.

Southern, E.M., 1975, Detection of specific sequences among DNA fragments separated by gel electrophoresis, *J. Mol. Biol.* **98**:503.

Spector, D.H., Hock, L., and Tamashiro, J.C., 1982, Cleavage maps for human cytomegalovirus DNA strain AD169 for restriction endonucleases EcoRI, BglII, and Hind III. *J. Virol.* **42**:558.

Stinski, M. F., 1976, Human cytomegalovirus: Glycoproteins associated with virions and dense bodies, *J. Virol.***19**:594.

Stinski, M.F., 1977, Synthesis of proteins and glycoproteins in cells infected with human cytomegalovirus, *J. Virol.* **23**:751.

Stinski, M.F., 1978, Sequence of protein synthesis in cells infected by human cytomegalovirus: Early and late virus-induced polypeptides, *J. Virol.* **26**:686.

Stinski, M.F., Mocarski, E.S., and Thomsen, D.R., 1979a, DNA of human cytomegalovirus: Size heterogeneity and defectiveness resulting from serial undiluted passage, *J. Virol.* **31**:231.

Stinski, M.F., Mocarski, E.S., Thomsen, D.R., and Urbanowski, M.L., 1979b, Membrane glycoproteins and antigens induced by human cytomegalovirus, *J. Gen. Virol.* **43**:119.

Stinski, M.F., Thomsen, D. R., and Wathen, M.W., 1980, Structure and function of the human cytomegalovirus genome, in: *The Human Herpesviruses*, (A. Nahmias, W. Dowdle, and R. Schinazi, eds.), pp. 72–84, Elsevier/North-Holland, New York.

Strand, B.C., and Aurelian, L., 1976, Proteins of herpesvirus type 2. I. Virion, nonvirion and antigenic polypeptides in infected cells, *Virology* **69**:438.

Talbot, P., and Almeida, J.D., 1977, Human cytomegalovirus: Purification of enveloped virions and dense bodies, *J. Gen. Virol.* **36**:345.

Tanaka, S., Furukawa, T., and Plotkin, S.A., 1975, Human cytomegalovirus-stimulated host cell DNA synthesis, *J. Virol.* **15**:297.

Tanaka, S., Otsuka, M., Ihara, S., Maeda, F., and Watanabe, Y., 1979, Induction of pre-early nuclear antigen(s) in HEL cells infected with human cytomegalovirus, *Microbiol. Immunol.* **23**:263.

Tanaka, J., Yabuki, Y., and Hatano, M., 1981, Evidence for early membrane antigens in cytomegalovirus-infected cells, *J. Gen. Virol.* **53**:157.

Taylor, A.T., Stafford, M.A., and Jones, O.W., 1972, Properties of thymidine kinase partially purified from human fetal and adult tissue, *J. Biol. Chem.* **247**:1930.

The, T.H., and Largenhuysen, M.M.A.C., 1972, Antibodies against membrane antigens of cytomegalovirus infected cells in sera of patients with a cytomegalovirus infection, *Clin. Exp. Immunol.* **11**:475.

Tocci, M.J., and St. Jeor, S.C., 1979a, Susceptibility of lymphoblastoid cells to infection with human cytomegalovirus, *Infect. Immun.* **23**:418.

Tocci, M.J., and St. Jeor, S.C., 1979b, Persistence and replication of the human cytomegalovirus genome in lymphoblastoid cells of B and T origin, *Virology* **96:**664.
Vonka, V., Anisimova, E., and Macek, M., 1976, Replication of cytomegalovirus in human epithelioid diploid cell line, *Arch. Virol.* **52:**283.
Wagner, M.J., and Summers, W.C., 1978, Structure of the joint region and the termini of the DNA of herpes simplex virus type 1, *J. Virol.* **27:**374.
Wathen, M.W., and Stinski, M.F., 1982, Temporal patterns of human cytomegalovirus transcription: Mapping the viral RNAs synthesized at immediate early, early, and late times after infection. *J. Virol.* **41:**462.
Wathen, M.W., Thomsen, D. R., and Stinski, M.F., 1981, Temporal regulation of human cytomegalovirus transcription at immediate early and early times after infection, *J. Virol.* **38:**446.
Weller, T.H., 1971, The cytomegalovirus: Ubiquitous agents with protein clinical manifestation, *N. Engl. J. Med.* **285:**267.
Weststrate, M.W., Geelen, J.L.M.C., and Van Der Noordaa, J., 1980, Human cytomegalovirus DNA: Physical maps for the restriction endonucleases BglII, HindIII and XbaI, *J. Gen. Virol.* **49:**1.
Wilkie, N.M., 1976, Physical maps for herpes simplex virus type 1 DNA for restriction endonucleases HindIII, HpaI, and XbaI, *J. Virol.* **20:**222.
Zavada, V., Erban, V., Rezacova, D., and V. Vonka, 1976, Thymidine-kinase in cytomegalovirus infected cells, *Arch. Virol.* **52:**333.

CHAPTER 3

Molecular Biology of Varicella–Zoster Virus

RICHARD W. HYMAN

I. INTRODUCTION

Varicella–zoster virus (VZV) is a human herpesvirus that causes two common clinical conditions: chicken pox (varicella) and shingles (herpes zoster). Virus isolated from patients with either disease is indistinguishable immunologically and biologically from virus isolated from patients with the other disease (reviewed by Taylor-Robinson and Caunt, 1972). Earlier experiments had established that vesicular fluid from patients with herpes zoster could cause varicella when injected into children. For example, Bruusgaard (1932) injected vesicular fluid from 5 herpes zoster patients into the arms of a total of 18 children; 4 of the 18 showed local vesicles and another 4 showed a more generalized vesicle spread. For these reasons, the earlier terms varicella virus and herpes zoster virus have been supplanted by the term varicella–zoster virus.

Most individuals contract chicken pox in childhood, and, once infected, a person appears to harbor VZV for life. The virus can remain without clinical manifestation for years, decades, or life. In a relatively small number of people (considering that virtually all adults harbor VZV), the virus reactivates to produce shingles. The immunological status of the person plays an important role in virus reactivation, as demonstrated by the fact that immunocompromised patients have a much higher risk

RICHARD W. HYMAN • Department of Microbiology and Cancer Research Center, The Pennsylvania State University College of Medicine, Hershey, Pennsylvania 17033.

of developing shingles than matched controls. However, beyond this fact, the molecular mechanism of long-term, inapparent VZV infection is unknown.

II. *IN VITRO* GROWTH PROBLEM

To study VZV in the laboratory, it would be extremely useful to grow the virus in cell culture, but therein lies a paradox. Even though VZV grows well in a child with chicken pox, the virus grows poorly in cell culture. The *in vitro* host range of VZV is very restricted. VZV grows, but poorly, in certain human cells (Weller *et al.*, 1958), e.g., human fetal diploid lung cells, human foreskin fibroblasts, human diploid fibroblasts, and melanoma cell lines (Grose *et al.*, 1979). VZV grows more poorly in human brain and human ganglion cells (Gilden *et al.*, 1978) and even more poorly in guinea pig embryo cells (Fioretti *et al.*, 1973; Harbour and Caunt, 1975).

In addition to poor *in vitro* growth, a second problem for laboratory manipulation of VZV is the highly cell-associated nature of the virus (cf. Taylor-Robinson and Caunt, 1972). As virus-related cytopathology advances in VZV-infected cells in culture, little or no infectivity is detectable in the medium (cf. Grose *et al.*, 1979). Figure 1 presents electron micrographs of stained thin sections of VZV-infected cells. The cells were harvested when they showed substantial cytopathic effects. Only a modest amount of VZV was observable in the infected cells. What VZV was observed appeared as a typical herpesvirus (Fig. 1). To obtain "cell-free" VZV, a parallel VZV-infected culture was thrice frozen and thawed, sonicated briefly, and subjected to low-speed centrifugation to remove cell debris. This virus preparation was "cell-free" in the sense that no viable cells were present. The VZV preparation was negatively stained and observed in the electron microscope (Fig. 2). Individual herpesviruslike particles were seen only very occasionally, and then usually without discernible envelopes (Fig. 2a). Rather, clumps of herpesviruslike particles were seen (Fig. 2b). These were interpreted to be VZV particles held together and occluded by host-cell membranes. In addition, larger membranous clumps were observed (Fig. 2c). These sometimes appeared to contain indistinct herpesviruslike particles.

These considerations lead to the final part of the *in vitro* growth problem. Despite the ubiquity of VZV as a pathogen, an estimated 2–3 $\times 10^6$ chicken pox cases occur per year in the United States (Preblud and D'Angelo, 1979), and even though chicken pox can be a life-threatening disease to immunocompromised children (cf. Feldman *et al.*, 1975), very few molecular studies have been published concerning VZV. Presumably, the paucity of data reflects, in part, the difficulties encountered in propagating VZV in cell culture.

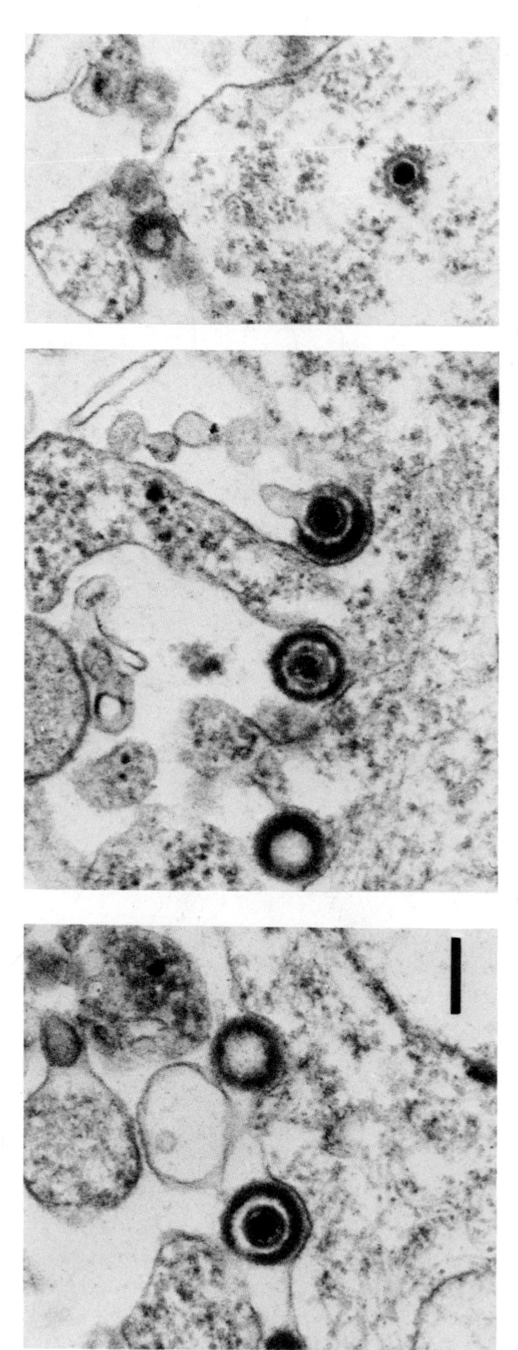

FIGURE 1. Electron micrographs of thin sections of VZV-infected cells. Varicella isolate (Kawaguchi)-infected whole human embryo diploid cells (Flow 5000) were fixed with glutaraldehyde and post-fixed with Dalton's chrome osmium. The cells were stained *en bloc* with 2% uranyl acetate in 35% ethanol and poststained with Reynold's lead citrate. Three examples are given. Scale bar: 100 nm.

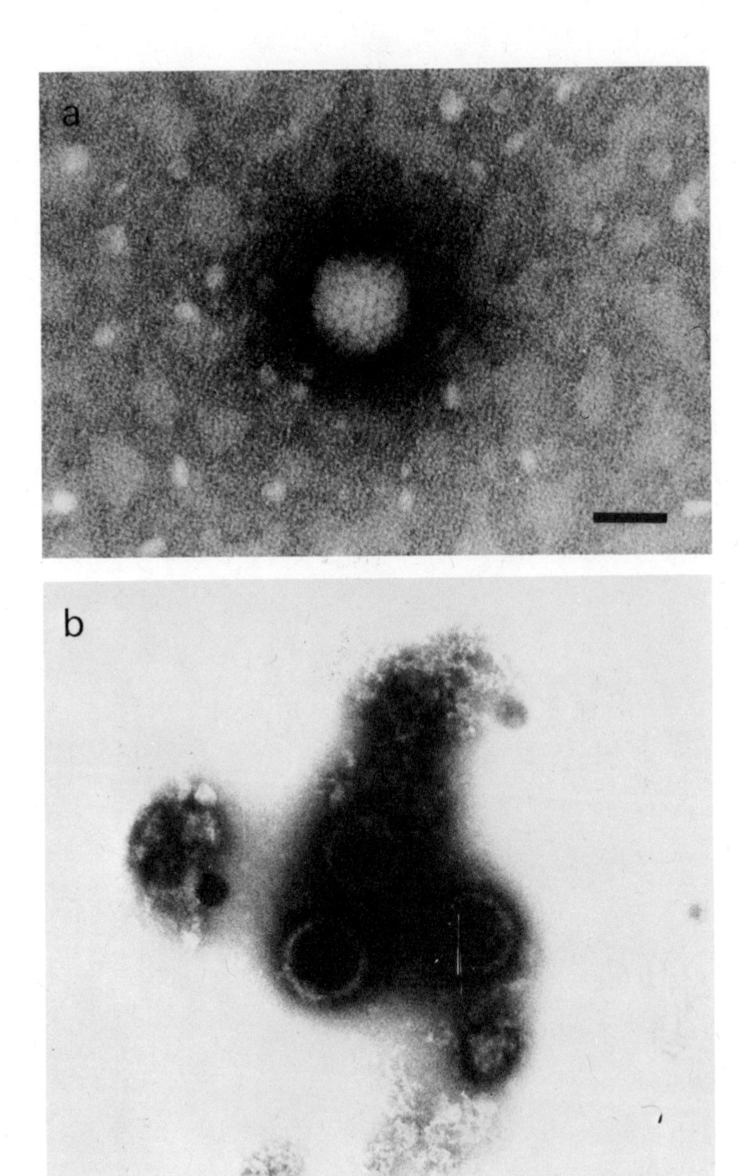

FIGURE 2. Electron micrographs of cell-free VZV preparations. In parallel with the VZV-infected cells in Fig. 1, VZV (Kawaguchi)-infected whole human embryo diploid cells were frozen and thawed three times and sonicated briefly. Cells debris was removed by low-speed centrifugation. This cell-free VZV preparation was negatively stained with 1% phosphotungstic acid, pH 5, and observed in the electron microscope. Scale bars: (a) 80 nm; (c) 100 nm.

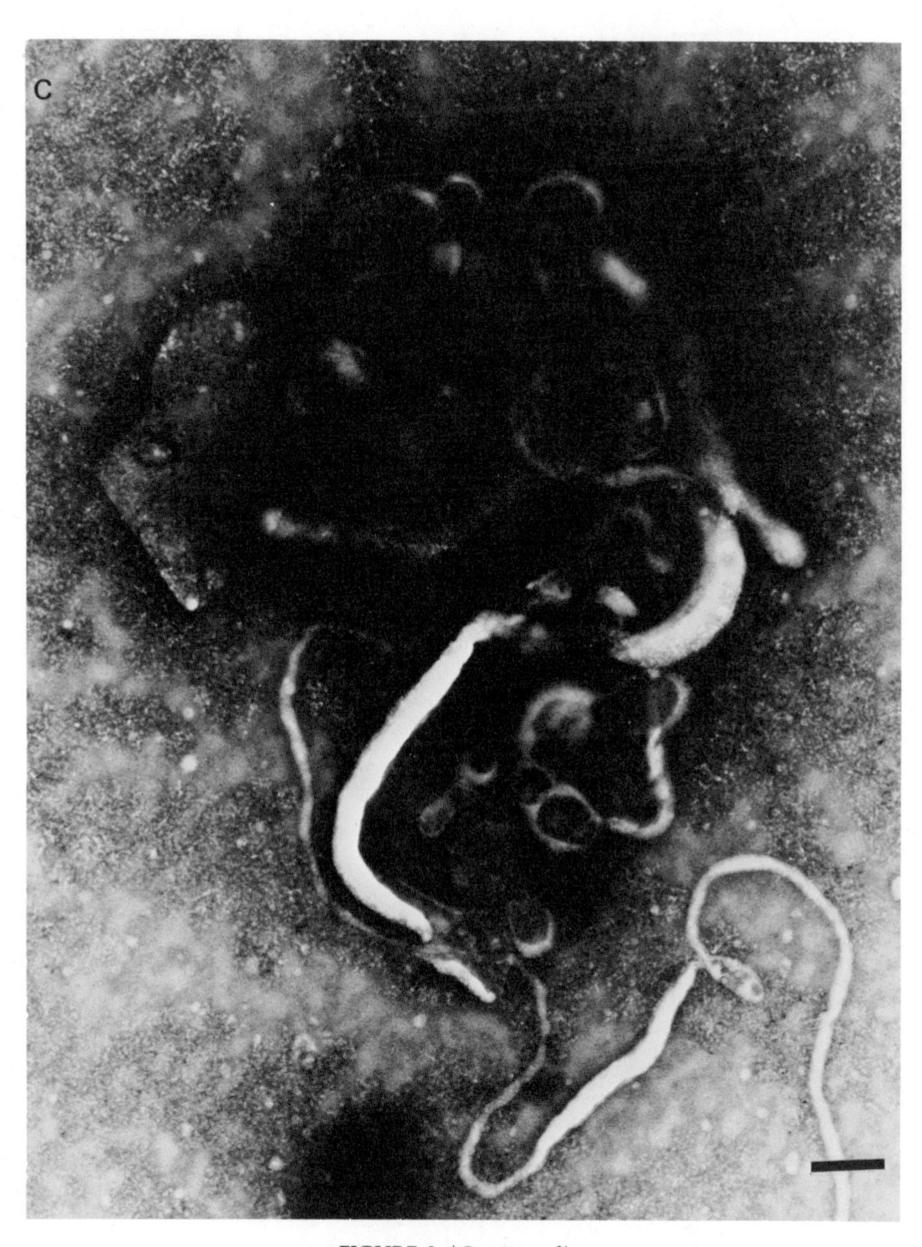

FIGURE 2. (*Continued*)

III. VARICELLA–ZOSTER-VIRUS-RELATED PROTEINS

Because of the cell-associated nature of VZV, it is difficult to purify VZV and to obtain virus unequivocally free of host-cell proteins. In the complete absence of genetic studies on VZV, no VZV-infected cell protein can be called VZV coded. Therefore, this section is titled VZV-related proteins. These proteins were first detected immunologically; e.g., Ito and Barron (1973) described surface antigens specific to VZV-infected cells.

Wolff (1978) radiolabeled VZV-infected cells with [^{35}S]methionine and then partially purified the virus by sucrose-gradient centrifugation. By polyacrylamide gel electrophoresis (PAGE) followed by fluorography, at least 31 proteins were identified. Their apparent molecular sizes ranged from 240 to 18 kilodaltons. Wolff (1978) immunoprecipitated proteins from partially purified VZV using a human serum with a complement-fixation titer of 1:80 against VZV. Of the originally identified 31 proteins, 14 were immunoprecipitated. Asano and Takahashi (1979) radiolabeled VZV-infected cells separately with (^{35}S]methionine or [^{14}C]glucosamine. The VZV-related proteins were immunoprecipitated with the use of antisera prepared in guinea pigs and green monkeys and subjected to PAGE and fluorography. The approximate neutralizing antibody titer of the sera was 1:1024. Thirty ^{35}S-labeled protein bands were observed; apparent molecular sizes ranged from approximately 145 to 22 kilodaltons. Three proteins precipitated by guinea pig antiserum were not precipitated by green monkey antiserum and vice versa. Ten ^{14}C-labeled glycoprotein bands were observed; apparent molecular sizes ranged from 105 to 56 kilodaltons. Grose (1980) has also examined VZV-related proteins and glycoproteins by PAGE and fluorography of immunoprecipitated proteins. Grose (1980) employed rabbit anti-VZV serum with a neutralizing antibody titer of 1:800 and a human convalescent serum with a titer greater than 1:512. Five glycoproteins with apparent molecular sizes of 118, 98, 88, 62, and 45 kilodaltons were identified in [^{3}H]glucosamine-labeled cells. Four of these five, excluding the 88-kilodalton glycoprotein, were identified using [^{3}H]fucose as the radiolabel.

Three technical conclusions can be reached from the work of Wolff (1978), Asano and Takahashi (1979), and Grose (1980): (1) Results from direct examinations of proteins radiolabeled in VZV-infected cells are not interpretable. VZV infection does not appear to shut off host protein synthesis tightly or promptly. (2) Partial purification of radiolabeled VZV is of questionable help because the virus is highly cell-associated. The protein patterns in the fluorographs of Wolff (1978) are greatly simplified after immunoprecipitation. (3) Given current techniques, immunoprecipitation helps significantly. The efficacy of the immunoprecipitation, however, is directly related to the type and strength of the antiserum. Asano and Takahashi (1979) achieved modestly different results with different antisera. There is a further caveat before a comparison of the authors' data can be undertaken. The authors used different concentrations of polyacrylamide and different cross-linking agents for the PAGE analyses. This variation could give rise to differences in apparent molecular weight for the same protein or glycoprotein. Nevertheless, certain agreements among the data can be noted. The authors detected prominent proteins with apparent molecular sizes of about 150, 70, and 35 kilodaltons. Less prominent proteins in the range of 90–110 kilodaltons were also detected. Asano and Takahashi (1979) and Grose (1980) detected five or six prominent glycoproteins. The apparent molecular weights are not in good agreement. However, disagreement is to be expected given the difference in PAGE conditions.

VZV-related proteins have also been detected as new enzymatic activities in VZV-infected cells. Several laboratories have reported the presence of a new thymidine kinase (deoxypyrimidine kinase) activity in VZV-infected cells (Dobersen *et al.*, 1976; Ogino *et al.*, 1977; Hackstadt and Mallavia, 1978). Miller *et al.* (1977) demonstrated that the spread of VZV infection in cell culture was reversibly inhibited by arabinofuranosylthymine (ARA-T), and Iltis *et al.* (1979) demonstrated the antiviral effect of 5-iodo-5′-amino-2′,5′-dideoxyuridine (AIU) on VZV replication *in vitro*. These results are in consonance with the interpretation that a VZV-induced thymidine kinase phosphorylates ARA-T and AIU and that the phosphorylated arabinose-containing bases block VZV DNA synthesis (cf. Aswell and Gentry, 1977). A new DNase activity has been reported in VZV-infected cells (Cheng *et al.*, 1979), as has a high salt-resistant DNA polymerase activity (Miller and Rapp, 1977; Mar *et al.*, 1978). May *et al.* (1977) previously found that VZV would not replicate in cell culture in the presence of phosphonoacetic acid. It is now well established that herpesvirus-induced DNA polymerases are sensitive to phosphonoacetate. Therefore, it is reasonable to assume that the salt-resistant DNA polymerase found in VZV-infected cells is VZV induced.

IV. VARICELLA–ZOSTER VIRUS DNA

Ludwig *et al.* (1972) first measured the CsCl buoyant density of VZV DNA using both purified virions and VZV-infected cells as starting materials for VZV DNA purification. The VZV DNA was subjected to isopycnic centrifugation in an analytical ultracentrifuge. Ludwig *et al.* (1972) investigated one virus isolated from varicella and a second isolated from herpes zoster. For both starting materials and both VZV DNAs, the measured buoyant density was 1.705 g/cm^3.

Richards *et al.* (1979) used the modified Hirt supernatant procedure (Rapp *et al.*, 1977) to purify VZV DNA for CsCl buoyant-density measurements. Herpes simplex virus type 1 (HSV-1) DNA was included in each gradient as an internal control. Figure 3 compares the buoyant dens-

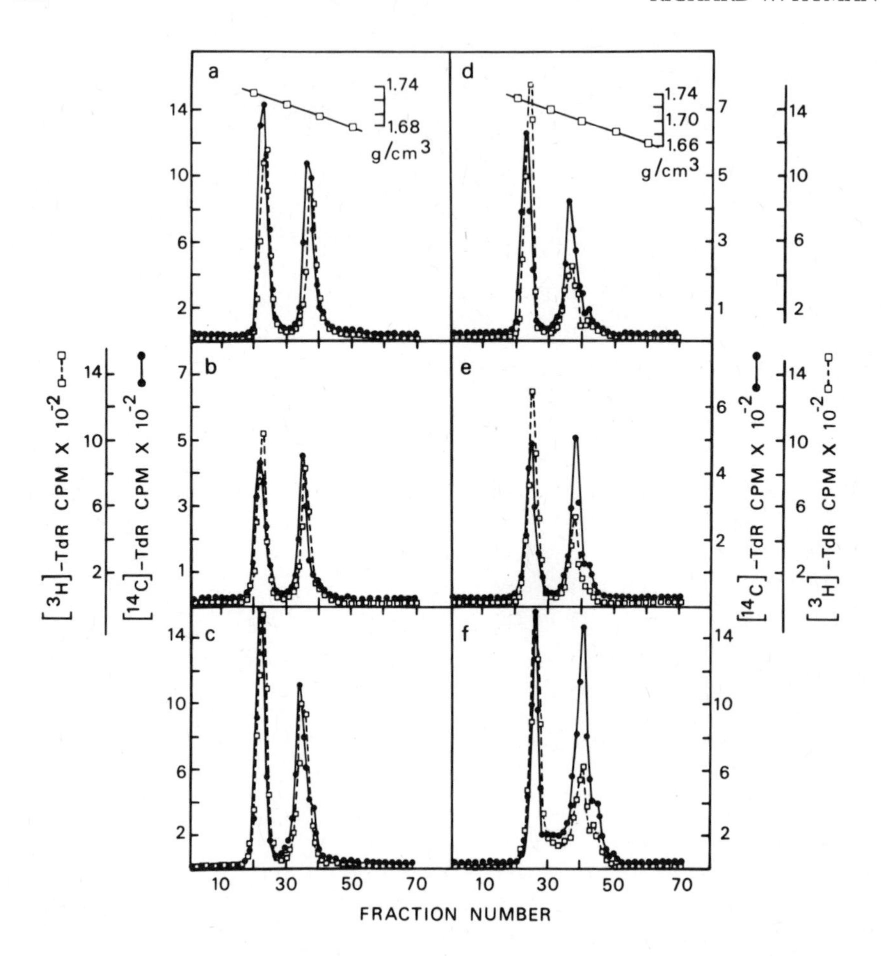

FIGURE 3. CsCl buoyant density analysis of VZV DNAs from varicella isolates. VZVs isolated from clinical varicella infections were propagated in Flow 5000 cells and independently labeled with [^{3}H]thymidine (TdR) or [^{14}C]-TdR, and the VZV DNAs were purified. HSV-1 (KOS) DNA was radiolabeled with either [^{14}C]- or [^{3}H]-TdR and purified from the Hirt supernatant of infected cells by CsCl density-gradient centrifugation. All gradient analyses represent artificial mixtures of ^{3}H-labeled and ^{14}C-labeled VZV DNAs, radiolabeled and purified prior to these reconstitution measurements. HSV-1 (KOS) [^{14}C]- and [^{3}H]-TdR DNAs were included in all reconstitution experiments as an internal control for isotope effect and, in each panel, are the DNAs of density 1.727 g/cm^{3} (leftmost bands). Isopycnic centrifugation was carried out at 35,000 rpm for 65 hr at 20°C in a Beckman 50.3 rotor. To avoid cluttering the figures, the density values are plotted only for the gradient at the top of each column. Density increases from right to left. A small amount of contaminating cell [^{14}C]-DNA is seen as the lightest (shoulder) DNA in (e) and (f), and cell [^{3}H]-DNA is seen in (f). (a) VZV (Oka) [^{3}H]-DNA and VZV (Oka) [^{14}C]-DNA; (b) VZV (Oka) [^{3}H]-DNA and VZV (Kawaguchi) [^{14}C]-DNA; (c) VZV (Kawaguchi) [^{3}H]-DNA and VZV (Oka) [^{14}C]-DNA; (d) VZV (Kawaguchi) [^{3}H]-DNA and VZV (Kawaguchi) [^{14}C]-DNA; (e) VZV (Kawaguchi) [^{3}H]-DNA and VZV (Rc) [^{14}C]-DNA; (f) VZV (Rc) [^{3}H]-DNA and VZV (Kawaghchi) [^{14}C]-DNA. (●) ^{14}C cpm; (□) ^{3}H cpm.

ities of VZV DNAs of varicella isolates. When ^{3}H-labeled VZV (Oka) DNA was compared to ^{14}C-labeled VZV (Oka) DNA, a one-fraction difference in band position was observed (Fig. 3a). Similarly, a one-fraction difference was observed between the ^{14}C- and ^{3}H-labeled HSV-1 DNAs (Fig. 3a). In both instances, the ^{14}C-labeled DNAs were heavier. The analogous result can be seen in Fig. 3b–d; the ^{14}C-labeled VZV DNAs were one fraction heavier than the ^{3}H-labeled VZV DNAs. Comparison of VZV (Kawaguchi) [^{3}H]-DNA and VZV (Rc) [^{14}C]-DNA (Fig. 3e) as well as the converse experiment (Fig. 3f) revealed indistinguishable DNA profiles. It can be concluded that the observed slight differences in CsCl buoyant density among the VZV DNAs from the varicella isolates were not larger than can be accounted for by isotope effect.

Comparisons of the buoyant densities of ^{14}C- and ^{3}H-labeled VZV DNAs from herpes zoster isolates are shown in Fig. 4. The mixture of VZV (Caufield) [^{3}H]-DNA and VZV (Caufield) [^{14}C]-DNA showed a banding profile in which the ^{14}C-labeled DNA was slightly heavier (Fig. 4a). Comparison of the buoyant densities of VZV (Caufield) and (Ludwig) DNAs (Fig. 4b and c) demonstrated that, in both cases, the ^{14}C-labeled DNAs were slightly heavier. The buoyant densities of the ^{3}H- and ^{14}C-labeled VZV (Ludwig) DNAs were indistinguishable (Fig. 4d). Comparison of VZV (Ludwig) [^{3}H]-DNA and VZV (Jab) [^{14}C]-DNA (Fig. 4e) indicated that the VZV (Jab) [^{14}C]-DNA was one-half fraction heavier than the VZV (Ludwig) (^{3}H]-DNA. In the converse experiment (Fig. 4f), buoyant-density analysis of VZV (Jab) [^{3}H]-DNA and VZV (Ludwig) [^{14}C]-DNA, the converse result was obtained; VZV (Ludwig) [^{14}C]-DNA was one fraction heavier (Fig. 4f). It can be concluded that the slight differences in CsCl buoyant density observed among the VZV DNAs from the herpes zoster isolates were not larger than can be accounted for by isotope effect.

The final part of the buoyant-density studies was a comparison of VZV DNA from varicella isolates with VZV DNA from herpes zoster isolates. Figure 5 shows that the ^{14}C-labeled DNA had the slightly greater buoyant density. The buoyant-density difference between VZV [^{14}C]-DNA and VZV [^{3}H]-DNA was never greater than the density difference observed between HSV-1 [^{3}H]-DNA and HSV-1 [^{14}C]-DNA within the same gradients (Fig. 5a and f–h). Thus, it can be concluded from all these buoyant-density measurements (Figs. 3–5) that the CsCl buoyant densities for VZV DNAs were indistinguishable for all clinical isolates.

The buoyant density in CsCl of all seven VZV DNAs was measured to be 1.705 g/cm^{3}, as determined by both refractive-index measurements and interpolations based on the gradient positions of HSV-1 DNA (density of 1.727 g/cm^{3}) and cell DNA (density of 1.695 g/cm^{3}). This value is in complete agreement with the values published by Ludwig *et al.* (1972) and Iltis *et al.* (1977). With the use of the equation presented by Schildkraut *et al.* (1962) relating the buoyant density of DNA to its guanine plus cytosine (G+C) content, the G+C content of VZV DNA is calculated to be 47% (Iltis *et al.*, 1977).

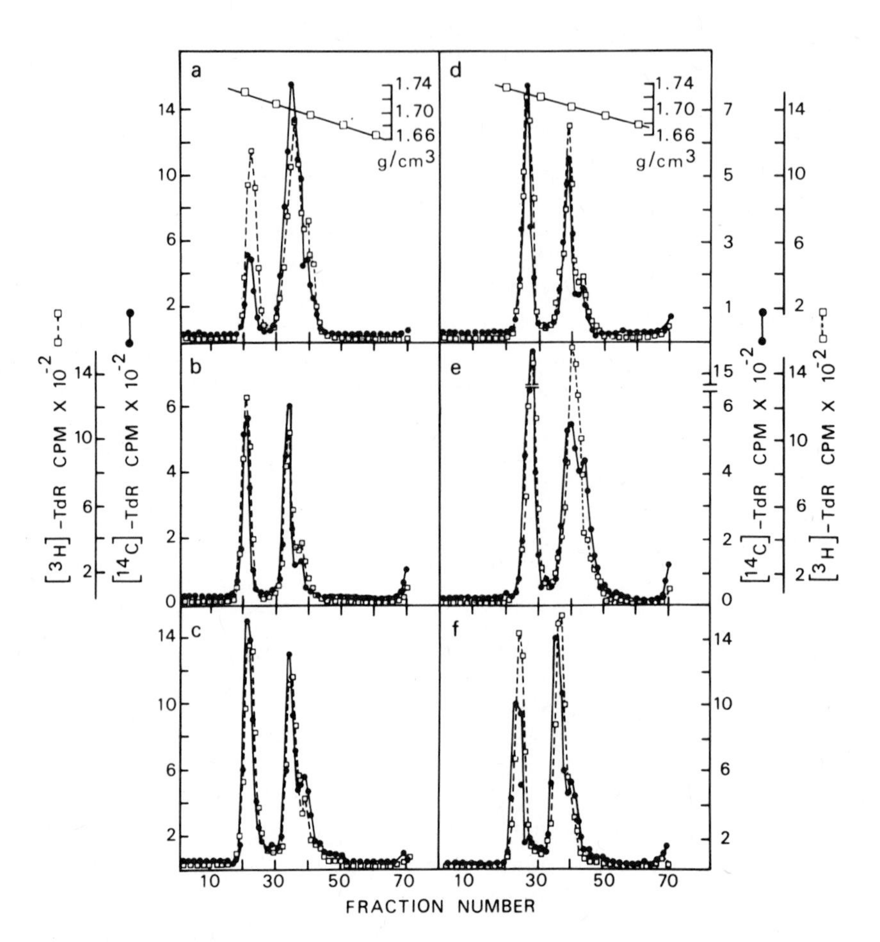

FIGURE 4. CsCl buoyant-density analysis of VZV DNAs from herpes zoster isolates. VZV isolates from clinical herpes zoster infections were propagated, and the VZV DNA was radiolabeled and purified. ^{3}H- and ^{14}C-labeled HSV-1 (KOS) DNAs were included in all gradients (leftmost DNA bands). Reconstitution CsCl density-gradient analysis of the various ^{3}H- and ^{14}C-labeled VZV DNAs was conducted as described in the Figure 3 caption. A small amount of contaminating cell DNA is seen as the lightest (shoulder) DNA of both isotopes in all panels. (a) VZV (Caufield) [^{3}H]-DNA and VZV (Caufield) [^{14}C]-DNA; (b) VZV (Caufield) [^{3}H]-DNA and VZV (Ludwig) [^{14}C]-DNA; (c) VZV (Ludwig) [^{3}H]-DNA and VZV (Caufield) [^{14}C]-DNA; (d) VZV (Ludwig) [^{3}H]-DNA and VZV (Ludwig) [^{14}C]-DNA; (e) VZV (Ludwig) [^{3}H]-DNA and VZV (Jab) [^{14}C]-DNA; (f) VZV (Jab) [^{3}H]-DNA and VZV (Ludwig) [^{14}C]-DNA. (●) ^{14}C cpm; (□) ^{3}H cpm.

Restriction-enzyme cleavage patterns have been published for VZV DNA (Oakes *et al.*, 1977; Richards *et al.*, 1979; Ecker and Hyman, 1981b). The single most important point in the restriction-enzyme cleavage patterns of VZV DNAs is that for each enzyme, the patterns are obviously very similar for the DNAs of all examined VZV isolates. For example, Fig. 6 presents a comparison of four restriction-enzyme cleavage patterns of the DNAs from VZV (Oka), the parental virus from which a live at-

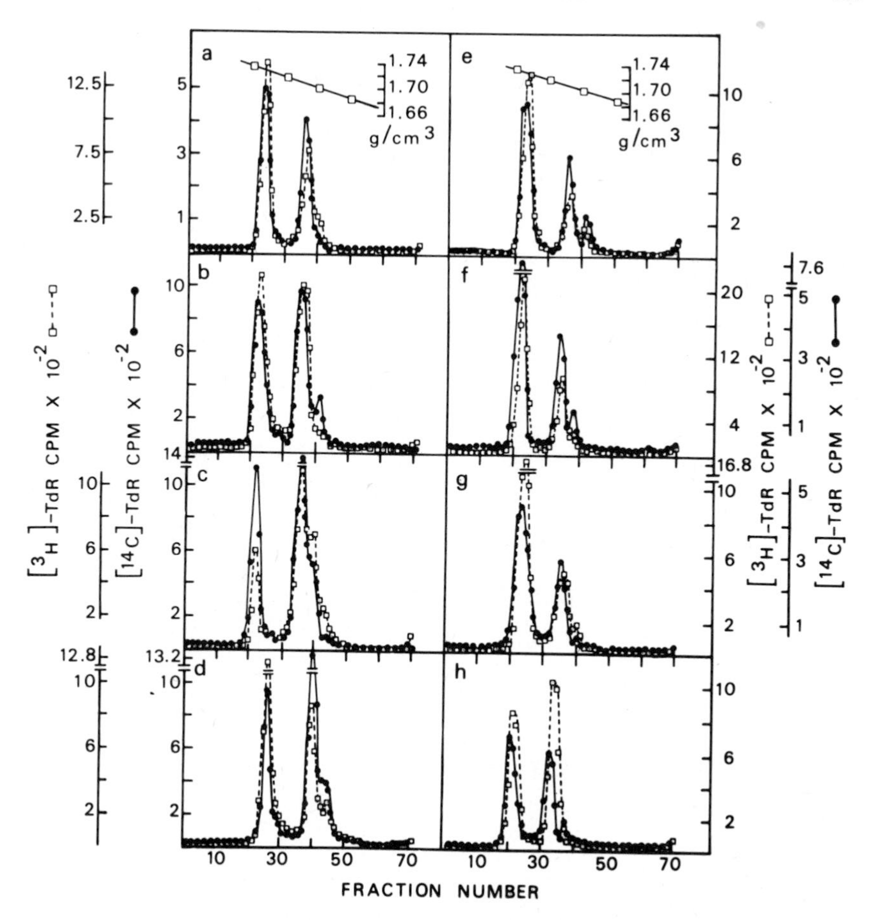

FIGURE 5. Comparison of the buoyant density of VZV DNAs from varicella isolates with the buoyant density of VZV DNAs from herpes zoster isolates. VZV DNAs were labeled with [^{14}C]- or [^{3}H]-TdR and purified. ^{3}H- and ^{14}C-labeled HSV-1 (KOS) DNAs were included in all gradients (leftmost DNA bands). CsCl buoyant-density centrifugation was conducted as described in the Fig. 3 caption. A small amount of contaminating cell DNA is seen as the lightest (shoulder) DNA for [^{3}H]-DNA in (a, c–e) and (g) and for [^{14}C]-DNA in (b–f) and (h). (a) VZV (Jab) [^{3}H]-DNA and VZV (Oka) [^{14}C]-DNA; (b) VZV (Oka) [^{3}H]-DNA and VZV (Jab) [^{14}C]-DNA; (c) VZV (Ludwig) [^{3}H]-DNA and VZV (Oka) [^{14}C]-DNA; (d) VZV (Oka) [^{3}H]-DNA and VZV (Ludwig) [^{14}C]-DNA: (e) VZV (Caufield) [^{3}H]-DNA and VZV (Ma) [^{14}C]-DNA; (f) VZV (Ma) [^{3}H]-DNA and VZV (Caufield) [^{14}C]-DNA; (g) VZV (Caufield) [^{3}H]-DNA and VZV (Kawaguchi) [^{14}C]-DNA; (h) VZV (Kawaguchi) [^{3}H]-DNA and VZV (Caufield) [^{14}C]-DNA. (●) ^{14}C cpm; (□) ^{3}H cpm. Reprinted from Richards *et al.* (1979), with the permission of ASM Publications.

tenuated vaccine has been derived (Takahashi *et al.*, 1975), and VZV (80-2), a recent herpes zoster isolate. The VZV DNA used in this comparison was extracted from purified virions and was not further selected by CsCl buoyant-density centrifugation. Therefore, the restriction-endonuclease cleavage patterns (Fig. 6) represent total virus DNAs. The fluorographs show the cleavage patterns of VZV (Oka) DNA (Fig. 6, lanes 1, 3, 5, 8)

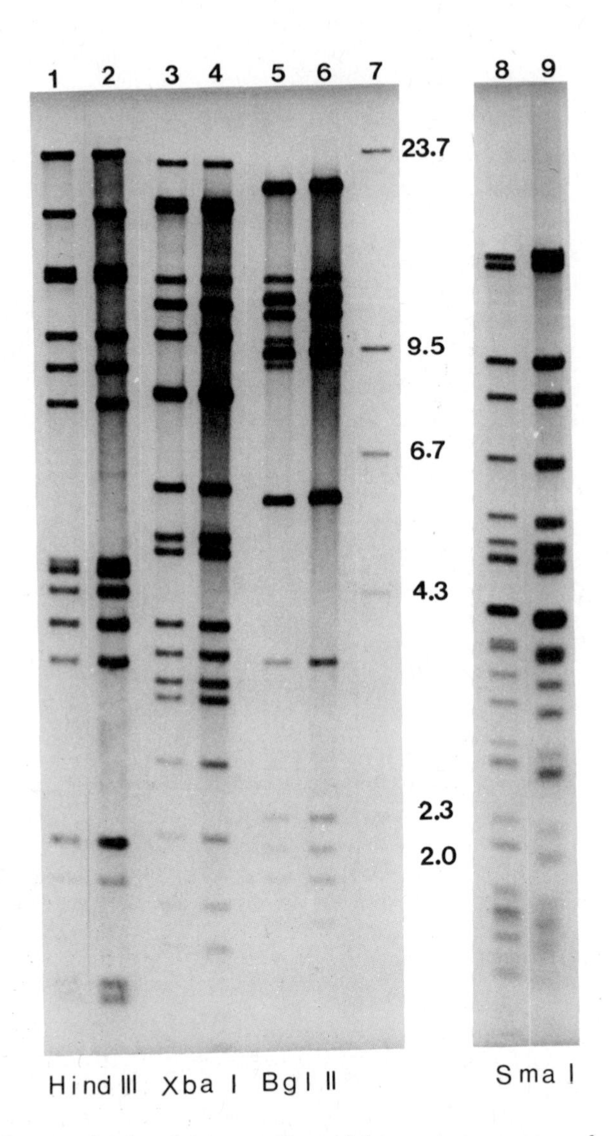

FIGURE 6. Fluorographs showing comparisons of the restriction-enzyme cleavage patterns of VZV (Oka) and VZV (80-2) DNAs. VZV (Oka) DNA (1, 3, 5, 8) was radiolabeled at passage 16; VZV (80-2) DNA (2, 4, 6, 9) was radiolabeled at passage 8. The VZV DNAs were purified as described (Ecker and Hyman, 1981b) and were digested by restriction enzymes *Hin*dIII (1, 2), *Xba*I (3, 4), *Bgl*II (5, 6), and *Sma*I (8, 9). Lane 7 contained λ DNA digested by *Hin*dIII. Sizes of the DNA fragments are given in kilobase pairs (kb). After electrophoresis, the agarose gels were dried and subjected to fluorography. Various exposures have been combined to achieve comparable contrast.

and VZV (80-2) DNA (Fig. 6, lanes 2, 4, 6, 9) using the enzymes *Hin*dIII (Fig. 6, lanes 1, 2), *Xba*I (Fig. 6, lanes 3, 4), *Bg1*II (Fig. 6, lanes 5, 6), and *Sma*I (Fig. 6, lanes 8,9). As can be clearly seen in Fig. 6, the four cleavage patterns were indistinguishable. Analogous results have been achieved with additional restriction enzymes and about a dozen VZV isolates (Oakes *et al.*, 1977; Richards *et al.*, 1979; Ecker and Hyman, 1981b).

Quantitation of the cleavage patterns of VZV DNA revealed molar and submolar fragments (Oakes *et al.*, 1977; Richards *et al.*, 1979). Submolar fragments are observed for the cleavage patterns of HSV-1 DNA, where the presence of submolar bands reflects major inversions within the HSV-1 genome (Hayward *et al.*, 1975; Skare *et al.*, 1975; Delius and Clements, 1976; Wilkie and Cortini, 1976; Skare and Summers, 1977; Roizman, 1979) and are also observed for Epstein–Barr virus (P3HR-1) DNA, where the presence of submolar bands reflects the heterogeneity of the virus (Sugden *et al.*, 1976; Delius and Bornkamm, 1978; Given and Kieff, 1978; Rymo and Forsblom, 1978). There are difficulties in interpreting the quantitative data on VZV DNA. Because VZV cannot be plaque-purified by current techniques, VZV is commonly propagated in cell culture by mixing VZV-infected cells with uninfected cells. There is little control of the multiplicity of infection. The effects on virus DNA following serial, undiluted passage of the virus have been studied for several herpesviruses (Bronson *et al.*, 1973; Ben-Porat *et al.*, 1974; Frenkel *et al.*, 1975; Wagner *et al.*, 1975; Cambell *et al.*, 1976; Graham *et al.*, 1978; Locker and Frenkel, 1979; Stinski *et al.*, 1979). The result of high-multiplicity passage is the synthesis of defective herpesvirus DNA. The defective herpesvirus DNA can have a different buoyant density than wild-type DNA, a changed restriction-enzyme cleavage pattern, and/or a lower molecular weight. Evidence has been presented (Ecker and Hyman, 1981b) that shows that as VZV is passed in culture, defective VZV DNA of lower buoyant density and altered *Hin*dIII cleavage pattern accumulates. For example, Fig. 7 presents a side-by-side comparison of VZV (Oka) DNA of very different *in vitro* passages employing the restriction enzymes *Hpa*I (lanes 2, 3), *Bgl*II (lanes 4, 5), *Kpn*I (lanes 6, 7), *Xba*I (lanes 8, 9), and *Hin*dIII (lanes 11, 12). Many additional bands were found in the DNA of the VZV (Oka) that has been passed extensively in cell culture. These additional bands were not found in the DNA of the VZV (Oka) that has been passed only sparingly in cell culture. Therefore, the presence of molar and submolar fragments in the restriction-enzyme cleavage patterns of VZV DNA could be the result of the presence of a heterogeneous population of mixed defective VZV DNAs or the result of structural idiosyncracies of the VZV genome or both.

The resolution of this problem has been achieved by the use of recombinant DNA. The *Eco*RI fragments of VZV DNA were cloned in the vector pACYC 184; the *Hin*dIII fragments in pBR322 (Ecker and Hyman, 1982). Appropriate double digestions, cross-hybridizations, and blot hybridizations established the restriction-enzyme cleavage maps. These

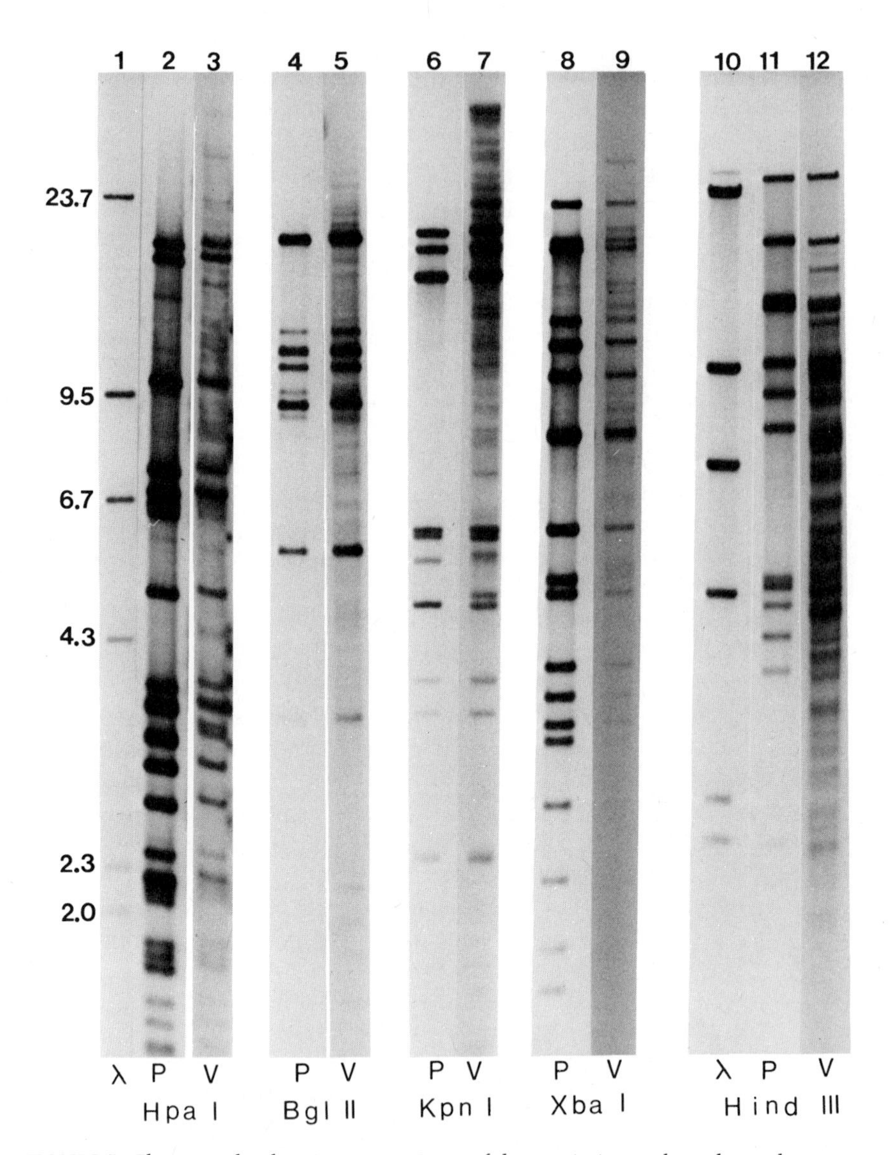

FIGURE 7. Fluorographs showing comparisons of the restriction-endonuclease cleavage patterns of total VZV (Oka) DNAs. Total tritium-labeled VZV (Oka) [parental virus (P)] DNA at passage 16 and VZV (Oka) [vaccine virus (V)] DNA at passage about 32 were prepared as described (Ecker and Hyman, 1981b). The VZV DNAs were cleaved with restriction enzymes *Hpa*I (2, 3), *Bgl*II (4, 5), *Kpn*I (6, 7), *Xba*I (8, 9), and *Hin*dIII (11, 12) and subjected to electrophoresis in agarose gels. The gels were dried, and fluorography was performed. Lanes (1) and (10) contained *Hin*dIII-cleaved λ DNA as standards. Sizes (kb) are given at left.

maps are presented in Fig. 8. The most important point to derive from Fig. 8 is that VZV DNA exists as a population of two isomers. The short unique (U_s) sequence region is found in two orientations with respect to the long unique (U_L) sequence region.

The size of VZV DNA has been measured by sucrose-gradient sedimentation and by electron microscopy. Figure 9 presents the neutral and alkaline sedimentation patterns of VZV DNA along with those of coliphage T_4 DNA as a standard. VZV DNA sedimented heterogeneously on an alkaline sucrose gradient (Fig. 9B). This unusual property is commonly observed for herpesvirus DNAs and reflects the presence of interruptions in the phosphodiester backbone of herpesvirus DNAs (Kieff *et al.*, 1971; Wilkie, 1973; Wilkie *et al.*, 1975; Hyman *et al.*, 1977; Roizman, 1979; Ecker and Hyman, 1981a). Dumas *et al.* (1980) measured the molecular length of VZV DNA relative to phage PM2 DNA using electron microscopy and calculated the molecular size of VZV DNA to be 80 megadaltons. This value is somewhat lower than earlier measurements based on inherently less accurate sucrose-gradient sedimentation data (Iltis *et al.*, 1977; Rapp *et al.*, 1977). An independent calculation of the molecular weight of VZV DNA can be derived from the sizes of the restriction-enzyme cleavage fragments (Table I) and the cleavage maps (Fig.

TABLE I. Molecular Sizes of Restriction-Enzyme Cleavage Fragments of VZV DNA

Fragment designation	Molecular mass (megadaltons)[a]	
	*Eco*RI	*Hin*dIII
A	10.8	17.0
B	9.4	11.0
C	9.0	8.6
D	8.3	8.0
E	8.25	6.2
F	7.7	5.5
G	5.8	4.6
H	5.7	3.0
I	5.3	2.95
J	5.25	2.8
K	4.9	2.5
L	4.6	2.2
M	3.4	1.25
N	3.0	0.9
O	2.6	0.7
P	1.2	0.65
Q[b]	0.6	0.4

[a] Derived from data published by Ecker and Hyman (1982). The nomenclature supersedes that used by Richards *et al.* (1979).
[b] Fragments less than or equal to 0.3 megadaltons have not been studied. In particular, this category includes the left molecular end of the *Eco*RI map (Fig. 8). The left end is composed of one or more small fragments.

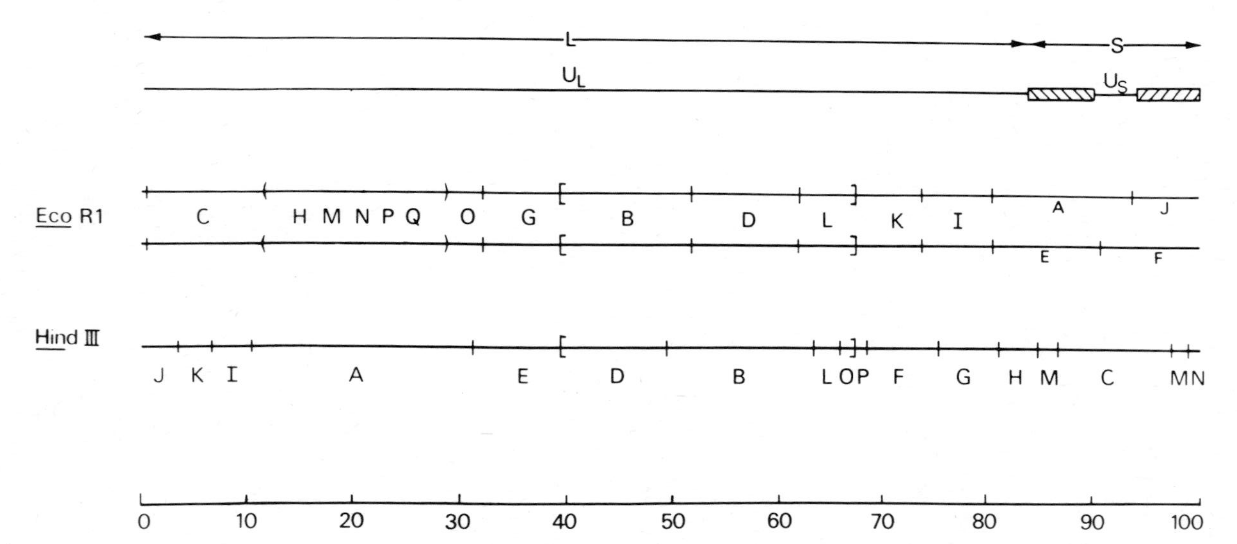

FIGURE 8. Restriction-enzyme cleavage maps of VZV DNA. VZV DNA is composed of two covalently linked segments, L and S, each of which contains unique sequences (U_L and U_S). The U_S sequence is bounded by inverted repeats (hatched boxes). The *Eco*RI (third and fourth lines) and *Hin*dIII (fifth line) maps are derived from published data (Ecker and Hyman, 1982). The current nomenclature (Ecker and Hyman, 1982) supersedes that of Richards *et al.* (1979). Fragments of molecular weight 0.4 md or less have not been positioned on the maps. The parentheses enclose fragments for which the relative order is not known. The brackets enclose fragments for which the relative order is known but the overall orientation has not yet been determined.

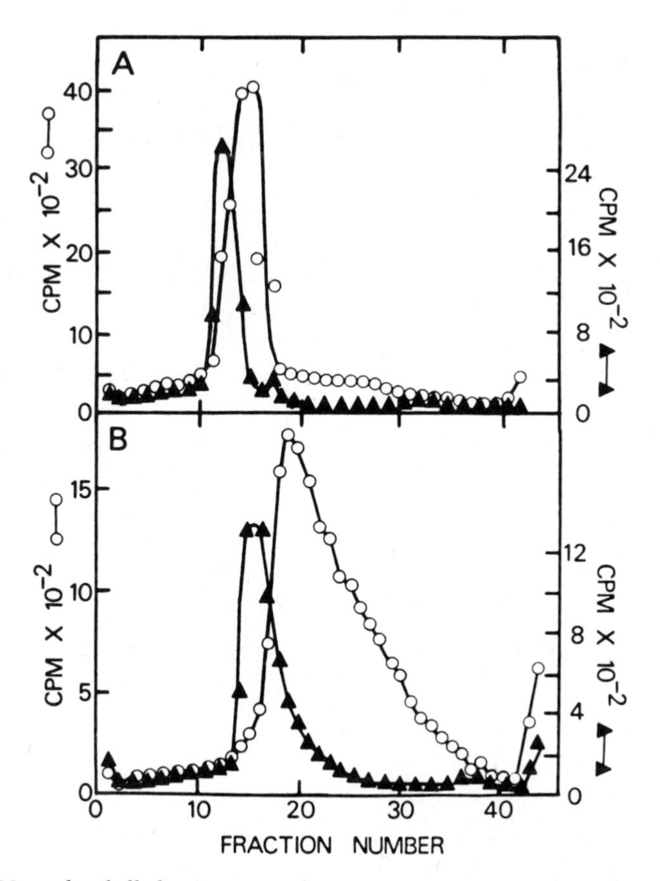

FIGURE 9. Neutral and alkaline sucrose sedimentation of VZV DNA. ^{3}H-labeled VZV (Batson) DNA, a herpes zoster isolate, was mixed with ^{14}C-labeled T_4 DNA. The mixture was sedimented in a neutral or alkaline sucrose gradient (Rapp *et al.*, 1977). Sedimentation is from right to left. (A) Neutral sucrose; (B) alkaline sucrose. (▲) ^{14}C-labeled T_4 DNA; (○) ^{3}H-labeled VZV DNA. Reprinted from Rapp *et al.* (1977), with the permission of S. Karger AG, Basel.

8). These calculations yield a molecular size of 80 ± 3 megadaltons, in agreement with the value published by Dumas *et al.* (1980).

V. CONCLUSIONS

Molecular studies of VZV have only recently been published. It is clear that work on VZV proteins and DNA is just beginning. The difficulties in growing VZV in cell culture have undoubtedly contributed to making VZV the least-studied human herpesvirus. The use of recombinant DNA techniques will partially overcome the difficulties. The highly cell-associated nature of VZV makes an unambiguous distinction between host-cell protein and VZV protein tendentious.

A clear and present conclusion from the preceding discussion is that a major challenge in current VZV research is biological in nature: to find the means to produce stable, high-titer, cell-free VZV. If that challenge can be met, the ambiguities in the interpretation of the molecular data concerning VZV should be obviated.

ACKNOWLEDGMENTS. I am grateful for the gifted collaboration of Mr. Joseph R. Ecker and Drs. Jeffrey P. Iltis, John E. Oakes, and James C. Richards. I thank Dr. Fred Rapp for his continuing interest and support. This work was supported by Grants CA 16498, CA 27503, and CA 18450 awarded by the National Cancer Institute and by Faculty Research Award FRA-158 from the American Cancer Society.

REFERENCES

Asano, Y., and Takahashi, M., 1979, Studies on the polypeptides of varicella–zoster (V-Z) virus. I. Detection of varicella–zoster virus polypeptides in infected cells, *Biken J.* **22**:81.

Aswell, J., and Gentry, G., 1977, Cell-dependent antiherpesviral activity of 5-methylarabinosylcytosine, an intracellular ara-T donor, *Ann. N.Y. Acad. Sci.* **284**:342.

Ben-Porat, T., Demarchi, J.M., and Kaplan, A., 1974, Characterization of defective interfering viral particles present in a population of pseudorabies virions, *Virology* **61**:29.

Bronson, D.L., Dreesman, G.R., Biswal, N., and Benyesh-Melnick, M., 1973, Defective virions of herpes simplex viruses, *Intervirology* **1**:141.

Bruusgaard, E., 1932, The mutual relation between zoster and varicella, *Br. J. Dermatol. Syph.* **44**:1.

Cambell, D., Kemp, M., Perdue, M., Randall, C., and Gentry, G., 1976, Equine herpesvirus *in vivo*: Cyclic production of a DNA density variant with repetitive sequence, *Virology* **69**:737.

Cheng, Y.-C., Tsou, T., Hackstadt, T., and Mallavia, L., 1979, Induction of thymidine kinase and DNase in varicella–zoster virus-infected cells and kinetic properties of the virus-induced thymidine kinase, *J. Virol.* **31**:172.

Delius, H., and Bornkamm, G., 1978, Heterogeneity of Epstein–Barr virus. III. Comparison of a transforming and a non-transforming virus by partial denaturation mapping of their DNAs, *J. Virol.* **27**:81.

Delius, H., and Clements, J.B., 1976, A partial denaturation map of herpes simplex virus type 1 DNA: Evidence for inversions of the unique DNA regions, *J. Gen. Virol.* **33**:125.

Dobersen, M., Jerkofsky, M., and Greer, S., 1976, Enzymatic basis for the selective inhibition of varicella–zoster virus by 5-halogenated analogues of deoxycytidine, *J. Virol.* **20**:478.

Dumas, A.M., Geelen, J.L.M., Maris, W., and van der Noordaa, J., 1980, Infectivity and molecular weight of varicella–zoster virus DNA, *J. Gen. Virol.* **47**:233.

Ecker, J.R., and Hyman, R.W., 1981a, Analysis of interruptions in the phosphodiester backbone of herpes simplex virus DNA, *Virology* **110**:213.

Ecker, J.R., and Hyman, R.W., 1981b, Varicella–zoster virus vaccine DNA differs from the parental virus DNA, *J. Virol.* **40**:314.

Ecker, J.R., and Hyman, R.W., 1982, Varicella–zoster virus DNA exists as two isomers *Proc. Natl. Acad. Sci. U.S.A.* **79**:156.

Feldman, S., Hughes, W.T., and Daniel, C.B., 1975, Varicella in children with cancer: Seventy-seven cases, *Pediatrics* **56**:388.

Fioretti, A., Iwasaki, Y., Furukawa, T., and Plotkin, S., 1973, The growth of varicella–zoster virus in guinea pig embryo cells, *Proc. Soc. Exp. Biol. Med.* **44**:340.

Frenkel, N.R., Jacob, R.J., Honess, R.W., Hayward, G.S., Locker, H., and Roizman, B., 1975, The anatomy of herpes simplex virus DNA. III. Characterization of defective DNA molecules and biological properties of virus populations containing them, *J. Virol.* **16:**153.

Gilden, D., Wroblewska, Z., Kindt, V., Warren, K., and Wolinsky, J., 1978, Varicella–zoster virus infection of human brain cells and ganglion cells in tissue culture, *Arch. Virol.* **56:**105.

Given, D., and Kieff, E., 1978, DNA of Epstein–Barr virus. IV. Linkage map of restriction enzyme fragments of the B95-8 and W91 strains of Epstein–Barr virus, *J. Virol.* **28:**524.

Graham, B., Bengali, Z., and Vande Woude, G., 1978, Physical map of the origin of defective DNA in herpes simplex virus type 1 DNA, *J. Virol.* **25:**878.

Grose, C., 1980, The synthesis of glycoprotein in human melanoma cells infected with varicella-zoster virus, *Virology* **101:**1.

Grose, C., Perrotta, D., Brunell, P., and Smith, G., 1979, Cell-free varicella–zoster virus in cultured human melanoma cells, *J. Gen. Virol.* **43:**15.

Hackstadt, T., and Mallavia, L., 1978, Deoxypyrimidine nucleoside metabolism in varicella–zoster virus-infected cells, *J. Virol.* **25:**510.

Harbour, D., and Caunt, A., 1975, Infection of guinea pig embryo cells with varicella–zoster virus, *Arch. Virol.* **49:**39.

Hayward, G.S., Jacob, R.J., Wadsworth, S.C., and Roizman, B., 1975, Anatomy of herpes simplex virus DNA: Evidence for four populations of molecules that differ in the relative orientations of their long and short components, *Proc. Natl. Acad. Sci. U.S.A.* **72:**4243.

Hyman, R.W., Oakes, J.E., and Kudler, L., 1977, *In vitro* repair of the pre-existing nicks and gaps in herpes simplex virus DNA, *Virology* **76:**286.

Iltis, J.P., Oakes, J.E., Hyman, R.W., and Rapp, F., 1977, Comparison of the DNAs of varicella–zoster viruses isolated from clinical cases of varicella and herpes zoster, *Virology* **82:**345.

Iltis, J., Lin, T.-S., Prusoff, W., and Rapp, F., 1979, Effect of 5-iodo-5′-amino-2′,5′-dideoxyuridine on varicella–zoster virus *in vitro, Antimicrob. Agents Chemother.* **16:**92.

Ito, M., and Barron, A., 1973, Surface antigens produced by herpesviruses: Varicella–zoster virus, *Infect. Immun.* **8:**48.

Kieff, E.D., Bachenheimer, S.L., and Roizman, B., 1971, Size, composition, and structure of the deoxyribonucleic acid of herpes simplex virus subtypes 1 and 2, *J. Virol.* **8:**125.

Locker, H., and Frenkel, N., 1979, Structure and origin of defective genomes contained in serially passaged herpes simplex virus type 1 (Justin), *J. Virol.* **29:**1065.

Ludwig, H., Haines, H.G., Biswal, N., and Benyesh-Melnick, M., 1972, The characterization of varicella–zoster virus DNA *J. Gen. Virol.* **14:**111.

Mar, E.-C., Huang, Y.-S., and Huang, E.-S., 1978, Purification and characterization of varicella–zoster virus-induced DNA polymerase, *J. Virol.* **26:**249.

May, D., Miller, R., and Rapp, F., 1977, The effect of phosphonoacetic acid on the *in vitro* replication of varicella–zoster virus, *Intervirology* **8:**83.

Miller, R., and Rapp, F., 1977, Varicella–zoster virus-induced DNA polymerase, *J. Gen. Virol.* **36:**515.

Miller, R., Iltis, J., and Rapp, F., 1977, Differential effect of arabinofuranosylthymine on the replication of human herpesviruses, *J. Virol.* **23:**679.

Oakes, J.E., Iltis, J. P., Hyman, R.W., and Rapp, F., 1977, Analyses by restriction enzyme cleavage of human varicella–zoster virus DNAs, *Virology* **82:**353.

Ogino, T., Otsuka, T., and Takahashi, M., 1977, Induction of deoxypyrimidine kinase activity in human embryonic lung cells infected with varicella–zoster virus, *J. Virol.* **21:**1232.

Preblud, S., and D'Angelo, L., 1979, Chicken pox in the United States, 1972–1977, *J. Infect. Dis.* **140:**257.

Rapp, F., Iltis, J., Oakes, J., and Hyman, R.W., 1977, A novel approach to study the DNA of herpes zoster virus, *Intervirology* **8:**272.

Richards, J., Hyman, R.W., and Rapp, F., 1979, Analysis of the DNAs from seven varicella–zoster virus isolates, *J. Virol.* **32:**812.

Roizman, B., 1979, The structure and isomerization of herpes simplex virus genomes, *Cell* **16:**481.

Rymo, L., and Forsblom, S., 1978, Cleavage of Epstein–Barr virus DNA by restriction endonucleases *Eco*RI, *Hin*dIII and *Bam*I, *Nucleic Acids Res.* **5:**1387.

Schildkraut, C.L., Marmur, J., and Doty, P., 1962, Determination of the base composition of deoxyribonucleic acid from its buoyant density in CsCl, *J. Mol. Biol.* **4:**430.

Skare, J., and Summers, W., 1977, Structure and function of herpesvirus genomes. II. *Eco*RI, *Xba*I and *Hin*dIII endonuclease cleavage sites on herpes simplex virus type 1 DNA, *Virology* **76:**581.

Skare, J., Summers, W., and Summers, W., 1975, Structure and function of herpesvirus genomes. I. Comparison of five HSV-1 and two HSV-2 strains by cleavage of their DNA with *Eco*RI restriction endonuclease, *J. Virol.* **15:**726.

Stinski, M., Mocarski, E., and Thomsen, D., 1979, DNA of human cytomegalovirus: Size heterogeneity and defectiveness resulting from serial undiluted passage, *J. Virol* **31:**231.

Sugden, B., Summers, W., and Klein, G., 1976, Nucleic acid renaturation and restriction endonuclease cleavage analyses show that the DNAs of a transforming and non-transforming strain of Epstein–Barr virus share approximately 90% of their nucleotide sequences, *J. Virol.* **18:**765.

Takahashi, M., Okuno, Y., Otsuka, T., Osame, J., Takamizawa, A., Sasada, T., and Kubo, T., 1975, Development of a live attenuated varicella vaccine, *Biken J.* **18:**25.

Taylor-Robinson, D., and Caunt, A.E., 1972, The varicella virus, *Virology Monographs*, Vol. 12, Springer-Verlag, Vienna.

Wagner, M., Skare, J., and Summers, W.C., 1975, Analysis of DNA of defective herpes simplex virus type 1 by restriction endonuclease cleavage and nucleic acid hybridization, *Cold Spring Harbor Symp. Quant. Biol.* **39:**683.

Weller, T., Witton, H., and Bell, E., 1958, The etiologic agents of varicella and herpes zoster: Isolation, propagation and cultural characteristics *in vitro*, *J. Exp. Med.* **108:**843.

Wilkie, N.M., 1973, The synthesis and substructure of herpesvirus DNA: The distribution of alkali-labile single strand interruptions in HSV-1 DNA, *J. Gen. Virol.* **21:**453.

Wilkie, N.M., and Cortini, R., 1976, Sequence arrangement in herpes simplex virus type 1 DNA: Identification of terminal fragments in restriction endonuclease digests and evidence for inversions in redundant and unique sequences, *J. Virol.* **20:**211.

Wilkie, N.M., Clements, J.B., MacNab, J.C.M., and Subak-Sharpe, J.H., 1975, The structure and biological properties of herpes simplex virus DNA, *Cold Spring Harbor Symp. Quant. Biol.* **39:**657.

Wolff, M., 1978, The proteins of varicella–zoster virus, *Med. Microbiol. Immunol.* **166:**21.

CHAPTER 4

Bovine Herpesviruses

HANNS LUDWIG

I. INTRODUCTION

The interest in bovine herpesviruses stems from their considerable economic importance, rather than from their contribution to the biology or molecular biology of Herpesviridae. Only recently has there been progress in studies on the basic properties of their genomes, antigens, proteins and on the immune response against members of this group.

Since previous reviews focused more or less on clinical aspects of herpesviruses from Bovidae (Kokles, 1967; Plowright, 1968; McKercher, 1973; Cilli and Castrucci, 1976; Gibbs and Rweyemamu, 1977a,b; Straub, 1978), this chapter will summarize the newer data on the molecular biology of these viruses and on the mechanisms by which they cause disease.

II. HISTORY

The exanthema coitale vesiculosum ("Bläschenausschlag") of cattle was well known in central Europe toward the end of the 19th century. Publications from Germany and other European countries assured "Bläschenausschlag" a prominent place in textbooks (Hutyra-Marek, 1910). The viral etiology of this disease was proposed in 1913 (Zwick and Gminder, 1913), and the filterability of the agent was proven some years later (Reisinger and Reimann, 1928). The virological investigations of Zwick and co-workers suggested that the virus of "Bläschenausschlag" was not the cause of abortions endemic in the cattle of small farms in villages of Hessia, since filter-sterilized material from the vagina had no abortigenic activity (Witte, 1933). Later reports on similar clinical pictures from sev-

HANNS LUDWIG • Institut für Virologie, Freie Universität Berlin, West Berlin, Germany.

eral other countries, including the United States, have contributed to the understanding of the disease caused by the virus of infectious pustular vulvovaginitis (IPV) (Kendrick *et al.*, 1958; McKercher, 1973).

The close relationship of IPV to the virus of infectious bovine rhinotracheitis (IBR) was demonstrated in animal experiments (Gillespie *et al.*, 1959). IBR virus infections were responsible for severe respiratory diseases in the cattle of feedlots in the United States (McKercher, 1973), and the causative agent named after the typical IBR (Miller, 1955; McKercher *et al.*, 1955) was isolated in the late 1950s (Madin *et al.*, 1956; Gründer *et al.*, 1960).

The biophysical characterization and pictures of its morphology were reported by Tousimis *et al.* (1958). The typical structures of intra- and extracellular IPV virus (Kubin and Klima, 1960) or IBR virus (Liess *et al.*, 1960) were demonstrated by electron microscopy and compared to that of herpes B virus (Reissig and Melnick, 1955). Thus, the IBR-IPV virus was the first bovine herpesvirus meeting morphological criteria of the herpesvirus group (Wildy *et al.*, 1960; Armstrong *et al.*, 1961).

The virus isolated from genital or respiratory organs was usually named IBR-IPV virus, and no fundamental differences between the genital and respiratory isolates emerged from biological experiments. Subsequently, viruses of the same type were isolated from cattle with conjunctivitis and even from aborted bovine fetuses (McKercher *et al.*, 1959; McKercher and Wada, 1964). The first report stating that this virus is also the causative agent of meningoencephalitis was by French (1962a,b), and it is now of interest to read that an Australian vaginal strain was neither encephalitic nor abortigenic (Snowdon, 1965).

Descriptions of isolates of IBR-IPV virus reported from other countries and their relationships to the disease were dealt with in the overview of Gibbs and Rweyemamu (1977a) and are of historical interest. It should be mentioned, however, that a restriction-endonuclease map of this virus has been determined only recently by Skare *et al.* (personal communication) and that the use of DNA restriction-endonuclease "fingerprinting" of the DNA—successfully practiced in herpes simplex viruses (Buchman *et al.*, 1978, 1979)—made it possible to differentiate IBR-IPV virus strains (Engels *et al.*, 1981a) and enabled us to answer pathogenetic questions concerning the role of different IBR-IPV virus strains from various clinical entities (Pauli *et al.*, 1981; Ludwig and Darai, unpublished).

The second well-established virus is the bovine herpes mammillitis (BHM) virus isolated from cases of mammillitis in Africa (Huygelen *et al.*, 1960a,b). It was first characterized to some extent by Martin *et al.* (1964, 1966) and is the cause of well-known lesions of the udder and teats (Hare, 1925). The virus was shown to cross-react with the prototype Allerton virus (Alexander *et al.*, 1957). Isolations and characterization of strains of BHM virus have been reported from several countries in Europe and North America (Deas and Johnston, 1966; Rweyemamu *et al.*, 1966; Yedloutschnig *et al.*, 1970; Castrucci *et al.*, 1972; Dilovsky *et al.*, 1974).

The finding of the immunological and genetic relationship between BHM virus and herpes simplex virus (Sterz *et al.*, 1973–1974) has stimulated further investigations on antigenic cross-reactivities between viruses of the herpes group (Pauli and Ludwig, 1977; Ludwig *et al.*, 1978a; Killington *et al.*, 1978; Norrild *et al.*, 1978b) and stimulated interest in the properties of its genome (Buchman and Roizman, 1978a,b).

Another herpesvirus associated with a characteristic clinicopathological disease of the wildebeest is the agent of African malignant catarrhal fever (MCF) (Plowright, 1968). The clinical picture of this disease is known worldwide, although herpesviruses could not always be defined as the etiological agents. The first reports on "bösartiges Katarrhalfieber des Rindes" date back to Götze and Liess (1929), who identified this disease in European countries (Hutyra-Marek, 1922) and compared it with the South African "snotsiekte" (Mettam, 1923). Clinical similarities and transmission to the rabbit raised the possibility that the disease is identical to Borna disease in cattle (Götze and Liess, 1929); this hypothesis has been ruled out by analysis of the agents (Ludwig and Becht, 1977). Numerous reports by Plowright and co-workers (Plowright *et al.*, 1960, 1975; Plowright, 1968; Patel and Edington, 1980), as well as others (Storz *et al.*, 1976; Ohshima *et al.*, 1977; Liggitt *et al.*, 1978; Straver.and van Bekkum, 1979), have contributed to the understanding of this highly interesting disease associated with lymphoproliferative changes in cattle. Molecular biological and immunological studies are required to clarify the role of a herpesvirus in this disease. By restriction-enzyme analysis of DNA from infected tissue cultures, it was recently possible to distinguish the virus of African MCF from isolates obtained from rather similar clinical entities in Europe and the United States (Ludwig and Darai, unpublished).

Numerous other herpesviral isolates from cattle and buffalo have been obtained from diseased as well as from apparently healthy animals (Bartha *et al.*, 1966; Liebermann *et al.*, 1967; Storz, 1968; Luther *et al.*, 1971; Mohanty *et al.*, 1971; Parks and Kendrick, 1973; Belák and Pálfi, 1974; for description of additional isolates, see Table 3 of Gibbs and Rweyemamu, 1977b). Some of the herpesviruses that were activated in normal tissue culture could be classified as IBR-IPV virus (Ludwig and Storz, 1973; Storz *et al.*, 1980) and specifically as IBR-like viruses (Pauli *et al.*, 1981). A few others from Europe and the United States comprise a new group that can be differentiated on the basis of the patterns of the fragments generated by cleavage of their DNAs with restriction endonucleases (see Section V.C).

Another herpesvirus (BHV-5), isolated from cases of "jaagsiekte" in sheep (Mackay, 1969; Malmquist *et al.*, 1972; De Villiers *et al.*, 1975; Verwoerd *et al.*, 1979), has also been recently characterized (De Villiers, 1979).

The Aujeszky's disease (pseudorabies) virus is actually also a common bovine herpesvirus and in fact was isolated first from fatally diseased cows and dogs (and not from pigs) by Aujeszky (1902). This virus will be mentioned in this chapter only for comparative purposes, since its natural host appears to be the pig.

III. CLASSIFICATION (TABLE I)

A first proposal to group IBR-IPV and BHM virus as bovid* herpesvirus (BHV) types 1 and 2 was generally accepted (Roizman *et al.*, 1973). The data collected in this review, especially the restriction-enzyme analysis of the DNAs (see also Sections V.A–E), justify separation of malignant catarrhal fever as group 3 and the forming of group 4 composed of various representative bovine herpesvirus isolates. The "jaagsiekte" virus is tentatively named group 5 virus and the goat herpesvirus group 6 virus. The reference strains for this classification are given in Tables I and IV. Only viruses for which serological data are available (see Section VI) and which cluster on the basis of restriction-endonuclease fingerprinting (see Sections V.A–E) were included in Table I. The abbreviations BHV-1 through BHV-6 will be used throughout this review.

Recent efforts to classify members of Herpesviridae according to the criteria proposed by the Herpesvirus Study Group (see Matthews, 1979; Roizman, 1982) would place BHV-1, -2, and -6 with the subfamily of Alphaherpesvirinae, whereas BHV-3, showing special affinity to the lymphoreticular system, shares properties of the Gammaherpesvirinae. The BHV-4 strains with their narrow host range and slow replicative cycle most probably belong to the Betaherpesvirinae. The restriction-endonuclease maps of BHV-3, -4, -5, and -6 DNAs are not yet available.

IV. MORPHOLOGY AND MORPHOGENESIS

The morphology of bovine herpesviruses is largely comparable to that of other members of this family (Roizman, 1978) like herpes simplex virus (HSV) (Watson *et al.*, 1964; Darlington and Moss, 1969), varicella–zoster virus (Almeida *et al.*, 1962), pseudorabies virus (Reissig and Kaplan, 1962), equine herpesviruses (Plummer and Waterson, 1963; Abodeely *et al.*, 1970; Ludwig *et al.*, 1971; O'Callaghan and Randall, 1976), and others. The viruses possess a regular icosahedral nucleocapsid with equilateral, triangular facets and 5 capsomeres on each edge. The axial symmetry is 5 : 3 : 2 as based on the presence of 5 and 6 coordinated capsomeres. From electron micrographs of negatively stained preparations of bovid herpesvirus 1 (BHV-1) (Watrach and Bahnemann, 1966; Bocciarelli *et al.*, 1966), the total number of capsomeres was calculated to be 162, as in the HSV capsids (Wildy *et al.*, 1960). By different techniques, the capsid of BHV-

* The term "bovid" was used to keep the scheme open for isolates from all Bovidae.

TABLE I. Classification of Bovine Herpesviruses

Virus groups[a]	Clinical entities and synonyms[b]	Natural host	Criteria[c]
BHV-1 (bovid herpesvirus 1)	"Bläschenausschlag"; exanthema coitale vesiculosum; infectious pustular vulvovaginitis (IPV)–infectious bovine rhinotracheitis (IBR)	Cattle	Serology[f]; DNA pattern
BHV-2 (bovid herpesvirus 2)	Allerton; bovine mammillitis; bovine herpes mammillitis (BHM); ps. lumpy skin disease	Cattle	Serology[g]; DNA pattern
BHV-3 (bovid herpesvirus 3)	"Bösartiges Katarrhalfieber"[d] (BKF) (African) malignant catarrhal fever (MCF); "snotsiekte"	Wildebeest	Serology[h]; DNA pattern
BHV-4 (bovid herpesvirus 4)	Associated with various clinical forms of disease (various strain abbreviations)	Cattle	Serology[i]; DNA pattern
BHV-5[e] (bovid herpesvirus 5)	Sheep pulmonary adenomatosis; "jaagsiekte"[l]; herpesvirus ovis infection	Sheep	Serology[j]; DNA analysis
BHV-6 (bovid herpesvirus 6)	Caprine herpesvirus infection	Goat	Serology[k]; DNA pattern

[a] These different herpesvirus groups, except BHV-1 and -6, do not cross-react, as shown by neutralization.
[b] The established names, together with abbreviations commonly used, are given.
[c] The serological cross-reactivity of isolates or data from restriction-endonuclease analysis of the DNA together with historical considerations are the basis for the above-proposed nomenclature.
[d] No herpesvirus has been unequivocally attributed to this term in European countries.
[e] This virus had already been proposed as bovid herpesvirus 4 (Roizman *et al.*, 1973), when the above various herpesvirus isolates from cattle were not yet characterized.
[f–k] Reference strains: [f]Strain LA (Madin *et al.*, 1956); [g]Strain TVA (Rweyemamu and Johnson, 1969); [h]Strain WC 11 (Plowright *et al.*, 1975); [i]Strain Movar 33/63 (Bartha *et al.*, 1966); [j]Strain JS-3 (De Villiers, 1979); [k]Strain E/CH (Mettler *et a;.*, 1965).
[l] The etiological relationship between BHV-5 and "jaagsiekte" remains uncertain (Verwoerd *et al.*, 1980).

1 was found to range from 95 to 110 nm; the capsomeres appear to be hollow cones with an axial hole of about 35 Å and measure 120 Å in length and 115 Å in width. Similarly to other herpesviruses (Roizman and Furlong, 1974), the nucleocapsid is surrounded by a tegument visible as an electron-dense zone (Valiček and Šmid, 1976).

The envelope consists of a bilayer with minute projections on its surface (Fig. 1). The virion envelope appears to be rather pleomorphic, especially after negative staining. The virion ranges in size up to 200 nm, sometimes incorporating two or more nucleocapsids (Fig. 1a). These features were determined in studies on BHV-1 strains originating from the various clinical entities or activated in cell cultures of normal tissues

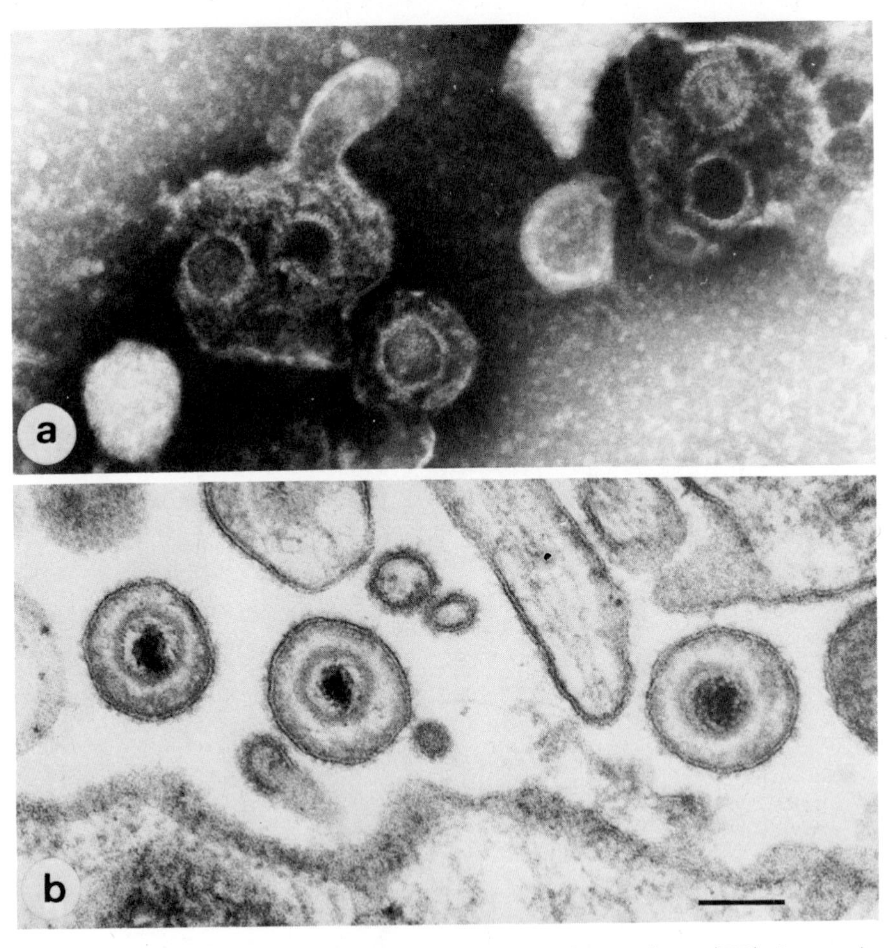

FIGURE 1. Fine structure of BHV-1 [infectious bovine rhinotracheitis (IBR) virus strain V101] after negative staining with phosphotungstic acid (a) and in thin sections (b). Note the small projections from the surface of the envelope. The inner layer of the envelope appears to be thickened and directly associated with the tegument material. A spindlelike structure comprising the DNA core is visible inside the hexagonal capsid. ×120,000. Scale bar: 100 nm. Courtesy of Hans Gelderblom.

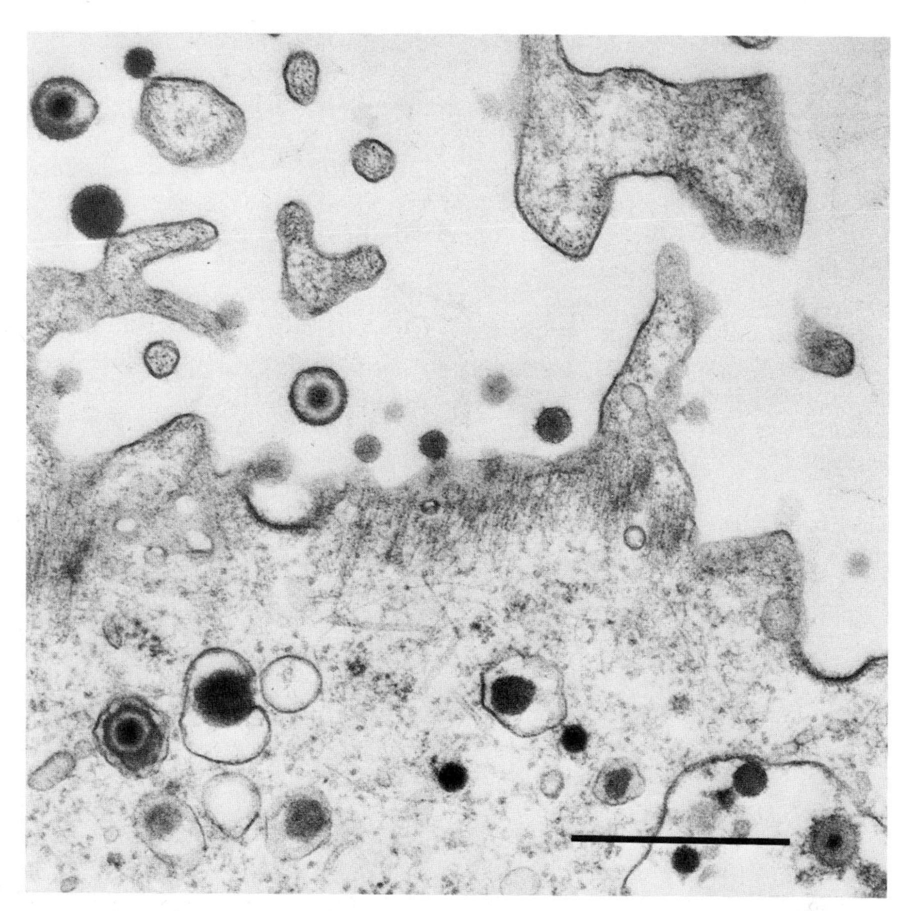

FIGURE 2. Thin section of a BHV-2 (strain TVA)-infected cell. The micrograph shows virions of various forms of dense bodies within vesicles and enveloped by membranes studded with surface projections. ×30,000. Scale bar: 1 μm. Courtesy of Hans Gelderblom.

(Armstrong *et al.*, 1961; Schulze *et al.*, 1964; Cruickshank and Berry, 1965; Watrach and Bahnemann, 1966; Bocciarelli *et al.*, 1966; Ludwig and Storz, 1973; Talens and Zee, 1976).

Reports on BHV-2 morphology indicate that the various isolates such as the Italian strain, the Minnesota isolate, or the TVA strain cannot be differentiated electron-microscopically from each other (Castrucci *et al.*, 1972; Breese and Dardiri, 1972; Sterz, 1973) or from the first isolate establishing this group (Martin *et al.*, 1966) (Fig. 2).

It is noteworthy that BHV-2-infected cell cultures produce low amounts of the typical enveloped virions, but significant amounts of non-enveloped structures similar to the cytomegalovirus dense bodies. This was also observed in cells infected with other herpesviruses (Fong *et al.*, 1973; Gershon *et al.*, 1973; Nii *et al.*, 1973). Immunoferritin tagging of BHV-2 specific antibodies confirmed the BHV-2 specificity of these viruslike particles detected with the TVA strain (Gelderblom and Ludwig, unpublished). Our observations are thus in accord with earlier reports by Šmid *et al.* (1971), who mentioned similar undefined, membrane-bound particles in cell cultures infected with another strain. From their electron-dense appearance, it might be deduced that these particles have selectively incorporated tegumentlike material.

There is little information concerning BHV-3 isolates from clinical cases of malignant catarrhal fever (Storz, 1968) and with respect to morphology (Patel and Edington, 1980) and morphogenesis (Armstrong *et al.*, 1964). Our studies on the attenuated, avirulent WC 11 strain (Plowright *et al.*, 1965) revealed that sufficient amounts of physical particles with typical herpesvirus morphology are produced (Fig. 3) (Gelderblom and Ludwig, unpublished) to permit further biochemical analysis of viral components (see Sections V.C and VI.).

In contrast to BHV-3, the herpesviruses now designated as BHV-4 have been classified predominantly on the basis of electron-microscopic studies. In addition, the morphogenesis of these viruses in tissue culture has been extensively studied by several laboratories (Bartha *et al.*, 1967 Liebermann *et al.*, 1967; Schulze *et al.*, 1967; Storz, 1968; Mohanty *et al.*, 1971; Mohanty, 1975). Some unusual forms detected in sections of virus-infected tissue-culture cells are shown in Fig. 4a–d. A few of the representative strains—all monitored by electron microscopy—produce enough cell-free virus to allow studies on the DNAs and the restriction-enzyme patterns that are the basis for designating these viruses as BHV-4 (see Sections III and V.C).

Viruses of the BHV-5 (Malmquist *et al.*, 1972; De Villiers *et al.*, 1975) and BHV-6 (Berrios and McKercher, 1975; Mettler *et al.*, 1979) groups have not been studied extensively by electron microscopy. They were identified in thin-tissue sections (see Fig. 32i). The caprine isolate was claimed to range in virion size between 120 and 150 nm, which is slightly smaller than HSV (Berrios and McKercher, 1975).

In several studies dealing with entry and morphogenesis of bovine herpesviruses, special emphasis has been placed on the BHV-1 group (Schulze *et al.*, 1964; Jasty and Chang, 1969; Zee and Talens, 1971, 1972). There are more similarities to, than diversities from, events that take place in the development of HSV. Evidence for the fusion of the viral envelope with the plasma membrane prior to entry of the nucleocapsid was demonstrated by Zee and Talens (1971). After replication of the virus, the assembled nucleocapsids normally acquire their envelopes from the inner lamella of the nuclear membrane, from cytoplasmic membranes, or from the plasma membrane itself (Jasty and Chang, 1971). In our studies using bovine fetal skin cells infected with the LA reference strain (IBR-like), apparent budding from the plasma membrane was not a rare event (Fig. 4a).

Morphological modifications of the envelope or the existence of a second, an inner envelope, have been reported by Valiček and Šmid (1976).

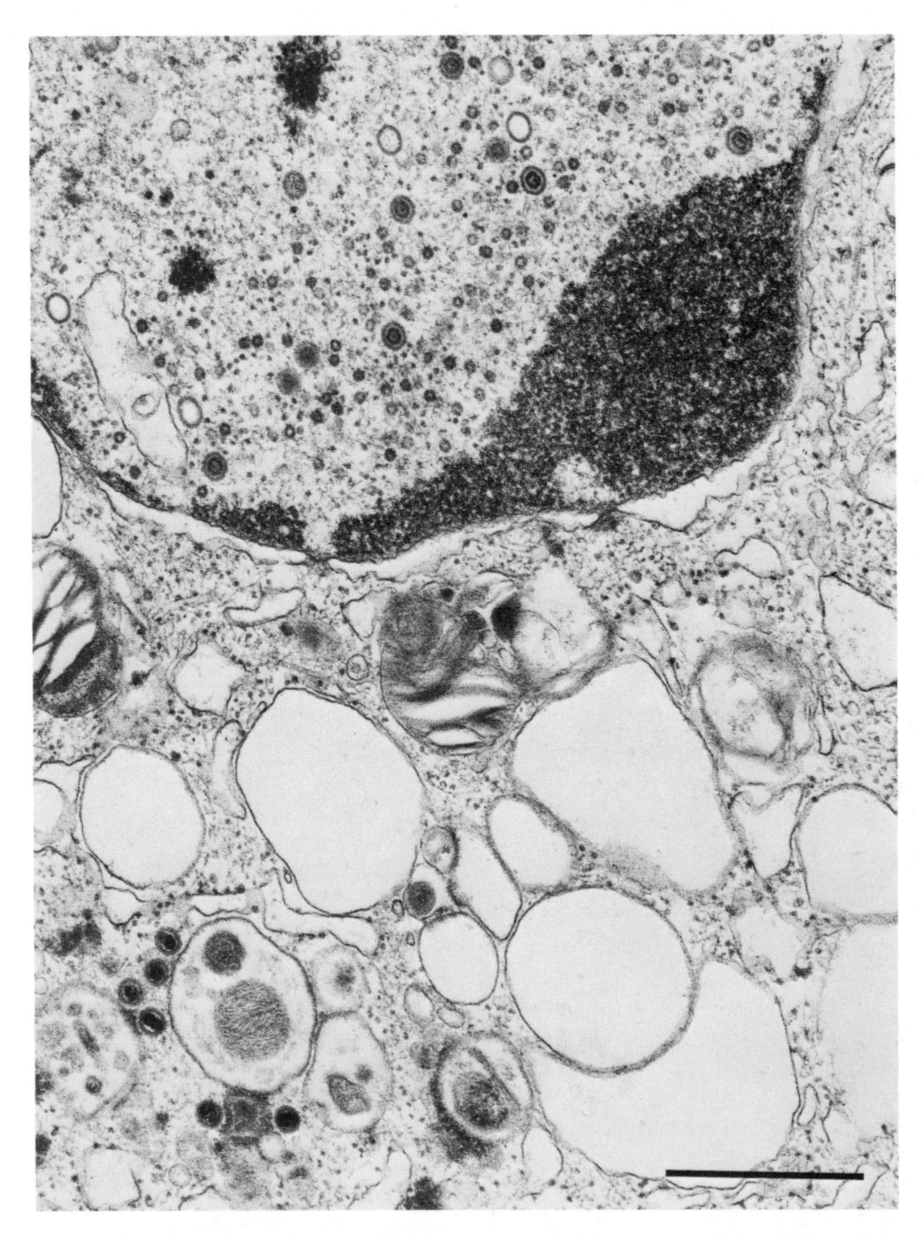

FIGURE 3. Thin section of a BHV-3 (malignant catarrhal fever, strain WC 11)-infected bovine fetal skin cell. Note numerous partially assembled nucleocapsids in the nucleus and the complete capsids—as judged from the typical electron-dense core structure—in the cytoplasm. ×27,000. Scale bar: 1 μ. Courtesy of Hans Gelderblom.

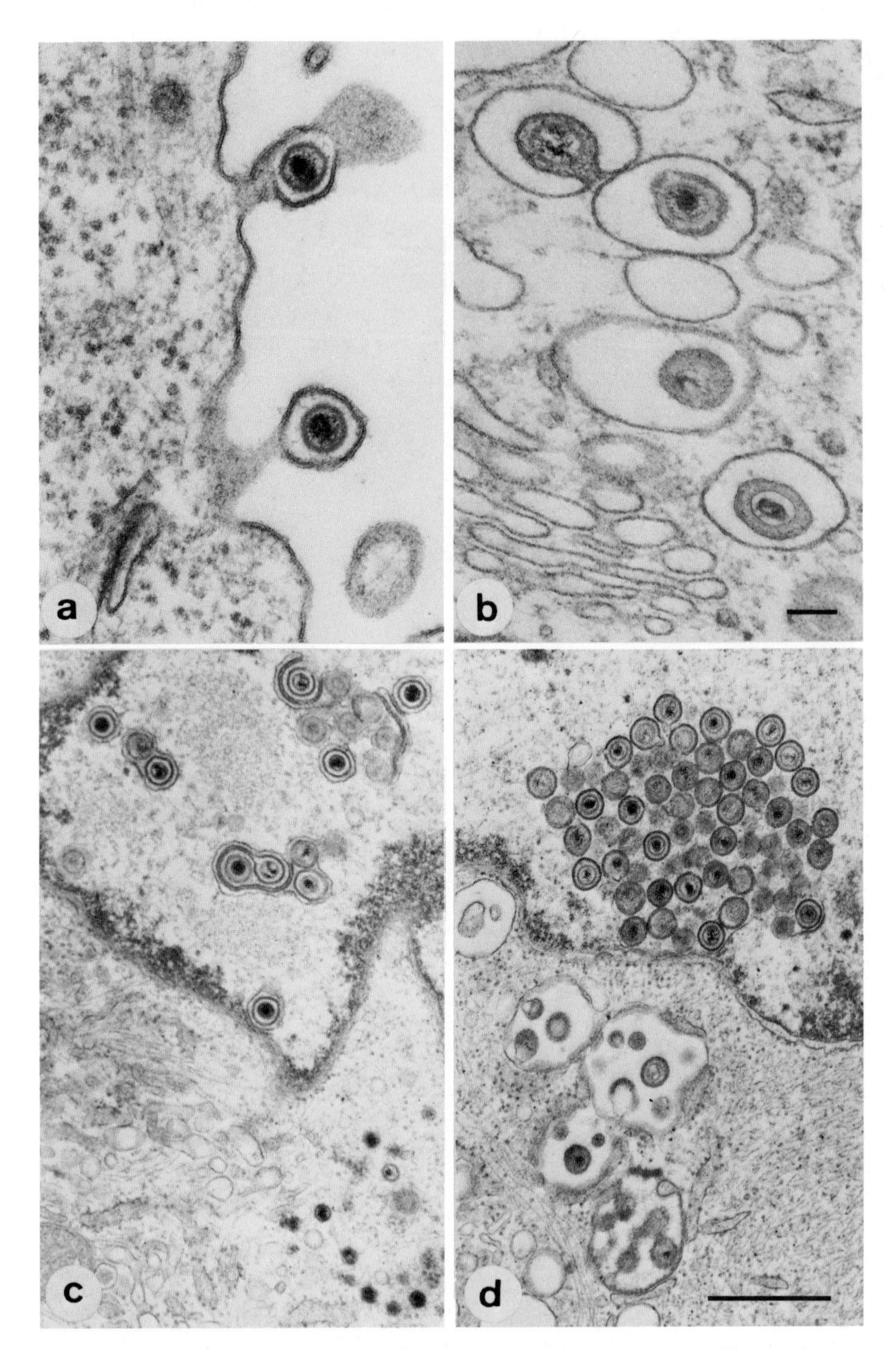

FIGURE 4. Peculiar features of maturation, morphogenesis, and egress of bovine herpesviruses as shown in thin sections of infected bovine fetal skin cells. Cells were fixed between 18 and 30 hr post infection and processed for electron microscopy as described by Geld-

Electron-dense material is often incorporated where envelopment of the capsid takes place by budding into endoplasmic reticulum, whereas particles enveloped in the nuclear membranes do not show much of this material. Such dense material can even bud by itself without a nucleocapsid into endoplasmic reticulum or directly into extracellular space, forming dense bodies. Similar structures have also been described in BHV-2 infected cells (Šmid *et al.*, 1971). Valiček and Šmid (1976) reported that particles in the perinuclear cisterna contained an electron-dense single membrane, while extracellular viruses possess the typical unit membrane with projections. Although the morphology and morphogenesis of the bovine herpesviruses are generally in line with the knowledge collected about other members of Herpesviridae, nothing is known about the localization of functionally important structural components.

V. GENETIC MATERIAL

A. Genome of Bovid Herpesvirus 1

The results of biophysical–chemical characterization of bovid herpesvirus 1 (BHV-1) DNA are summarized in Table II. The DNA has a guanosine plus cytosine (G + C) content of 72 moles%—similar to that of pseudorabies (PsR) virus. Various isolates from different clinical entities, including reference strains from the United States and Europe, did not significantly differ in G + C content, as computed from buoyant densities obtained by analytical ultracentrifugation. The viral DNA can be easily separated from the bulk of cellular DNA with the characteristic shoulder of satellite DNA (Fig. 5A and B) by centrifugation of extracts of infected cell lysates (Ludwig and Storz, 1973). After shearing, the BHV-1 DNA, like sheared herpes simplex virus 1 (HSV-1) DNA, forms a rather homogeneous peak, when centrifuged to equilibrium in an analytical ultracentrifuge (Ludwig, 1972a). This pattern differs significantly from that obtained with equine herpesvirus DNA (Borgen and Ludwig, 1974) or with *Herpesvirus saimiri* DNA (Fleckenstein *et al.*, 1975) because inhomogeneity in base composition of these DNAs results in physical separation of DNA fragments with low and high G + C content.

---

erblom *et al.*, (1974). (a) Two capsids of BHV-1 (strain V101) appear to be enveloped at the plasma membrane. The egress seems to be similar to retrovirus maturation (Gelderblom *et al.*, 1972). (b) One of four BHV-4 (strain ÜT) particles appears to be in the process of budding into the endoplasmic reticulum. ×70,000. Scale bar: 100 nm. (c) Atypical nuclear and cytoplasmic distribution of enveloped and naked BHV-2 (strain TVA) capsids. (d) Aggregate of apparently mature, enveloped particles of BHV-4 (strain Movar 33/63) within the nucleus. These may, however, be contained within an indentation of the cytoplasm into the nucleus. ×18,000. Scale bar: 500 nm. Courtesy of Hans Gelderblom.

TABLE II. Physicochemical Properties of DNAs from Bovine Herpesviruses[a]

Group	Buoyant density in CsCl (g/cm^3)	Melting point (T_M °C) in 0.1 × SSC	G + C (%)	Molecular weight	DNA analysis: Performed? (Strain differences?)	DNA analysis: Physical map?
BHV-1	1.730[f,g]	85.6[f]	71.5[f]	88×10^6 [b,h]	Yes (Yes)[h–j]	Yes[h]
BHV-2	1.723[k,l]	NT	63.5[k,l]	88×10^6 [b,m]	Yes (NT)[m,n]	Yes[m]
BHV-3	NT	NT	—	—	Yes (NT)[n]	—
BHV-4	NT	NT	—	76×10^6 [c,n]	Yes (Yes)[n]	—
BHV-5	1.706[o]	NT	46.5[o]	67×10^6 [d,o]	NT	—
BHV-6	1.729[p]	84.5[p]	70	84×10^6 [e,p]	Yes (NT)[p]	—

[a] (NT) Not tested; (—) not known.
[b] Calculated from the DNA fragments after restriction-endonuclease cleavage.
[c] Estimated from the DNA fragments in reference to known DNA fragments.
[d] Calculated from S_{20} values.
[e] Calculated from electron–micrographic DNA measurements in reference to PM_2 DNA. The indices give the references for the analysis and the reference strains used (for strains, see also Tables III and IV).
[f–p] References: [f]Graham *et al.* (1972) (Plummer isolate), Ludwig (1972a) (several strains); [g]W.C. Russel and Crawford (1964); [h]Skare *et al.* (in prep.) (strains K_{22} and Cooper); [i]Engels *et al.* (1981a) (various isolates); [j]Pauli *et al.* (1981) (various isolates); [k]Martin *et al.* (1966) (BMV); [l]Sterz *et al.* (1973–1974) (TVA strain); [m]Buchman and Roizman (1978a,b) (BMV strain); [n]Ludwig and Darai (unpublished) (WC 11; strains Movar 33/63; ÜT; 66-P-347; DN-599); [o]De Villiers (1979) (JS-3 strain); [p]Engels *et al.* (unpublished) (goat isolate E/CH).

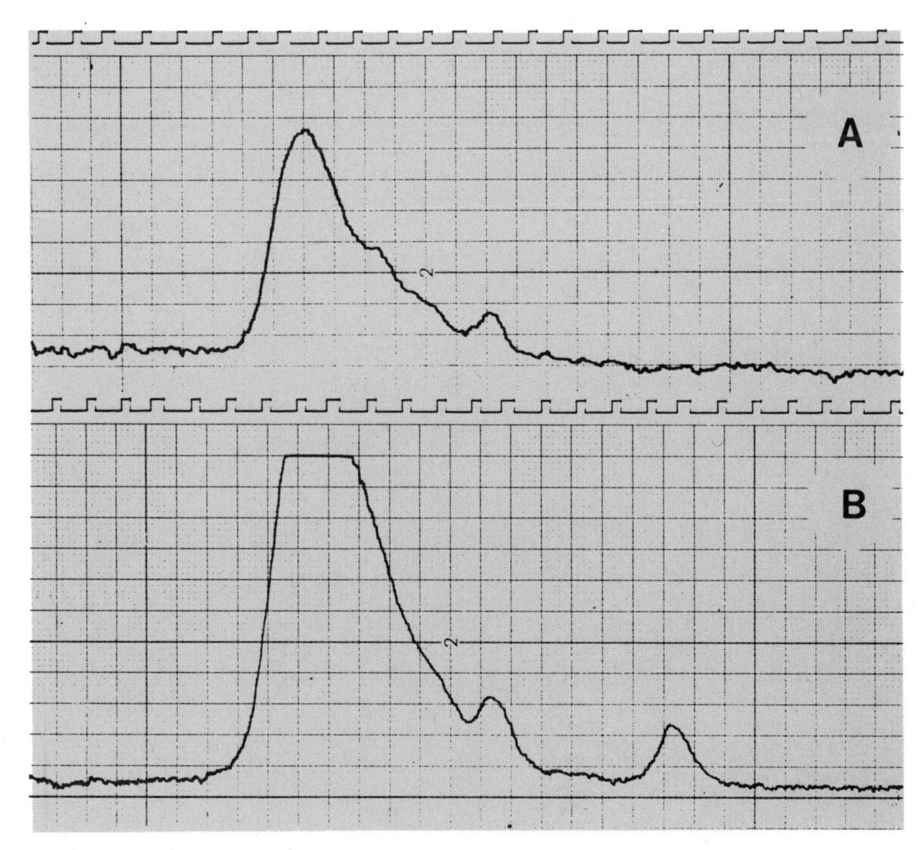

Figure 5. Analytical ultracentrifugation of DNA contained in BHV-1-infected (B) and mock-infected (A) bovine cell lysates in CsCl. Note the satellite DNA in normal bovine cells (A). From Ludwig and Storz (1973).

No obvious internal heterogeneity in base composition was found from the comparisons of the melting profiles of BHV-1 DNA with that of HSV or of PsR virus DNAs (Graham *et al.*, 1972). In this respect, the profile differed significantly from that of other herpesviruses (Fleckenstein *et al.*, 1975).

The restriction-endonuclease maps of BHV-1 DNA were determined by Skare and colleagues (Skare *et al.*, 1975; Skare, personal communication) and are shown in Fig. 6. The sequence arrangement of BHV-1 DNA resembles that of PsR virus (Stevely, 1977). Specifically, the small component (mol. wt. 22×10^6) inverts relative to the large component (mol. wt. 66×10^6), which has one unique orientation. The molecular weight of BHV-1 DNA is 88×10^6.

The genome of BHV-1 does not show significant homology to that of HSV-1, HSV-2, PsR virus, or BHV-2 when tested either by DNA–DNA filter hybridizations or by hybridization of viral DNAs with infected cell

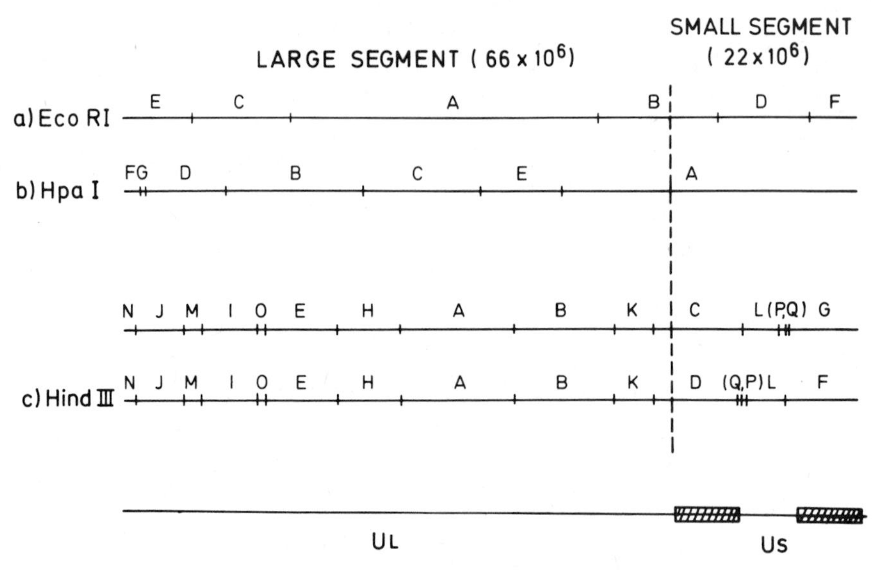

FIGURE 6. Physical maps and sequence arrangement of BHV-1 (strain K22) DNA. (U_L, Long unique region); (U_S, short unique region). Data from J. Farley, J. Skare, and I. B. Skare (in prep.) and J. Skare (in prep.).

RNAs (Ludwig *et al.*, 1972a; Ludwig, 1972b; Bronson *et al.*, 1972; Sterz *et al.*, 1973–1974).

There is little information on the replication of BHV-1 DNA. The finding that ribonucleotides are present in newly synthesized BHV-1 DNA (Babiuk and Rouse, 1976) is in accord with similar results obtained with other herpesvirus DNAs (Hirsch and Vonka, 1974; Murray and Biswal, 1974).

1. Differentiation of BHV-1 Strains by DNA Restriction-Enzyme Analysis

Although it was commonly believed that no significant biological differences exist between BHV-1 strains originating from the respiratory and genital tracts (Bowling *et al.*, 1969; McKercher, 1973), clear variations were recently demonstrated in infectious bovine rhinotracheitis–infectious pustular vulvovaginitis (IBR-IBV) virus strains (Engels *et al.*, 1980, 1981a). This was possible by the use of restriction-endonuclease "fingerprinting" of virus DNA, a technique that has been successfully applied in epidemiological studies of HSV and PsR virus infections (Buchman *et al.*, 1978, 1979; Ludwig *et al.*, 1982). It was demonstrated for the first time that not only fresh isolates and attenuated strains but also various isolates obtained from clearly different clinical entities could be assigned to an IBR-like or IPV-like BHV-1 group (Engels *et al.*, 1980, 1981a; Pauli *et al.*, 1981). One of the clearly discriminating enzymes is *Hpa*I (Figs. 7 and 8). Isolates from IBR, conjunctivitis, encephalitis, and abortion and the BHV-1 isolates activated from normal tissue-culture cells (Ludwig and Storz, 1973; Storz *et al.*, 1980) all fall into the IBR-like virus group—and this appears to be of major epidemiological importance—whereas isolates from typical "Bläschenausschlag" (IPV) or even attenuated and high-passage vaccine strains originating from IPV showed the IPV-like DNA pattern (Table III and Figs. 7 and 8).

A BHV-1 strain isolated from a case of IPV retained its characteristic overall IPV-like *Hpa*I cleavage sites after approximately 460 passages in bovine cell culture (Geilhausen, personal communication; Pauli *et al.*, 1981). Further analysis with other restriction enzymes, however, revealed that one fragment in the DNA of the high-passage strain was significantly smaller than in the original strain (Engels *et al.*, 1980). Figures 7 and 8 summarize some of the results of analyses with those DNAs. As could be expected, differences among individual BHV-1 strains could

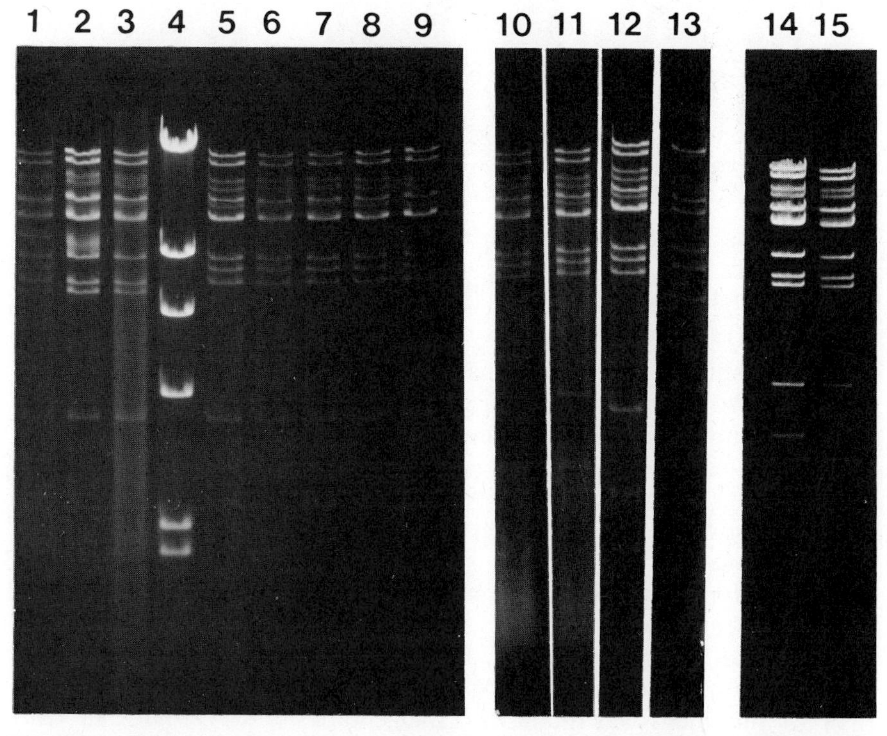

FIGURE 7. Electrophoretic patterns of *Hin*dIII restriction-endonuclease digests of DNAs of BHV-1 strains (see also Table III). (1) IPV isolate; (2) IPV isolate passaged, B3; (3) IPV isolate passaged, B2; (4) lambda phage (marker); (5) IBR/IPV isolate V 101; (6) IBR isolate LAE: (7) IBR strain LA; IBR isolate Gi-5; (9) IBR isolate BFN-1H; (10) IBR isolate Gi-1; (11) IBR isolate Gi-2; (12) IBR isolate Gi-4; (13) BHV-2 strain TVA; (14) IPV isolate B1; (15) IPV isolate passaged, B4. Electrophoresis was done in 0.5% agarose slab gels, 350 × 200 × 2 mm at 90 V and 4°C for 18 hr (Darai and Ludwig, unpublished).

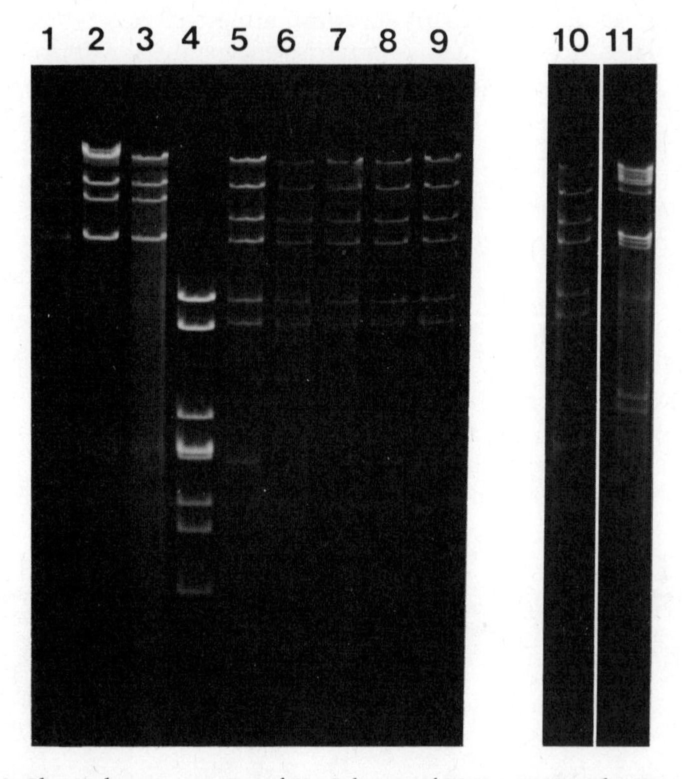

FIGURE 8. Electrophoretic patterns of *Hpa*I digests of BHV-1 DNAs. The DNAs in lanes 1–10 correspond to those in Fig. 7; lane 11, BHV-2 DNA. Electrophoretic analyses were done as described in the Fig. 7 caption.

be readily shown with additional restriction enzymes (Engels *et al.*, 1981a). These conclusions were confirmed by other groups (Rziha, personal communication; Herring, personal communication).

B. Genome of Bovid Herpesvirus 2

Characteristics of the DNA are summarized in Table II and shown in Figs. 8–10 and 13. The viral DNA has a G + C content of 63–64 moles%. (Martin *et al.*, 1966; Sterz *et al.*, 1973–1974). The DNA of BHV-2 has a surprisingly high degree (15%) of homology with HSV-1 DNA, but less than 6% homology was found with the genomes of BHV-1 and PsR virus (Fig. 11). Thus, BHV-2 was the first nonhuman herpesvirus closely related to HSV-1 (Sterz *et al.*, 1973–1974). This finding stimulated interest in the genome structure of BHV-2. The sequence arrangement and restriction-endonuclease maps determined by Buchman and Roizman (1978a,b) revealed another characteristic similarity to HSV. Both herpesviruses possess inverted repetitions at the end of the molecules and at the junction

TABLE III. Grouping of BHV-1 Strains as IBR-Like or IPV-Like Viruses as Based on Restriction-Endonuclease Cleavage of the DNAs[a]

Strain	IBR-like site of isolation	Ref. No.[b]	Strain	IPV-like site of isolation	Ref. No.[b]
LA	Nasal cavity	1	K_{22}	Vagina	9
Montagne de Moutier	Lung	2	Oberhüningen	Vagina, pharynx	2
V 101	Preputium	3	Wabu	Preputium	2
Conj.	Conjunctiva	4	B_1	Vagina, TC passage 10	10
LAE	Brain	5	B_2	TC passage 260	10
Burroughs	Aborted fetus	6	B_3	TC passage 310	10
Gi-1 to Gi-5	TC, bovine fetal spleen	7	B_4	TC passage 460, Bayferon®	10
BFN-1H	TC, bovine fetal kidney, thyroid, and testes	8			

[a] From Darai and Ludwig (unpublished), Engels *et al.* (1980, 1981), and Pauli *et al.* (1981a). Significant differences were seen in fragment distribution after digestion with *Eco*RI, *Hpa*I, and *Hin*dIII, which allowed this kind of grouping. The DNAs of the strains (references 1, 3–7, 9, and 10) did not significantly differ in GC- content (Ludwig, 1972a). (TC) Tissue culture.

[b] References: (1) Madin *et al.* (1956); (2) Engels *et al.* (1981a); (3) Ludwig *et al.* (1969); (4) Hughes *et al.* (1964); (5) Barenfus *et al.* (1963); (6) obtained from McKercher; (7) Ludwig and Storz (1973); (8) Storz *et al.* (1980); (9) Kendrick *et al.* (1958); (10) obtained from Bayer, Leverkusen.

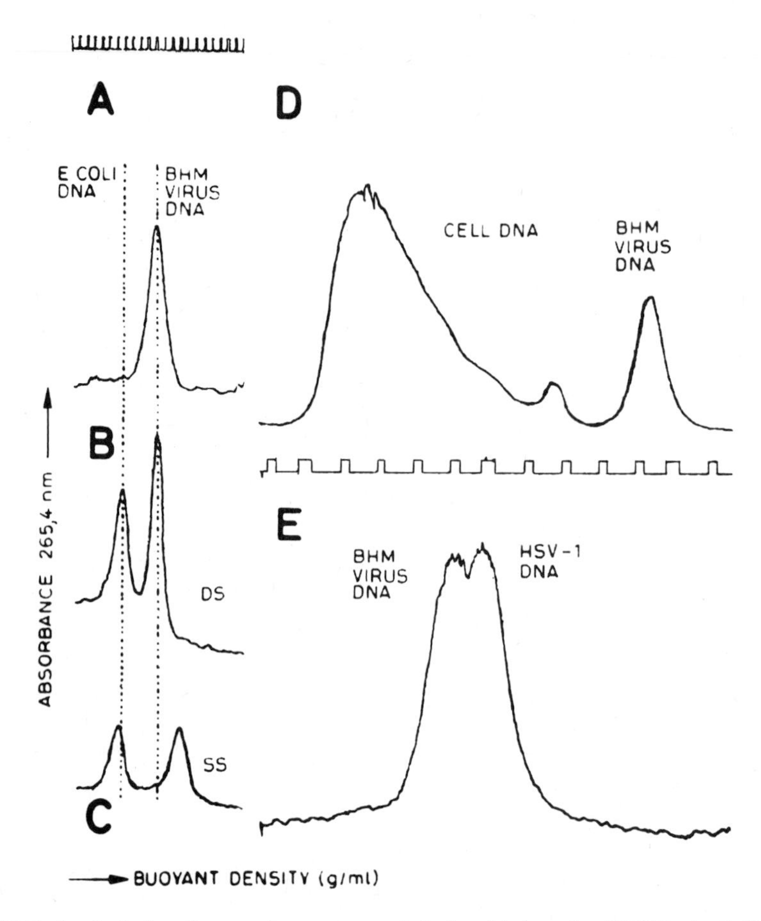

FIGURE 9. Analytical sedimentation patterns of viral and infected cell DNAs in CsCl density gradients. (A) Native, double-stranded (DS) BHV-2 DNA; (B) native (DS) BHV-2 DNA with *Escherichia coli* DNA; (C) same as in (B), but BHV-2 DNA denatured [single-stranded (SS)]; (D) DNA extracted from lysates of BHV-2-infected bovine cells; (E) native (DS) BHV-2 and HSV-1 DNAs (Sterz *et al.*, 1973–1974). (BHM) Bovine herpes mammillitis.

between the long and short components (see Fig. 10). The homology and common architectural similarity of BHV-2 and HSV genomes raised the question of a common ancestor, since the genomes of both viruses clearly differ from those of other herpesviruses (Fig. 10) such as PsR virus (Stevely, 1977), BHV-1 (Skare *et al.*, personal communication) and *Herpesvirus saimiri* (Fleckenstein *et al.*, 1975, 1978). In view of the strong serological cross-reactivity of BHV-2 and HSV-1 (Sterz *et al.*, 1973–1974), encountered in glycoprotein gA/B (Pauli and Ludwig, 1977; Norrild *et al.*, 1978b; Eberle and Courtney, 1980), which maps at 0.3–0.4 map unit on HSV-1 DNA (Ruyechan *et al.*, 1979), one might speculate that this locus has been conserved at least in part during the evolutionary divergence of these viruses.

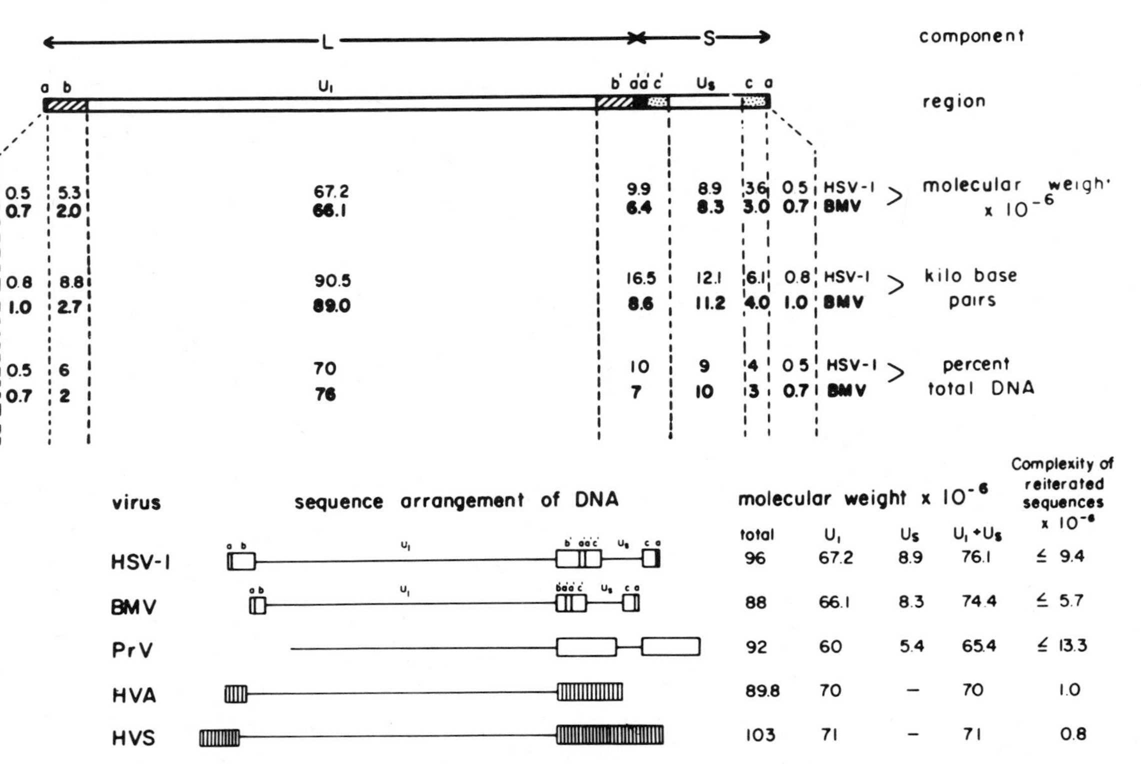

FIGURE 10. *Top:* Comparison of HSV-1 and BHV-2 DNAs. The results of BHV-2 DNA analyses are given in boldface. Sizes of the regions are in molecular weight ($\times 10^6$) of double-stranded DNA and in kilo base pairs. *Bottom:* Comparison of the sequence arrangements of different herpesviral genomes including BHV-2. BHV-1 would be similar to PsR virus. From Buchman and Roizman (1978b).

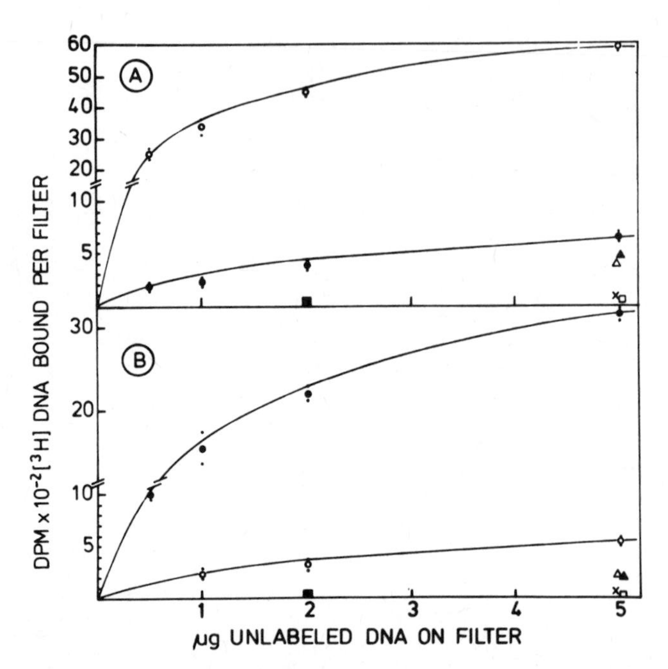

FIGURE 11. Homology of BHV-2 DNA with HSV-1 DNA and other DNAs as measured by filter hybridization. (A) [^{3}H]-HSV-1 DNA, 7400 dpm, was hybridized to DNA immobilized on filters: HSV-1 (○), BHV-2 (●), BHV-1 (△), PsR virus (▲), *E. coli* (□), rabbit kidney cells (■), and bovine kidney cells (×). (B) [^{3}H]-BHV-2 DNA, 4800 dpm, was hybridized to the DNAs listed under (A) (Sterz *et al.*, 1973–1974).

C. DNAs of Bovid Herpesviruses 3 and 4

No investigations on the genetic material of these bovine herpesviruses have been reported, but recent determinations of a few biophysical parameters of the purified DNAs have unequivocally grouped these isolates (Table IV). The high-passage avirulent strain (WC 11) of malignant catarrhal fever virus and the virulent strain (C500) have an estimated molecular weight of approximately 80×10^6, similar to that of isolates of the "Movar type." The reference strains of malignant catarrhal fever and "Movar-type" viruses (see Table IV) can be readily differentiated by restriction-endonuclease "fingerprinting" into the BHV-3 and BHV-4 groups (see Table I and Fig. 12).

The cleavage pattern shows not only that malignant catarrhal fever viruses are completely different from the "Movar-type" viruses but also that the European and the American strains of BHV-4 are closely related by DNA patterns. The American strains (66-P-347 and DN-599) have virtually identical cleavage sites for *Eco*RI and *Hin*dIII enzymes, but differ slightly from the Hungarian and the German reference strains (Table IV and Fig. 12).

TABLE IV. Characterization and Grouping of BHV-3, BHV-4, and BHV-6 Strains on the Basis of Restriction Endonuclease Patterns of Their DNAs[a]

Group	Reference strain	Authors	Site of isolation	Estimated number of cleavage sites with restriction enzymes		
				*Eco*RI	*Hin*dIII	Estimated identity
BHV-3	C500, virulent MCF	Plowright *et al.* (1965)	—	—	—	—
	WC 11, avirulent MCF	Plowright *et al.* (1965)	TC passage 50	—	15	
BHV-4	Movar 33/63	Bartha *et al.* (1966)	Ocular discharge	10	12	90%
	ÜT	Überschär, Tübingen (unpublished)	TC isolate	10	12	
	DN-599	Mohanty *et al.* (1971)	Nasal discharge	10	12	100%
	66-P-347	Storz (1968)	TC, spleen cells	10	12	
BHV-6	McK/US	Saito *et al.* (1974)	Organ pool	1	—	90%
	E/CH	Mettler *et al.* (1979)	Organ pool	1	—	

[a] These data were obtained from purified or partially purified virions. For DNA extraction, the virions were lysed with sodium dodecyl sulfate and submitted to proteinase K digestion and phenol treatment. After digestion with bacterial enzymes, the DNA fragments were analyzed on 0.5–0.8% agarose gels in reference to λ DNA digests (Darai and Ludwig, unpublished). The reference strains were kindly supplied to H. Ludwig in 1971–1972 and 1980 by Drs. Bartha (Budapest), Edington (London), Mohanty (College Park, Maryland), Storz (Fort Collins), and Überschär (Tübingen) and were propagated on the mycoplasma-free GBK cell line.

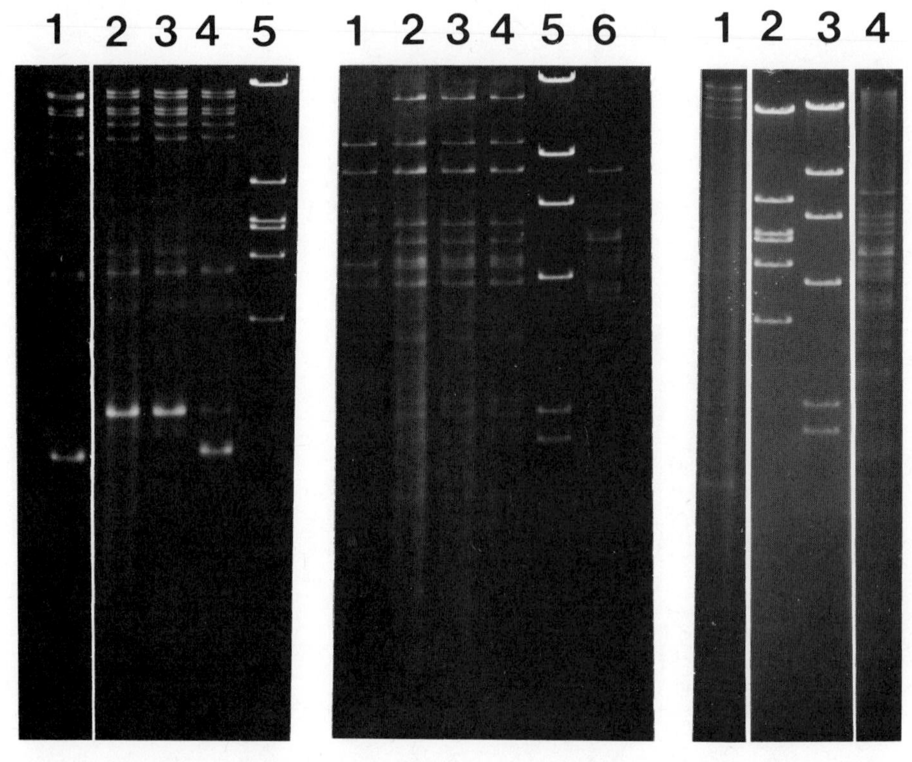

FIGURE 12. Electrophoretic patterns of BHV-3 and BHV-4 DNAs digested with restriction endonucleases. (A) BHV-4 strain Movar (1), strain DN599 (2), strain 66-P-347 (3), and strain ÜT (4), and (5)Lambda phage (marker) were all digested with *Eco*RI. (B) Lanes (1–4) correspond to (A); lambda phage (marker) (5) and BHV-3 strain WC 11 (6) digested with *Hin*dIII. (C) BHV-3 strain C500 (1) and lambda phage (2) digested with *Eco*RI; lambda phage (marker) (3); BHV-3 strain C500 (4) digested with *Hin*dIII. Electrophoresis was done as described in the Fig. 7 caption.

D. DNA of Bovid Herpesvirus 5

Analyses of the DNA of "jaagsiekte" virus have been performed by De Villiers (1979) and are given in Table II. The molecular weight of the DNA appears to be remarkably low (67×10^6). In its G+C content, the genome of this virus resembles those of Marek's disease virus (Lee *et al.*, 1971) and varicella–zoster virus (Ludwig *et al.*, 1972b).

E. DNA of Bovid Herpesvirus 6

The goat herpesvirus was isolated first by Saito *et al.* (1974) in the United States and more recently also in Europe (Mettler *et al.*, 1979). The viruses belonging to this group we provisionally assigned the designation

BHV-6, although some serological cross-reactivity with BHV-1 is apparent (see also Section VI.C). The cleavage patterns of its DNA do not correspond to those of any of the other known herpesviruses. It is of general interest that *Eco*RI or *Hin*dIII restriction enzymes cleave the DNA only at one site (Fig. 13A), whereas treatment with *Bst*EII gave at least ten fragments. This observation differentiates the genome of this virus from representative genomes of other bovine herpesviruses and also from those of the equine herpesviruses and PsR viruses (Fig. 13B). Preliminary ex-

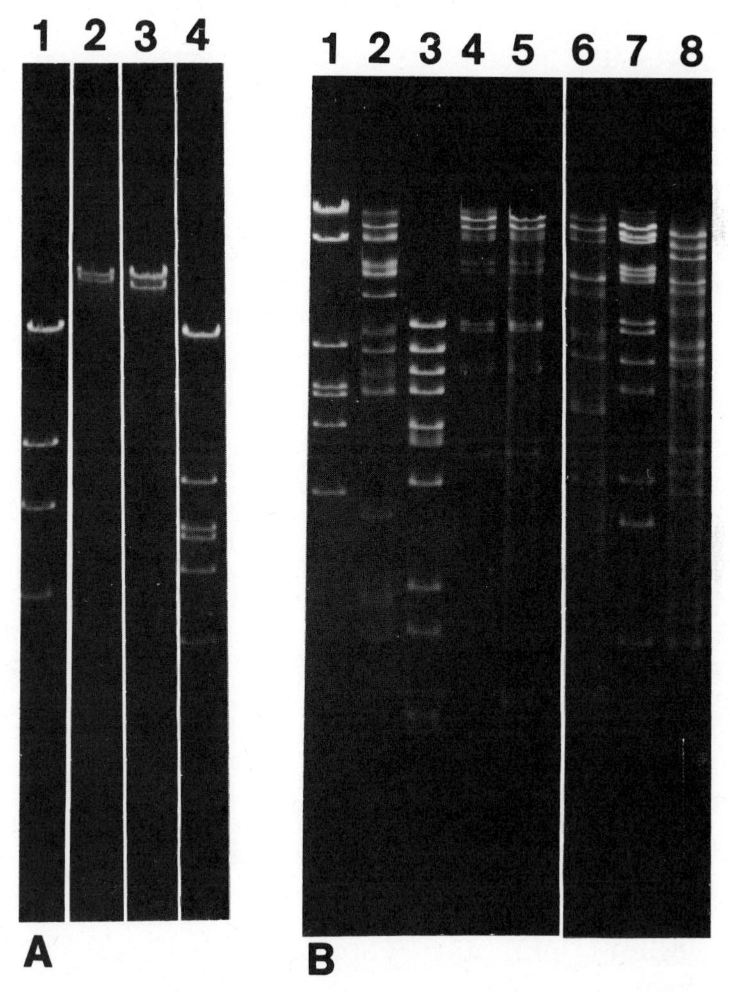

FIGURE 13. Comparison of electrophoretic patterns of BHV-6 DNA with those of other DNAs digested with restriction endonucleases. (A) (1) Lambda phage (*Hin*dIII); (2) BHV-6 (*Hin*dIII); (3) BHV-6 (*Eco*RI); (4) lambda phage (*Eco*RI). (B) (1) Lambda phage DNA undigested and mixed with *Eco*RI-digested DNA (marker); (2) BHV-6; (3) lambda phage; (4) BHV-1 strain LA; (5) BHV-1 strain B2; (6) BHV-2 strain TVA; (7) EHV-1; (8) PsR virus strain DEK. All DNAs were digested with *Bst*EII. Electrophoresis was done as described in the Fig. 7 caption.

periments revealed that the DNA has a buoyant density of 1.729 g/ml and a melting point of 84.5°C, which corresponds to a G+C content of 70%. The molecular weight of the DNA calculated from electron-microscopic measurements using PM_2 phage DNA as standard was 84 ± 4 × 10^6, which agrees with calculations based on restriction-enzyme analyses (Engels *et al.*, 1981b).

VI. PROTEINS AND ANTIGENS

A. Virus-Specific Proteins

Studies on bovine herpesvirus specific proteins are not nearly as advanced as those on other lytic herpesviruses such as herpes simplex virus (HSV) (Roizman and Furlong, 1974) or pseudorabies (PsR) virus (Stevely, 1975). Only recently were the bovid herpesvirus 1 (BHV-1) specific proteins described (Sklyanskaya *et al.*, 1977; Pastoret *et al.*, 1980).

From this virus, 21 structural polypeptide bands were separated by denaturing polyacrylamide gel electrophoresis (PAGE) from purified virus, and 10 of these appeared to contain glycoproteins. The nonglycosylated viral polypeptide No. 7 with an apparent molecular weight of 99,000 seems to reflect the major structural component, since it accounted for approximately 15% of the total protein (Pastoret, 1981). It is noteworthy that slight differences between proteins of infectious bovine rhinotracheitis (IBR) virus and infectious pustular vulvovaginitis (IPV) virus strains could be detected (Fig. 14).

Comparisons of the structural proteins of BHV-2 with those of four other herpesviruses (Killington *et al.*, 1977) revealed that many BHV-2 polypeptides correspond to those of HSV with respect to electrophoretic mobility and that BHV-2 glycosylated polypeptides cluster in three molecular-weight ranges.

Information on proteins specified by all the other bovine herpesviruses (groups BHV-3 to -6) is still lacking. This may be due in part to their slow growth and in part to their tendency to remain strongly cell-associated. These characteristics make labeling of viral proteins and purification of enveloped particles relatively difficult.

B. Serological Identification of Strains and Individual Strain Differences

The neutralization test is still the preferred method for identification of bovine herpesvirus strains. It has also been used to identify strain differences with more or less success. (Restriction-enzyme analysis of the

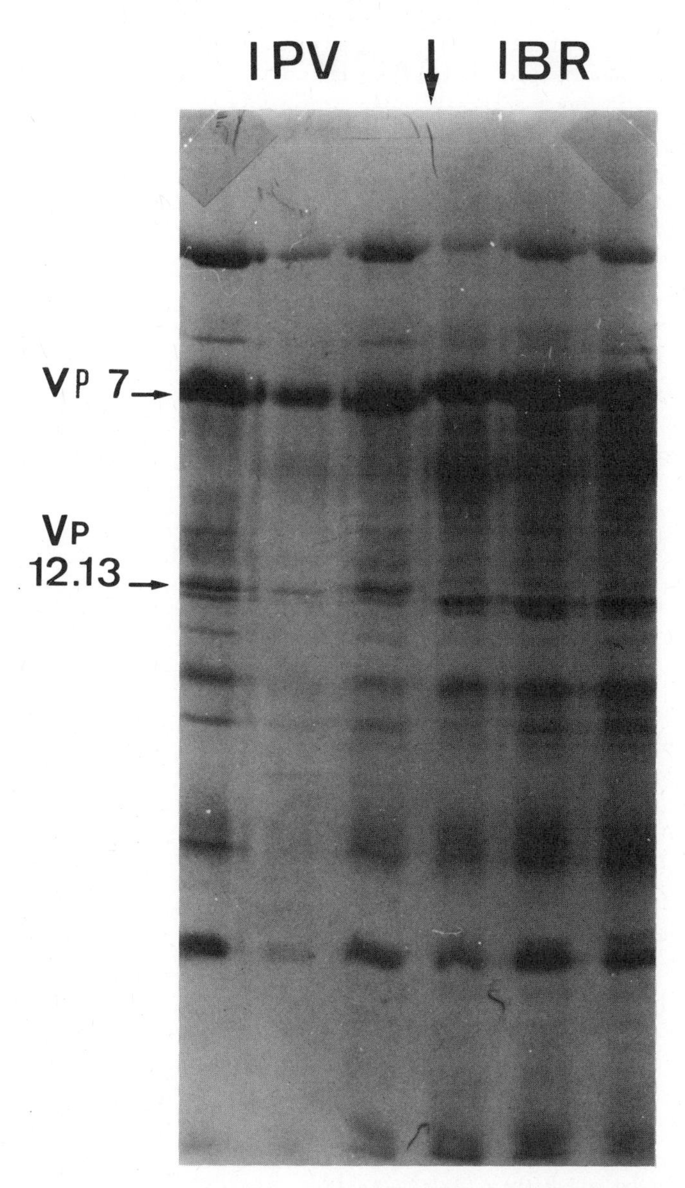

FIGURE 14. Electrophoresis patterns of the proteins of three different strains each of IBR virus or IPV virus in denaturing polyacrylamide gels. Note molecular weights of differences between proteins of the two groups of viruses. Proteins were labeled with a [^{3}H]-amino acid mixture. Courtesy of P.-P. Pastoret.

DNA might be a better technique.) Other serological methods such as immunoprecipitation or complement-fixation tests have been used for the identification of different groups (Table V).

Since BHV-1 is preferentially isolated from the respiratory or the genital tract, it has frequently been compared with HSV-1 and HSV-2. The human viruses differ immunologically (Schneweis, 1962) and biochemically (Ludwig *et al.*, 1972a). Several reports conclude that BHV-1 isolates from different clinical entities of diseased animals cannot be differentiated on the basis of neutralization tests (Gillespie *et al.*, 1959; Bowling *et al.*, 1969). Other investigators reported antigenic variation among IBR and IPV viruses (Buening and Gratzek, 1967; House, 1972; Crandell, 1973; Potgieter and Maré, 1974). It is of particular interest that an encephalitic strain could be differentiated from others by neutralization (Bagust, 1972). However, while they may differ antigenically from each other, there is no definitive evidence that BHV-1 viruses comprise distinct groups differentiable with respect to antigenic properties.

The BHV-2 strains, including isolates from Africa, Britain, Italy, and North America, are strongly related immunologically (Rweyemamu and Johnson, 1969; Breese and Dardiri, 1972; Sterz and Ludwig, 1974; Castrucci *et al.*, 1975). Minor variations among strains from different countries have been detected by cross-immunity tests made in cattle (Cilli and Castrucci, 1976).

Not much is known about the immunological relationship among BHV-3 strains. The isolates from different places in Africa appear to be immunologically homogeneous (Plowright, 1968). Serological identification is usually done by neutralization (Mushi and Plowright, 1979; Russel, 1979), immunodiffusion (Rossiter, 1980), immunofluorescence microscopy (Ferris *et al.*, 1976), or complement-fixation tests (Rossiter and Jessett, 1980).

Virus isolates believed to belong, on the basis of analyses of their DNAs, to the BHV-4 group have been identified as herpesviruses by virological tests other than serology. Numerous isolates that might belong to this group are summarized in Table III of the review by Gibbs and Rweyemamu (1977b). In some instances, cross-reactivities among strains have been shown, and these seem to belong to the "Movar-type" viruses (Bartha *et al.*, 1967). Since these viruses appear to be weakly immunogenic, very little is known concerning the antigens of the members of this group.

Serological identity of BHV-5 isolates obtained from cases of "jaagsiekte" in sheep and a Scottish isolate was demonstrated by neutralization tests (Verwoerd *et al.*, 1979); however, further immunological studies on these viruses were not reported.

The two goat isolates classified, as BHV-6 (Saito *et al.*, 1974; Mettler *et al.*, 1979) on the basis of serological analyses, have not yet been tested for identity.

TABLE V. Antigenic Diversity and Cross-Reactivities[a]

Group	Reference strains	Strain differences tested by:		Cross-reactivity tested by antisera to BHV-					
		Serology	DNA analysis	1	2	3	4	5	6
BHV-1	LA, K_{22}	Might exist	Yes	+ + + +	–	–	–	–	+ +
BHV-2	TVA	Yes	NT	–	+ + + +	–	NT	NT	–
BHV-3	WC 11; C500	Possible	NT	–	–	+ + + +	NT	NT	–
BHV-4	Movar 33/63; 66-P-347	Exist	Yes	–	–	–	+ + + +	NT	–
BHV-5		No	NT	–	NT	–	NT	+ + + +	NT
BHV-6	^{12}E/CH; McK/US	Yes	NT	+ +	–	–	NT	NT	+ + + +

[a] This table was compiled from published reports and is based in part on results of neutralization (Engels and Ludwig, unpublished) and immunoelectrophoretic and fluorescent antibody tests (Ludwig and Leiskau, unpublished). The following antisera with homologous neutralizing antibody titers (75% reduction of 100 plaques or foci) were used; anti-BHV-1 (1:180) and anti-BHV-2 (1:120) hyperimmune sera from cattle, anti-BHV-3 (1:40) serum from calves (kindly supplied by Dr. Edington, London), and anti-BHV-6 (1:160) hyperimmune serum from goats. Anti-BHV-4 and -5 sera were not available. (–) Negative; (+ through + + + +) estimated antigenic cross-reactivity from low to total; (NT) not tested.

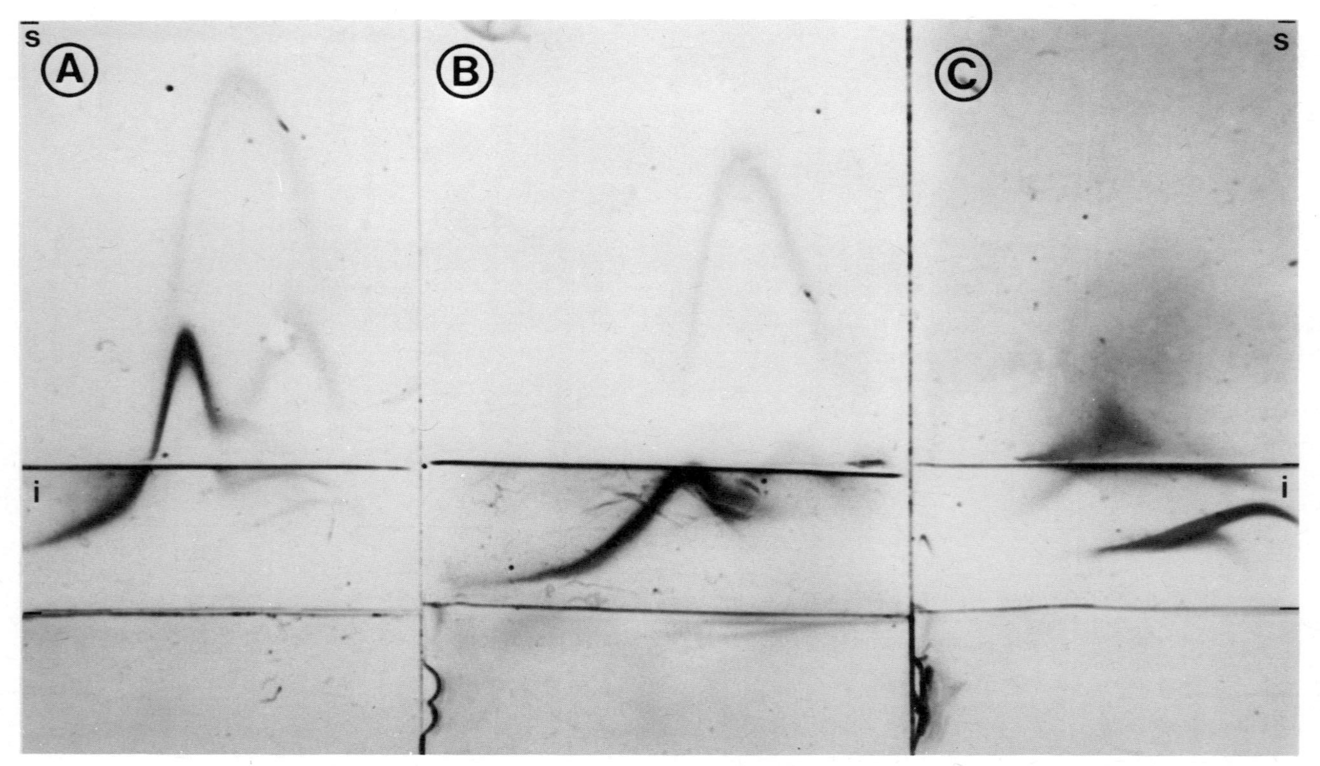

FIGURE 15. Antigenic relationship of BHV-1 and BHV-6 as shown by crossed immunoelectrophoresis using the intermediate gel technique (for details, see Norrild *et al.*, 1978b). BHV-1- (A, B) or BHV-6- (C) infected cell antigens were subjected to electrophoresis into the second-dimension gels (s) containing 15 μl/cm^2 each of antibodies directed against BHV-1 (A, B) or BHV-6 (C). The intermediate gels (i) contained 20 μl/cm^2 of preimmune serum (A, C) or goat anti BHV-6 hyperimmune serum (B). Data from Engels *et al.* (1981b).

C. Virus-Specific Antigens

Knowledge of antigens of bovine-herpesvirus-infected cells is still fragmentary, and almost nothing has been reported on the major immunogenic components of the viruses. It is certainly of great interest to localize viral antigens within the physical structure of the virion and to determine which antigens induce protection from disease. Studies dealing with these questions are in progress for HSV, Epstein–Barr virus (EBV) and PsR virus (Vestergaard, 1973; Honess *et al.*, 1974; Honess and Watson, 1977; Pauli and Ludwig, 1977; Norrild *et al.*, 1978b; Spear, 1980; Zur Hausen, 1980; Pauli *et al.*, 1982). The studies on bovine herpesvirus antigens are less advanced.

Different BHV-1 strains show the same overall antigen composition when investigated by immunoelectrophoresis. One major and five minor antigens can be identified (Fig. 15A). By immunoprecipitation two major immunogenic compounds (Fig. 16), one of which appears primarily in-

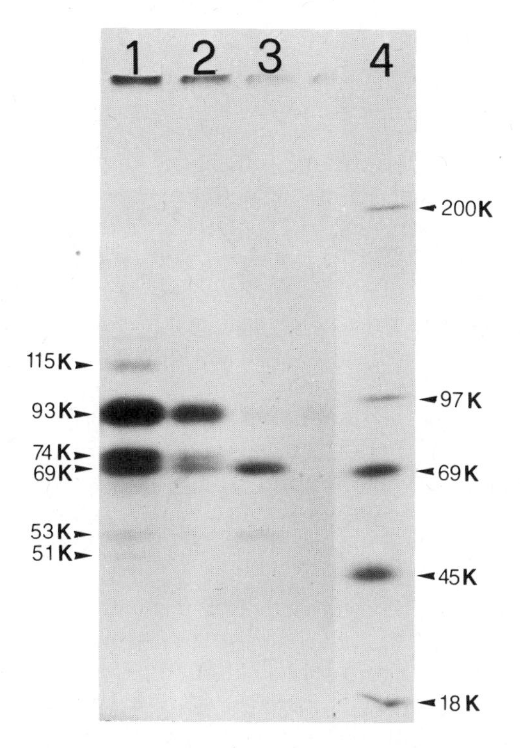

FIGURE 16. Analysis of major immunogenic components of BHV-1. ^{3}H-glucosamine labeled BHV-1 infected cell lysates were immunoprecipitated using rabbit hyperimmune sera and *Staphylococcus aureus* and analyzed by denaturing PAGE. Lane 1, serum directed against infected cells: neutralizing titer 1:585; lanes 2 and 3, sera directed against defined individual precipitation bands obtained by crossed immunoelectrophoresis: neutralizing titers of 1:500 and <1:4, respectively; lane 4, ^{14}C labeled marker proteins. Courtesy of Jens-Peter Gregersen.

volved in neutralization, have recently been identified (Pauli and Aguilar-Setién, 1980; Gregersen, in prep).

There is more information on BHV-2, possibly because this virus has attracted more interest with its antigenic relationship to HSV-1 and -2 (Sterz *et al.*, 1973–1974; Norrild *et al.*, 1978b). Analyses of BHV-2 by immunoelectrophoresis yielded evidence for at least six defined antigens (Fig. 17D). Analyses of immunoprecipitation by denaturing PAGE showed that homologous reconvalescent or hyperimmune sera react with several viral glycoproteins (data not shown).

BHV-3 specific antigens have been identified by immunoprecipitation techniques (Rossiter, 1980), but detailed analyses of the viral antigens remain unreported. In studies using DNA inhibitors, two early antigens were differentiated from late antigens (Rossiter *et al.*, 1978).

Detailed analyses on BHV-4 and -5 antigens have not been reported.

BHV-6 (the goat herpesvirus)-infected cell extracts precipitated with hyperimmune goat serum yielded one major and several minor antigens (see Fig. 15C) (Engels and Ludwig, unpublished).

D. Antigenic Diversity of Bovine Herpesviruses and Immunological Cross-Reactivities with Other Herpesviruses

Studies on bovine herpesviruses have the advantage that antisera can be made in their natural host inoculated with infected bovine cells grown and maintained in bovine serum. Specific antisera, which do not recognize host components, can be obtained despite the genetic differences among cattle (Sterz *et al.*, 1973–1974; Ludwig and Pauli, unpublished results). These sera are invaluable reagents because they enable clarification of the antigenic relationships and diversities among bovine herpesviruses. Although such reagents have not been intensively studied, the few reports together with our own data indicate that the BHV-1, BHV-2, BHV-3, BHV-4, and BHV-5 viruses are not related immunologically (Table V) (compare also Fig. 19), and cross-neutralization is generally not demonstrable. This was shown for BHV-1 and -2 (Martin *et al.*, 1966; Castrucci *et al.*, 1972; Sterz *et al.*, 1973–1974), BHV-4 (Bartha *et al.*, 1967; Mohanty *et al.*, 1971), and BHV-5 (Verwoerd *et al.*, 1979). Neutralization tests of BHV-3 have been performed only in the homologous system (Rossiter *et al.*, 1977; P.H. Russel, 1979). BHV-6 and BHV-1 are exceptions and are neutralized by heterologous antisera (Saito *et al.*, 1974; Mettler *et al.*, 1979). The cattle (BHV-1) and the goat (BHV-6) herpesviruses share common immunologically determined sites on the virion surface. In each case, the homologous complement-independent neutralization with immunoglobulin (IgG) is considerably stronger than in the heterologous reaction. This serological relationship extends to several antigens of both viruses as measured by immunoelectrophoresis (see Fig. 15B). The goat virus, however, does not cross-react with other bovine herpesviruses or with HSV-1, HSV-2, B virus, or the canine and feline herpesviruses (Engels and Ludwig, unpublished) (Fig. 19).

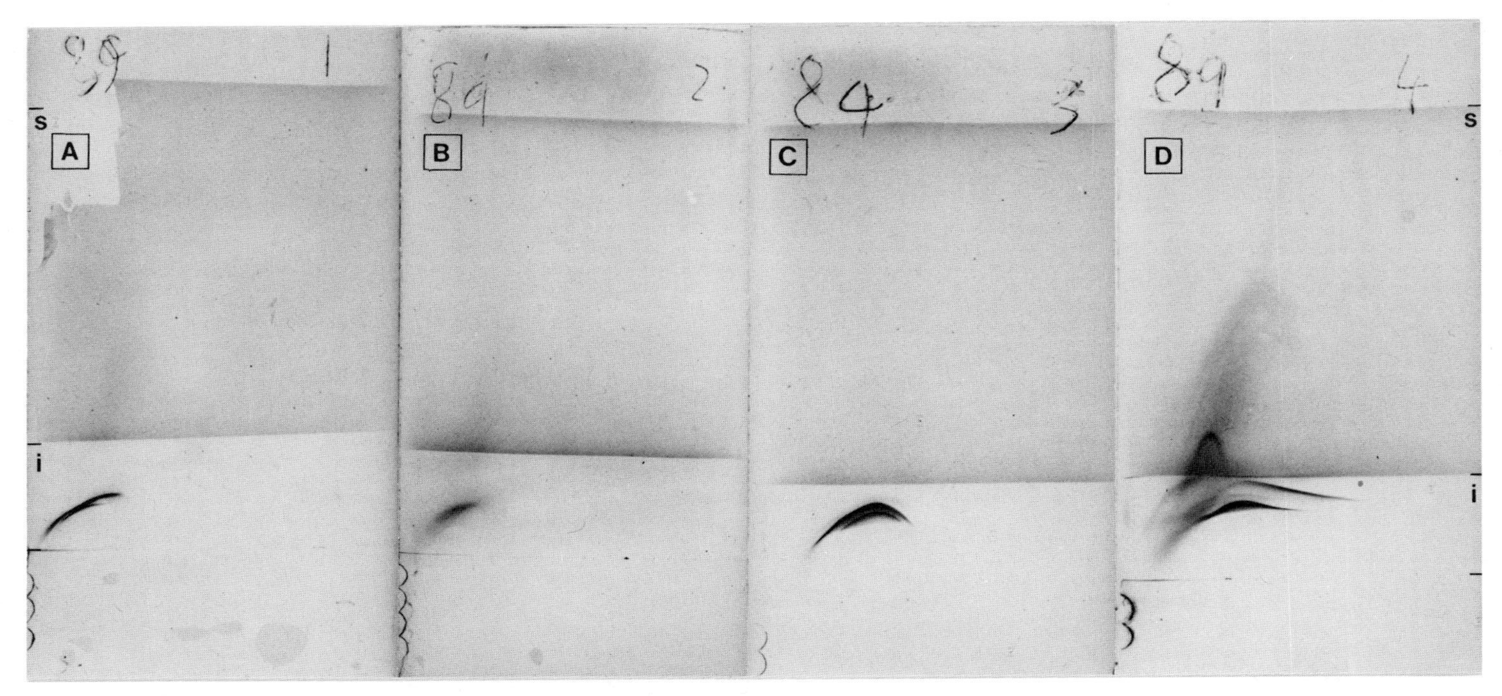

FIGURE 17. Antigenic relationship of BHV-2 with HSV-1, HSV-2, and B virus as detected by crossed immunoelectrophoresis. The experimental conditions were similar to those described in the Fig. 15 caption. Various amounts of infected-cell antigens were subjected to electrophoresis into second-dimension gels (s), containing 18 μl/cm² of bovine antibodies directed against BHV-2. Blank agarose intermediate gels (i) were used for better resolution of the precipitates. HSV-1 (A) and B virus (C) cross-react more strongly than HSV-2. (A) HSV-1 antigen; (B) HSV-2 antigen; (C) B virus antigen; (D) BHV-2 antigen.

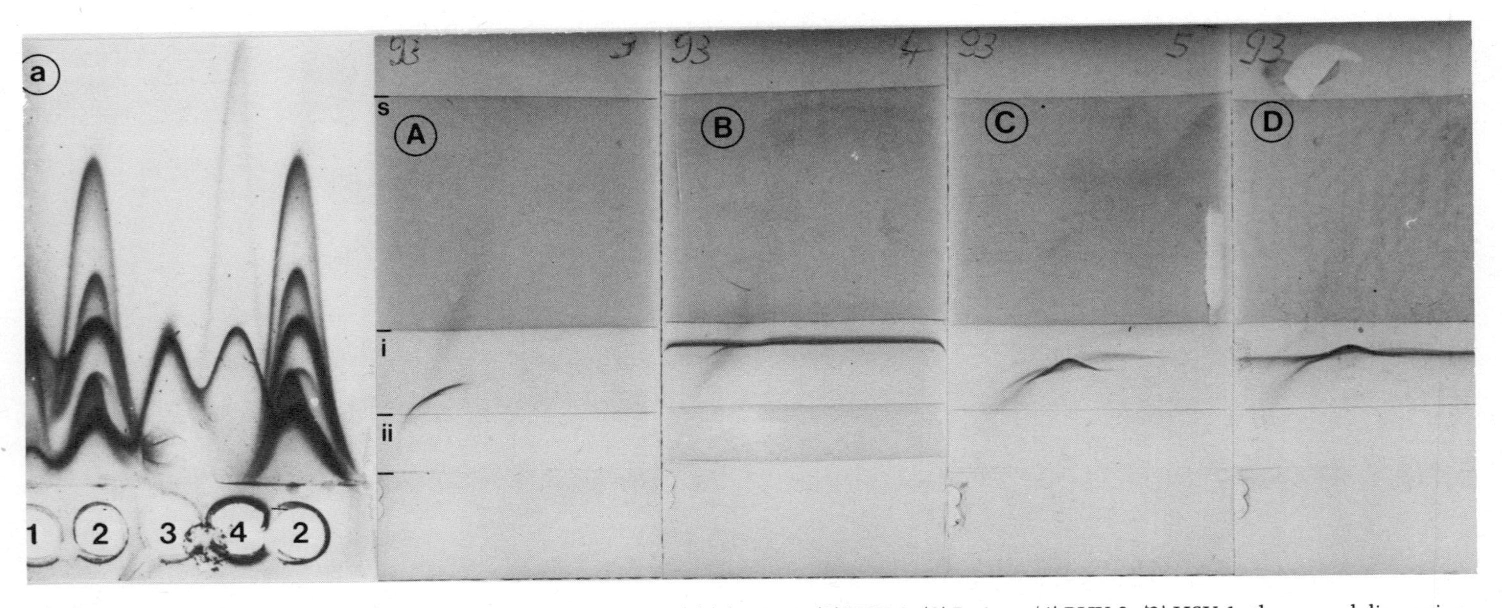

FIGURE 18. Antigenic cross-reactivity of different herpesviruses: (a) (1) HSV-2; (2) HSV-1; (3) B virus; (4) BHV-2; (2) HSV-1; the second dimension gel contains polyvalent antibodies directed against HSV-1; shown by fused rocket immunoelectrophoresis. (A–D) the partial identity of the common antigen between BHV-2 and HSV-1 as demonstrated by crossed-line immunoelectrophoresis. Antigens of cells infected with HSV-1 (A, B) and BHV-2 (C, D) were subjected to electrophoresis in the first dimension (1, 5 hr at 10 V/cm). The second dimension gels (s) contained antibodies to BHV-2 ($15/\mu l/cm^2$); the first intermediate gels (i) were blank gels; the second intermediate gels (ii) were either blank gels (A, C) or contained BHV-2 (B) or HSV-1 (D) antigens. Electrophoresis was performed for 18 hr at 2 V/cm. Data from Ludwig *et al.* (1978b).

There have been reports of slight cross-reactivities between BHV-1 and BHV-3 using fluorescent-antibody techniques and other tests (Rossiter *et al.*, 1977). Complement-fixation tests (Sterz *et al.*, 1973–1974) and recent fluorescent-antibody studies (Ludwig and Leiskau, unpublished data) have failed to yield detectable cross-reactivity between BHV-1 and BHV-3.

The complexity of immunological cross-reactivity among herpesviruses in general has been discussed extensively by Hampar and Martos (1973), and their comments are still relevant. Investigations on bovine herpesviruses are consistent with their conclusions. Specifically, the immunological cross-reactivity between bovine herpesviruses and viruses of other species is not overwhelming. Sparse information indicates that BHV-1 and equine abortion virus (equine herpesvirus type 1) do cross-react (Blue and Plummer, 1973) and that BHV-1 and Marek's disease virus, as well as EBV, share common antigenic determinants (Evans *et al.*, 1972).

There exists, however, a clear immunological relationship among BHV-2, HSV, and B virus, detected first by our group (Sterz *et al.*, 1973–1974; Ludwig, 1976; Pauli and Ludwig, 1977; Ludwig *et al.*, 1978a; Norrild *et al.*, 1978b) and now also confirmed for HSV and BHV-2 by others (Killington *et al.*, 1977; Engels, 1979; Poli *et al.*, 1980; Yeo *et al.*, 1981; Castrucci *et al.*, 1981). This unique interspecies specific serological relationship is discussed in more detail below.

1. Nature of the Common Antigen between BHV-2 and Other Herpesviruses

Virion surface structures are involved in cross-reactivity between HSV and B virus, whereas common antigenic determinants seem not to be accessible on the BHV-2 virion. This conclusion is based on neutralization (Sterz *et al.*, 1973–1974; Ludwig *et al.*, 1978a) and immunoprecipitation tests (Pauli *et al.*, 1979).

By several techniques such as the immunodiffusion test (Sterz *et al.*, 1973–1974), the indirect precipitation test (Pauli and Ludwig, 1977), and immunoelectrophoresis combined with denaturing PAGE analysis of the precipitate (Norrild *et al.*, 1978b), it could be demonstrated that the common antigen involves proteins of BHV-2, HSV-1 and -2, and B virus. The results of an electrophoretic analysis are shown in Fig. 17.

The antigens recognized by heterologous sera in HSV-1- and BHV-2-infected cells and virions share only partial identity, as demonstrated by fused rocket and also by crossed-line immunoelectrophoresis (Fig. 18a, A–D).

The common antigenic sites are located on a glycoprotein in HSV-1 and in BHV-2 (Fig. 20). The HSV glycoprotein with an apparent molecular weight of 125,000 is part of glycoprotein region I of Pauli and Ludwig (1977); it was designated as antigen 11 by Vestergaard (1973) and as virion proteins 7 and 8.5 by Heine *et al.* (1974) and renamed as gly-

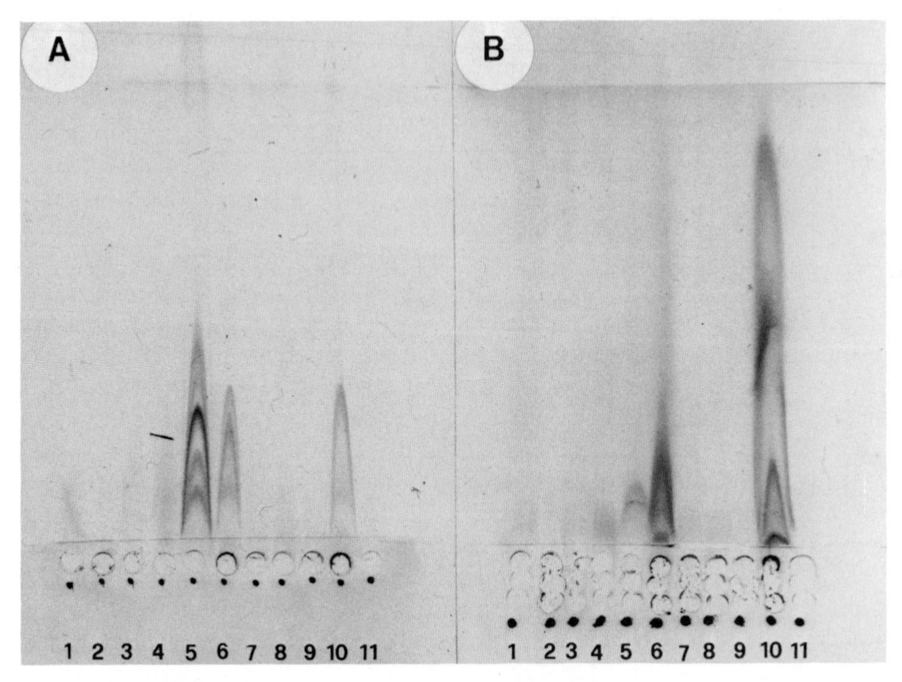

FIGURE 19. Rocket immunoelectrophoresis of antigens from infected cells in antisera directed against BHV-1 (A) or BHV-6(B). Lysates, 10 μl (A) or 30 μl (B) per well, were subjected to electrophoresis into the second dimension gel containing 15 $1/cm^2$ of antiserum. The antigens were prepared from cells infected with BHV-3 (strain C 500) (1), BHV-3 (strain WC 11) (2), BHV-4 (strain Movar) (3), PsR virus (4), BHV-1 (strain LA) (5), BHV-6 (strain E/CH) (6), BHV-2 (strain TVA) (7), HSV-1 (8), EHV-1 (9), and BHV-6 (strain McK/US) (10), (11) Mock-infected GBK cells.

coproteins gA and gB by Spear *et al.* (1978). It is now apparent that gA and gB are products of a single gene, gA/B (Pereira *et al.*, 1981). The identification is based on experiments involving human HSV convalescent sera and cerebrospinal fluid that reacted specifically with this glycoprotein (Pauli and Ludwig, 1977). Because of the cross-reactivity of glycoprotein gA/B with the corresponding BHV-2 glycoprotein, the hyperimmune sera to BHV-2 appear to be a useful tool to analyze the major HSV-1 glycoprotein. BHV-2 specific antisera do not cross-react with glycoprotein region II of HSV (Pauli and Ludwig, 1977). This glycoprotein region is recognized by anti-band II serum directed against HSV proteins (Sim and Watson, 1973) and contains glycoprotein D (Cohen *et al.*, 1978; Spear, 1980).

Cross-neutralization experiments revealed that HSV was inactivated by the BHV-2 specific antiserum in the presence of complement. However, anti-HSV-1 serum did not significantly neutralize BHV-2 even in the presence of complement (Fig. 25) (Pauli *et al.*, 1979). Monoclonal

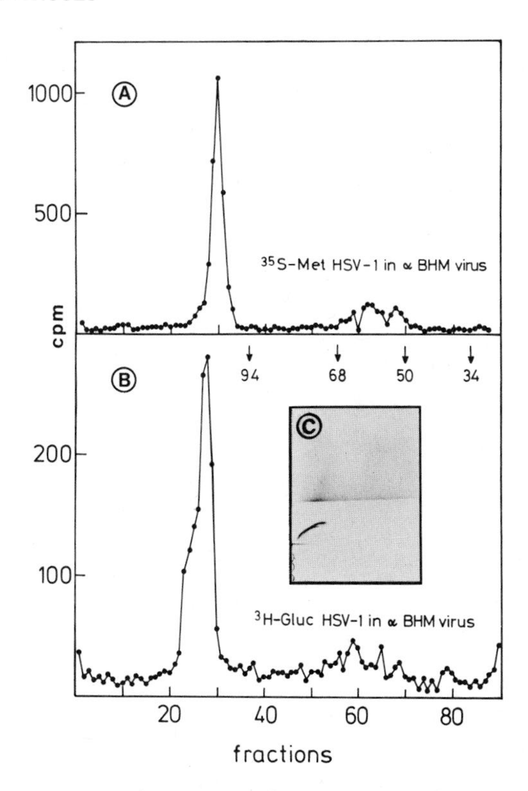

FIGURE 20. Characterization of HSV-1 specific proteins (A–C) sharing common antigenic determinants with BHV-2. The immunoprecipitate (C) obtained after crossed immunoelectrophoresis with anti-BHV-2 serum (α BHM virus) was analyzed by denaturing PAGE according to procedures reported elsewhere (Pauli and Ludwig, 1977; Norrild *et al.*, 1978b). The arrows indicate the molecular-weight markers (semliki forest virus proteins) incorporated in the same gels. The HSV-1 glycoproteins in (B) were identified as gA/B. Data from Ludwig *et al.* (1978b).

antibodies (H233; HD1) directed against HSV-1 or -2, similarly did not inactivate BHV-2 (Lewis, Pereira, and Ludwig, unpublished).

Neutralization of HSV-1 by decomplemented antisera directed against the glycoprotein gA/B complex (Courtney and Powell, 1975) might occur over different antigenic sites as in the HSV-1/BHV-2 system.

Further studies have shown that BHV-2 specific antibodies bind to the surface of HSV-1-infected cells. However, anti-HSV sera do not react with BHV-2 infected cells (Fig. 22). This phenomenon was quantitated by measuring radioactively labeled protein A bound to cells incubated with the homologous or the heterologous sera (Fig. 23). The antigenic determinants on HSV-1-infected cells available for homologous antibodies have been reported earlier in some detail (Pauli and Ludwig, 1977; Norrild *et al.*, 1978a).

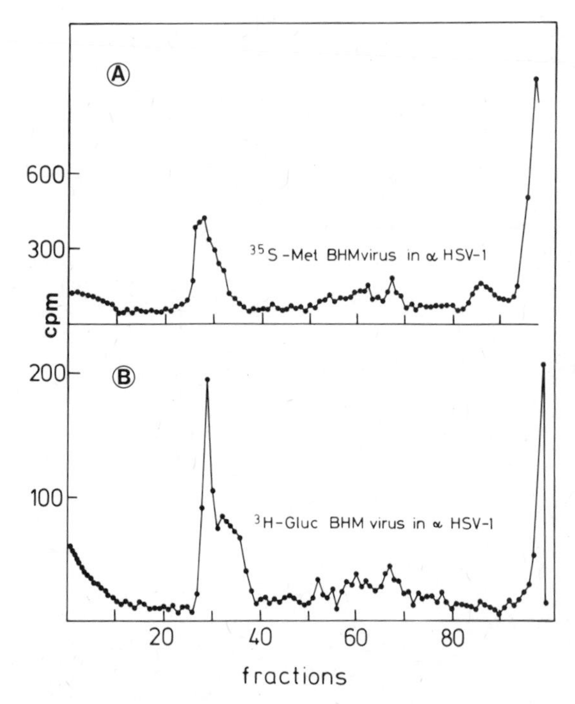

FIGURE 21. Characterization of BHV-2 specific proteins (A, B) sharing common antigenic determinants with HSV-1. Crossed immunoelectrophoresis was performed with anti-HSV-1 serum (αHSV-1) as described in the Fig. 20 caption.

The one-way reaction of accessibility of the common antigen on the surface of cells infected with one or the other herpesvirus seems likewise to hold true for the virion (Fig. 24), as demonstrated by immunoelectron microscopy (Gelderblom *et al.*, 1980).

The demonstration of a common antigen in the human and bovine herpesviruses has stimulated further research on its molecular structure and localization in the lipid bilayer. Charge-shift immunoelectrophoretic analyses (Fig. 26) support the serological findings that the antigen is on the surface of the virion, inasmuch as the HSV-1 antigen seems to be strongly amphiphilic in that it binds considerable amounts of nonionic detergent. In the BHV-2 system, however, the antigen binds little or no detergent. This indicates, as based on other membrane studies (Bhakdi *et al.*, 1977; Helenius and Simons, 1977), that in HSV-1 the antigen is an integral part of the membrane, whereas in BHV-2 it appears to be localized on the internal side of the virion envelope (Ludwig *et al.*, 1978b).

Another outstanding property of BHV-2 and HSV is their ability to cross-protect against a lethal outcome of disease. The relevant results (Sterz, Ludwig, and Rott, unpublished results), which were mentioned in several reports (Sterz *et al.*, 1973–1974; Ludwig, 1976; Ludwig *et al.*,

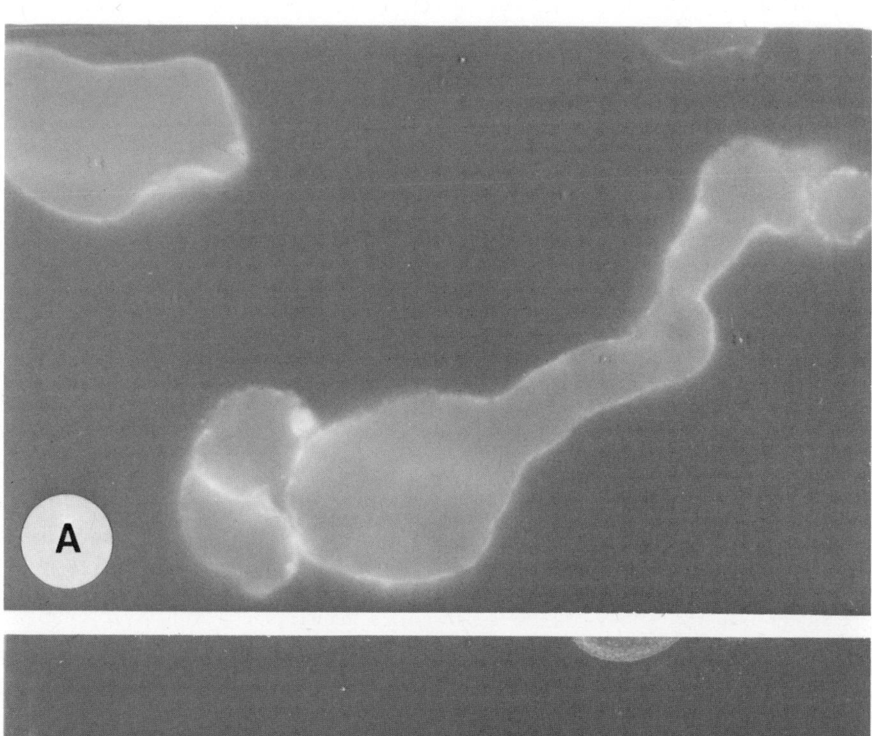

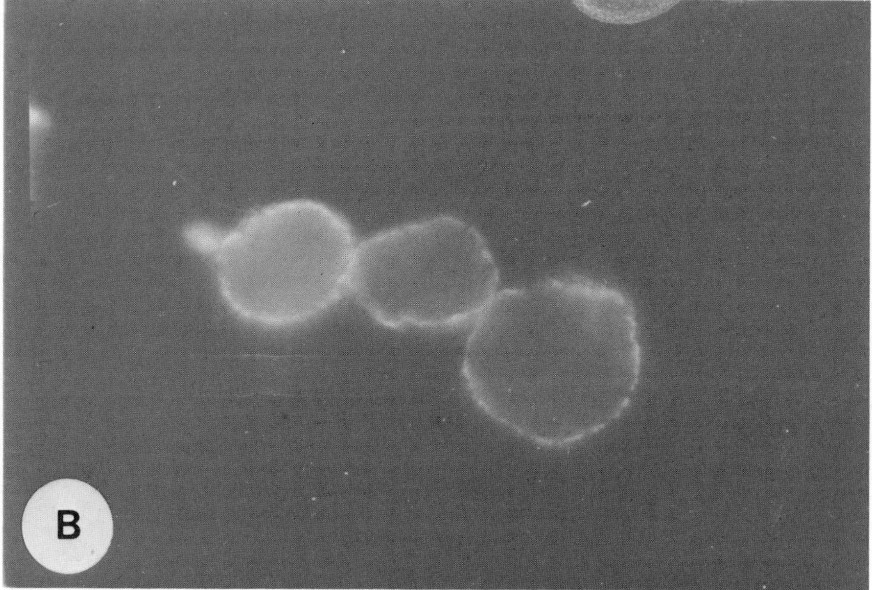

FIGURE 22. Surface immunofluorescence of infected cells demonstrating common antigenic determinants in BHV-2 and HSV-1. BHV-2-infected bovine embryonic skin cells forming polykaryocytes (A) and HSV-1-infected primary rabbit kidney cells (B) were treated with bovine BHV-2 antibodies in an indirect fluorescent-antibody test. No cell-surface staining was observed when (A) was treated with human or rabbit anti-HSV-1 antibodies (not shown), demonstrating the one-way reaction (Ludwig *et al.*, 1978b).

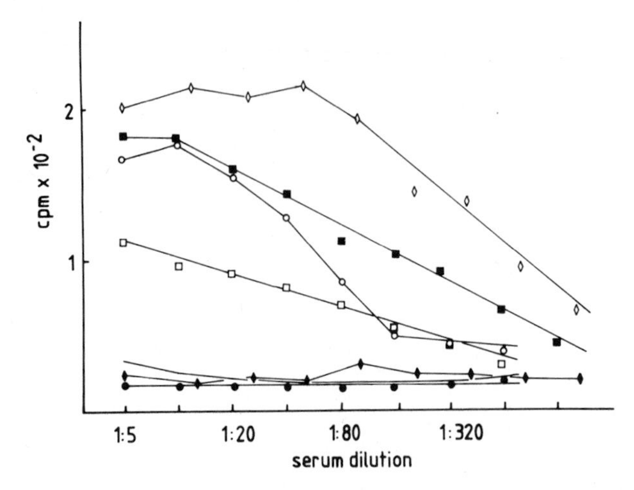

FIGURE 23. Common viral antigenic sites on the surface of BHV-2- and HSV-1-infected cells as demonstrated by binding of antibodies. Rabbit kidney cells (RK) or bovine embryonic fibroblasts (BEF) were grown in microtiter plates and infected with 10 plaque-forming units per cell. Between 18 and 24 hr postinfection, the binding of IgG was measured with the help of ^{125}I-labeled protein A. Open symbols represent RK infected with HSV-1; solid symbols, BEF infected with BHV-2. These cells were treated with rabbit anti-HSV-1 serum (◇, ◆), rabbit anti-BHV-2 serum (□, ■), and human serum containing antibodies to HSV (○, ●) (Pauli *et al.*, 1979).

1978a), are summarized in Table VI. The data clearly show that immunizations with BHV-2 or with membranes of BHV-2-infected cells significantly prevent encephalitis induced by HSV in mice. This outstanding immunogenic capacity of BHV-2 has recently also been confirmed in a reversed experimental approach, when cattle, as the natural host for BHV-2, were preimmunized with HSV and challenged with BHV-2. The disease was considerably milder than in control animals (Castrucci *et al.*, 1981).

On the basis of the antigenic relationship of HSV and BHV-2, Honess and Watson (1977) grouped these viruses in one "neutroseron," which,

TABLE VI. Protection of Mice with BHV-2 (Strain TVA) against a Lethal Encephalitis Induced by HSV-1 (Strain KOS) or HSV-2 (Strain 196)[a]

Preimmunization with:	LD_{50} determination after challenge with:	
	HSV-1	HSV-2
Cell-free BHV-2	5.0×10^{0}	1×10^{-2}
Cell medium	1×10^{-3}	5×10^{-3}
Purified membranes from BHV-2-infected cells	5×10^{0}	1×10^{-3}
Purified membranes from mock-infected cells	8×10^{-2}	2.5×10^{-3}

[a] The mice were preimmunized by intraperitoneal and intracranial inoculation. Challenge virus was given intracerebrally. Similar results were obtained when preimmunization was done subcutaneously and intraperitoneally.

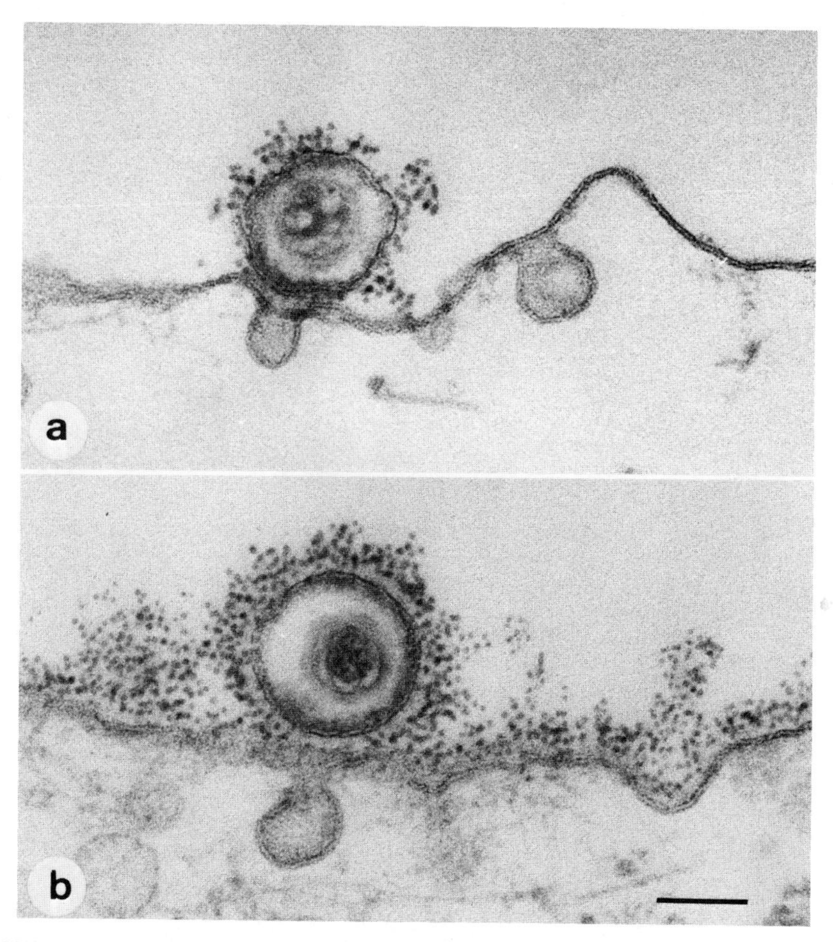

FIGURE 24. Cross-reactivity of BHV-2 and HSV-1 as demonstrated by immunoelectron microscopy. HSV-1 (strain KOS)-infected cells were prefixed with glutaraldehyde and reacted with (b) uninfected cell-absorbed homologous polyvalent antibodies prepared in the rabbit (Pauli and Ludwig, 1977) and (a) uninfected cell-absorbed heterologous polyvalent antibodies directed against BHV-2 (strain TVA) prepared in a heifer (Ludwig, 1976). Details of the preparation and staining with anti-IgG ferritin conjugate were reported elsewhere (Gelderblom *et al.*, 1980). ×120,000. Scale bar: 100 nm. Courtesy of Hans Gelderblom.

according to our findings, would not hold for BHV-2 and the B virus, since these two viruses do not cross-neutralize. Nevertheless, such a common antigen of four distinct herpesviruses from different species, involving a major glycoprotein of the viruses, might represent a vestige of a common ancestor, and will certainly be of interest for further studies on the evolution of herpesviruses (Nahmias, 1972).

It is of interest to note in connection with the preceding discussion of common antigenic determinant sites on a viral glycoprotein that DNA binding proteins of several herpesviruses, including those of BHV-2 and

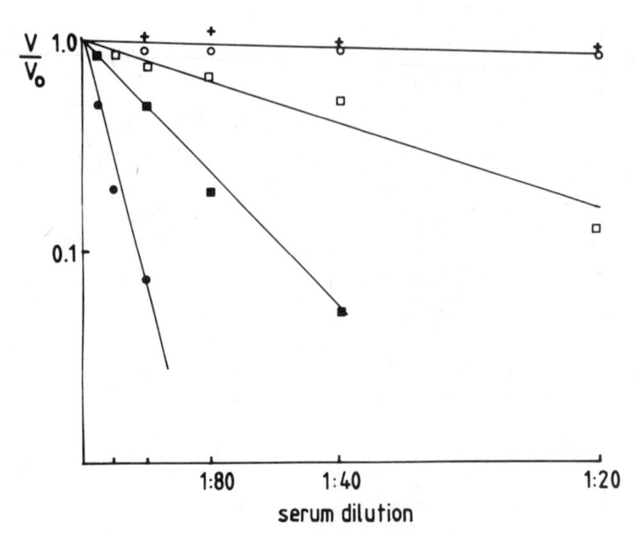

FIGURE 25. Detection of common virion antigens on BHV-2 and HSV-1 by complement-requiring neutralization. Approximately 100 plaque-forming units were incubated (1 hr at 37°C) with serum diluted as indicated. Active or inactivated guinea pig serum was added and incubation continued for 1 hr at 37°C. The results are shown in terms of fraction (V/V_0) surviving neutralizations. The reaction mixtures were: HSV-1 (●) and BHV-2 (○) plus rabbit anti-HSV-1 serum, plus inactivated C′; HSV-1 (□) and BHV-2 (■) plus rabbit anti-BHV-2 serum, plus active C′; HSV-1 (+) plus rabbit anti-BHV-2 serum, plus inactivated C′ (Pauli *et al.*, 1979).

HSV-1, appear to be immunologically related. These proteins are not glycosylated (Yeo *et al.*, 1981).

VII. PROPERTIES OF THE VIRUSES

A. *In Vitro*

1. Growth Curves and Cytopathic Effects

The bovine herpesviruses vary considerably in their growth properties (see Table VII). Bovid herpesvirus 1 (BHV-1) and BHV-6 replicate quickly in a one-step growth curve (Fig. 27), reaching maximal titers of 10^9 infectious particles/ml approximately 1 day postinfection. The ratio of physical to infectious particles at that time approaches 100 (Fig. 28). Both viruses have a broad host-cell range. BHV-2 attains lower titers, replicates more slowly, and has a more restricted host-cell range. BHV-3 resembles in many aspects the lymphotropic herpesviruses of other species (Edington *et al.*, 1979; Patel and Edington, 1980). Only the laboratory strains can be easily assayed for infectivity in bovine cells (see Fig. 27). BHV-4 strains replicate considerably slower than the lytic BHV-1 and -2 and best in cells of bovine origin. Not much information exists on growth properties of BHV-5. Defective particles, which might affect

TABLE VII. Properties of Bovine Herpesviruses *in Vitro*[a]

Group	Host range	Optimal cell system	Kind of cytopathic effect[b]	Assay system	Average titers (infectious units/ml)	Time post-infection	Transforming ability or "persistent" infection?	Ref. Nos.[c]
BHV-1	Broad	Bovine fetal skin	R	Plaque assay	10^9	20 hr	Yes	1–4
BHV-2	Relatively broad	Bovine fetal skin	F!!	Plaque assay	10^{5-6}	30 hr	Yes	2, 5, 6
BHV-3	Narrow	Bovine fetal cells	F	$TCID_{50}$ immunofluorescent plaque assay	10^{5-6}	5 day	—	7–9, 13
BHV-4	Narrow	Bovine cells	R	$TCID_{50}$	10^{3-4}	4 day	—	10–13
BHV-5	Narrow	Sheep cells	R/F	$TCID_{50}$	10^6	3 day	Yes	14–16
BHV-6	Broad	Bovine fetal lung	R	Plaque assay	10^8	18 hr	—	13, 17

[a] The table gives a generalized summary of some biological properties of these viruses as compiled from the literature and based on our own unpublished results.

[b] (R) Rounding up; (F) fusion of infected cells.

[c] References: (1) Stevens and Groman (1963); (2) Gibbs and Rweyemamu (1977a,b); (3) Michalski and Hsiung (1975); (4) Geder *et al.* (1980); (5) Sterz and Ludwig (1972); (6) Rweyemamu *et al.* (1972); (7) Ferris *et al.* (1976); (8) Rossiter *et al.* (1978); (9) P.H. Russel (1980); (10) Bartha *et al.* (1966); (11) Storz (1968); (12) Mohanty (1975); (13) Engels, Leiskau, and Ludwig (unpublished results); (14) MacKay (1969); (15) DeVilliers *et al.* (1975); (16) Martin *et al.* (1976); (17) Berrios and McKercher (1975).

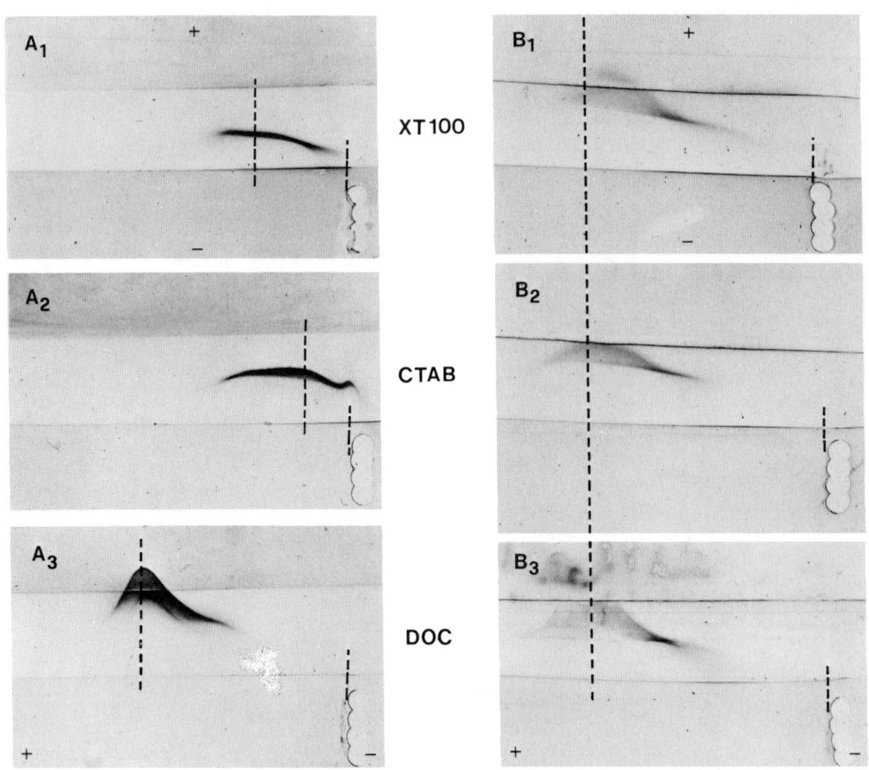

FIGURE 26. Charge-shift crossed immunoelectrophoresis of herpesvirus antigens (Ludwig *et al.*, 1978b). Amphiphilic, "intrinsic" membrane proteins bound to Triton X-100 (TX100) micelles, show an anodal shift when treated with deoxycholate (DOC) and a cathodal shift when treated with cetyltrimethylammoniumbromide (CTAB). No shift is found with hydrophilic proteins, which do not bind significant amounts of detergent (Bhakdi *et al.*, 1977); Helenius and Simons, 1977). The antigens of BHV-2 (B_{1-3}) and HSV-1 (A_{1-3}) carrying common antigenic determinants were subjected to electrophoresis in one dimension and then in the second dimension into heterologous antisera. The HSV-1 antigen appears to be an intrinsic, the BHV-2 antigen an extrinsic, protein.

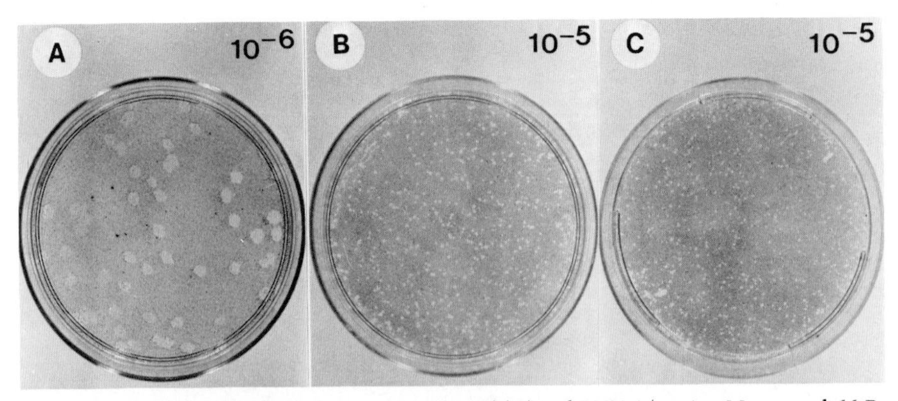

FIGURE 27. Plaque test of BHV-1 (strain Cooper) (A) and BHV-4 (strains Movar and 66-P-347) (B, C) on bovine fetal skin cells. Monolayers were strained with crystal violet at days 4 (A) and 10 (B, C) after infection. Courtesy of Johannes Storz.

some of the biological properties of bovine herpesviruses, have not been investigated. The growth properties are also reflected in the plaque morphology. BHV-1, -2, and -6 usually show large plaques of 2–3 mm in about 3 days, whereas BHV-4 strains give rise to plaques of less than 2 mm in approximately 8 days (Fig. 27a). The delay in development is also seen with the microplaques formed by BHV-3 (Ludwig and Storz, unpublished results). Generally speaking, the bovine herpesviruses prefer to grow in cultures of cells of bovine origin in contrast to other lytic herpesviruses

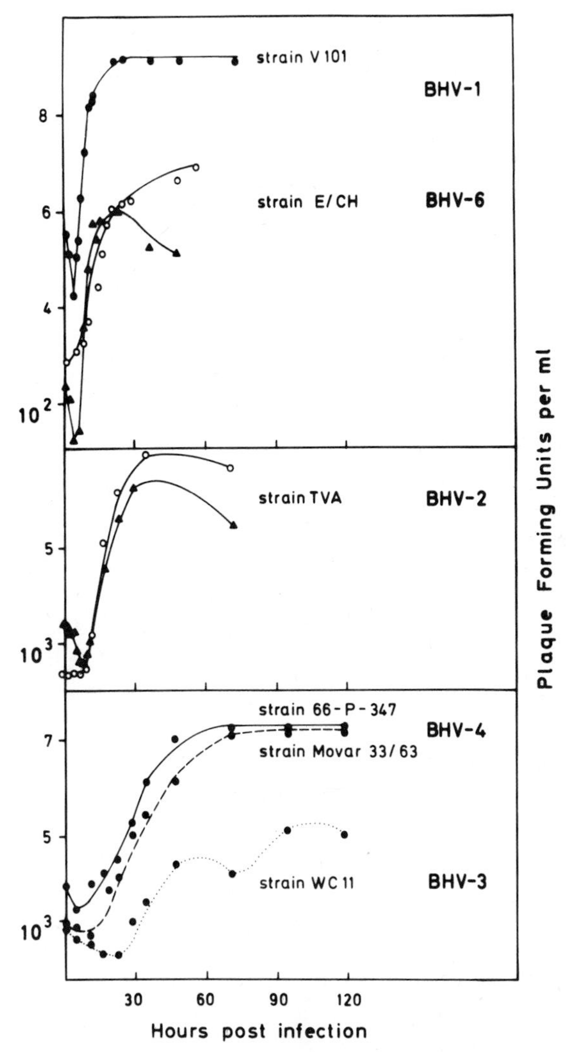

FIGURE 27. (*Continued*). Growth curves of bovine herpesviruses on bovine cells. Note the relatively rapid replication of BHV-1 and BHV-6 and the rather slow growth of BHV-3 and BHV-4 strains. BHV-3 was assayed by a fluorescent focus test. For BHV-2 and BHV-6, open and closed symbols refer to extracellular and intracellular virus, respectively. The curves were redrawn from or kindly provided by: Törner (1974), BHV-1; Sterz and Ludwig (1972), BHV-2; Leiskau (unpublished), BHV-3 and -4; Engels (unpublished), BHV-6.

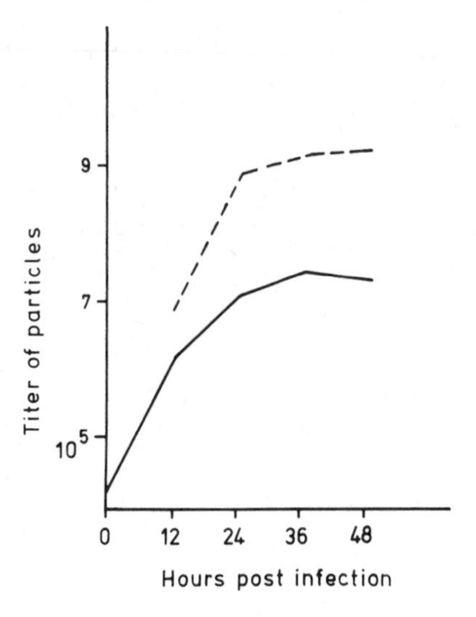

FIGURE 28. Production of infectious (——) and physical (– – –) virus particles of strain LA (BHV-1) in MDBK cells. Courtesy of P.-P. Pastoret.

such as herpes simplex virus (HSV), B virus, or pseudorabies (PsR) virus (Kaplan, 1973).

With respect to the social behavior of infected cells (Ejercito *et al.*, 1968), bovine herpesviruses clearly fall into two categories. BHV-1 and -6 lead to rounding of cells, whereas all the BHV-2 strains fuse tissue-culture cells. BHV-3 likewise induces polykaryocytosis, whereas the BHV-4 strains cause slow degeneration of cells, and clumps of rounded cells detach from the monolayer (Table VII).

The various type-specific cell-surface alterations induced by BHV-1 (rounding) and BHV-2 (fusion) using the same cell-culture system are shown in Fig. 29 by scanning electron microscopy. Related to the rounding of cells is an increase in number of microvilli, whereas a smoothing on the cell surface can be seen in cells undergoing fusion. Similar observations were described for HSV-infected HEp-2 cells (Schlehofer *et al.*, 1979).

Fusion of bovine cells, especially by BHV-2 strains, cannot be inhibited by convalescent or hyperimmune sera (unpublished results). This had been shown for PsR virus, which also induces extensive polykaryocytosis (Ludwig *et al.*, 1974), and for HSV-1 (Hoggan and Roizman, 1959). Although "fusion from within" appears to be the prominent mode of cell fusion in BHV-2-infected cells, nothing is known about the "fusion factor" and the molecular events that play a role. More extensive studies on cell fusion have been reported for other virus cell systems including the HSV and PsR virus systems (Roizman, 1962; Gallaher *et al.*, 1973; Ludwig and Rott, 1975; Knowles and Person, 1976; Schmidt *et al.*, 1976; Compans and Klenk, 1979; Ruyechan *et al.*, 1979; Spear, 1980).

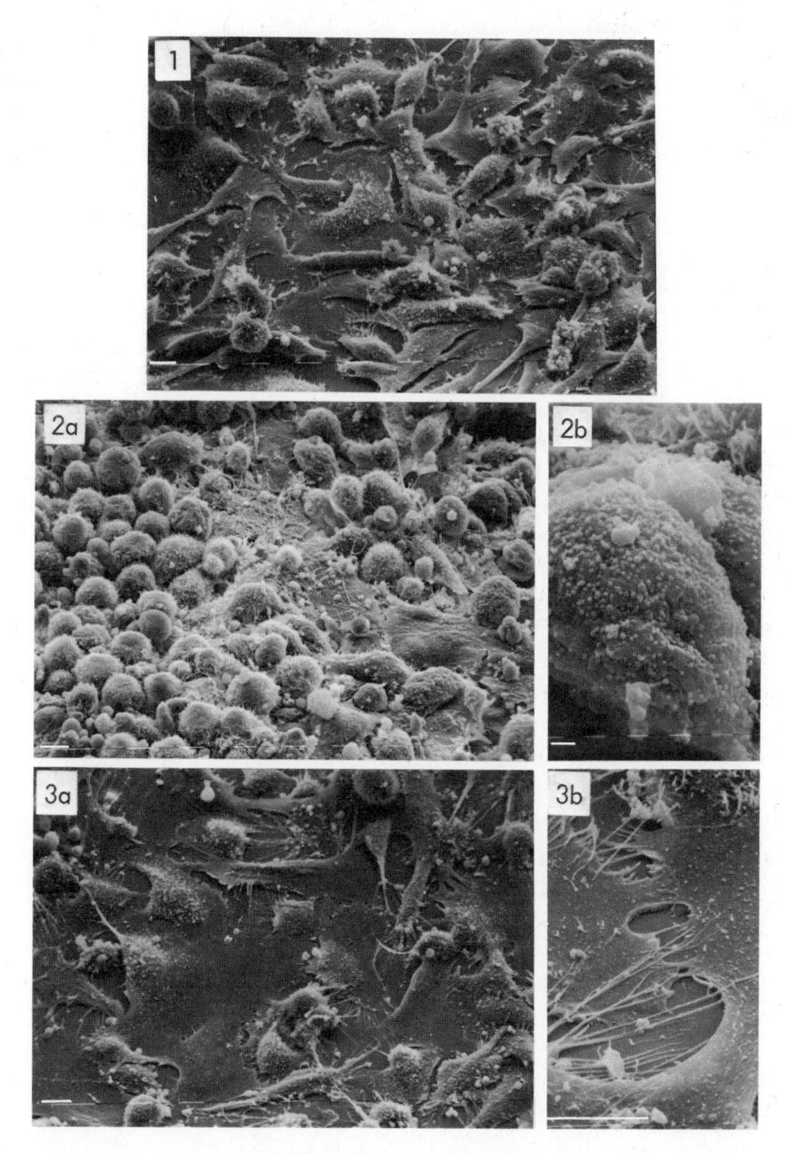

FIGURE 29. Virus-specific surface alterations of bovine fetal lung fibroblasts after infection with bovine herpesviruses (BHV-1, BHV-2) visualized in scanning electron microscopy. (1) Mock-infected cells. (2a) BHV-1 infection induces rounding of cells (cells prepared 18 hr postinfection) and leads to an increase in the number of microvilli. (2b) The areas free of microvilli appear to be packed with large amounts of released virus particles adhering to the surface of the plasma membranes. Similar results were obtained with HSV-1-infected HEp-2 cells (Schlehofer *et al.*, 1979) and PsR-virus-infected cells (Bijok *et al.*, 1979). (3a) Cell fusion induced by BHV-2 infection (cells prepared 24 hr postinfection). The cell surface is smoothed, virus particles are not discernible at the membrane. (3b) Magnified area between two cells. Scale bars: (1, 2a, 3a, b) 10 μm; (2b) 1 μm. Courtesy H. Zeichardt, J. R. Schlehofer, A. Hennig, and K. O. Habermehl.

Studies on the growth of bovine herpesviruses in organ cultures were reported on tracheal rings infected with BHV-1 (Shroyer and Easterday, 1968; Chia and Savan, 1974a,b) and on bovine teat skin infected with BHV-2 (James and Povey, 1973). Infected organ cultures may contribute valuable information on cell tropism and the microepidemiology of bovine herpesviruses.

2. Transforming Ability and Persistent Infection

Among the known bovine herpesviruses, only BHV-1 has been reported to transform hamster cells (Michalski and Hsiung, 1975) and mouse embryo fibroblasts (Geder *et al.*, 1980). The BHV-1 used in the latter studies was also able to cause a persistent infection of human kidney cancer cells; a cell line expressing infectious bovine rhinotracheitis (IBR)-virus-specific antigens was also established (Geder *et al.*, 1979).

The activation of BHV-1 from normal bovine fetal spleen cell cultures (Ludwig and Storz, 1973; Storz *et al.*, 1980) supports the view that BHV-1 is able to persist in an as yet unknown state in bovine tissues and that the virus can be activated by growing the cells in culture. It is of considerable interest that BHV-1 was also recovered from spleen cells of normal ferrets (Porter *et al.*, 1975). These independent findings with quite different animal species indicate that the genome of at least some BHV-1 strains remains present in a repressed form in cells of the reticuloendothelial system. DNA analysis of some of these strains revealed that they can be grouped with the IBR-like virus strains (Pauli *et al.*, 1981a).

BHV-2 was reported to persist in its host in a kind of steady-state carrier state that appears to differ from the typical persistent infection found with BHV-1 (Rweyemamu *et al.*, 1972).

Although not yet proved, BHV-3 and BHV-5 might be able to transform cells *in vitro* since both viruses are associated with proliferative processes *in vivo* (Plowright, 1968; Coetzee *et al.*, 1976; Liggitt *et al.*, 1978).

B. *In Vivo*

1. Viremia and Virus Spread

Dissemination of bovine herpesviruses can occur hematogenously and neurally (Fenner *et al.*, 1974). Of the known bovine viruses, only BHV-3 is known to persist in a viremic stage for prolonged intervals. The association of this virus with whole blood has already been recognized in the early days of malignant catarrhal fever research, and transfer of whole blood is still practiced to transmit the disease (Götze and Liess, 1929; Plowright, 1968; Straver and van Bekkum, 1979).

In BHV-1 infections, and especially in experimental infection, the virus can be excreted from the portal of entry (Fig. 30A). It was also recovered from buffy coat cells, but not from whole blood samples. Antigen was found in mononuclear cells, which correlates with *in vitro* investigations (Narita *et al.*, 1978; Nyaga and McKercher, 1980).

In the BHV-2 infections, a viremic stage was not unequivocally demonstrated, although the virus was disseminated in experimental infections, indicating such a spread (Martin *et al.*, 1969; Ludwig, 1976; Gibbs and Rweyemamu, 1977b). The failure to detect viremia might reflect the concentration of virus in the blood; it may be below the threshold levels necessary for its isolation.

Penetration of the central nervous system (CNS) was observed with BHV-1 and -3. Spread to the CNS may be via routes similar to those used by other neurotropic viruses (Johnson and Mims, 1968; Wolinsky and Johnson, 1980), that is, by either the neural route—which is now assumed to be more probable—or the hematogenous route. In the case of the BHV-1 infections occurring primarily on the nasal mucosa, the virus could be transmitted by way of the nerve fibers extending from the olfactory mucosa to the olfactory bulb into the brain.

Ascending infection through peripheral nerves, trigeminal ganglionitis, and nonsuppurative encephalitis have been demonstrated (Bagust and Clark, 1972; Narita *et al.*, 1976). Infection of the vulva with the IBR-LA strain resulted in lesions of the sacrolumbar spinal cord and ganglia, as

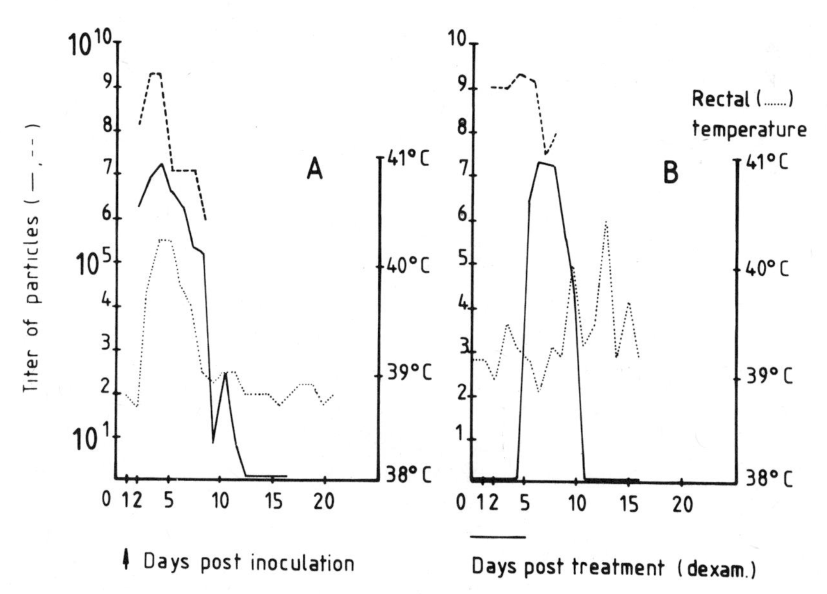

FIGURE 30. Excretion of BHV-1 [physical (– – –) and infectious (——) particles] after intranasal inoculation (A) and reexcretion after dexamethasone treatment (B) of bovine animals (Pastoret, 1981).

well as in the trigeminal ganglia, suggesting neural and hematogenous spread (Narita *et al.*, 1978).

2. Target Cells

Though gross pathological changes have been described in several bovine herpesvirus infections, the identity of the cells that propagate the virus *in vivo* is not known. In the respiratory tract, BHV-1 has a dramatic effect on the tracheal mucosa demonstrated by scanning electron microscopy (Allen and Msolla, 1980). The first effect in the ciliated epithelium cells was significant loss of the microvilli 4 days after infection, resulting in large areas lacking the cilia. These ciliated cells appeared to be able to propagate the virus as deduced from the presence of herpesvirus particles. The severe loss of microvilli accelerated invasion by other microbial agents.

In the genital tract, BHV-1 also infects the epithelial cells. Virus can reach epithelial regions in the semen channels and the testes of the male animal and cause degeneration with polykaryocytosis of spermiocytes (Steck *et al.*, 1969). In the CNS, virus infection was demonstrated where ganglia, neuroglia and Schwann cells support virus replication (Narita *et al.*, 1978).

BHV-2 destroys cells in the dermis, preferentially those of the stratum germinativum and spinosum, and causes confluent syncytial masses (Martin *et al.*, 1969; Cilli and Castrucci, 1976; Gibbs and Rweyemamu, 1977b).

BHV-3 is found in medium-sized lymphocytes after experimental infection (Patel and Edington, 1980). In natural infection, infectivity is associated with whole blood cells (Plowright, 1968), but can also be detected with corneal and turbinate cells (Mushi *et al.*, 1980).

The cells that support the multiplication of BHV-4 and -5 in the infected animal have not been identified precisely. During recent studies on BHV-6 infection, it became obvious that the virus seems to preferentially destroy epithelium cells of the digestive tract (Mettler *et al.*, 1979).

3. Latency and Reactivation

The ability to remain latent is a general property of herpesviruses (Rapp and Jerkofsky, 1973). Latent infection of cattle with BHV-1 was reported by numerous groups (Snowdon, 1965; Saxegaard, 1968; Sheffy and Davis, 1972; Bitsch, 1973; Davies and Carmichael, 1973; Davies and Duncan, 1974; Darcel and Dorward, 1975; Gibbs *et al.*, 1975; Castrucci *et al.*, 1980a; Pastoret, 1981). The major finding was that BHV-1 could be easily and very quickly induced to multiply and be excreted from experimentally and naturally infected animals after treatment with corticosteroids (Fig. 30B). Dexamethasone was used to test cattle for latent BHV-1 infection by stimulating virus multiplication and excretion. We have shown that BHV-1 could be activated from a significant number of healthy fetuses (Ludwig and Storz, 1973; Storz *et al.*, 1980).

Although BHV-2 is assumed to remain latent, attempts to reactivate the virus have not produced consistent results (Martin *et al.*, 1975; Probert and Povey, 1975; Castrucci *et al.*, 1980b).

Clinically inapparent infection of wildebeest with BHV-3 and induction of virus excretion by betamethasone injection have been reported (Rweyemamu *et al.*, 1974).

Although no specific studies on BHV-4 viruses have been reported, the reactivation of these viruses by other diseases associated with inflammatory processes or even with proliferative processes points to their ability to remain present in the animal in a latent form.

Aspects of latency and recrudescence of clinical disease will also be discussed in connection with immune response and other defense mechanisms (see Section VIII), since very likely a balance between the immune status and the appearance of excretion of virus exists.

4. Association with Cell Proliferation

Although none of the bovine herpesviruses is known to be directly involved in oncogenic processes, there are reports that BHV-1 was isolated in cases of tumors of the eye (Taylor and Hanks, 1969; B. Epstein, 1972). The *in vitro* transforming activity of this virus was reported (Michalski and Hsiung, 1975; Geder *et al.*, 1980).

Recent studies on the nature of acute lymphoproliferative changes in rabbits after inoculation with BHV-3 have suggested a relationship between virus infection and proliferation of lymphoblastoid cells (Edington *et al.*, 1979). Although not detected in the altered tissues, viral antigen was present in lymphocytes on cultivation of explants (Patel and Edington, 1980). A similar disease pattern and demonstration of BHV-3 replication have also been achieved in calves (Edington, personal communication).

Although the association of BHV-5 with the adenomatous–proliferative processes in the lungs of "jaagsiekte" sheep is known, there is little convincing evidence that this virus is the etiological agent (De Villiers *et al.*, 1975; Martin *et al.*, 1976; Hiepe and Ilchmann, 1978).

VIII. IMMUNE RESPONSE AND OTHER DEFENSE MECHANISMS

A. Humoral Immune Response

Although bovine fetuses can produce both IgM and IgG antibodies after the 5th month of gestation, the bovine fetal sera seldom contain virus-specific antibodies, presumably because fetal infection is rare and the structure of the placenta precludes the passage of maternal antibodies to the fetus. After birth, colostral antibodies can be resorbed optimally during the first 12 hr, with resorption ceasing by 36 hr. The antibodies acquired in this fashion can persist in the calf until the 3rd month post-partum (Owen *et al.*, 1968; Kendrick, 1973; Osburn, 1973; Schultz, 1973).

Under normal conditions, the placenta seems to protect the fetus against infection by herpesviruses. Fetal infection with bovid herpesvirus 1 (BHV-1) did not occur even in cases where virus could be isolated from the placenta. If the fetus becomes infected, the virus is viremically spread to all tissues, and this is followed by peracute fetal disease. The pattern of disease is similar in colostrum-deprived neonates.

Most of the investigations on humoral immune response in the adult were concerned with diagnosis or epidemiology. Numerous reports were published documenting the production of antibody to BHV-1, -2, -3, -4, and -6 by a variety of techniques such as neutralization, complement fixation, fluorescent antibody, direct and indirect hemagglutination, immunodiffusion, immunoelectrophoresis, and enzyme-linked immunosorbent assay tests. References and details can be found in Kaminjolo *et al.* (1969), Sterz *et al.* (1973–1974), Saito *et al.* (1974), Törner (1974), Gibbs and Rweyemamu (1977a,b), Kahrs (1977), Aguilar-Setién *et al.* (1980), Herring *et al.* (1980), and Rossiter *et al.* (1980). Several reports have dealt with the sequence of appearance of different immunoglobulin classes (Rossi and Kiesel, 1976), the role of complement-requiring antibodies (Potgieter, 1975), and the role of humoral defense parameters in the recovery from infection (Rouse and Babiuk, 1978). The differences in requirement of complement for neutralization by sera taken early or late after infection have been used as a sensitive indicator system to recognize infection (Potgieter, 1975). From *in vitro* experiments, different mechanisms have been proposed to explain how the humoral immune response serves to inhibit virus spread and ultimately to eliminate infected cells. Mechanisms such as direct lysis of infected cells in the presence of antibody and complement or antibody-dependent cytolysis with and without complement mediated by a variety of white blood cells were described by Rouse and Babiuk (1978).

Considerable attention has also been given to the role of IgA antibodies against BHV-1, locally produced in the respiratory and genital tracts. Such antibodies protected by neutralizing reinfecting virus (Asso and Le Jan, 1977; Bier *et al.*, 1977; Whitmore and Archbald, 1977), but could not mediate the direct killing of virus-infected cells.

The humoral immune response in BHV-2 (Martin, 1973; Sterz *et al.*, 1973–1974; Ludwig, 1976; Cilli and Castrucci, 1976) as well as in BHV-6 infections (Saito *et al.*, 1974; Mettler *et al.*, 1979) generally follows the pattern of anti-BHV-1 response. These viruses induce a pronounced antibody response in the natural host and in experimentally infected animals. The situation with BHV-4 is different in that only antibodies of low titers are detectable after natural (Bartha *et al.*, 1967; Potgieter and Aldridge, 1977) or experimental infection (Mohanty, 1973). The development of a cell-culture-adapted BHV-3 strain (WC 11) permitted demonstration of anti-BHV-3 neutralizing antibodies in a variety of ungulates (Reid *et al.*, 1975). Further studies are required to clarify whether the generally weak humoral immune responses in BHV-3, -4, and -5 infections correlate in some way with the more or less strong cell association of these viruses and, in the case of BHV-4 infections, with a relatively mild clinical disease.

B. Cell-Mediated Immunity

In a series of elegant experiments, Babiuk and Rouse drew attention to cellular defense mechanisms in BHV-1-infected cattle (Rouse and Babiuk, 1978; Babiuk and Rouse, 1979). They found a virus-specific lymphocyte stimulation, essentially confirming earlier reports of Davies and Carmichael (1973), and demonstrated direct killing of virus-infected cells mediated by T cells (Rouse and Babiuk, 1977). They also reported that immune peripheral-blood lymphocytes could interact with virus-infected cells, preventing intercellular virus spread. The containment of virus infection could reflect direct killing of virus-infected cells or the release of interferon from virus-stimulated T lymphocytes (Babiuk and Rouse, 1976).

Cellular immunity can also be detected by the delayed hypersensitivity skin test in BHV-1-infected cattle. Experience with this test had been gained in other virus infections of cattle (Morein and Moreno-Lopez, 1973). The test provides a useful measure of the immune status of BHV-1-infected animals (Darcel and Dorward, 1972) and was successfully applied to detect virus carriers (Fig. 31) and latently infected animals that lacked neutralizing antibodies (Aguilar-Setién *et al.*, 1978, 1979; Pastoret, 1981).

Cellular immunity to BHV-3 infection has been studied primarily in the rabbit system. Lymphocyte transformation was depressed from the onset of disease, and no stimulation *in vitro* with viral antigens could be detected. An immunosuppressive factor in lethally infected rabbits has been reported but not confirmed (Wilks and Rossiter, 1978; P.H. Russel, 1980). To our knowledge, there are no reports on the role of cellular immunity in bovine infections with BHV-3, BHV-4, BHV-5, and BHV-6.

C. Interferon

Interferon was demonstrated in nasal secretions in BHV-1-infected bovines as early as 5 hr postinfection and persisted until 8 days postinfection. Titers of interferon were higher in nasal secretions than in blood

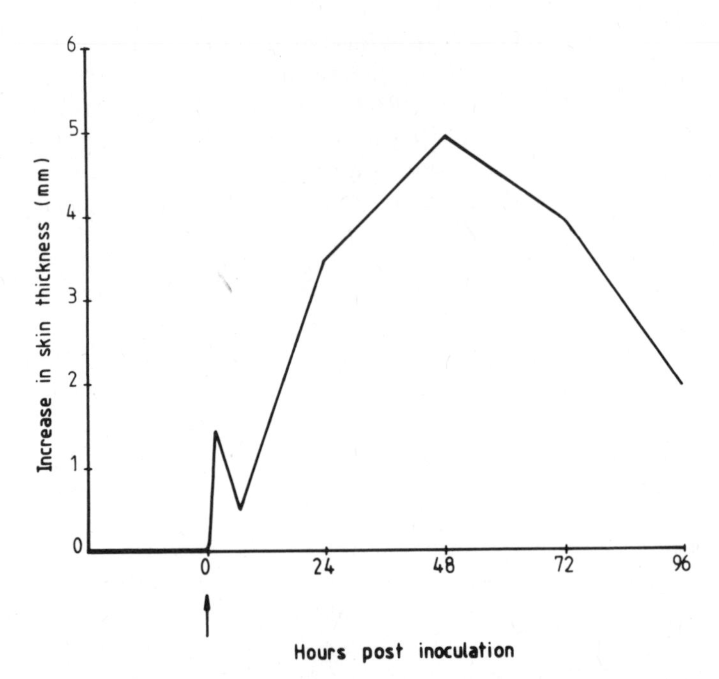

FIGURE 31. Delayed hypersensitivity skin test. After intradermal inoculation of BHV-1 specific antigen (↑), a nonspecific followed by a specific increase in skin thickness can be measured in the BHV-1-infected animal (Pastoret, 1981).

(Ahl and Straub, 1971; Todd *et al.*, 1972; Zygraich *et al.*, 1974; Cummins and Rosenquist, 1977, 1979).

Immune bovine interferon was induced *in vitro*, and it has been suggested that T-lymphocyte populations provided the antigen-specific step and that physical contact with macrophages enhanced interferon levels *in vitro* (Babiuk and Rouse, 1976). It has also been reported that bovine immune interferon enhanced cell-mediated cytotoxicity in the antibody-dependent cellular cytotoxicity (ADCC) test (Wardley *et al.*, 1976).

D. Role of Host Defense Mechanisms

In the neonate, humoral immunity, especially virus-neutralizing antibodies, is thought to play a major protecting role against BHV-1 infections. This is not the case in BHV-3 infections, inasmuch as a protective effect of colostral antibodies could not be demonstrated (Plowright, 1968).

In the adult bovine, an interplay of humoral, cellular, and other defense mechanisms, such as, for example, interferon induction, is thought to play a significant role in viral recrudescence and recovery from overt disease (Rouse and Babiuk, 1978). Moreover, it appears that BHV-1, -2, and -3 can persist in the presence of circulating antibodies (Plowright, 1968; Cilli and Castrucci, 1976; Gibbs and Rweyemamu, 1977a,b) and give rise to recrudescent infection.

The immunosuppressive situation often seen in cattle either during pregnancy, under the influence of stress factors such as transport or crowding of animals, or experimentally by application of immunosuppressive drugs has shed some light on the role of different immune mechanisms in controlling herpesvirus pathogenesis.

BHV-1 has been reported to remain latent in cattle following primary infection (Davies and Carmichael, 1973; Pastoret *et al.*, 1980), but the ability to establish latency varies from one virus strain to another (Narita *et al.*, 1980). As a general rule, treatment of infected, asymptomatic cattle with immunosuppressive drugs results in reactivation of latent virus identified on the basis of biological markers and restriction-endonuclease cleavage sites. Of particular interest is the observation that animals with the lowest immune status—reflected by neutralizing antibodies (ADCC) and blastogenesis response—excreted the highest levels of virus (Pastoret *et al.*, 1980). There seemed to be a correlation between the level of the immune response and the duration of virus shedding. Since the immune response can control inter- and extracellular virus spread, repeated reactivations of latent virus would act as a natural booster of immunity.

With respect to BHV-2, it has been reported that primary infection rendered cattle immune for several years, which could explain disease outbreaks at intervals of approximately 5 years (Rweyemamu, 1969; Gibbs *et al.*, 1972). In contrast to easy reactivation of BHV-1, virus could not be recovered following dexamethasone treatment of asymptomatic animals that had been infected with BHV-2 (Castrucci *et al.*, 1980b).

Animals surviving BHV-3 infection appear to develop a lifelong resistance to reinfection (Götze, 1930) and are resistant to challenge by high doses (10^6 $TCID_{50}$) of virulent virus (Plowright, 1968).

IX. DISEASES, EXPERIMENTAL INFECTIONS, AND PATHOLOGY

The clinical features of infections with different bovine herpesviruses were described in Section II. The characteristics of the different clinical expressions by viruses of the same group will be summarized in this section. Detailed reviews of the individual diseases emphasizing one or the other aspect of the clinical picture prevalent in different parts of the world have been published elsewhere (Plowright, 1968; McKercher, 1973; Cilli and Castrucci, 1976; Gibbs and Rweyemamu, 1977a,b; Kahrs, 1977; Hiepe and Ilchmann, 1978; Straub, 1978).

A. Bovid Herpesvirus 1 Infections

These infections are characterized by a variety of clinical manifestations that overlap to some extent. For the sake of clarity, the different forms will be described briefly as follows:

Infectious bovine rhinotracheitis (IBR) is the most prominent and best-known disease and is associated with fever and more or less severe affliction of the respiratory tract. Pathological pictures vary from mere inflammations of the mucosae to severe alterations in the tissue of the respiratory tract. Complicated cases are often the consequences of infection with other viruses and bacteria. *Keratoconjunctivitis* is a frequent consequence of IBR infection either by itself or in conjunction with respiratory illness.

Encephalitis is a rather rare form of bovid herpesvirus 1 (BHV-1)-induced disease. It can be found in the calf, in which Aujeszky's disease has been ruled out (Barenfus *et al.*, 1963; Wellemans *et al.*, 1976). Encephalitis in adult bovines was first reported from Australia (French, 1962a,b). Sporadic outbreaks have been observed all over the world. Strains of BHV-1 may differ considerably in their affinities for CNS tissue (McKercher *et al.*, 1970).

BHV-1 also causes *abortions* (McKercher and Wada, 1964; Afshar, 1965). Conjunctivitis frequently precedes abortion, suggesting that the same agent is responsible for both. An association of abortion with intramuscular application of modified live vaccine was soon recognized. Fetuses that were infected *in utero* or after birth have shown systemic infections that are frequently fatal. The focal necrotic lesions observed in liver, kidneys, lymph nodes, and the digestive tract seem to be rather characteristic and are reminiscent of those described in goats infected with BHV-6 (see Section IX.F).

The *enteritis* associated with BHV-1 infection is an unusual disease and may result from systemic infection (Gratzek *et al.*, 1966; Burkhardt and Paulsen, 1978).

Another classic syndrome is *infectious pustular vulvovaginitis*, also known as "Bläschenausschlag" in German-speaking countries. In the bull, it expresses itself as a *balanoposthitis*. Semen can be contaminated by virus (Spradbrow, 1968). The pathological lesions are usually pustules followed by necrotic erosions in the vaginal or penile mucosa (Steck *et al.*, 1969; Bwangamoi and Kaminjolo, 1971; Straub, 1978). The lesions of this kind of disease heal within days, but the animals have been reported to remain virus carriers. Therefore, the virus can be disseminated by infected asymptomatic bulls either in natural breeding or through artificial insemination.

Although BHV-1 usually infects only bovine animals (see Gibbs and Rweyemamu, 1977a), isolations have also been reported from pigs (Saxegaard and Onstad, 1967; Derbyshire and Caplan, 1976) and even from minks and ferrets (Porter *et al.*, 1975).

BHV-1 causes disease in experimentally inoculated cattle. Smaller laboratory animals such as mice, rats, and guinea pigs appear to be refractory to the virus (see Gibbs and Rweyemamu, 1977a). The rabbit has been reported to show signs of disease after inoculation with BHV-1 (Armstrong *et al.*, 1961) and has been suggested as an experimental model (Lupton *et al.*, 1980).

B. Bovid Herpesvirus 2 Infections

Bovine herpes mammillitis (BHM) was reviewed by several investigators (Martin, 1973; Sterz, 1973; Cilli and Castrucci, 1976; Gibbs and Rweyemamu, 1977b). In European countries, BHV-2 causes mammillitis with severe skin gangrene of the bovine udder. In the African form (Allerton disease), a mild febrile reaction is followed by the appearance of skin nodules. These kinds of scattered nodules can also be observed after experimental infection of animals. Clinical BHM is more limited to the teats and the udder, where severe inflammations of the skin occur. Clinical forms of stomatitis and facial infection occurring preferentially in calves with mammillitis have also been associated with BHV-2. Although the clinical symptoms that may accompany BHM are severe, mortality resulting from infection is negligible.

The pathological alterations result from the inflammation of the dermis and are the consequence of diffuse inflammatory foci or edema, preceding the typical degeneration of cells in the different layers of the epidermis. More detailed information on this dermatotropic virus infection is presented in numerous reports summarized by Cilli and Castrucci (1976).

Experimental infection with BHV-2 was tested in sheep, goats, and pigs, which showed mild skin reactions. Furthermore, BHV-2 infection was used in a model system in rabbits, guinea pigs, and mice to induce dermal reactions. For further details about experimental infections, see Cilli and Castrucci (1976) and Gibbs and Rweyemamu (1977b). Humans are not susceptible to infection (Pepper *et al.*, 1966; Sterz and Ludwig, unpublished).

C. Bovid Herpesvirus 3 Infections

In Europe, four clinical forms of BHV-3 disease have been reported. These are the peracute form with generalized disease, the intestinal form, the head and eye form, and a light form characterized by catarrhal symptoms (Götze, 1930). These clinical expressions have often been attributed to diseases associated with sheep (Götze and Liess, 1930; Bürki *et al.*, 1972; Buxton and Reid, 1980).

In the words of Plowright (1968) and Rossiter *et al.* (1977), the African form of

> malignant catarrhal fever (MCF) is an acute, generalized disease of cattle and domestic buffalos, characterized by high fever, and severe inflammatory and degenerative lesions in the mucosa of the upper respiratory tract and throughout the alimentary tract. Ophthalmia is frequent and manifested inter alia by corneal opacity which is at first peripheral but extends centripetally, often leading to blindness. Enlargement of lymph nodes occurs in the majority of African cases, perhaps less frequently elsewhere. The disease has a worldwide distribution, but its incidence is generally sporadic and low, with no particular

reference to age, sex, or breed. The mortality is extremely high, usually greater than 95%, and death commonly occurs 5 to 12 days after onset of pyrexia.

The disease outside Africa is usually associated with sheep and, whilst it is clinico-pathologically indistinguishable from the wildebeest-derived form, no aetiological agent has yet been isolated and characterized. It is remarkable that no cytological or electron microscopic evidence of a herpesvirus infection is present in the tissue of cattle infected by wildebeest-derived or sheep-associated strains of the virus. In this respect, MCF shows a considerable resemblance to infectious mononucleosis of man, now known to be caused by the Epstein-Barr (EB) herpesvirus.

The basic pathology is characterized by lymphoreticular-cell proliferation, angiitis, and cell infiltrations (Dahme and Weiss, 1978; Pallaske, 1960; Ohshima *et al.*, 1977). In rabbits, focal necrosis of paracortical areas of lymph nodes followed by a progressive destruction of the cortex of the thymus was reported, suggesting involvement of T cells in the disease (Edington *et al.*, 1979).

In addition to the experimental infection of the host animal and sheep, the rabbit has become a useful model to study this disease (Götze and Liess, 1929; Edington *et al.*, 1979; Straver and van Bekkum, 1979; Edington and Plowright, 1980; Buxton and Reid, 1980).

D. Bovid Herpesvirus 4 Infections

Viruses grouped together as BHV-4 have been isolated from different clinical diseases, although it is not known whether they are etiologically involved in the clinical entities from which they were isolated. Four viruses (strains Movar, ÜT, 66-P-347, and DN-599), used as reference strains in our laboratory, are closely related by restriction-endonuclease cleavage patterns (see Fig. 12) and form the basis of this group. Three of the viruses were associated with disease of the upper respiratory tract and with conjunctivitis (compare Table IV).

Some isolates from the respiratory tract were placed together on the basis of weak serological cross-reactivity (P.C. Smith, 1976). The "Movar-type" viruses were assumed to participate in various clinical entities. Recently, such a virus was also isolated from an aborted fetus (Reed *et al.*, 1979).

There are numerous other isolates (see Gibbs and Rweyemamu, 1977b) that resemble the grouped viruses discussed above with respect to *in vitro* and *in vivo* properties. These isolates were found to be associated with bronchopneumonia, metritis, lymphosarcoma, skin afflictions, vaginitis, altered lymph nodes, and tumors of the urinary bladder and the rumen. Other isolates were obtained from asymptomatic animals. Although their classification is unknown, they could be passenger viruses activated by disease or in tissue culture. Activation of herpesviruses associated with diseases in other species has been reviewed by Stevens (1978).

E. Bovid Herpesvirus 5 Infections

Herpesviruses have been recovered from tumor-bearing animals in the course of "jaagsiekte" or contagious pulmonary adenomatosis disease of sheep (Wandera, 1971; W. Smith and Mackay, 1969; De Villiers *et al.*, 1975). Although the etiological role of the viruses in the initiation of the disease has not been proven, the involvement of BHV-5 has been suggested (Malmquist *et al.*, 1972). The incubation period for this disease may range from a few months to several years. The disease affects sheep of all ages, but the characteristic manifestations have been seen mainly in adults. Seromucose rhinitis is an early indication; the disease progresses slowly and eventually leads to a severe pulmonary syndrome often complicated by secondary infections.

Histologically affected lungs can show solitary adenomata or more confluent areas in the lung parenchyma. The alterations are characterized by a proliferation of the flat alveolar epithelium cells, which change to cylindrical tumor cells. Detailed descriptions of the pathology of the disease were reported by De Villiers *et al.* (1975) and by Hiepe and Ilchmann (1978).

De Villiers *et al.* (1975) reported failure to reproduce the disease by experimental inoculation of animals. More recently, however, Martin *et al.* (1976) reported that a tumor suspension from the lung of a natural case induced a clinical picture of "jaagsiekte" in one sheep 6 months after inoculation. The role of the herpesvirus is unclear inasmuch as a C-type retrovirus has also been reported to be present in altered lung material (Perk *et al.*, 1974).

F. Bovid Herpesvirus 6 Infections

The two outbreaks attributed to BHV-6 are rather similar in their disease pattern and pathology (Saito *et al.*, 1974; Mettler *et al.*, 1979) and clinically differentiable from BHV-1 infections in goats (Mohanty *et al.*, 1971). In the first case, kids in a goat herd in California had difficulties in breathing and showed general weakness without further major involvement of the respiratory tract, but had no diarrhea. The disease increased in severity and animals died (Saito *et al.*, 1974). In the other case, goat kids in Bregaglia Valley (Switzerland) developed conjunctivitis, rhinitis, and erosions in the mouth (Fig. 32a). In some animals, petechial bleedings in the skin were recognized. A high mortality rate was observed (Mettler *et al.*, 1979).

Gross pathological lesions were largely confined to the gastrointestinal tract with ulcerations and necrosis in the mucosa of the rumen, abomasum, cecum, and colon (Fig. 32b–f). Fresh focuslike necrosis with signs of inflammation predominantly in the colon was evident in all animals. The lungs showed an enlargement of the interstitium and typical

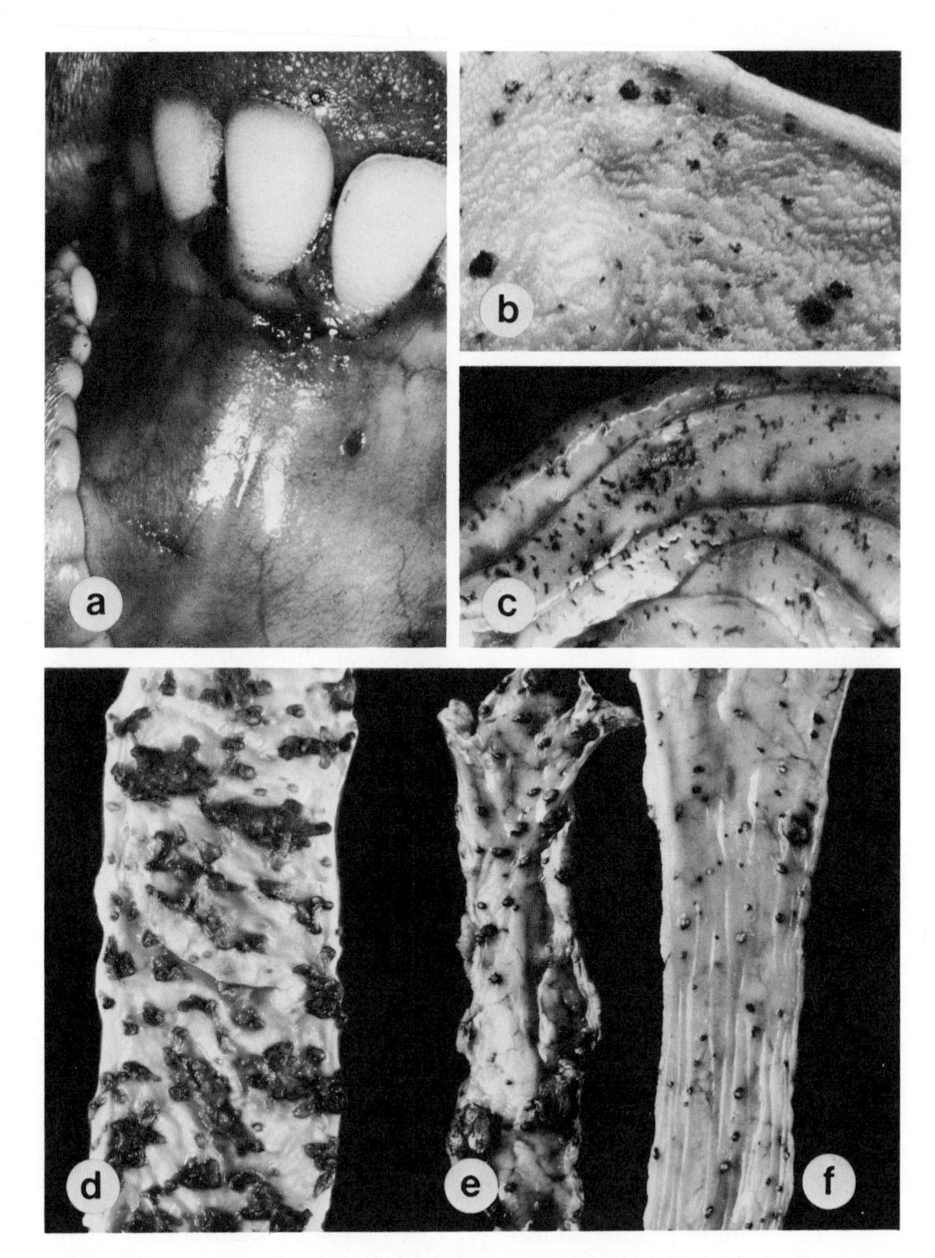

FIGURE 32. Pathology of a natural BHV-6 infection in the goat. (a) Erosions in the mucosa of the mouth. Focal necrotic alterations in the rumen (b), abomasum (c), cecum (d), colon (e), and rectum (f), demonstrating the peracute generalized infection. Histopathology of such a fresh lesion showing intranuclear inclusion bodies (g) and inflammatory reactions in the cecum (h). History of the corresponding BHV-6 tissue culture isolate (E/CH) in a bovine embryonic lung cell (i), with all stages of virus replication, as demonstrated by electron microscopy. From Mettler *et al.* (1979).

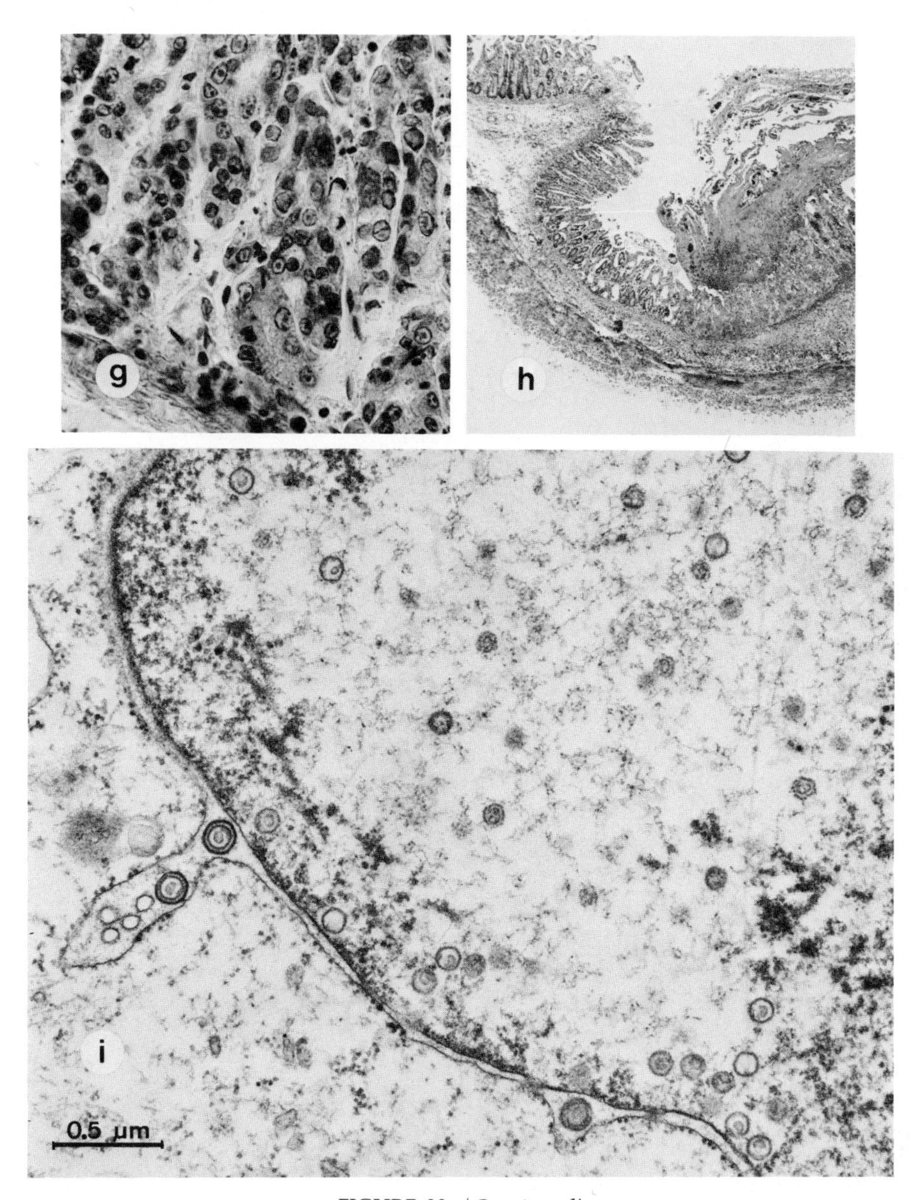

FIGURE 32. (*Continued*)

round-cell infiltrations (Fig. 32h). Histological examination revealed intranuclear inclusion bodies in necrotic areas (Fig. 32g).

Experimental infection with severe disease could be demonstrated in adult goats as well as in kids. Pathogenicity was lacking in lambs and calves (Berrios *et al.*, 1975).

The isolates from the two outbreaks could not be differentiated by analyses of their DNAs (see Fig. 13) (Engels, Herrmann, and Ludwig, unpublished results).

X. PATHOGENESIS

In general, the diseases that result from infection with BHV-1, -2, -4, or -6 appear to be the consequences of destruction resulting from virus replication, whereas proliferation of cells and most probably immunopathological events are prominent features in animals infected with BHV-3 and -5.

Recent data on the molecular biology of clinical isolates strongly suggest that BHV-1 consists of two groups of viruses differing in their genomes and that each group is responsible for a distinct set of clinically recognized diseases (see Section V.A.1). Further studies and detailed genetic analyses of the BHV-1 genome will be necessary to correlate genetic and biological information with pathogenesis.

After primary infection in which BHV-1 multiplies at the portal of entry in the respiratory and genital tracts, the virus appears to be transported by monocytes or other white blood cells via the bloodstream to other target organs (Nyaga and McKercher, 1980). Whether each IBR-like virus has the potential to cause all the known clinical manifestations of infection such as encephalitis, keratoconjunctivitis, rhinotracheitis, abortion, and enteritis or whether the apparent organ tropism is determined by other factors is unclear. Latency is probably established by viruses that are transported intraaxonally from the portal of entry to sacral or cervical ganglia (Narita *et al.*, 1978, 1980; Pastoret *et al.*, 1980). The latent virus may be reactivated by stress and disseminated by centrifugal spread. In these respects, the processes involved in BHV-1 latency may resemble those of other herpesviruses (Goodpasture, 1925; Tenser, 1980).

The vast affliction of several organs showing necrotic foci—a sign of acute virus replication—is seen in BHV-6 infections, which bear some similarities to experimental BHV-1 infections of newborn calves. From the two single reports (Saito *et al.*, 1974; Mettler *et al.*, 1979), it becomes evident that the goat herpesvirus causes severe disease only in the kid, with its still weak immune defense. The spread of virus to various organs and the destruction of cells in the course of replication—demonstrated by the pathological examination—would easily explain the pathogenesis of the disease and the death of the animals.

The pathogenesis of BHV-4 infections meets all the criteria of an organ-specific lytic infection in an immunologically weakened organism by an acute disease. It seems that the viruses placed in this group are usually nonpathogenic, latent agents that are reactivated by unrelated disease processes *in vivo* and by cultivation of tissues *in vitro*.

The clinical manifestations of BHV-2 infections—whether generalized or local—may depend on the virus, the immune status, or the genotype of the animals. All this has not yet been investigated. The selectivity that this virus shows for replication in the stratum germinativum of the skin, however, cannot be overlooked. When virus reaches these skin layers via the blood or directly from the irritated skin, it infects and multiplies primarily in the growing cells of the germinative layers. The syncytia formed in this layer resemble those induced in tissue culture. The lesions in the target organ and the host reaction to this inflammatory process constitute the typical clinical picture of the disease (Cilli and Castrucci, 1976; Gibbs and Rweyemamu, 1977b). The pathogenesis is certainly influenced by hormonal changes and by depression of cellular immunity because the disease occurs most often in dairy cows and only occasionally in pregnant cows. In typical herpes mammillitis, the tissues afflicted are the fine and intensively innervated skin of the udder and teats. BHV-2 appears to have evolved a high affinity for the cells of the mammary gland.

Herpesviruses have not been unequivocally identified as the etiological agent of European malignant catarrhal fever even though the clinical disease resembles that of the wildebeest infected with BHV-3. Our knowledge concerning the pathogenesis of BHV-3 infections in the wildebeest has been supplemented by the results of experimental studies in rabbits (Plowright, 1968; Edington *et al.*, 1979). In typical infections, whether in the rabbit or in the calf, no specific cytopathic lesions due to virus replication have been reported. The prevailing picture is that of proliferation of reticuloendothelial cells. These cells invade many tissues and seem to be associated with infectious virus. In fatally diseased rabbits, virus was found attached to medium-sized lymphocytes rather than proliferative lymphoblastoid cells. Focal necrosis of lymphocytes was observed in organs of the lymphoreticular system and particularly in the thymus, but it remains to be established that the cells carrying the virus are the T lymphocytes (Patel and Edington, 1980). Although the pathogenesis of the disease resulting from natural infection is far from being understood, some of its features resemble those of Epstein–Barr virus infections in humans (M.A. Epstein *et al.*, 1964) and especially those of *Herpesvirus sylvilagus* infections in rabbits (Wenger and Hinze, 1974).

Most obscure of all bovine herpesvirus infections is the pathogenesis of BHV-5 infection and the association of this virus with "jaagsiekte." Although virus can be isolated from lung cell cultures derived from sheep with pulmonary adenomatosis, no direct involvement of virus in the tumor events has been demonstrated (Perk *et al.*, 1974; De Villiers *et al.*, 1975). It is conceivable that the disease results from interaction of a herpesvirus with a retrovirus, as has been reported in other systems (Hsiung and Fong, 1980), or that as yet unidentified factors determine the outcome of viral infection as in Burkitt's lymphoma (Zur Hausen *et al.*, 1970).

XI. DIAGNOSIS

Bovid herpesvirus 1 (BHV-1), BHV-2, and BHV-3 infections can be diagnosed clinically, although various other diseases have to be ruled out (Gibbs and Rweyemamu, 1977a,b; Straub, 1978). Unambiguous diagnosis,

TABLE VIII. Diagnosis of Bovine Herpesvirus Infections

Virus group (animals)	Clinically?	Histologically?	Differential diagnosis[a]	Virus isolation?	Antigen demonstration[b]?	Serologically?	Molecular biologically?
BHV-1 (cattle)	Yes	Eventually	MCF, BHM, MD/BVD, parainfluenza, various bacterial infections	Yes	Eventually	Yes	Yes
BHV-2 (cattle)	Yes	Eventually	VSV infection, IBR/IPV, cow pox, pseudo-cow pox	Yes	—	Yes	Yes
BHV-3 (cattle, sheep)	Yes	Yes	MD/BVD, IBR/IPV, BT, BD	Eventually	Eventually	Yes	Yes
BHV-4 (cattle)	Unclear	—	MCF, MD/BVD, IBR/IPV	Yes	—		Yes
BHV-5 (sheep)	Unclear	Yes	Maedi, bacterial infections	Eventually	—	Eventually	
BHV-6 (goat)	Unclear	Eventually	BHV-1 infection in goats	Yes	Eventually	Yes, but cross-reactivity with BHV-1	Yes

[a] (MCF) Malignant catarrhal fever; (BHM) bovine herpes mammillitis; (MD/BVD) mucosal disease/bovine virus diarrhea; (VSV) vesicular stomatitis virus; (IBR/IPV) infectious bovine rhinotracheitis/ infectious pustular vulvovaginitis; (BT) bluetongue; (BD) Borna disease. Further details on some of these diseases, which are relevant for the clinician and practitioner, can be found in veterinary textbooks (Rolle and Mayr, 1978).

[b] Here the possibility is mentioned of performing a quick diagnosis (in approximately 6 hr) on the basis of antigen detection in nasal secretions, ocular discharge, vaginal mucosal discharge, etc., with fluorescent-antibody tests, ELISA, or radioimmunoassays, as successfully done in diagnosis of viral diseases in man (Halonen and Habermehl, personal communication).

however, requires virus isolation and identification. This can be done for viruses of all six groups. Some difficulties are encountered with BHV-3; transfer of whole blood to bovine animals is still a good indicator system for virulent malignant catarrhal fever virus.

As shown in this chapter, restriction-enzyme analyses of viral DNAs have been used successfully to characterize BHV-3, -4, and -6 strains and for diagnosis and differentiation of viral strains. This technique will certainly be of great help in identifying bovine herpesviruses and in tracing their spread. Quick diagnosis of bovine herpesvirus infections by radioimmunoassay or enzyme-linked immunosorbent assay (ELISA)—especially in diseases in which virus-specific antigen is present in secretions and discharge material—is a matter of time. The effectiveness of the currently available diagnostic procedures is summarized schematically in Table VIII (see also Sections VI, VII, and IX).

XII. EPIDEMIOLOGY

The epidemiology of bovid herpesvirus 1 (BHV-1) infection depends on virus, age and immune status of the animals, and other factors determined by the environment. Rhinotracheitis and, eventually, abortion are the most serious sequelae of BHV-1 infection (Kahrs, 1977). Infection and transmission of BHV-1 in cattle are facilitated in feedlots, because of close contact among animals. On the other hand, the classic pustular vulvovaginitis, a major problem in natural breeding (Witte, 1933; Steck *et al.*, 1969), should be reduced in incidence by widespread application of artificial insemination.

Reduction and eradication of BHV-1-associated diseases are an important goal of cattle husbandry. Because BHV-1 remains latent, it may be difficult to eradicate. The conventional approach to prevention of BHV-1 infection is by extensive vaccination with live, modified live, and inactivated vaccines. Until now, no defined subunit vaccines have been produced, but attempts to stimulate antibody production by inoculation of extracts containing virion envelope antigens yielded promising results (Darcel *et al.*, 1979). It is conceivable that nucleic-acid-free immunogenic components of BHV-1-infected cells may induce stable immunity.

The generalized form of BHV-2-induced disease in South Africa has the highest incidence during summer months and is confined to humid areas, preferentially along river banks. This suggests transmission by insects (Weiss, 1968). The epidemiology of BHV-2 infections leading to mammillitis was studied during the yearly outbreaks in the United Kingdom, where the disease is endemic in some areas. Transmission is thought to occur less by contact than by biting flies or by mechanical vectors such as milking machines. Hygenic control measures as well as the use of inactivated BHV-2 vaccines seem to prevent disease (Gibbs and Rweyemamu, 1977b).

The BHV-3 infection of wildebeest (African malignant catarrhal fever) can be transmitted to other ruminants as well. Malignant catarrhal fever as a clinicopathological entity is known worldwide, although the etiological agent may vary from one geographic area to another. Epidemiological studies on the widespread African form have been reported, but the mode of transmission of the virus is still not clear (Plowright, 1968; Rweyemamu *et al.*, 1974). In contrast, the European form appears to be localized to single animals and occurs sporadically, and it has been thought for many years that sheep are a major factor in the epidemiology of the disease (Götze and Liess, 1930; Selman *et al.*, 1974). Vaccination against (African) malignant catarrhal fever has been advocated (Plowright *et al.*, 1975), but was only partly successful even in a rabbit model (Rweyemamu *et al.*, 1976; Edington and Plowright, 1980).

The problems associated with prevention of herpesvirus infections in animals by immunization with the different kinds of vaccines were extensively discussed by Biggs (1975) and Plowright (1978).

The epidemiology of BHV-4 infections is not well understood. Localized disease outbreaks have yielded several isolates [e.g. strain Movar 33/63 (Bartha *et al.*, 1966)]. In addition, virus was activated from individually diseased animals [strain 66-P-347 (Storz, 1968)] and even from healthy animals (strain ÜT; tissue-culture isolate). The mode of transmission of these viruses is unclear. The epidemiology of BHV-4 infection resembles that of malignant catarrhal fever (Storz, 1968; Liebermann *et al.*, 1967; Gibbs and Rweyemamu, 1977b).

The mode of spread of sheep pulmonary adenomatosis ("jaagsiekte") is unknown. It should be emphasized that efforts to transmit the disease were successful with intratracheal injections of lung homogenates, cell cultures, or disrupted cells, but not with isolated virus (BHV-5). Although this finding could suggest that transmission of the tumor occurs in nature, the epidemiology of the infectious agent remains obscure (Verwoerd *et al.*, 1978).

Our knowledge concerning the epidemiology of BHV-6 infections in goats is based on studies of the two outbreaks cited earlier in the text (Saito *et al.*, 1974; Mettler *et al.*, 1979). The disease is highly contagious and virulent in young goats; it is reminiscent of fatal pseudorabies infection in piglets (Baskerville *et al.*, 1973), of BHV-1 infection of calves (Kendrick, 1973), and of canine herpesvirus infection in puppies (Carmichael *et al.*, 1969). The epidemiology of BHV-6 infections and the relationship of this group of viruses to BHV-1 remain to be elucidated.

XIII. CONCLUDING REMARKS

Bovine herpesvirus infections are of considerable economic importance. The most significant of these diseases are those associated with bovid herpesvirus 1 (BHV-1), followed by those caused by BHV-2 or BHV-

3 depending on geographic location. BHV-4 infections cause sporadic losses, whereas BHV-5 and -6 infections are problems in sheep and goat husbandry.

Knowledge of the biology of some of these viruses was obtained more from field studies than from laboratory experimentation. For example, the worldwide campaign to vaccinate cattle against BHV-1 infection with live virus resulted in the maintenance of the vaccine strain in the animal population as well as losses through abortion. Since BHV-1 has the potential to remain latent, the vaccine virus may remain and contribute to infection in nature.

Given the economic importance of these viruses, the impetus for fundamental studies on these viruses is obvious. For example, studies on the viral genomes may enable epidemiological studies based on fingerprinting of the DNA. It may also be possible to assign specific viral functions such as virulence and immunogenicity to specific genes in the viral genome. In addition, studies on the structure and function of bovine herpesviruses may enrich our knowledge concerning other herpesviruses and of viruses in general. It is important to note that bovine herpesviruses resemble the human and primate herpesviruses in many aspects of their biology and molecular biology. For example, the highly lytic BHV-1, -6, and -2 resemble herpes simplex and B viruses. The slow-growing viruses of the BHV-4 group resemble cytomegaloviruses, and the lymphoproliferative-disease-associated BHV-3 and the human-associated BHV-5 resemble Epstein–Barr virus, *Herpesvirus saimiri*, and *H. ateles*.

Several lines of future research are already evident: It will be a major challenge to establish the sequence arrangements and functional maps of the genomes of the six bovine herpesviruses. Because of the relatively short reproductive cycle and high virus yields, BHV-1 and BHV-6 are likely to be studied first. Another interesting question is whether BHV-6, the goat herpesvirus, represents a natural variant of BHV-1 with a restricted host range. Interest in BHV-2 stems from its special dermatotropism and from its ability to fuse cells *in vitro* and *in vivo*. In addition, this is the only bovine virus that shares common antigenic determinants with major immunogenic components of human (herpes simplex) and primate (B) herpesviruses. The BHV-3 rabbit system is a very useful model for the study of the pathogenesis of lymphoproliferative diseases. The role of BHV-4 viruses as passenger viruses and their participation in disease processes will almost certainly attract further attention. Finally, the elucidation of the role of BHV-5 in lung adenomatosis is an exciting challenge.

Another potential field of interest concerns the interaction of bovine herpesviruses with their hosts, particularly from the point of view of immune responses and mechanisms. A unique advantage of large animals is that they can provide large quantities of immunocompetent cells, serum, and even tissues frequently over an interval of many years. The cells can be grown in the animal's own serum—an important consideration in certain types of studies. Why should a cow not provide milk and meat and in addition compete in the laboratory with the mouse?

ACKNOWLEDGMENTS. I wish to thank my colleagues and co-workers G. Pauli, G. Darai, M. Engels, H. Gelderblom, P.-P. Pastoret, T. Leiskau, and B. Heppner for help and generosity to incorporate unpublished results in this review. The supply of re- and preprints by numerous scientists throughout the world has been of invaluable importance. I am indebted to Jan G. Lewis for editorial contributions, to M. Hartmann for secretarial assistance, Meike Ludwig for continuous encouragement, and Johannes Storz for his informative discussions about topics of this review. Finally, I would like to thank Dr. J.L. Melnick and Dr. h.c. R. Rott for scientific support given during my stays in their departments, where initial studies covered in this review were done. Some results reported in this review were supported by grants from the "Deutsche Forschungsgemeinschaft" and the "Bundesministerium fur Forschung und Technologie."

REFERENCES

Abodeely, R.A., Lawson, L.A., and Randall, Ch.C., 1970, Morphology and entry of enveloped and deenveloped equine abortion (herpes) virus, *J. Virol.* **5**:513.

Afshar, A., 1965, Virus diseases associated with bovine abortion and infertility, *Vet. Bull.* **35**:735.

Aguilar-Setién, A., Pastoret, P.-P., Burtonboy, G., and Schoenaers, F., 1978, Test d'hypersensibilité retardé au virus de la rhinotrachéite infectieuse bovine (Bovid Herpesvirus 1) avec du virus purifié, *Ann. Méd. Vét.* **122**:193.

Aguilar-Setién, A., Pastoret, P.-P., Burtonboy, G., Coignouil, F., Jetteur, P., and Schoenaers, F., 1979, Communauté antigénique entre le virus de la rhinotrachéite infectieuse bovine (Bovid Herpesvirus 1, BHV 1) et le virus de la maladie d'Aujeszky (suis herpesvirus 1, SHV 1) démontrée, chez le bovin, par un test d'hypersensibilité retardée, *Ann Méd. Vét.* **123**:55.

Aguilar-Setién, A., Pastoret, P.-P, and Schoenaers, F., 1980, L'immunité envers le virus de la rhinotrachéite infectieuse bovine (Bovid Herpesvirus 1), *Ann. Méd. Vét.*, **124**:103.

Ahl, R., and Straub, O.C., 1971, Die lokale Interferonbildung im Respirations- und Genitaltrakt nach experimenteller Infektion mit Rhinotracheitis (IBR)- und Bläschenausschlag (IPV)-Virus, *Dtsch. Tieraerztl. Wochenschr.* **78**:653.

Alexander, R.A., Plowright, W., and Haig, D.A., 1957, Cytopathogenic agents associated with lumpy-skin disease of cattle, *Bull. Epizoot. Dis. Afr.* **5**:489.

Allan, E.M., and Msolla, P.M., 1980, Scanning electron microscopy of the tracheal epithelium of calves inoculated with bovine herpesvirus 1, *Res. Vet. Sci.* **29**:325.

Almeida, J.D., Howatson, A.F., and Williams, M.C., 1962, Morphology of varicella (chicken pox) virus, *Virology* **16**:353.

Armstrong, J.A., Pereira, H.G., and Andrewes, C.H., 1961, Observations on the virus of infectious bovine rhinotracheitis, and its affinity with the herpesvirus group, *Virology* **14**:276.

Armstrong, J.A., Plowright, W., and Macadam, R.F., 1964, Observations on the virus of bovine malignant catarrhal fever in cell cultures, in: *3rd European Regional Conference on Electronmicroscopy*, Vol. B, pp. 363–364, Prague.

Asso, J., and Le Jan, C., 1977, Viral infections of the respiratory tract of calves: Local immunity, *Vet. Sci. Commun.* **1**:297.

Aujeszky, A., 1902, Uber eine neue Infektionskrankheit bei Haustieren, *Zentralblatt Bakteriol. I: Orig.* **32**:353.

Babiuk, L.A., and Rouse, B.T., 1976, Ribonucleotides in infectious bovine rhinotracheitis virus DNA, *J. Gen. Virol.* **31**:221.

Babiuk, L.A., and Rouse, B.T., 1979, Immune control of herpesvirus latency, *Can. J. Microbiol.* **25**:267.

Bagust, T.J., 1972, Comparison of the biological, biophysical and antigenic properties of four strains of infectious bovine rhinotracheitis herpesvirus, *J. Comp. Pathol.* **82**:365.

Bagust, T.J., and Clark, L., 1972, Pathogenesis of meningo-encephalitis produced in calves by infectious bovine rhino-tracheitis herpesvirus, *J. Comp. Pathol.* **82**:375.

Barenfus, M., Delli Quadri, C.A., McIntyre, R.W., and Schroeder, R.J., 1963, Isolation of infectious bovine rhinotracheitis virus from calves with meningoencephalitis, *J. Am. Vet. Med. Assoc.* **143**:725.

Bartha, A., Juhász, M., and Liebermann, H., 1966, Isolation of a bovine herpesvirus from calves with respiratory disease and keratoconjunctivitis, *Acta Vet. Acad. Sci. Hung.* **16**:357.

Bartha, A., Juhász, M., Liebermann, H., Hantschel, H., and Schulze, P., 1967, Isolierung und Eigenschaften eines bovinen Herpesvirus von einem Kalb mit respiratorischer Krankheit und Keratokonjunktivitis, *Arch. Exp. Veterinaermed.* **21**:615.

Baskerville, A., McFerran, J.B., and Dow, C., 1973, Aujeszky's disease in pigs, *Vet. Bull. (Weybridge)* **43**:465.

Belák, S., and Pálfi, V., 1974, Characterization of a herpesvirus isolated from spontaneously degenerated bovine kidney culture, *Acta Vet. Hung.* **24**:249.

Berrios, P.E., and McKercher, D.G., 1975, Characterization of a caprine herpesvirus, *Am. J. Vet. Res.* **36**:1755.

Berrios, P.E., McKercher, D.G., and Knight, H.D., 1975, Pathogenicity of a caprine herpesvirus, *Am. J. Vet. Res.* **36**:1763.

Bhakdi, S., Bhakdi-Lehnen, B., and Bjerrum, O.J., 1977, Detection of amphiphilic proteins and peptides in complex mixtures: Charge-shift crossed immuno-electrophoresis and two-dimensional charge-shift electrophoresis, *Biochim. Biophys. Acta* **470**:35.

Bier, P.I., Hall, C.E., Duncan, J.R., and Winter, A.J., 1977, Measurement of immunoglobulins in reproductive tract fluids of bulls, *Vet. Microbiol.* **2**:1.

Biggs, P.M., 1975, Vaccination against oncogenic herpesviruses—a review, in: *Oncogenesis and Herpesviruses II/2* (G. de Thé, M.A. Epstein, and H. zur Hausen, eds.), p. 317, IARC Sci. Publ. No. 11, Lyon.

Bijok, U., Dimpfel, W., Habermann, E., and Ludwig, H., 1979, Normal and pseudorabies virus infected primary nerve cell cultures in scanning electron microscopy, *Med. Microbiol. Immunol.* **167**:117.

Bitsch, V., 1973, Infectious bovine rhinotracheitis virus infection in bulls, with special reference to preputial infection, *Appl. Microbiol.* **26**:337.

Blue, W.T., and Plummer, G., 1973, Antigenic relationship among four herpesviruses, *Infect. Immun.* **7**:1000.

Bocciarelli, D.S., Orfei, Z., Mondino, G., and Persechino, A., 1966, The core of a bovine herpes virus, *Virology* **30**:58.

Borgen, H.C., and Ludwig, H., 1974, Equine Herpesvirus 1: Biological and biophysical comparison of two viruses from different clinical entities, *Intervirology* **4**:189.

Bowling, C.P., Goodheart, C.R., and Plummer, G., 1969, Oral and genital bovine herpesviruses, *J. Virol.* **3**:95.

Breese, S.S., Jr., and Dardiri, A.H., 1972, Electron microscopic characterization of a bovine herpes virus from Minnesota, *J. Gen. Virol.* **15**:69.

Bronson, D.L., Graham, B.J., Ludwig, H., Benyesh-Melnick, M., and Biswal, N., 1972, Studies on the relatedness of herpesviruses through DNA–RNA hybridization, *Biochim. Biophys. Acta* **259**:24.

Buchman, T.G., and Roizman, B., 1978a, Anatomy of bovine mammillitis DNA. I. Restriction endonuclease maps of four populations of molecules that differ in the relative orientation of their long and short components, *J. Virol.* **25**:395.

Buchman, T.G., and Roizman, B., 1978b, Anatomy of bovine mammillitis DNA. II. Size and arrangements of deoxynucleotide sequences, *J. Virol.* **27:**239.

Buchman, T.G., Roizman, B., Adams, G., and Stover, B.H., 1978, Restriction endonuclease fingerprinting of herpes simplex virus DNA: A novel epidemiological tool applied to a nosocomial outbreak, *J. Infect. Dis.* **138:**488.

Buchman, T.G., Roizman, B., and Nahmias, A.J., 1979, Demonstration of exogenous genital reinfection with herpes simplex virus type 2 by restriction endonuclease fingerprinting of viral DNA, *J. Infect. Dis.* **140:**295.

Buening, G.M., and Gratzek, J.B., 1967, Comparison of selected characteristics of four strains of infectious bovine rhinotracheitis virus, *Am. J. Vet. Res.* **28:**1257.

Burkhardt, E., and Paulsen, J., 1978, Nachweis von bovinem Herpesvirus 1 (IBR/IPV) bei Rindern mit Affektionen des Verdauungstraktes, *Berl. Muench. Tieraerztl. Wochenschr.* **91:**480.

Bürki, F., Schlerka, G., and Sibalin, M., 1972, Untersuchungen auf bösartiges Katarrhalfieber und bovine Virusdiarrhoe in Gebirgsgegenden Österreichs. I. Versuche zum Nachweis des Virus des bovine malignant catarrh, *Wien. Tieraerztl. Monatsschr.* **59:**307.

Buxton, D., and Reid, H.W., 1980, Transmission of malignant catarrhal fever to rabbits, *Vet. Rec.* **106:**243.

Bwangamoi, O., and Kaminjolo, J.S., 1971, Isolation of IBR/IPV virus from the semen and skin lesions of bulls at Kabete, Kenya, *Zentralbl. Veterinaermed. Reihe B* **18:**262.

Carmichael, L.E., Barnes, F.D., and Percy, D.H., 1969, Temperature as a factor in resistance of young puppies to canine herpesvirus, *J. Infect. Dis.* **120:**669.

Castrucci, G., Pedini, B., Cilli, V., and Arancia, G., 1972, Characterisation of viral agent resembling bovine herpes mammillitis virus, *Vet. Rec.* **90:**325.

Castrucci, G., Martin, W.B., Pedini, B., Cilli, V., and Ranucci, S., 1975, A comparison in calves of the antigenicity of three strains of bovid herpesvirus 2, *Res. Vet. Sci.* **18:**208.

Castrucci, G., Wada, E.M., Ranucci, S., Frigeri, F., Cilli, V., Pedini, B., Tesei, B., and Arush, M.A., 1980a, Reactivation of latent infection by infectious bovine rhinotracheitis virus in calves, *Microbiologica* **3:**307.

Castrucci, G., Frigeri, F., Cilli, V., Tesei, B., Arush, A.M., Pedini, B., Ranucci, S., and Rampichini, L., 1980b, Attempts to reactivate Bovid herpesvirus-2 in experimentally infected calves, *Am. J. Vet. Res.* **41:**1889.

Castrucci, G., Ranucci, S., Ferrari, M., Frigeri, F., Cilli, V., and Cassai, E., 1981, A study in calves of an immunologic relationship between herpes simplex virus and bovid herpesvirus 2, *Comp. Immunol. Microbiol. Infect. Dis.* **4:**1

Chia, W.-K., and Savan, M., 1974a, Pathogenesis of infectious bovine rhinotracheitis (IBR) virus infection in bovine fetal organ cultures, *Can. J. Microbiol.* **20:**839.

Chia, W.-K., and Savan, M., 1974b, Electron microscopic observations on infectious bovine rhinotracheitis virus in bovine fetal tracheal organ cultures, *Arch. Gesamte. Virusforsch.* **45:**185.

Cilli, V., and Castrucci, G., 1976, Infection of cattle with Bovid herpesvirus 2, *Fol. Vet. Lat.* **6:**1.

Coetzee, S., Els, H.J., and Verwoerd, D.W., 1976, Transmission of jaagsiekte (ovine pulmonary adenomatosis) by means of a permanent epithelial cell line established from affected lungs, *Onderstepoort J. Vet. Res.* **43:**133.

Cohen, G.H., Katze, M., Hydrean-Stern, C., and Eisenberg, R.J., 1978, Type-common CP-1 antigen of herpes simplex virus is associated with a 59,000-molecular-weight envelope glycoprotein, *J. Virol.* **27:**172.

Compans, R. W., and Klenk, H.-D., 1979, Viral membranes, in: *Comprehensive Virology,* Vol. 13 (H. Fraenkel-Conrat and R.R. Wagner, eds.), pp. 293–407, Plenum Press, New York.

Courtney, R.J., and Powell, K.L., 1975, Immunological and biochemical characterization of polypeptides induced by herpes simplex virus types 1 and 2, *IARC Sci. Publ.* **11:**63.

Crandell, R.A., 1973, Antigenic variations among infectious bovine rhinotracheitis viruses, *Proc. U.S. Anim. Health Assoc.* **76:**480.

Cruickshank, J.G., and Berry, D.M., 1965, Morphology of infectious bovine rhinotracheitis virus, *Virology* **25:**481.

Cummins, J.M., and Rosenquist, B.D., 1977, Effect of hydrocortisone on the interferon response of calves infected with infectious bovine rhinotracheitis virus, *Am. J. Vet. Res.* **38:**1163.

Cummins, J.M., and Rosenquist, B.D., 1979, Leukocyte changes and interferon production in calves injected with hydrocortisone and infected with infectious bovine rhinotracheitis virus, *Am. J. Vet. Res.* **40:**238.

Dahme, E., and Weiss, E., 1978, *Grundriss der speziellen pathologischen Anatomie der Haustiere,* 2nd ed., Enke, Stuttgart.

Darcel Le, Q.C., and Dorward, W.J., 1972, Skin reactivity and infectious bovine rhinotracheitis, *Can. Vet. J.* **13:**100.

Darcel Le, Q.C., and Dorward, W.J., 1975, Recovery of infectious bovine rhinotracheitis virus following corticosteroid treatment of vaccinated animals, *Can. Vet. J.* **16:**87.

Darcel Le, Q.C., Bradley, J.A., and Mitchell, D., 1979, Immune response of cattle to antigens obtained from bovine herpesvirus 1-infected tissue culture, *Can. J. Comp. Med.* **43:**288.

Darlington, R.W., and Moss, L.H., III, 1969, The envelope of herpesvirus, *Prog. Med. Virol.* **11:**16.

Davies, D.H., and Carmichael, L.E., 1973, Role of cell-mediated immunity in the recovery of cattle from primary and recurrent infections with infectious bovine rhinotracheitis virus, *Infect. Immun.* **8:**510.

Davies, D.H., and Duncan, J.R., 1974, Pathogenesis of recurrent infectious bovine rhinotracheitis virus induced in calves by treatment with corticosteroids, *Cornell Vet.* **64:**340.

Deas, D.W., and Johnston, W.S., 1966, An outbreak of an ulcerative skin condition of the udder and teats of dairy cattle in the East of Scotland, *Vet. Rec.* **78:**828.

Derbyshire, J.B., and Caplan, B.A., 1976, The isolation and characterization of a strain of infectious bovine rhinotracheitis virus from stillbirth in swine, *Can. J. Comp. Med.* **40:**252.

De Villiers, E.-M., 1979, Purification of the JS-3 isolate of *Herpesvirus ovis* (bovid herpesvirus 4) and some properties of its DNA, *J. Virol.* **32:**705.

De Villiers, E.-M., Els, H.J., and Verwoerd, D.W., 1975, Characteristics of an ovine herpesvirus associated with pulmonary adenomatosis (jaagsiekte) in sheep, *S. Afr. J. Med. Sci.* **40:**165.

Dilovsky, M., Tekerlekov, P., and Hadjiev, G., 1974, Isolation of the bovine herpes mammillitis virus in Bulgaria, *Vet. Sci.* **11:**60.

Eberle, R., and Courtney, R.J., 1980, gA and gB glycoproteins of herpes simplex virus type 1: two forms of a single polypeptide, *J. Virol.* **36:**665.

Edington, N., and Plowright, W., 1980, The protection of rabbits against the herpesvirus of malignant catarrhal fever by inactivated vaccines, *Res. Vet. Sci.* **28:**384.

Edington, N., Patel, J., Russel, P.H., and Plowright, W., 1979, The nature of the acute lymphoid proliferation in rabbits infected with the herpes virus of bovine malignant catarrhal fever, *Eur. J. Cancer* **15:**1515.

Ejercito, P.M., Kieff, E.D., and Roizman, B., 1968, Characterization of herpes simplex virus strains differing in their effects on social behaviour of infected cells, *J. Gen. Virol.* **2:**357.

Engels, M., 1979, Seroepizootologishe Untersuchung über das Vorkommen des bovinen Herpes Mammillitis Virus in der Schweiz, Vet. med. dissertation, Universität Zürich.

Engels, M., Ludwig, H., and Mulder, C., 1980, Mapping of genomes from attenuated strains of infectious bovine rhinotracheitis virus and infectious pustular vulvovaginitis virus, 5th Cold Spring Harbor Meeting on Herpesviruses, Cold Spring Harbor, New York, August 1980, p. 59.

Engels, M., Steck, F., and Wyler, R., 1981a, Comparison of the genomes of infectious bovine rhinotracheitis and infectious pustular vulvovaginitis virus strains by restriction endonuclease analysis: Brief report, *Arch. Virol.* **67:**169.

Engels, M., Darai, G., Gelderblom, H., and Ludwig, H., 1981b, Properties of the goat herpesvirus, International Workshop on Herpesvirus, Bologna, Italy, July 1981, p. 20.

Epstein, B., 1972, Virus herpético del bovino en carcinoma celular escamosos de ojos, *Rev. Med. Vet. (Buenos Aires)* **53:**105.

Epstein, M.A., Achong, B.G., and Barr, Y.M., 1964, Virus particles in cultured lymphoblasts from Burkitt's lymphoma, *Lancet* **1**(7335)**:**702.

Evans, D.L., Barnett, J.W., Bowen, J.M., and Dmochowski, L., 1972, Antigenic relationship between the herpesviruses of infectious bovine rhinotracheitis, Marek's disease and Burkitt's lymphoma, *J. Virol.* **10:**277.

Fenner, F., McAuslan, B.R., Mims, C.A., Sambrook, J., and White, D.O., 1974, *The Biology of Animal Viruses*, 2nd. ed. Academic Press, New York, San Francisco, London.

Ferris, D.H., Hamdy, F.M., and Dardiri, A.H., 1976, Detection of African malignant catarrhal fever virus antigens in cell cultures by immunofluorescence, *Vet. Microbiol.* **1:**437.

Fleckenstein, B., Bornkamm, G.W., and Ludwig, H., 1975, Repetitive sequences in complete and defective genomes of *Herpesvirus saimiri*, *J. Virol.* **15:**398.

Fleckenstein, B., Bornkamm, G.W., Mulder, C., Werner, F.-J., Daniel, M.D., Falk, L.A., and Delius, H., 1978, *Herpesvirus ateles* DNA and its homology with *Herpesvirus saimiri* nucleic acid, *J. Virol.* **25:**361.

Fong, C.K.Y., Tenser, R.B., Hsiung, G.D., and Gross, P.A., 1973, Ultrastructural studies on the envelopment and release of guinea pig herpes-like virus in cultured cells, *Virology* **52:**468.

French, E.L., 1962a, A specific virus encephalitis in calves: Isolation and characterization of the causal agent, *Aust. Vet. J.* **38:**216.

French, E.L., 1962b, Relationship between infectious bovine rhinotracheitis (IBR) virus and a virus isolated from calves with encephalitis, *Aust. Vet. J.* **38:**555.

Gallaher, W.R., Levitan, D.B., and Blough, H.A., 1973, Effect of 2-deoxy-D-glucose on cell fusion induced by Newcastle disease and herpes simplex viruses, *Virology* **55:**193.

Geder, L., Ladda, R.L., Kreider, J.W., Sanford, E.J., and Rapp, F., 1979, Properties of human epitheloid cells established *in vitro* by a herpesvirus [IBRV (HMC)] isolated from cytomegalovirus-transformed human cells, *J. Natl. Cancer Inst.* **63:**1313.

Geder, L., Kap, J.L., Dawson, M.S., Hyman, R.W., Maliniak, R.M., and Rapp, F., 1980, Properties of mouse embryo fibroblasts transformed *in vitro* by infectious bovine rhinotracheitis virus, *J. Natl. Cancer Inst.* **65:**441.

Gelderblom, H., Bauer, H., Bolognesi, D.P., and Frank H., 1972, Morphogenese und Aufbau von RNS-Tumorviren: Elektronenoptische Untersuchungen an Virus-Partikeln vom C-Typ, *Zentralbl. Bakteriol. Parasitenkd. Infektionskr. Hyg. Abt. 1: Orig. Reihe A* **220:**79.

Gelderblom, H., Ogura, H., and Bauer, H., 1974, On the occurrence of oncornavirus-like particles in HeLa cells, *Cytobiology* **8:**339.

Gelderblom, H., Pauli, G., Schäuble, H., and Ludwig, H., 1980, Detection of a cross-reacting antigen between herpes simplex virus type 1 and bovine herpes mammillitis virus by immunoelectron microscopy, *Electron Microsc.* **2:**482.

Gershon, A., Cosio, L., and Brunell, P.A., 1973, Observation on the growth of varicella-zoster virus in human diploid cells, *J. Gen. Virol.* **18:**21.

Gibbs, E.P.J., and Rweyemamu, M.M., 1977a, Bovine herpesviruses. Part I. Bovine herpesvirus 1, *Vet. Bull.* **47:**317.

Gibbs, E.P.J., and Rweyemamu, M.M., 1977b, Bovine herpesviruses. Part II. Bovine herpesviruses 2 and 3, *Vet. Bull.* **47:**411.

Gibbs, E.P.J., Johnson, R.H., and Osborne, A.D., 1972, Field observations on the epidemiology of bovine herpes mammillitis, *Vet. Rec.* **91:**395.

Gibbs, E.P.J., Pitzolis, G., and Lawman, M.J.P., 1975, Use of corticosteroids to isolate IBR virus from cattle in Cyprus after respiratory disease and ataxia, *Vet. Rec.* **96:**464.

Gillespie, J.H., McEntee, K., Kendrick, J.W., and Wagner, W.C., 1959, Comparison of infectious pustular vulvovaginitis virus with infectious bovine rhinotracheitis virus, *Cornell Vet.* **49:**288.

Goodpasture, E.W., 1925, The axis-cylinder of peripheral nerves as portals of entry to the central nervous system for the virus of herpes simplex in experimentally infected rabbits *Am. J. Pathol.* **1:**11.

Götze, R., 1930, Untersuchungen über das bösartige Katarrhalfieber des Rindes. III. Mitt., *Dtsch. Tieraerztl. Wochenschr.* **38**:487.
Götze, R., and Liess, J., 1929, Erfolgreiche Übertragungsversuche des bösartigen Katarrhalfiebers von Rind zu Rind: Identität mit der Südafrikanischen Snotsiekte, *Dtsch. Tieraerztl. Wochenschr.* **37**:433.
Götze, R., and Liess, J., 1930, Untersuchungen über das bösartige Katarrhalfieber des Rindes: Schafe als Überträger, *Dtsch. tieraerztl. Wochenschr.* **38**:194.
Graham, B.J., Ludwig, H., Bronson, D.L., Benyesh-Melnick, M., and Biswal, N., 1972, Physicochemical properties of the DNA of herpes viruses, *Biochim. Biophys. Acta* **259**:13.
Gratzek, J.B., Peter, C.P., and Ramsey, F.K., 1966, Isolation and characterization of a strain of infectious rhinotracheitis virus associated with enteritis in cattle: Isolation, serologic characterization and induction of the experimental disease, *Am. J. Vet. Res.* **27**:1567.
Gründer, H.-D., Reuleaux, I.-R., and Lieb, B., 1960, Feststellung der virusbedingten Rhinotracheitis infectiosa des Rindes. I. Herkunft und Isolierung der Virus, *Dtsch. Tieraerztl. Wochenschr.* **67**:514.
Hampar, B., and Martos, L.M., 1973, Immunological relationships, in: *The Herpesviruses* (A.S. Kaplan, ed.), pp. 221–259, Academic Press, New York, San Francisco, London.
Hare, T., 1925, Staphylococcal and Streptococcal dermatitis of the udder in dairy cows, *Vet. Rec.* **5**:943.
Heine, J.W., Honess, R.W., Cassai, E., and Roizman, B., 1974, Proteins specified by herpes simplex virus. XII. The virion polypeptides of type 1 strains, *J. Virol.* **14**:640.
Helenius, A., and Simons, K., 1977, Charge shift electrophoresis: Simple method for distinguishing between amphiphilic and hydrophilic proteins in detergent solution, *Proc. Natl. Acad. Sci. U.S.A.* **74**:529.
Herring, A.J., Nettleton, P.F., and Burrells, C., 1980, A micro-enzyme-linked immunosorbent assay for the detection of antibodies to infectious bovine rhinotracheitis virus, *Vet. Rec.* **107**:155.
Hiepe, T., and Ilchmann, G., 1978, Lungenadenomatose der Schafe, in: *Handbuch der Virusinfektionen bei Tieren,* Vol. VI/2 (H. Röhrer, ed.), pp. 795–815, VEB Gustav Fischer, Jena.
Hirsch, J., and Vonka, V., 1974, Ribonucleotides linked to DNA of herpes simplex virus type 1, *J. Virol.* **13**:1162.
Hoggan, M.D., and Roizman, B., 1959, The isolation and properties of a variant of herpes simplex producing multinucleated giant cells in monolayer cultures in the presence of antibody, *Am. J. Hyg.* **70**:208.
Honess, R.W., and Watson, D.H., 1977, Unity and diversity in the herpesviruses, *J. Gen. Virol.* **37**:15.
Honess, R.W., Powell, K.L., Robinson, D.J., Sim, C., and Watson, D.H., 1974, Type specific and type common antigens in cells infected with herpes simplex virus type 1 and on the surface of naked and enveloped particles of the virus, *J. Gen. Virol.* **22**:159.
House, J.A., 1972, Bovine herpesvirus IBR-IPV: Strain differences, *Cornell Vet.* **62**:431.
Hsiung, G.D., and Fong, C.K.Y., 1980, Interaction of herpesviruses and retroviruses, in: *Oncogenic Herpesviruses,* Vol. 2 (F. Rapp, ed.), pp. 61–86, CRC Press, Boca Raton, Florida.
Hughes, J.P., Harvey, J.O., and Wada, M., 1964, Keratoconjunctivitis associated with infectious bovine rhinotracheitis, *J. Am. Vet. Med. Assoc.* **145**:32.
Hutyra-Marek, 1910, *Lehrbuch der speziellen Pathologie und Therapie der Haustiere,* 3rd ed., Vol. 1, S. 354, Gustav Fischer-Verlag, Jena.
Hutyra-Marek, 1922, *Spezielle Pathologie und Therapie der Haustiere,* 6th ed., Vol. I, S. 480, Gustav Fischer-Verlag, Jena.
Huygelen, C., Thienpont, D., Dekeyser, P.J., and Vandervelden, M., 1960a, Allerton virus, a cytopathogenic agent associated with lumpy skin disease. I. Inoculation of animals with tissue culture passaged virus, *Zentralbl. Veterinaermed.* **7**:754.
Huygelen, C., Thienpont, D., and Vandervelden, M., 1960b, Isolation of a cytopathogenic agent from skin lesions of cattle, *Nature (London)* **186**:979.

James, Z.H., and Povey, R.C., 1973, Organ culture of bovine teat skin and its application to the study of herpes mammillitis and pseudocowpox infections, *Res. Vet. Sci.* **15:**40.

Jasty, V., and Chang, P.W., 1969, Infectious bovine rhinotracheitis virus in bovine kidney cells: Sequence of viral production, cellular changes, and localization of viral nucleic acid and protein, *Am. J. Vet. Res.* **30:**1325.

Jasty, V., and Chang, P.W., 1971, Envelopment of infectious bovine rhinotracheitis viral particles in bovine kidney cell cultures: An electron microscopic study, *Am. J. Vet. Res.* **32:**1945.

Johnson, R.T., and Mims, C.A., 1968, Pathogenesis of viral infections in the nervous system, *N. Engl. J. Med.* **278:**23.

Kahrs, R.F., 1977, Infectious bovine rhinotracheitis: A review and update, *J. Am. Vet. Med. Assoc.* **171:**1055.

Kaminjolo, J.S., Paulsen, J., and Ludwig, H., 1969, Serologische Untersuchungen über die Verteilung von Antikörpern gegen die Viren der Mucosal Disease, der infektiösen Rhinotracheitis bzw. pustulären Vulvovaginitis und der Parainfluenza-3 bei Rindern in Hessen, *Berl. Muench. Tieraerztl. Wochenschr.* **82:**161.

Kaplan, A.S., (ed.), 1973, *The Herpesviruses,* Academic Press, New York, San Francisco, London.

Kendrick, J.W., 1973, Effects of the infectious bovine rhinotracheitis virus on the fetus, *J. Am. Vet. Med. Assoc.* **163:**852.

Kendrick, J.W., Gillespie, J.H., and McEntee, K., 1958, Infectious pustular vulvovaginitis of cattle, *Cornell Vet.* **48:**458.

Killington, R.A., Yeo, J., Honess, R.W., Watson, D.H., Duncan, B.E., Halliburton, I.W., and Mumford, J., 1977, Comparative analysis of the proteins and antigens of five herpesviruses, *J. Gen. Virol.* **37:**297.

Killington, R.A., Randall, R.E., Yeo, J., Honess, R.W., Halliburton, I.W., and Watson, D.H., 1978, Observations on antigenic relatedness between viruses of the herpes simplex "neutroseron," in: *Oncogenesis and Herpesviruses III/1* (G. de Thé, W. Henle, and F. Rapp, eds.), pp. 185–194, IARC Sci. Publ. No. 24, Lyon.

Knowles, R.W., and Person, S., 1976, Effects of 2-deoxyglucose, glucosamine, and mannose on cell fusion and the glycoproteins of herpes simplex virus, *J. Virol.* **18:**644.

Kokles, R., 1967, Die infektiöse Rhinotracheitis und das Coitalexanthem des Rindes, in: *Handbuch der Virusinfektionen bei Tieren,* Vol. II (H. Röhrer, ed.), pp. 901–960, VEB Gustav Fischer, Jena.

Kubin, G., and Klima, J., 1960, Experimentelle Untersuchungen über das Virus des Bläschenausschlages des Rindes, *Wien. Tieraerztl. Monatsschr.* **47:**741.

Lee, L.F., Kieff, E.D., Bachenheimer, S.L., Roizman, B., Spear, P.G., Burmester, B.R., and Nazerian, K., 1971, Size and composition of Marek's disease virus deoxyribonucleic acid, *J. Virol.* **7:**289.

Liebermann, H., Schulze, P., Kokles, R., and Hantschel, H., 1967, Isolierung und Identifizierung eines weiteren neuartigen bovinen Herpesvirus, *Arch. Exp. Veterinaermed.* **21:**761.

Liess, B., Knocke, K.-W., and Schimmelpfennig, H., 1960, Feststellung der virusbedingten Rhinotracheitis infectiosa des Rindes. II. Kulturelle, serologische und elektronenoptische Charakterisierung des Virus unter Berücksichtigung der in Gewebekulturen auftretenden zytopathischen Veränderungen, *Dtsch. Tieraerztl. Wochenschr.* **67:**551.

Liggitt, H.D., DeMartini, J.C., McChesney, A.E., Pierson, R.E., and Storz, J., 1978, Experimental transmission of malignant catarrhal fever in cattle, gross and histopathologic changes, *Am. J. Vet. Res.* **39:**1249.

Ludwig, H., 1972a, Untersuchungen am genetischen Material von Herpesviren. I. Biophysikalisch-chemische Charakterisierung von Herpesvirus-Desoxyribonucleinsäuren, *Med. Microbiol. Immunol.* **157:**186.

Ludwig, H., 1972b, Untersuchungen am genetischen Material von Herpesviren. II. Genetische Verwandtschaft verschiedener Herpesviren, *Med. Microbiol. Immunol.* **157:**212.

Ludwig, H., 1976, Bovine herpes mammillitis (BHM) virus and its relationship to other herpesviruses, Proceedings of the XXth World Veterinary Congress, Thessaloniki, July 6–12, 1975, pp. 1318–1319.

Ludwig, H., and Becht, H., 1977, Borna disease and its causative agent (Workshop, Würzburg 1975), in: *Slow Virus Infections of the Central Nervous System: Investigational Approaches to Etiology and Pathogenesis of These Diseases* (V. ter Meulen and M. Katz, eds.), pp. 75–83, Springer-Verlag, New York.

Ludwig, H., and Rott, R., 1975, Effect of 2-deoxy-D-glucose on herpesvirus-induced inhibition of cellular DNA synthesis, *J. Virol.* **16:**217.

Ludwig, H., and Storz, J., 1973, Activation of herpesvirus from normal bovine fetal spleen cells after prolonged cultivation, *Med. Microbiol. Immunol.* **158:**209.

Ludwig, H., Paulsen, J., and Kaminjolo, J.S., 1969, Kälbercornea-Zellen in Kultur. I. Eigenschaften einer fibroblasten- und einer epithelartigen Zellinie und deren Eignung für die Untersuchung einiger tierpathogener Viren, *Z. Med. Mikrobiol. Immunol.* **155:**133.

Ludwig, H., Biswal, N., Bryans, J.T., and McCombs, R.M., 1971, Some properties of the DNA from a new equine herpesvirus, *Virology* **45:**534.

Ludwig, H.O., Biswal, N., and Benyesh-Melnick, M., 1972a, Studies on the relatedness of herpesviruses through DNA–DNA hybridization, *Virology* **49:**95.

Ludwig, H., Haines, H.G., Biswal, N., and Benyesh-Melnick, M., 1972b, The characterization of varicella–zoster virus DNA, *J. Gen. Virol.* **14:**111.

Ludwig, H., Becht, H., and Rott, R., 1974, Inhibition of herpes virus-induced cell fusion by Concanavalin A, antisera, and 2-deoxy-D-glucose, *J. Virol.* **14:**307.

Ludwig, H., Pauli, G., Norrild, B., Vestergaard, B.F., and Daniel, M.D., 1978a, Immunological characterization of a common antigen present in herpes simplex virus, bovine mammillitis virus and *Herpesvirus simiae* (B virus), in: *Oncogenesis and Herpesviruses III/1* (G. de Thé, W. Henle, and F. Rapp, eds.), p. 235, IARC Sci. Publ. No. 24, Lyon.

Ludwig, H., Pauli, G., Bhakdi, S., and Norrild, B., 1978b, Properties of the common antigen between bovine herpes mammillitis (BHM) virus and herpes simplex virus (HSV), Herpesvirus Workshop, Cambridge, 1978, p. 53.

Ludwig, H., Heppner, B., and Herrmann, S., 1982, The genomes of different field isolates of Aujeszky's Disease virus, in: *Aujeszky's Disease* (G. Wittmann and S.A. Hall, eds.), pp. 15–20, Martinus Nijhoff Publishers, The Hague.

Lupton, H.W., Barnes, H.J., and Reed, D.E., 1980, Evaluation of the rabbit as a laboratory model for infectious bovine rhinotracheitis virus infection, *Cornell Vet.* **70:**77.

Luther, P.D., Bradley, P.G., and Haig, D.A., 1971, The isolation and characterization of a herpesvirus from calf kidney cell cultures, *Res. Vet. Sci.* **12:**496.

Mackay, J.M.K., 1969, Tissue culture studies of sheep pulmonary adenomatosis (jaagsiekte). II. Transmission of cytopathic effects to normal cultures, *J. Comp. Pathol.* **79:**147.

Madin, S.H., York, C.J., and McKercher, D.G., 1956, Isolation of the infectious bovine rhinotracheitis virus, *Science* **124:**721.

Malmquist, W.A., Krauss, H.H., Moulton, J.E., and Wandera, J.G., 1972, Morphologic study of virus-infected lung cell cultures from sheep pulmonary adenomatosis (jaagsiekte), *Lab. Invest.* **26:**528.

Martin, W.B., 1973, Bovine mammillitis: Epizootiologic and immunologic features, *J. Am. Vet. Med. Assoc.* **163:**915.

Martin, W.B., Martin, B., and Lauder, I.M., 1964, Ulceration of cows' teats caused by a virus, *Vet. Rec.* **76:**15.

Martin, W.B., Hay, D., Crawford, L.V., LeBouvier, G.L., and Crawford, E.M., 1966, Characteristics of bovine mammillitis virus, *J. Gen. Microbiol.* **45:**325.

Martin, W.B., James, Z.H., Lauder, I.M., Murray, M., and Pirie, H.M., 1969, Pathogenesis of bovine mammillitis virus infection in cattle, *Am. J. Vet. Res.* **30:**2151.

Martin, W.B., Wells, P.W., Lauder, I.M., and Martin, B., 1975, Features of the epidemiology of bovine mammillitis in Britain, Proceedings of the 20th World Veterinary Congress, Thessaloniki, Vol. 2, p. 1307.

Martin, W.B., Scott, F.M.M., Sharp, J.M., and Angus, K.W., 1976, Experimental production of sheep pulmonary adenomatosis, *Nature (London)* **264:**183.

Matthews, R.E.F., 1979, Classification and nomenclature of viruses, *Intervirology* **12:**165.
McKercher, D.G., 1973, Viruses of other vertebrates, in: *The Herpesviruses* (A.S. Kaplan, ed.), pp. 427–493, Academic Press, New York, San Francisco, London.
McKercher, D.G., and Wada, E.M., 1964, The virus of infectious bovine rhinotracheitis as a cause of abortion in cattle, *J. Am. Vet. Med. Assoc.* **144:**136.
McKercher, D.G., Moulton, J.E., and Jasper, D.E., 1955, Virus and virus-like cattle disease entities new to California, *Proc. U.S. Livestock Sanit. Assoc.* **58:**260.
McKercher, D.G., Saito, J.K., Wada, E.M., and Straub, O., 1959, Current status of the newer virus disease of cattle, *62nd Annu. Proc. U.S. Livestock Sanit. Assoc.*, 1958, p. 136.
McKercher, D.G., Bibrack, B., and Richards, W.P.C., 1970, Effects of the infectious bovine rhinotracheitis virus on the central nervous system of cattle, *J. Am. Vet. Med. Assoc.* **156:**1460.
Mettam, R.W.M., 1923, Snotsiekte in cattle, 9th and 10th Reps. Dir. Vet. Educ. Res. Union of S. Africa, p. 395.
Mettler, F., Engels, M., Wild, P., und Bivetti, A., 1979, Herpesvirus-Infektion bei Zicklein in der Schweiz, *Schweiz. Arch. Tierheilkd.* **121:**662.
Michalski, F., and Hsiung, G.D., 1975, Malignant transformation of hamster cells following infections with a bovine herpesvirus (infectious bovine rhinotracheitis virus), *Proc. Soc. Exp. Biol. Med.* **148:**891.
Miller, N.J., 1955, Infectious necrotic rhinotracheitis of cattle, *J. Am. Vet. Med. Assoc.* **126:**463.
Mohanty, S.B., 1973, New herpesviral and rhinoviral respiratory infections, *J. Am. Vet. Med. Assoc.* **163:**855.
Mohanty, S.B., 1975, Immunoferritin and immune electron microscopic study of bovine herpesvirus strain DN-599, *Am. J. Vet. Res.* **36:**319.
Mohanty, S.B., Hammond, R.C., and Lillie, M.G., 1971, A new bovine herpesvirus and its effect on experimentally infected calves, *Arch. Gesamte Virusforsch.* **34:**394.
Morein, B., and Moreno-Lopez, J., 1973, Skin hypersensitivity to parainfluenza-3 virus in cattle, *Zentralbl. Veterinaermed. Reihe B* **20:**540.
Murray, B.K., and Biswal, N., 1974, Synthesis of herpes simplex virus type 1 (HSV-1) DNA in isolated nuclei. II. Covalent linkage of RNA to nascent viral DNA, *Intervirology* **4:**14.
Mushi, E.Z., and Plowright, W., 1979, A microtitre technique for the assay of malignant catarrhal fever virus and neutralising antibody, *Res. Vet. Sci.* **27:**230.
Mushi, E.Z., Karstad, L., and Jessett, D.M., 1980, Isolation of bovine malignant catarrhal fever virus from ocular and nasal secretions of wildebeest calves, *Res. Vet. Sci.* **29:**168.
Nahmias, A.J., 1972, Herpesviruses from fish to man—a search for pathobiological unity, *Pathobiol. Annu.* **2:**153.
Narita, M., Inui, S., Namba, K., and Shimizu, Y., 1976, Trigeminal ganglionitis and encephalitis in calves intranasally inoculated with infectious bovine rhinotracheitis virus, *J. Comp. Pathol.* **86:**93.
Narita, M., Inui, S., Namba, K., and Shimizu, Y., 1978, Neural changes in calves intravaginally inoculated with infectious bovine rhinotracheitis virus, *J. Comp. Pathol.* **88:**381.
Narita, M., Inui, S., Namba, K., and Shimizu, Y., 1980, Neural changes in vaccinated calves challenge exposed with virulent infectious bovine rhinotracheitis virus, *Am. J. Vet. Res.* **41:**1995.
Nii, S., Katsume, I., and Ono, K., 1973, Dense bodies in duck embryo cells infected with turkey herpesvirus, *Biken J.* **16:**111.
Norrild, B., Bjerrum, O.J., Ludwig, H., and Vestergaard, B.F., 1978a, Analysis of herpes simplex virus type 1 antigens exposed on the surface of infected tissue culture cells, *Virology* **87:**307.
Norrild, B., Ludwig, H., and Rott, R., 1978b, Identification of a common antigen of herpes simplex virus, bovine herpes mammillitis virus, and B virus, *J. Virol.* **26:**712.
Nyaga, P.N., and McKercher, D.G., 1980, Pathogenesis of bovine herpesvirus-1 (BHV-1) infections: Interactions of the virus with peripheral bovine blood cellular components, *Comp. Immunol. Microbiol. Infect. Dis.* **2:**587.

O'Callaghan, D.J., and Randall, C.C., 1976, Molecular anatomy of herpesviruses: Recent studies, *Prog. Med. Virol.* **22:**152.

Ohshima, K., Miura, S., Numakunai, S., Sato, T., and Yasuda, J., 1977, Pathological studies on bovine malignant catarrhal fever with special reference to intranuclear structures in parietal cells, *Jpn. J. Vet. Sci.* **39:**293.

Osburn, B.I., 1973, Immune responsiveness of the fetus and neonate, *J. Am. Vet. Med. Assoc.* **163:**801.

Owen, N.V., Chow, T.L., and Molello, J.A., 1968, Infectious bovine rhinotracheitis: Correlation of fetal and placental lesions with viral isolations, *Am. J. Vet. Res.* **29:**1959.

Pallaske, G., 1960, *Pathologische Histologie,* 2nd ed., VEB Gustav Fischer, Jena.

Parks, J.B., and Kendrick, J.W., 1973, The isolation and partial characterization of a herpesvirus from a case of bovine metritis, *Arch. Gesamte Virusforsch.* **41:**211.

Pastoret, P.-P., 1981, Le virus de la rhinotracheite infectieux bovine (bovid herpesvirus 1): Aspects biologiques et moleculaires, Thesis, Université de Liége (in press).

Pastoret, P.-P., Burtonboy, G., Aguilar-Setién, A., Godart, M., Lamy, M.E., and Schoenaers, F., 1980, Comparison between strains of infectious bovine rhinotracheitis virus (bovid herpesvirus 1) from respiratory and genital origins, using polyacrylamide gel electrophoresis of structural proteins, *Vet. Microbiol.* **5:**187.

Patel, J.R., and Edington, N., 1980, The detection of the herpesvirus of bovine malignant catarrhal fever in rabbit lymphocytes *in vivo* and *in vitro, J. Gen. Virol.* **48:**437.

Pauli, G., and Aguilar-Setién, A., 1980, The major immunogenic components of IBR virus, International Conference on Human Herpesviruses; Workshop: Animal Models for Human Herpesvirus Infection (Hsiung, Ludwig), Atlanta, Georgia, March 17–21, 1980.

Pauli, G., and Ludwig, H., 1977, Immunoprecipitation of herpes simplex virus type 1 antigens with different antisera and human cerebrospinal fluids, *Arch. Virol.* **53:**139.

Pauli, G., Norrild, B., and Ludwig, H., 1979, Expression of common antigenic determinants of HSV-1 and BHM virus infected cells, Fourth Cold Spring Harbor Meeting on Herpesviruses, Cold Spring Harbor, New York, August 28–Sept. 2, 1979, p. 68.

Pauli, G., Darai, G., Storz, J., and Ludwig, H., 1981, IBR-IPV viruses: Genome structure and disease, *Med. Microbiol. Immunol.* **169:**129.

Pauli, G., Bund, K., and Podesta, B., 1982, Antigenic components of Aujeszky's Disease virus, in: *Aujeszky's Disease* (G. Wittmann and S.A. Hall, eds.), pp. 23–27, Martinus Nijhoff Publishers, the Hague.

Pepper, T.A., Stafford, L.P., Johnson, R.H., and Osborne, A.D., 1966, Bovine ulcerative mammillitis caused by a herpes-virus, *Vet. Rec.* **78:**569.

Pereira, L., Dondero, D., Norrild, B., and Roizman, B., 1981, Differential immunologic reactivity and processing of glycoproteins gA and gB of herpes simplex viruses 1 and 2 made in VERO and Hep-2 cells, *Proc. Natl. Acad. Sci. USA,* **78:**5202.

Perk, K., Michalides, R., Spiegelman, S., and Schlom, J., 1974, Biochemical and morphologic evidence for the presence of an RNA tumor virus in pulmonary carcinoma of sheep (jaagsiekte), *J. Natl. Cancer Inst.* **53:**131.

Plowright, W., 1968, Malignant catarrhal fever, *J. Am. Vet. Med. Assoc.* **152:**795.

Plowright, W., 1978, Vaccination against diseases associated with herpesvirus infections in animals: A review, in: *Oncogenesis and Herpesviruses III/2* (G. de Thé, W. Henle, and F. Rapp, eds.), p. 965, IARC Sci. Publ. No. 24, Lyon.

Plowright, W., Ferris, R.D., and Scott, G.R., 1960, Blue wildebeest and the aetiological agent of bovine malignant catarrhal fever, *Nature (London)* **188:**1167.

Plowright, W., Macadam, R.F., and Armstrong, J.A., 1965, Growth and characterization of the virus of bovine malignant catarrhal fever in East Africa, *J. Gen. Microbiol.* **39:**253.

Plowright, W., Herniman, K.A.J., Jesset, D.M., Kalunda, M., and Rampton, C.S., 1975, Immunisation of cattle against the herpesvirus of malignant catarrhal fever: Failure of inactivated culture vaccines with adjuvants, *Res. Vet. Sci.* **19:**159.

Plummer, G., and Waterson, A.P., 1963, Equine herpes viruses, *Virology* **19:**437.

Poli, G., Ponti, W., Uberti, F., and Balsari, A., 1980, Serological comparison of bovid herpesvirus 2 and herpes simplex virus by reciprocal neutralization kinetic studies, *Comp. Immunol. Microbiol. Infect. Dis.* **3:**509.

Porter, D.D., Larsen, A.E., and Cox, N.A., 1975, Isolation of infectious bovine rhinotracheitis virus from Mustelidae, *J. Clin. Microbiol.* **1:**112.

Potgieter, L.N.D., 1975, The influence of complement on the neutralization of infectious bovine rhinotracheitis virus by globulins derived from early and late bovine antisera, *Can. J. Comp. Med.* **39:**427.

Potgieter, L.N.D., and Aldridge, P.L., 1977, Frequency of occurence of viruses associated with respiratory tract disease of cattle in Oklahoma: Serologic survey for bovine herpesvirus DN 599 *Am. J. Vet. Res.* **38:**1243.

Potgieter, L.N.D., and Marë, D.J., 1974, Differentiation of strains of infectious bovine rhinotracheitis virus by neutralization kinetics with late 19 s rabbit antibodies, *Infect. Immun.* **10:**520.

Probert, M., and Povey, R.C., 1975, Experimental studies concerning the possibility of a latent carrier state in bovine herpes mammillitis (BHM), *Arch. Virol.* **48:**29.

Rapp, F., and Jerkofsky, M.A., 1973, Persistent and latent infections, in: *The Herpesviruses* (A.S. Kaplan, ed.), pp. 271–289, Academic Press, New York, San Francisco, London.

Reed, D.E., Langpap, T.J., and Bergeland, M.E., 1979, Bovine abortion associated with mixed Movar 33/63 type herpesvirus and bovine viral diarrhea virus infection, *Cornell Vet.* **69:**54.

Reid, H.W., Plowright, W., and Rowe, L.W., 1975, Neutralising antibody to herpesviruses derived from wildebeest and hartebeest in wild animals in East Africa, *Res. Vet. Sci.* **18:**269.

Reisinger, L., and Reimann, H., 1928, Beitrag zur Ätiologie des Bläschenausschlages der Rinder, *Wien. Tieraerztl. Monatsschr.* **15:**249.

Reissig, M., and Kaplan, A.S., 1962, The morphology of noninfectious pseudorabies virus produced by cells treated with 5-fluorouracil, *Virology* **16:**1.

Reissig, M., and Melnick, J.L., 1955, The cellular changes produced in tissue cultures by herpes B virus correlated with the concurrent multiplication of the virus, *J. Exp. Med.* **101:**341.

Roizman, B., 1962, Polykaryocytosis, *Cold Spring Harbor Symp. Quant. Biol.* **27:**327.

Roizman, B., 1978, The herpesviruses, in: *The Molecular Biology of Animal Viruses*, Vol. 2 (D.P. Nayak, ed.), pp. 769–848, Marcel Dekker, New York and Basel.

Roizman, B., 1982, The family Herpesviridae: General description, taxonomy and classification, in: *The Viruses, Vol A, Herpesviruses* (B. Roizman, ed.), Plenum Press, New York.

Roizman, B., and Furlong, D., 1974, The replication of herpesviruses, in: *Comprehensive Virology*, Vol. 3 (H. Fraenkel-Conrat and R.R. Wagner, eds.), pp. 229–403, Plenum Press, New York.

Roizman, B., Bartha, A., Biggs, P.M., Carmichael, L.E., Granoff, A., Hampar, B., Kaplan, A.S., Meléndez, L.V., Munk, K., Nahmias, A., Plummer, G., Rajcani, J., Rapp, F., Terni, M., de Thé, G., Watson, D.H., and Wildy, P., 1973, Provisional labels for herpesviruses, *J. Gen. Virol.* **20:**417.

Rolle, M., and Mayr, A., 1978, *Mikrobiologie, Infektions-und Seuchenlehre,* 4th ed., Enke, Stuttgart.

Rossi, C.R., and Kiesel, G.K., 1976, Antibody class and complement requirement of neutralizing antibodies in the primary and secondary antibody response of cattle to infectious bovine rhinotracheitis virus vaccine, *Arch. Virol.* **51:**191.

Rossiter, P.B., 1980, Antigens and antibodies of malignant catarrhal fever herpesvirus detected by immunodiffusion and counter-immunoelectrophoresis, *Vet. Microbiol.* **5:**205.

Rossiter, P.B., and Jessett, D.M., 1980, A complement fixation test for antigens of and antibodies to malignant catarrhal fever virus, *Res. Vet. Sci.* **28:**228.

Rossiter, P.B., Mushi, E.Z., and Plowright, W., 1977, The development of antibodies in rabbits and cattle infected experimentally with an African strain of malignant catarrhal fever virus, *Vet. Microbiol.* **2:**57.

Rossiter, P.B., Mushi, E.Z., and Plowright, W., 1978, Antibody response in cattle and rabbits to early antigens of malignant catarrhal fever virus in cultured cells, *Res. Vet. Sci.* **25:**207.

Rossiter, P.B., Jessett, D.M., and Mushi, E.Z., 1980, Antibodies to malignant catarrhal fever virus antigens in the sera of normal and naturally infected cattle in Kenya, *Res. Vet. Sci.* **29:**235.

Rouse, B.T., and Babiuk, L.A., 1977, The direct antiviral cytotoxicity by bovine lymphocytes is not restricted by genetic incompatibility of lymphocytes and target cells, *J. Immunol.* **118:**618.

Rouse, B.T., and Babiuk, L.A., 1978, Mechanisms of recovery from herpesvirus infections—a review, *Can. J. Comp. Med.* **42:**414.

Russel, P.H., 1979, Malignant catarrhal fever virus—plaque assay and enhanced neutralisation in hypotonic medium, *Vet. Microbiol.* **4:**29.

Russel, P.H., 1980, The *in vitro* response to phytohaemagglutinin during malignant catarrhal fever of rabbits and cattle, *Res. Vet. Sci.* **28:**39.

Russel, W.C., and Crawford, L.V., 1964, Properties of the nucleic acids from some herpes group viruses, *Virology* **22:**288.

Ruyechan, W.T., Morse, L.S., Knipe, D.M., and Roizman, B., 1979, Molecular genetics of herpes simplex virus. II. Mapping of the major viral glycoproteins and of the genetic loci specifying the social behavior of infected cells, *J. Virol.* **29:**677.

Rweyemamu, M.M., 1969, Bovine herpes mammillitis and skin gangrene of the bovine udder, *Vet. Rec.* **85:**697.

Rweyemamu, M.M., and Johnson, R.H., 1969, A serological comparison of seven strains of bovine herpes mammillitis virus, *Res. Vet. Sci.* **10:**102.

Rweyemamu, M.M., Johnson, R.H., and Tutt, J.B., 1966, Some observations on herpes virus mammillitis of bovine animals, *Vet. Rec.* **79:**810.

Rweyemamu, M.M., Johnson, R.H., and Gibbs, E.P.J., 1972, Studies on cell cultures persistently infected with bovine herpes mammillitis virus: The possible role of deionised water in inducing a carrier status, *Br. Vet. J.* **128:**611.

Rweyemamu, M.M., Karstad, L., Mushi, E.Z., Otema, J.C., Jessett, D.M., Rowe, L., Drevemo, S., and Grootenhuis, J.G., 1974, Malignant catarrhal fever virus in nasal secretions of wildebeest: A probable mechanism for virus transmission, *J. Wildl. Dis.* **10:**478.

Rweyemamu, M.M., Mushi, E.Z., Rowe, L., and Karstad, L., 1976, Persistent infection of cattle with the herpesvirus of malignant catarrhal fever and observations on the pathogenesis of the disease, *Br. Vet. J.* **132:**393.

Saito, J.K., Gribble, D.H., Berrios, P.E., Knight, H.D., and McKercher, D.G., 1974, A new herpesvirus isolate from goats: Preliminary report, *Am. J. Vet. Res.* **35:**847.

Saxegaard, F., 1968, Serological investigations of bulls subclinically infected with infectious pustular vulvovaginitis virus (IPV virus), *Nord. Veterinaermed.* **20:**28.

Saxegaard, F., and Onstad, O., 1967, Isolation and identification of IBR-IPV virus from cases of vaginitis and balanitis in swine and from healthy swine, *Nord. Veterinaermed.* **19:**54.

Schlehofer, J.R., Hampl, H., and Habermehl, K.-O., 1979, Differences in the morphology of herpes simplex virus infected cells. I. Comparative scanning and transmission electron microscopic studies on HSV-1 infected Hep-2 and chick embryo fibroblast cells, *J. Gen. Virol.* **44:**433.

Schmidt, M.F.G., Schwarz, R.T., and Ludwig, H., 1976, Fluorosugars inhibit biological properties of different enveloped viruses, *J. Virol.* **18:**819.

Schneweis, K.E., 1962, Serologische Untersuchungen zur Typendifferenzierung des Herpesvirus hominis, *Z. Immunitaetsforsch. Exp. Ther.* **124:**24.

Schultz, R.D., 1973, Developmental aspects of the fetal bovine immune response: A review, *Cornell Vet.* **63:**507.

Schulze, P., Hahnefeld, H., Hantschel, H., and Hahnefeld, E., 1964, Die Feinstruktur der Viren der infektiösen bovinen Rhinotracheitis und des Bläschenausschlages des Rindes (Exanthema coitale vesiculosum bovis), *Arch. Exp. Veterinaermed.* **18:**417.

Schulze, P., Liebermann, H., Hantschel, H., Bartha, A., and Juhász, M., 1967, Licht- und eletronenmikroskopische Untersuchungen an einem neuartigen bovinen Herpesvirus, *Arch. Exp. Veterinaermed.* **21:**747.

Selman, I.E., Wiseman, A., Murray, M., and Wright, N.G., 1974, A clinico-pathological study of bovine malignant catarrhal fever in Great Britain, *Vet. Rec.* **94:**483.

Sheffy, B.E., and Davis, D.H., 1972, Reactivation of a bovine rhinotracheitis virus after corticosteroid treatment, *Proc. Soc. Exp. Biol. Med.* **140:**974.
Shroyer, E.L., and Easterday, B., 1968, Growth of infectious bovine rhinotracheitis virus in organ cultures, *Am. J. Vet. Res.* **29:**1355.
Sim, C., and Watson, D.H., 1973, The role of type specific and crossreacting structural antigens in the neutralisation of herpes simplex types 1 and 2, *J. Gen. Virol.* **19:**217.
Skare, J., Summers, W.P., and Summers, W.C., 1975, Structure and function of herpesvirus genomes. I. Comparison of five HSV-1 and two HSV-2 strains by cleavage of their DNA with *Eco*RI restriction endonuclease, *J. Virol.* **15:**726.
Sklyanskaya, E.I., Itkin, Z.B., Gofman, Y.P., and Kaverin, N.V., 1977, Structural proteins of infectious bovine rhinotracheitis virus, *Acta Virol.* **21:**273.
Šmid, B., Valiček, L., and Pleva, V., 1971, Bovine herpes mammillitis (BHM) virus: Electron microscopy of the virus in tissue culture of primary bovine kidney cells, *Zentralbl. Veterinaermed. Reihe B* **18:**1.
Smith, P.C., 1976, The bovine herpesviruses: An overview, *Proc. 80th Annu. Conf. U.S. Anim. Health Assoc.*, p. 149.
Smith, W., and Mackay, J.M.K., 1969, Morphological observations on a virus associated with sheep pulmonary adenomatosis (jaagsiekte), *J. Comp. Pathol.* **79:**421.
Snowdon, W.A., 1965, The IBR-IPV virus: Reaction to infection and intermittent recovery of virus from experimentally infected cattle, *Aust. Vet. J.* **41:**135.
Spear, P.G., 1980, Composition and organization of herpesvirus virions and properties of some of the structural proteins, in: *Oncogenic Herpesviruses*, Vol. 1 (F. Rapp, ed.), pp. 53–84, CRC Press, Boca Raton, Florida.
Spear, P.G., Sarmiento, M., and Manservigi, R., 1978, The structural proteins and glycoproteins of herpesviruses: A review, in: *Oncogenesis and Herpesviruses III/1* (G. de Thé, W. Henle, and F. Rapp, eds.), pp. 157–167, IARC Sci. Publ. No. 24, Lyon.
Spradbrow, P.B., 1968, The isolation of infectious bovine rhinotracheitis virus from bovine semen, *Aust. Vet. J.* **44:**418.
Steck, F., Raaflaub, W., König, H., and Ludwig, H., 1969, Nachweis von IBR-IPV-Virus, Klinik und Pathologie bei zwei Ausbrüchen von Bläschenseuche, *Schweiz. Arch. Tierheilkd.* **111:**13.
Sterz, H., 1973, Biologische und physikalisch-chemische Eigenschaften des Virus der bovinen Herpes Mammillitis, Vet. med. dissertation, Justus Liebig-Universität, Giessen.
Sterz, H., and Ludwig, H., 1972, Plaque test and biological properties of bovine herpes mammillitis (BHM) virus, *Zentralbl. Veterinaermed. Reihe B* **19:**473.
Sterz, H., and Ludwig, H., 1974, Über die Bedeutung des Virus der bovinen Herpes Mammillitis (BHM), *Fortschr. Veterinaermed.* **20:**231.
Sterz, H., Ludwig, H., and Rott, R., 1973–1974, Immunologic and genetic relationship between herpes simplex virus and bovine herpes mammillitis virus, *Intervirology* **2:**1.
Stevely, W.S., 1975, Virus-induced proteins in pseudorabies-infected cells, II. Proteins of the virion and nucleocapsid, *J. Virol.* **16:**944.
Stevely, W.S., 1977, Inverted repetition in the chromosome of pseudorabies virus, *J. Virol.* **22:**232.
Stevens, J.G., 1978, Persistent, chronic, and latent infections by herpesviruses: A review, in: *Oncogenesis and Herpesviruses III/2* (G. de Thé, W. Henle, and F. Rapp, eds.), pp. 675–685, IARC Sci. Publ. No. 24, Lyon.
Stevens, J.G., and Groman, N.B., 1963, Properties of infectious bovine rhinotracheitis virus in a quantitated virus-cell culture system, *Am. J. Vet. Res.* **24:**1158.
Storz, J., 1968, Comments on malignant catarrhal fever, *J. Am. Vet. Med. Assoc.* **152:**804.
Storz, J., Okuna, N., McChesney, A.E., and Pierson, R.E., 1976, Virologic studies on cattle with naturally occuring and experimentally induced malignant catarrhal fever, *Am. J. Vet. Res.* **37:**875.
Storz, J., Ludwig, H., and Rott, R., 1980, Aktivierung des boviden Herpesvirus 1 aus foetalem Gewebe, *Proceedings of the XIth International Congress on Diseases of Cattle,* Tel-Aviv, pp. 471–473, Bregman Press, Haifa.

Straub, O.C., 1978, *Bovine Herpesvirusinfektionen*, VEB Gustav Fischer, Jena.

Straver, P.J., and van Bekkum, J.G., 1979, Isolation of malignant catarrhal fever virus from a European bison (*Bos bonasus*) in a zoological garden, *Res. Vet. Sci.* **26:**165.

Talens, L.T., and Zee, Y.C., 1976, Purification and buoyant density of infectious bovine rhinotracheitis virus, *Proc. Soc. Exp. Biol. Med.* **151:**132.

Taylor, R.L., and Hanks, M.A., 1969, Viral isolations from bovine eye tumors, *Am. J. Vet. Res.* **30:**1885.

Tenser, R.B., 1980, Control of herpesvirus infections, in: *Oncogenic Herpesviruses*, Vol. 2 (F. Rapp, ed.), pp. 87–115, CRC Press, Boca Raton, Florida.

Todd, J.D., Volenec, F.J., and Paton, I.M., 1972, Interferon in nasal secretions and sera of calves after intranasal administration of avirulent infectious bovine rhinotracheitis virus; association of interferon in nasal secretions with early resistance to challenge with virulent virus, *Infect. Immun.* **5:**699.

Törner, M., 1974, Indirekte Hämagglutinationsreaktion zum serologischen Nachweis von Herpesviren, Vet. Med. dissertation, Justus Liebig-Universität, Giessen.

Tousimis, A.J., Howells, W.V., Griffin, T.P., Porter, R.P., Cheatham, W.J., and Maurer, F.D., 1958, Biophysical characterization of infectious bovine rhinotracheitis virus, *Proc. Soc. Exp. Biol. Med.* **99:**614.

Valiček, L., and Šmid, B., 1976, Envelopment and the envelopes of infectious bovine rhinotracheitis virus in ultrathin sections, *Arch. Virol.* **51:**131.

Verwoerd, D.W., de Villiers, E.-M., and Coetzee, S., 1978, On the etiological role of *Herpesvirus ovis* in jaagsiekte, in: *Oncogenesis and Herpesviruses III/2* (G. de Thé, W. Henle, and F. Rapp, eds.), pp. 869–873, IARC Sci. Publ. No. 24, Lyon.

Verwoerd, D.W., de Villiers, E.-M., and Tustin, R.C., 1980, Aetiology of jaagsiekte: Experimental transmission to lambs by means of cultured cells and cell homogenates, *Onderstepoort J. Vet. Res.* **47:**13.

Verwoerd, D.W., Meyer-Scharrer, E., Brockman, J., and de Villiers, E.M., 1979, The serological relationship of herpesvirus ovis to other herpesviruses and its possible involvement in the aetiology of jaagsiekte, *Onderstepoort J. Vet. Res.* **46:**61.

Vestergaard, B.F., 1973, Crossed immunoelectrophoretic characterization of *Herpesvirus hominis* type 1 and 2 antigens, *Acta Pathol. Microbiol. Scand.* **81:**808.

Wandera, J.G., 1971, Sheep pulmonary adenomatosis, in: *Advances in Veterinary Science and Comparative Medicine*, Vol. 15 (C.A. Brandley and C.E. Cornelius, eds.), pp. 251–283, Academic Press, New York.

Wardley, R.C., Rouse, B.T., and Babiuk, L.A., 1976, Antibody dependent cytotoxicity mediated by neutrophils: A possible mechanism of antiviral defence, *J. Reticuloendothel. Soc.* **19:**323.

Watrach, A.M., and Bahnemann, H., 1966, The structure of infectious bovine rhinotracheitis virus, *Arch. Gesamte Virusforsch.* **18:**1.

Watson, D.H., Wildy, P., and Russel, W.C., 1964, Quantitative electron microscopy studies on the growth of herpes virus using techniques of negative staining and ultramicrotomy, *Virology* **24:**523.

Weiss, K.E., 1968, Lumpy skin disease virus, in: *Virology Monographs*, Vol. 3, pp. 111–131, Springer-Verlag, Vienna and New York.

Wellemans, G., Pastoret, P.-P., and Gouffaux, M., 1976, Maladie d'Aujeszky dans un rassemblement de jeunes bovins, *Ann. Méd. Vét.* **120:**196.

Wenger, D.L., and Hinze, H.C., 1974, Virus-host-cell relationship of *Herpesvirus sylvilagus* with cottontail rabbit leukocytes, *Int. J. Cancer* **14:**567.

Whitmore, H.L., and Archbald, L.F., 1977, Demonstration and quantitation of immunoglobulins in bovine serum, follicular fluid, and uterine and vaginal secretions with reference to bovine viral diarrhea and infectious bovine rhinotracheitits, *Am. J. Vet. Res.* **38:**445.

Wildy, P., Russel, W.C., and Horne, R.W., 1960, The morphology of herpes virus, *Virology* **12:**204.

Wilks, C.R., and Rossiter, P.B., 1978, An immunosuppressive factor in serum of rabbits lethally infected with the herpesvirus of bovine malignant catarrhal fever, *J. Infect. Dis.* **137**:403.

Witte, J., 1933, Untersuchungen über den Bläschenausschlag (Exanthema pustulosum coitale) des Rindes, *Z. Hyg. Infektionskrank.* **44**:163.

Wolinsky, J.S., and Johnson, R.T., 1980, Role of viruses in chronic neurological diseases, in: *Comprehensive Virology*, Vol. 16 (H. Fraenkel-Conrat and R.R. Wagner, eds.), p. 257, Plenum Press, New York.

Yedloutschnig, R.J., Breese, S.S., Jr., Hess, W.R., Dardiri, A.H., Taylor, W.D., Barnes, D.M., Page, R.W., and Ruebke, H.J., 1970, Bovine herpes mammillitis-like disease diagnosed in the United States, *Proc. U.S. Anim. Health Assoc.* **74**:208.

Yeo, J., Killington, R.A. Watson, D.H., and Powell, K.L., 1981, Studies on cross-reactive antigens in the herpesviruses, *Virology* **108**:256.

Zee, Y.C., and Talens, L., 1971, Entry of infectious bovine rhinotracheitis virus into cells, *J. Gen. Virol.* **11**:59.

Zee, Y.C., and Talens, L., 1972, Electron microscopic studies on the development of infectious bovine rhinotracheitis virus in bovine kidney cells, *J. Gen. Virol.* **17**:333.

Zur Hausen, H., 1980, The role of Epstein–Barr virus in Burkitt's lymphoma and nasopharyngeal carcinoma, in: *Oncogenic Herpesviruses*, Vol. 2 (F. Rapp, ed.), pp. 13–24, CRC Press, Boca Raton, Florida.

Zur Hausen, H., Schulte-Holthausen, H., Klein, G., Henle, W., Henle, G., Clifford, P., and Santesson, L., 1970, EBV DNA in biopsies of Burkitt tumours and anaplastic carcinomas of the nasopharynx, *Nature* (*London*) **228**:1056.

Zwick, W., and Gminder, J.G., 1913, Untersuchungen über den Bläschenausschlag (Exanthema vesiculosum coitale) der Rinder, *Berl. Tieraerztl. Wochenschr.* **29**:637.

Zygraich, N., Huygelen, C., and Vascoboinic, E., 1974, Vaccination of calves against infectious bovine rhinotracheitis using a temperature sensitive mutant, 13th International Congress of the IABS, Budapest, 1973, Part B, *Dev. Biol. Stand.* **26**:8.

CHAPTER 5

The Equine Herpesviruses

DENNIS J. O'CALLAGHAN, GLENN A. GENTRY, AND CHARLES C. RANDALL

I. INTRODUCTION

The equine herpesviruses comprise a diverse group of three antigenically distinct biological agents of protean manifestation in the horse, causing a variety of natural infections that vary from the subclinical to fatal generalized disease. Equine herpesvirus type 1 (EHV-1), also known as equine abortion virus (EAV), or equine rhinopneumonitis virus, is a major cause of respiratory disease and abortion in the horse. This agent, in particular, appears to be a model system for *in vitro* and *in vivo* study of disease, persistent infection, biochemistry of viral infection, and biochemical and oncogenic transformation. This review will therefore emphasize certain biological features of this equine herpesvirus, the best studied of the group. Equine herpesvirus type 2 (EHV-2), or equine cytomegalovirus (ECMV), is a ubiquitous, loosely defined, antigenically heterogeneous, usually slowly growing group of viruses, causing no known disease. Equine herpesvirus type 3 (EHV-3), equine coital exanthema (ECE) virus, is the causative agent of a relatively mild progenital exanthema of both mare and stallion.

These agents, classified as equine herpesviruses (Plummer *et al.*, 1973; Mathews, 1979), have been the subject of recent review articles (Studdert, 1974; O'Callaghan *et al.*, 1978). As a consequence, the references for this presentation have been arbitrarily chosen. Each virus and

DENNIS J. O'CALLAGHAN, GLENN A. GENTRY, AND CHARLES C. RANDALL • Department of Microbiology, University of Mississippi Medical Center, Jackson, Mississippi 39216. This chapter is dedicated to the memory of former colleagues Everett C. Bracken and Robert W. Darlington.

its biological and molecular parameters will be described more or less separately.

II. CLASSIFICATION AND CLINICAL FEATURES

A. Equine Herpesvirus Type 1

1. Historical Background

In historical perspective, the study of the first disease attributed to an equine herpesvirus (EHV-1) began at the University of Kentucky Agriculture Experiment Station in Lexington. Strategically located in central Kentucky, with the highest concentration of thoroughbreds in the world, and aided by cooperative owners and well-trained veterinarians, the Kentucky workers have had and continue to have an exceptional opportunity to study diseases of the horse.

Abortion of the equine fetus due to a possible specific viral infection was first described as a clinical entity by Dimock and Edwards (1932) and in the same year (1932) was produced in mares by inoculation of material from aborted fetuses. Further studies established the viral etiology and defined the clinical manifestations and the definitive pathological lesions in the aborted foal (Dimock and Edwards, 1933; Dimock and Edwards, 1936; Dimock, 1940).

Several European workers were the first to associate viral abortion with respiratory disease. Manninger and Csontos (1941), in Hungary, inoculated pregnant mares with bacteriologically sterile filtrates from aborted fetuses with lesions of viral abortion and observed that these animals and contacts developed a clinical picture resembling that of mild influenza. They proposed that viral abortion was secondary to equine influenza infections. Their collaborator, Salyi (1942), studied the fetal abortion material and showed that the gross and microscopic lesions were identical with those reported in Kentucky. Further studies by Manninger (1949)concluded that viral abortion was due to infection of pregnant mares by equine influenza virus. Kress (1941) further suggested that the abortion virus was pneumotropic because of the occurrence of bronchopneumonia in horses that had been in contact with aborted mares and fetuses.

The respiratory infection associated with equine abortion virus (EAV) was first studied experimentally by Doll *et al.* (1954b), and the symptomatology produced in inoculated young horses was similar to that described for equine influenza. The evidence from this study showed that EAV was the etiological agent of an epizootic respiratory disease of young horses. It remained for Doll and associates to prove that several putative influenza virus isolates were identical with EAV (Doll *et al.*, 1954a; Doll and Kintner, 1954; Doll and Wallace, 1954). In another study, Doll *et al.* (1957)

suggested that the agent previously known as EAV should be regarded as a respiratory virus, because the principal histological lesions in young horses and in aborted foals occur in the respiratory tract. Accordingly, the authors designated the disease caused by the virus as viral rhinopneumonitis and the agent as equine rhinopneumonitis virus.

A major hindrance to the study of experimental abortion was the inability to reproduce the disease consistently on inoculation of pregnant mares, since only about 20% of such mares abort after such inoculation. This obstacle was overcome by Doll (1953),who showed that after direct inoculation of uterus or fetus, 100% of fetuses could be aborted with virus infection, with lesions typical of the natural infection.

2. Cultivation and Host Range

Investigation of EHV-1 infections was greatly handicapped by inability of workers to cultivate the agent in any animal system except the horse. An encouraging approach was begun by Anderson and Goodpasture (1942), who showed that suckling hamsters inoculated with extracts of aborted fetal tissue developed focal liver necrosis and that the involved parenchymal cells contained typical intranuclear inclusions. The infection, however, could not be maintained beyond three serial passages.

The virus was first cultivated in tissue culture (Maitland type) in equine fetal lung and spleen (Randall *et al.*, 1953), in domestic cat tissue (Randall *et al.*, 1954b), in tissues of dog, hamster, and rabbit (Randall, 1954a), and in HeLa cell monolayers (Randall, 1954b). Subsequently, use of monolayers of equine kidney (Shimizu *et al.*, 1957, 1959), swine and sheep kidney (McCollum *et al.*, 1962), and L-cells (Randall and Lawson, 1962) was reported for the propagation of the virus.

Concomitant with these studies, Doll *et al.* (1953) reported the adaptation of EHV-1 to the Syrian hamster, and Randall *et al.* (1954a) demonstrated viremia in the same infected host. In another study, Randall and Bracken (1957) correlated the development of hepatic inclusions with the log-phase growth cycle of the virus, in both blood and liver.

As previously thoroughly reviewed (O'Callaghan *et al.*, 1978), it has been shown by many workers that EHV-1 has a wide host range, with regard to both animal and tissue-culture systems. EHV-1 appears to offer a relatively unique system to herpesvirus workers in that the agent replicates with one-step growth kinetics *in vivo* in the Syrian hamster (Randall and Bracken, 1957; O'Callaghan *et al.*, 1972b) and *in vitro* in the L-M cell line (Darlington and James, 1966; O'Callaghan *et al.*, 1968b).

3. Serotypes

The three members of the equine herpes group have been compared serologically without the detection of any significant relationship and are considered antigenically distinct (Plummer *et al.*, 1973). Gutekunst *et al.* (1978), however, showed that EHV-1 and EHV-3 have certain common antigens not shared by EHV-2. Further discussion of antigenic comparisons will be included in Section II.C.

Even though EHV-1 has practically a worldwide distribution (McCollum, 1966; Matumoto *et al.*, 1965; Studdert, 1974), only one serotype is recognized (Burrows, 1970; Burrows and Goodridge, 1973). Two subtypes, however, have been claimed on the basis of neutralization tests (Shimizu *et al.*, 1959; Mayr *et al.*, 1965; Burrows, 1968). Subtype 1 is typified by the Kentucky D strain (Doll and Wallace, 1954) and subtype 2 by the Japanese H-45 strain (Shimizu *et al.*, 1959). Mayr *et al.* (1965) found that some European isolates of the virus were more closely related to the Army 183 than to the Ky-D strain and classified them as subtype 2 strains. In this latter work, the isolates appear not to have been subjected to critical comparison with the Japanese subtype 2 (Burrows and Goodridge, 1973). It remains to be shown that the differentiation into subtypes has any real significance. Burrows (1970) has pointed out that strains of both subtypes have been associated with respiratory and abortion syndromes and that both are widely distributed. It is claimed that Japanese respiratory and abortigenic isolates are serologically identical, but differ from the Kentucky D strains of EHV-1 by cross-neutralization. Thus, there are Japanese and American strains (Kawakami *et al.*, 1962).

It is commonly accepted (without proof) that all rhinopneumonitis strains, other than those modified by passage in tissue culture or animals, have the same disease-producing potential. There has been some indication, however, that the clinical picture varies in some countries. Although respiratory infection is relatively common in Japan, infection by EHV-1 as a cause of abortion appears to be rare (Kawakami and Shimizu, 1978), and abortion occurred in Britain only in association with newly imported mares. This situation led Miller (1966) to postulate a difference in virulence between respiratory and abortion strains.

With this difference in mind, Burrows and Goodridge (1973) investigated a number of isolates thought to represent the two clinical conditions, designating abortion virus isolates as fetal (F) strains and those obtained from respiratory infections as R strains. Differences between the two groups, both *in vivo* and *in vitro,* were demonstrated. The majority of F strains produced plaques in a variety of cells, while most of the R group did not. In tissue cultures supporting growth of both, the F designates grew to somewhat higher titers, and in foal trachea organ cultures, the peak concentration was 10-fold greater with the F strains. *In vivo,* F infected the horses more readily, grew better in the nasopharynx, was present in the buffy coat of blood, was excreted more readily, and elicited a better neutralizing-antibody response. In Australia, F and R isolates exhibited clear differences in plaque size and host-cell range (Studdert and Blackney, 1979).

4. Clinical Conditions Associated with EHV-1 Infections of Horses

As discussed below, infections of horses with EHV-1 are associated with several clinically distinguishable entities: (1) respiratory disease; (2) abortion; and (3) neurological disease.

a. *Respiratory Disease and Abortion*

As discussed above, the concept that equine viral abortion and an influenzalike (respiratory) disease are associated was advanced by a number of workers (Manninger and Csontos, 1941; Manninger, 1949; Kress, 1941, 1944). Doll and Kintner (1954) clarified this problem by showing that the putative equine influenza virus of Jones *et al.* (1948) was, in fact, EHV-1.

To clarify the problem of the respiratory-disease complex in horses, a type A influenza virus has been identified as one of the causes of epizootic respiratory disease in horses, and the use of the term "equine influenza" is restricted to disease caused by influenza virus (Sovinova *et al.*, 1958).

Infection by EHV-1 is evidenced in pregnant mares by abortion and in young horses—which have little or no immunity—by signs mainly localized to the respiratory tract. The respiratory diseases in question are among the most important viral diseases of horses from an economic standpoint. In young horses (foals) contracting the infection for the first time, the symptoms can usually be attributed to involvement of the upper respiratory system. Infection of perinatal foals is much more severe, leading to a fatal viral bronchopneumonia (Bryans, 1969). Appropriate clinical descriptions of the two conditions of interest have been more or less summarized in the discussion to follow (Doll *et al.*, 1954b; Doll 1962; Doll and Bryans, 1963a; Bryans, 1980, 1981).

Occult respiratory disease usually occurs in immunologically naïve young horses (foals), characteristically as an epidemic disease of farm populations after weaning and in the fall and winter months of the first year of life. Signs of acute infection include fever up to 106°F, viremia, pharyngitis, tracheobronchitis, and neutropenia. Nasal discharge is conspicuous during the febrile period. Necrosis of the lymphoreticular tissue of the upper respiratory tract and respiratory mucosa occurs accompanied by acute inflammation, and large amounts of virus are shed to the external environment. Virus may be recovered from foals up to 9 days and from the nasopharynx of experimentally infected mares (which show no sign of infection) for similar periods of time. At this stage, secondary bacterial infection is likely to occur, which, if untreated, may result in abscess formation in the lymphoid tissue of the pharynx, regional lymph node adenitis, and heavy mucopurulent discharge. As a consequence, foals may also develop pneumonia or severe chronic respiratory infection as sequelae of rhinopneumonitis.

Infection by EHV-1 does not, unfortunately, result in permanent immunity of the respiratory tract. Such immunity persists for only a few months. Horses may be reinfected naturally (respiratory route), or by experimental infection, every 3–6 months throughout life. After the first experience, reinfection by way of the respiratory tract results in production of virus, but usually without signs of clinical disease.

Pregnant mares, because of past exposure, generally have sufficient immunity so that infection is not observable. However, the occurrence of infection in pregnant mares may become evident after a delay of 3 weeks to 3 months by abortion of a virus-infected fetus with typical lesions (which will be discussed in detail in Section II.A.4.a.i) or by the birth of an infected foal, which dies within hours to several days with severe bronchopneumonia and pulmonary edema. Mares usually show no premonitory signs, and abortion happens rather suddenly without untoward results to the mare. Abortion is usually a disease of late pregnancy, 86% occurring from the 9th month to full term (11 months). The aborted fetus apparently dies from suffocation consequent to separation of the placenta and inability to respire after delivery because of extensive pulmonary edema and viral pneumonitis. Rarely, mares abort a second time after an interval of 2 years or more (Dimock *et al.*, 1942).

A previously unrecognized aspect of neonatal foal disease associated with prenatal infection by EHV-1 has been reported by Bryans *et al.* (1977). In a series of cases, some foals were ill at birth, but most appeared in good condition. Illness was usually manifest in the 1st week of life with respiratory distress, marked infection of conjunctival and buccal membranes, weakness, and diarrhea. At autopsy, all foals exhibited interstitial pneumonia and hypoplasia of spleen and thymus. Viral inclusions were not found, but EHV-1 was isolated from three of nine cases. The authors postulated that the evidence, though speculative, suggests that the disease is caused by infection of the fetus late in gestation by EHV-1.

With regard to diagnosis of EHV-1 infections (Bryans, 1980), the most reliable diagnostic procedures for determination of EHV-1-caused abortion involves histological and virological examination of autopsy material. In horses and foals with respiratory signs of infection, virus may be cultured from the nasopharynx and from the buffy coat of the blood during the acute febrile period. During the 2nd week of the respiratory infection, serological evidence for EHV-1 infection may be attained by demonstrating a rise in either complement-fixing (CF) or virus-neutralizing (VN) antibodies. There appears to be no satisfactory method for serological diagnosis of abortigenic disease because of the prevalence of high-titer antibody in mares (Bryans, 1968).

i. Pathology of Respiratory Disease and Abortion. Dimock and Edwards (1936) were the first to relate the disease occurring in fetuses to an epizootic abortion of mares. In this and subsequent studies, Dimock (1940) and Dimock *et al.* (1942) described the gross and microscopic pathological lesions that have come to be considered pathognomonic of equine viral abortion. This pioneer work was confirmed and enlarged on by a number of investigators (Salyi, 1942; Westerfield and Dimock, 1946; Jeleff, 1957; Kawakami *et al.*, 1959; Prickett, 1970a).

Virus abortion ordinarily occurs from the 7th month to term with foals being stillborn and usually in a good state of preservation, or they may be alive at birth, but die soon afterward. In summary, the classic and most consistent gross lesions occurring in aborted foals include petechial hemorrhages of serosal surfaces, respiratory mucosa, and conjunctiva, markedly edematous lungs and voluminous pleural effusion, focal necrosis of liver, and splenic enlargement. Microscopically, the principal lesions are bronchiolitis and bronchitis, with characteristic necrosis of epithelium, bronchopneumonia, or even diffuse pneumonitis and focal necrosis of spleen and liver. Conspicuous and characteristic herpes-type intranuclear inclusions are present in affected cells.

Equine viral abortion is ordinarily a disease of late pregnancy, but may occur as early as the 5th month and may be induced experimentally as early as the 3rd month (Doll *et al.*, 1955). Prickett (1970a), in the most complete study to date of the pathology of experimental EHV-1 infections of the fetus, foal, and pregnant mare, has made a number of interesting observations. In all instances, whether naturally occurring or experimentally produced, the gross and microscopic lesions before the 6th month are the same and show no evidence of local tissue reaction (inflammation), which is a prominent feature of infection of older fetuses (late pregnancy). In the early abortions, the fetus is severely autolyzed and there is diffuse scattering of typical inclusions. Between the 6th and 7th months, a mixture of lesions is evident; both those in early and late abortions (from the 7th month until term) are seen. The pathological features vary with the age of the fetus and may be related to the maturation of the fetal immune system.

Experimental inoculation by the respiratory route of susceptible young horses and those with some degree of immunity produced distinctly different results, and the pathological findings correlated with the results of a serological study of the affected animals. Susceptible horses without antibodies developed bronchopneumonia characterized by conspicuous polymorphonuclear infiltration of small bronchi and brochioles with necrosis of respiratory epithelium, with typical intranuclear inclusions, peribronchiolar and perivascular infiltration of round cells, and serofibrinous exudate into alveoli. In addition, necrosis and intranuclear inclusions were observed in bronchial lymph nodes. In contrast, animals having past experience with the virus showed marked regional lymph-node hyperplasia without necrosis or inclusion-body formation.

Experimentally infected mares were sacrificed after aborting EHV-1-infected fetuses, and the tissues were studied. No gross lesions were evident. Microscopic lesions in the mares consisted of perivascular infiltration by lymphocytes, plasma cells, and eosinophiles of the subendometrial lamina propria and interglandular connective tissue. No inclusion bodies or microscopic lesions other than those described were observed in the various tissues of the mare, though in all cases the regional lymph nodes were undergoing marked hyperplasia. The placental tissue showed no significant gross abnormalities. The chorionic villi exhibited fine necrosis and low-grade inflammation. Edema of the intermicrovillar spaces between chorionic epithelium and the endometrium was widespread, indicating placental and endometrial separation prior to abortion.

ii. Epidemiology, Pathogenesis, and Immunity. These areas cannot be rigidly separated and will be discussed together; they concern studies of pathogenesis (Bryans, 1969; Bryans and Prickett, 1970), immunological relationships (Doll *et al.*, 1955; Doll, 1961; Doll and Bryans, 1962a,b, 1963a–c; Bryans, 1968, 1969, 1980), and epidemiology (Doll and Bryans, 1963a).

All epidemiological evidence suggests that EHV-1 spreads chiefly from one horse to another by inhalation of airborne secretions from the respiratory tract, but also may be disseminated by contaminated articles, feed and water, and the products of conception in viral abortion. The annual epizootics that occur in young horses serve as a source of virus for the annual reinfection of adult horses, which tend to maintain sufficient resistance to prevent most abortions. Farms on which these epizootics are annual among highly susceptible young horses have no or only a very few abortions. Those farms that have experienced serious outbreaks of abortion have escaped epizootics in young horses for several years or have kept young horses isolated from adults, thus preventing spread. Escaping frequent reinfection allows resistance to wane, thus maximizing opportunity for a serious outbreak of abortion.

Immunity against reinfection of the respiratory tract is of short duration, lasting from 3 to 5 months, but immunity that protects against abortion is of longer duration. Horses become infected early in life and are reinfected annually or more often. Horses may be reinfected experimentally or by natural exposure, and these reinfections may occur every 3–6 months or longer throughout the life of the horse. Most mares (but not all), as a consequence of natural occurring reinfection of the respiratory tract, have a relatively high level of resistance against abortion, presumably because of sufficient VN antibody. Approximately 30% of mares exposed to virus in the respiratory tract may be reinfected 3 months later.

There appears to be enchanced susceptibility to abortion from fetal rhinopneumonitis from 8 months to full term, approximately 95% of abortions occurring at this time. Epizootics of abortion have greater frequency from January through April, with an incidence of approximately 90%. The seasonal periodicity of respiratory epidemics and of abortion, at least in Kentucky, appears to have a definite relationship to the age of the fetus and the restricted breeding season of thoroughbred horses.

Foals are born during a restricted period of the year, raised during the summer and early fall, usually in dispersed groups, which results in a population of susceptible foals at weaning time. Since they are usually weaned as a group during the fall and held in fairly closely confined quarters, it is understandable that the exposure maximizes the spread of respiratory infection and an epizootic of rhinopneumonitis occurrs. Despite much work, the source of virus to initiate such an epizootic remains to be discovered. The situation suggests natural reservoirs or a high incidence of carriers among adult horses. In this regard, it may be significant that outbreaks of respiratory infection caused by EHV-1 appeared to have been activated by vaccination of isolated horses with African horse sickness virus (Erasmus, 1966).

The pattern of infection in young horses, the long incubation period for abortion, and the seasonal breeding of mares create a situation that favors abortion in the late stages of gestation. A majority of mares are exposed during a period of gestation that combined with an incubation period of 30–90 days places them in the period of 8 months to full term, apparently the optimal time for abortion from fetal infection. It is interesting that natural or artifically induced infection of mares at 60–120 days of gestation does not result in abortion.

As indicated previously, immunity to natural or induced respiratory infection is of relatively short duration, and the nasopharyx may be repeatedly infected without febrile response or signs of infection. Protection against abortion is longer lasting but inconsistent. Abortion in a few mares occurs naturally, or after injection of virus, despite high-titer antibody.

The pathogenesis of abortigenic disease is still an unsolved problem. It has always been assumed that virus is transported from the respiratory mucosa to the fetus by way of the bloodstream. Bryans (1969) made a study of this problem in pregnant mares. Infection was instituted either by intranasopharyngeal spray or by subcutaneous inoculation, and samples of blood and nasapharynx were cultured in tissue culture. The Bryans study showed that in horses infected by the natural route (nasopharynx), there was a relationship between the level of serum VN antibody titer and immunity to infectivity. Bryans also showed that viremia developed (in the presence of VN antibody) associated with leukocytes only, and not with red blood cells or serum, and persisted in some cases for as long as 3 weeks. No animal with a VN antibody titer of 2.0 or higher ($\log_{10}$ of reciprocal of serum VN antibody titer) at the time of exposure of the nasopharyx subsequently developed viremia, infection of the upper respiratory tract, or abortigenic infection. On the other hand, all animals inoculated subcutaneously developed viremia regardless of antibody titer, and some pregnant mares delivered normal foals, while others aborted an infected fetus. In this case, there is no correlation between serum VN antibody and protective immunity to abortion. Bryans (1969), as a result of this study, has postulated that animals naturally infected, with a sufficiently high protective level of serum VN antibody (2.0 or higher), have developed immunity against abortigenic disease that results from a barrier to infection located on or in the mucosa or nasapharyngeal accessory structures. Also, according to Bryans, high-titer antibody might protect against tissue damage and release of infected cells to the circulation and subsequent infection of the fetus with EHV-1.

This provocative work should stimulate further study. It would be interesting to determine the type(s) of white blood cell infected and whether viral replication occurs. Darlington (1978) has shown that EHV-1 replicates in white blood cells of immunologically naïve foals but not on reinfection. As expected, the white blood cells of adult horses show no evidence of replication.

Control by vaccination of respiratory and other diseases in horses, particularly abortigenic infections of the fetus, has been a long-sought goal of many laboratories, the work of which has been reviewed by Bryans (1978). Accordingly, chemically inactivated virus derived from infected horse fetuses was not satisfactory because of untoward side effects; neither was inactivated hamster passage virus, because of lack of potency. Attenuated live virus vaccines may have reduced the incidence of abortion, but were suspect because of the tendency to cause virus abortion in the recipients. Recent work of Bryans and Allen (1982) describes the effective use of a chemically inactivated virulent tissue culture strain of EHV-1 to reduce the incidence of abortion 4-fold.

New information indicates change in the molecular epidemiology of abortion. Allen *et al.* (1983b) have shown by restriction endonuclease analyses the rapid emergence of 1 B as the new dominant strain, identified in less than 5% of abortion before 1980 and 53% in 1982.

b. Neurological Disease

As summarized by Bryans (1980), a neurological syndrome was first observed in horses in association with EHV-1 viral abortion in mares. It has since been shown to occur in horses of either sex of any age as well as in pregnant mares with or without abortigenic infection. The clinical condition may be experimentally produced in horses within 6–12 days after subcutaneous or intranasal inoculation with EHV-1. Clinical signs may include paresis, ataxia, weakness, and paralysis, without disturbance of consciousness and usually without fever. The severity of the disease varies considerably, and the prognosis is good in those nonrecumbent.

Manninger (1949) was the first to report the occurrence of neurological disease in horses; during the course of studies on equine abortion, he noted that several mares inoculated with EHV-1 from aborted fetuses developed myelitis. It remained for Saxegaard (1966) to incriminate EHV-1 infection yet again with paralytic disease. In this study, EHV-1 was isolated from the central nervous system (CNS) of two mares and a stallion with paralysis. Each mare developed the disease about 4 weeks after aborting a fetus from which EHV-1 was isolated. Other reports have since appeared involving studies of neurological disease associated with EHV-1 infections. According to these reports, ataxia and paralysis occurred during natural outbreaks of rhinopneumonitis or abortion or both (Jackson and Kendrick, 1971; Dalsgaard, 1970; Little and Thorsen, 1974; Petzoldt *et al.*, 1972; Charlton *et al.*, 1976; Dinter and Klingeborn, 1976). The condition has also been reproduced by inoculation of virulent virus (Jackson and Kendrick, 1971; Little and Thorsen, 1974; Jackson *et al.*, 1977). Prickett (1970b), however, was unable to induce paralysis (or any other signs referable to the CNS) by direct intracerebral inoculation of EHV-1, and there was no evidence of infection at autopsy.

The histopathology seems to vary somewhat depending on the report. In the studies of Jackson and Kendrick (1971) and Jackson *et al.* (1977), vasculitis involving small arteries and veins and nervous-tissue degeneration were the principal lesions. The authors noted the lack of the usual indications of encephalomyelitis. Virus was not isolated at autopsy in either study. A more severe lesion was described by Charlton *et al.* (1976). In this study, it was found that during an outbreak of EHV-1 abortion, two pregnant mares developed severe neurological disease and at autopsy EHV-1 was cultured from brain and lung. Histologically, both animals showed severe disseminated meningoencephalitis characterized by necrotizing vasculitis and focal malacia in brain and spinal cord. Perivascular cuffing of lymphocytes, histiocytes, neutrophiles, and occasionally giant cells was noted. No intranuclear inclusions were detected in the CNS lesions, but typical inclusions occurred in necrotic foci in the thyroid.

Circumstantial evidence indicates that some strains of EHV-1 are more neurotropic than others. Apparently, none of the strains associated with outbreaks of CNS disease have been compared, particularly with this potential in mind.

B. Equine Herpesvirus Type 2

As far as is known, despite much research, this group of herpesviruses does not cause disease in horses. The members are antigenically heterogeneous by VN-antibody tests (Erasmus, 1970; Plummer *et al.*, 1969a, 1973). Mumford and Thompson (1978) also found major differences between slow-growing and fast-growing isolates by VN-antibody testing, but by indirect immunofluorescence and CF, they demonstrated a common antigen among all the many isolates tested. Among these, the L-K strain of Plummer and Waterson (1963) may be considered the prototype. Recently, Wharton *et al.* (1981) have compared the genomic properties of equine cytomegalovirus and other cytomegaloviruses.

Studdert (1974), O'Callaghan *et al.* (1978), and Bryans (1968) have largely reviewed the many contributions to the field. In brief, the many candidate viruses have a limited host range, and with the exception of cultivation in primary cat and rabbit kidney cells, growth is restricted to equine cells. Most investigators have reported that most isolates of EHV-2 from representative parts of the world were slow-growing, requiring up to 28 days to produce cytopathic effect (CPE) in suitable cell cultures. A few, however, are fast-growing, producing CPE in 24 hr or less (Karpas, 1966; Mumford and Thompson, 1978). Because of slow growth of the majority of strains, a tendency to be cell-associated, and a proclivity to form conspicuous intranuclear inclusions, these agents have been termed equine cytomegaloviruses (Wharton *et al.*, 1981). The members are ubiquitous, having been isolated from uninoculated primary equine cell cultures from a variety of equine tissues (including the buffy coat) and also from a variety of sites from normal horses and other horses afflicted with a variety of clinical syndromes (Studdert, 1974).

Infection occurs early in life, probably by inhalation of contagious material from the respiratory tracts of other horses. The high incidence of infection is associated with long-lasting viral persistence and continuous viral shedding. EHV-2, however, has not been isolated from fetal tissue and appears not to cross the placental barrier (Studdert, 1974).In some ways, EHV-2 might well be considered a member of the indigenous flora; it certainly appears to have less potential for causing disease than *Escherichia coli*, for example.

C. Equine Herpesvirus Type 3

This most recent addition to the equine herpesviruses was discovered independently in the United States (Bryans, 1968), in Canada (Girard *et al.*, 1968), and in Australia (Pascoe *et al.*, 1968). It has been designated as the third member of the equine herpesvirus triad (Roizman, 1973; Plummer *et al.*, 1973; Mathews, 1979), is the etiological agent of a relatively innocuous genital disease commonly known as equine coital exthanthema (ECE), and appears to be a typical herpesvirus (Studdert, 1974; O'Callaghan *et al.*, 1978; Atherton *et al.*, 1980, 1981 and 1982).

All EHV-3 strains isolated from four different countries were antigenically identical with the two strains of Bryans and Allen (1973) by cross-neutralization. Also, all horse ECE strains examined by Burrows (1973) were indistingishable serologically, but a strain of donkey coital exanthema virus was antigenically different from the horse viruses and from other equine herpesvirus strains.

Replication of EHV-3 is restricted to cells of equine origin (Bryans and Allen, 1973), with the exception of rabbit kidney cells (Girard *et al.*, 1968). Cytolysis of infected cells and the production of intranuclear inclusions is noted in 24 hr or less, with release of virus from the equine transitional carcinoma cell occurring slowly with low efficiency; less than 10% of total virus is recovered in the extracellular fluid (Allen and Bryans, 1977). However, with the Australian strains (grown in primary equine fetal kidney), the growth curves for cell-associated and free virus were reversed (Studdert, 1974).

As previously noted, the three equine herpesviruses are antigenically distinct and do not cross-neutralize, but EHV-1 and EHV-3 do show some common antigens detectable by direct and indirect immunofluorescence (IF), immunodiffusion, and CF in studies in ponies (Gutekunst, 1978). Mumford and Thompson (1978), however, showed no relationship (as measured by indirect IF and CF) among the three herpesviruses. Also, in another study, Burki *et al.* (1973) showed that EHV-1 and EHV-3 were not related by CF.

1. Clinical Conditions

As a clinical entity, ECE has been described in the European literature as early as the 19th century, but with unknown etiology (Studdert, 1974). The clinical course of natural infections and experimental ECE caused by EHV-3 has been reviewed in some detail (Bryans and Allen, 1973; Studdert, 1974; Bryans, 1980). According to Studdert, after experimental inoculation into appropriate areas of the external genitalia of horses, the incubation period is about 48 hour, but after coitus, the incubation period may be as long as 10 days before lesions develop. They appear initially as small vesicles or papules and progress rapidly to pustules (probably due in part to superinfection by streptococci) and within days may become frank ulcers with narrow ethrythematous borders of varying size, and may coalesce to form lesions several centimeters or more in diameter and 0.5 cm deep (Studdert, 1974). Healing occurs in about 2 weeks, leaving white patches that resolve in 3–4 weeks (Bryans and Allen, 1973).

The histopathology is more or less characteristic of herpes lesions, which generally show ballooning degeneration, lysis, and sloughing of necrotic epithelium, and leave ulcers filled with fibrin and acute exudate that extends superficially into the surrounding tissue. Characteristic intranuclear inclusions are present in the epithelial cells at the ulcer margins. Virus may be isolated from experimental lesions for up to 9 days.

ECE is ordinarily an acute but mild exanthematous disease, usually localized to the external genitalia, and in most cases without systemic signs of illness (Bryans and Allen, 1973; Pascoe *et al.*, 1968). It is worth noting that the natural disease is not limited to the external genitalia, and lesions may appear as well on the conjunctivae, lips, external nares, and nasal mucosa (Krogsrud and Onstad, 1971). After intranasal inoculation, horses may also develop a mild infection as evidenced by development of low-grade fever for several days in some animals. The virus could be cultured from the nasopharynx of all these animals for 6–8 days (Bryans and Allen, 1973). Initially, none of the experimental animals had VN antibody, but at 21 days of convalescence, all had significant VN-antibody titers; none showed signs of progenital disease. EHV-3 apparently never causes abortion; even direct inoculation into the fetus does not result in infection (Bryans and Allen, 1973).

It is interesting that EHV-1 may also cause a very mild progenital disease with lesions that are more superficial than those caused by EHV-

3 (Bryans and Allen, 1973). Histologically, the experimental lesions showed superficial necrosis and sloughing of epithelium that contained conspicuous intranuclear inclusions so typical of herpesvirus eruptions. This type of genital disease also occurs naturally (Petzoldt, 1970; Virat *et al.*, 1972) and following experimental infection with EHV-1 (Turner *et al.*, 1970; Bryans and Allen, 1973). The incidence is probably low (Bryans, 1981, personal communication); a natural outbreak has never been recognized in Kentucky.

2. Epidemiology

ECE caused by EHV-3 is usually venereally transmitted, but may also be spread by means other than coitus; the method of such transmission is unknown. Krogsrud and Onstad (1971) observe transmission in an outbreak without venereal spread. It has become apparent that Equidae may develop antibody without signs of clinical infection and are capable of transmitting the virus to others with or without coitus. Burrows and Goodridge (1973) reported that a number of ponies (both mares and stallions) had acquired infection (as judged by VN antibody) without clinical signs, with and without coital activity. They also reported that the virus can be recovered for weeks after a single vulvar lesion and again from a similar lesion in the same location in the same animal 8 months later. They also presented evidence that the virus is capable of persisting from season to season in both stallion and mare. Whether this represents true persistence or reactivation of latency is not known.

The general population of horses must have considerable contact with EHV-3 judging from the data of Bagust *et al.* (1972). In this study, beginning with 2-year-old horses (which have very low titers of antibody), there is a steady increase to age 8, when 52% have significant antibody titers.

Finally, it would appear that there are several possible clinical forms of ECE, caused variously (as discussed) by EHV-1, EHV-3, and the donkey strain of Burrows (1973). The lesions produced by these agents probably are not clinically identifiable (etiologically) and must depend for identification on the laboratory. It should be considered that since the pony, donkey, and horse are different species, different subtypes or serotypes may have evolved with them.

III. STRUCTURE AND MOLECULAR ANATOMY OF EQUINE HERPESVIRUS VIRIONS AND THEIR COMPONENTS

A. Morphology

Morphologically, the equine herpesviruses are typical of members of the Herpetoviridae family and possess an internal DNA-containing core enclosed within an icosahedral capsid to form a nucleocapsid; this structure is surrounded by a loose outer envelope that is derived from the modified nuclear membrane of the cell (see O'Callaghan and Randall, 1976) (Fig. 1). Virions of all three types of equine herpesviruses have been reported to range in size from 150 to 198 nm (Arhelger *et al.*, 1963; Darlington and Randall, 1963; Darlington and Moss, 1968, 1969; Abodeely *et al.*, 1970; O'Callaghan *et al.*, 1968a, 1978), but 150–170 nm is the most frequently observed size (Ludwig *et al.*, 1971; Bryans and Allen, 1973; O'Callaghan *et al.*, 1978; Wharton *et al.*, 1981).

The three EHV types are morphologically indistinguishable from each other and are similar in morphology to herpesviruses of other species. The capsid of all three EHV types has been examined by electron-microscopic methods and shown to be approximately 100 nm in diameter and to be comprised of 162 capsomers arranged to form an icosadeltahedron that exhibits 2-, 3-, and 5-fold symmetry (Darlington and Randall, 1963; Darlington and James, 1966; O'Callaghan *et al.*, 1968b, 1978; Abodeely *et al.*, 1970; Ludwig *et al.*, 1971; Bryans and Allen, 1973; Wharton *et al.*, 1981). The capsomers appear to be arranged as 12 pentameric capsomers, located at the vertices, and 150 hexameric capsomers that have a centrally located hole or channel and are located in the faces of the $T=16$ structure (Abodeely *et al.*, 1970; O'Callaghan and Randall, 1976). Some evidence from examination of the capsids of EHV-1 and other herpesviruses suggests that 2-nm intercapsomeric fibrils link adjacent capsomers (Vernon *et al.*, 1974; Palmer *et al.*, 1975).

The core of the EHV virions has been shown to contain the viral genome in a form that is significantly resistant to DNase (Perdue *et al.*, 1976) and that exhibits the morphological features of a toroid structure as proposed by Roizman and co-workers (Furlong *et al.*, 1972; Roizman and Furlong, 1974) and Nazerian (1974). This electron-dense toroid structure in EHV-1 virions appears to measure approximately 59 nm high with an inside diameter of 16 nm and an outside diameter of 64 nm and to contain a less electron-dense central cylinder or bar structure (approximately 13 nm in diameter) around which the linear viral DNA molecule is "spooled" or wound in strands with regular spacing (Perdue *et al.*, 1975, 1976; O'Callaghan and Randall, 1976; Perdue and O'Callaghan, unpublished).

The area between the capsid and the envelope of EHV virions appears as layers of amorphous material and does not exhibit the morphology of a unit membrane. This area has been termed the tegument (Roizman and Furlong, 1974) and has been shown to vary in size for different members of the Herpetoviridae family (see Roizman and Furlong, 1974; O'Callaghan and Randall, 1976; Spear and Roizman, 1980). There is evidence that specific proteins of EHV-1 virions are located within this intravirion compartment (see below), and differences observed in overall virion diameter and volume among EHV virions of the same type or of different types appear to be due mostly to differences in the size of the tegument.

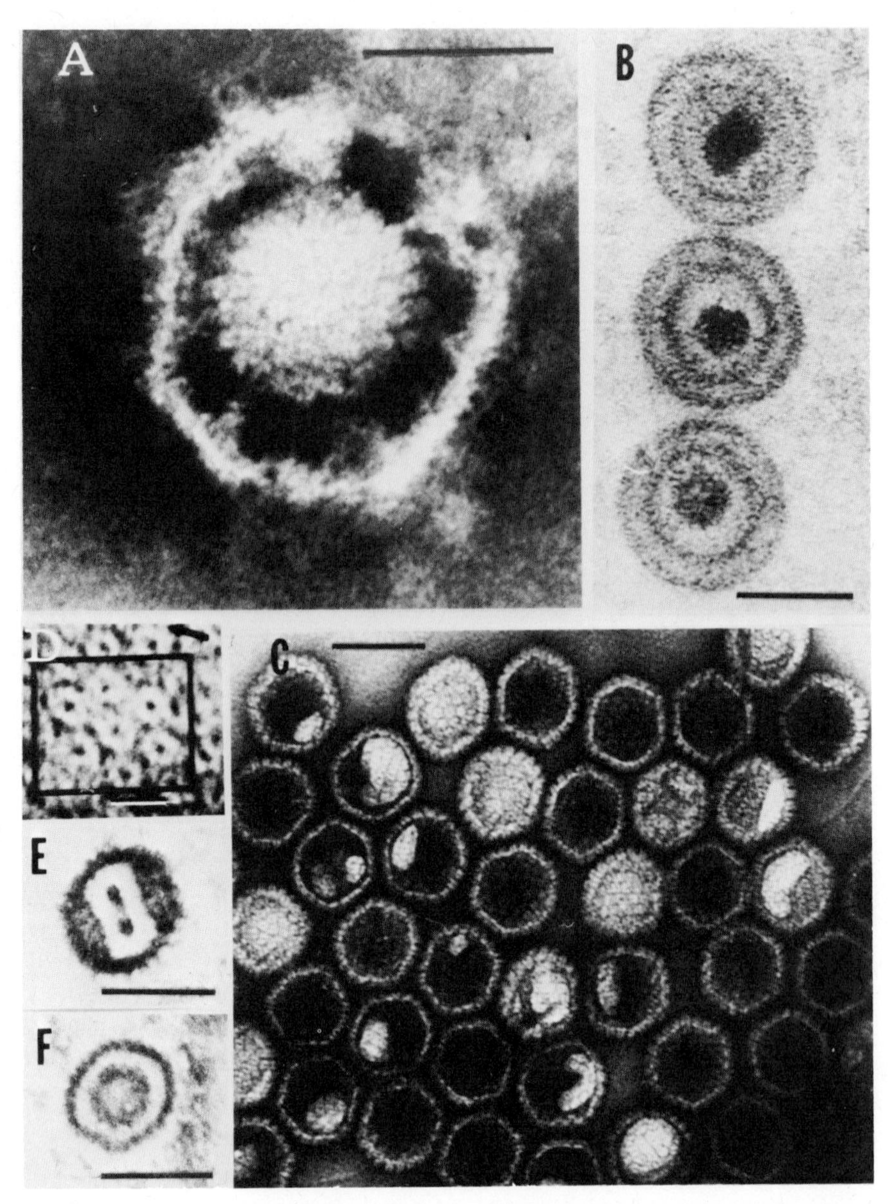

FIGURE 1. Structure of equine herpesviruses as typified by EHV-1. (A) Pseudoreplica of EHV-1 virion stained with phosphotungstic acid showing envelope, tegument, and icosahedral capsid. (B) Cross section of EHV-1 virions illustrating densely staining, centrally located core; stained with uranyl acetate (UA). (C) Pseudoreplica showing UA-stained nucleocapsids. (D) Pseudoreplica of UA-stained capsid illustrating hexameric arrangement of capsomeres and centrally located hole of each capsomere. (E, F) Cross sections of nucleocapsid at different angles of sectioning. Core structure appears in (E) as barlike structure and in (F) as spherical body with less electron-dense center.

The envelope is the outer, triple-layered membranous structure that surrounds the nucleocapsid and is composed of a lipid bilayer and associated proteins (see Spear and Roizman, 1980). The envelope of the EHV virion exhibits morphological and structural properties typical of those described for other herpesviruses: (1) it is similar to other membranes in permeability and sensitivity to detergents and lipid solvents (Abodeely *et al.*, 1970, 1971; Bryans and Allen, 1973; O'Callaghan *et al.*, 1978; Wharton *et al.*, 1981; Wharton and O'Callaghan, unpublished); (2) it contains lipids, viral glycoproteins, and phosphoproteins (see below); and (3) it is derived from the nuclear membrane of the infected cell by the budding of mature nucleocapsids that have interacted with the inner lamina of the membrane or by a mechanism in which convolutions of nuclear membrane invaginate into the nuclear space and react with mature capsids (Darlington and James, 1966; Abodeely *et al.*, 1970; O'Callaghan and Randall, 1976; O'Callaghan *et al.*, 1978; Wharton *et al.*, 1981).

B. Properties of Equine Herpesvirus DNAs

1. Conformation, Size, and Base Composition

Of the three equine herpesviruses, the genome of EHV-1 is the best characterized. Results of early studies by Randall and co-workers (Randall and Bracken, 1957; Randall *et al.*, 1959; Gentry and Randall, 1960; Gentry *et al.*, 1960, 1961) employing chemical, staining, and histological as well as isotopic analyses of EHV-1-infected hamster hepatocyte nuclei and viral inclusions indicated that EHV-1 was a DNA-containing virus. Direct evidence that the EHV-1 genome is DNA was provided by Darlington and Randall (1963), who used chemical analysis to show that DNA comprises approximately 9% of the dry particle weight of EHV-1 virions and that the DNA has a guanosine plus cytosine (G + C) content of approximately 56 moles%. These findings were confirmed by Russell and Crawford (1964).Soehner *et al.* (1965) published the first detailed investigation of EHV-1 DNA and showed that the DNA has a buoyant density of 1.716 g/cm^3 and a melting temperature of 51.4°C and that these values correspond to a G + C content of 57 moles%. These workers showed that the molecule has a size of approximately 93 megadaltons (md) by three independent methods that involved sedimentation coefficient (approximately 52 S) and bandwidth measurements in the analytical ultracentrifuge and determination of the contour length of viral DNA molecules in the electron microscope. This value has been confirmed for the DNA of several strains of EHV-1 propagated in the hamster as well as in cell-culture systems (Plummer and Waterson, 1963; O'Callaghan *et al.*, 1968a, 1972a; Plummer *et al.*, 1969a,b, 1973). Recently, the size of the genome of the KyA tissue culture (L-M cell) strain was shown by restriction-enzyme analysis (O'Callaghan *et al.*, 1981a; Henry *et al.*, 1981) and electron-microscopic measurements (Ruyechan *et al.*, 1982) to be 92 md (Table I).

Initial studies by Russell and Crawford (1964) reported that EHV-2 DNA has a sedimentation coefficient of 49 S, an apparent molecular size of 84 md, and a buoyant density of 1.715 g/cm^3, which corresponds to a G+C content of 56 moles%. However, it should be noted that in this same study, these workers reported herpes simplex virus (HSV) DNA to be only 68 md in size; none of these values has withstood the test of time. Plummer and co-workers (1969a,b), however, obtained a density of 1.717 g/cm^3, which indicated a G+C content of 58 moles%. Recent studies in our laboratory (Wharton *et al.*, 1981; O'Callaghan *et al.*, 1981a; Ruyechan, Staczek, and O'Callaghan, in prep.) of the genome of this equine cytomegalovirus (ECMV) have employed a variety of experimental approaches and have yielded results significantly different from those of the initial reports. CsCl isopyknic analysis in the analytical ultracentrifuge of viral DNA extracted and purified from virions, nucleocapsids, or infected cells (Hirt fractionation) gave an average density of 1.7165 g/cm^3, which corresponded to a G+C content of 57.7 moles%. Sedimentation analyses of the viral DNA in neutral sucrose gradients using phage and herpesvirus DNAs as markers indicated that the sedimentation coefficient of this DNA is approximately 61.8 S. This value equated to a molecular weight of 120–129 md when used in the more reliable formulas for equating molecular weight and sedimentation of large DNA molecules (Wharton *et al.*, 1981). Determination of the molecular weight by restriction-enzyme methods employing several enzymes and by measurement

TABLE I. Properties of Equine Herpesvirus DNAs[a]

Property	EHV-1	EHV-2[b]	EHV-3[b]
Conformation	Linear, 2×	Linear, 2×	Linear, 2×
Molecular weight			
Bandwidth measurements	92 md	ND	ND
Sedimentation analysis	93.8 md	112–122 md	92–103 md
Restriction-enzyme analysis	92 md	126 md	96.2 md
Electron-microscopic measurement	90–94 md	120 md	96–100 md
G+C (moles %)	57%	57.7–58.0%	66–67.9%
Buoyant density (CsCl)	1.716 g/cm^3	1.7165 g/cm^3	1.725–1.727 g/cm^3
Sedimentation coefficient	49–55 S	61.8 S	55.4 S
Fragmentation in alkalai[c]	Yes	Yes	Yes
Isomeric arrangement	Two isomers; S region can invert	ND	Two isomers; S region can invert
Infectivity by transfection?	Yes[d]	Yes[e]	Yes[f]

[a] References are cited in the text. [b](ND) Not determined.
[c] Sedimentation profiles of EHV DNA treated with alkali indicate the presence of single-stranded regions (gaps) within the molecule.
[d] Allen and Randall (1978). [e]Wharton and O'Callaghan (unpublished). [f]Allen (personal communication); Atherton *et al.* (1982).

of intact DNA molecules in the electron microscope confirmed that ECMV DNA molecules have a size of approximately 126 md (range 120–126 md). The value of 126 md for ECMV DNA in our studies (Wharton *et al.*, 1981; O'Callaghan *et al.*, 1981a; Ruyechan, Staczek, and O'Callaghan, in prep.) exceeds that of EHV-1 DNA by 30%, and this is consistent with the finding that the molecular weights of cytomegalovirus genomes are significantly greater than those of the rapidly cytocidal group of herpesviruses such as HSV-1 and HSV-2 (Table I).

Ludwig *et al.* (1971) demonstrated that EHV-3 DNA is double-stranded and reported a density of 1.725 g/cm^3, which corresponds to a G + C content of 66 moles%. Our lack of knowledge about the properties of this herpesvirus DNA has been due, at least in part, to the difficulty in obtaining large quantities of viral particles, since the virus replicates only in cells of equine origin at low to moderate titer and remains cell-associated (see O'Callaghan *et al.*, 1978). Recently, however, methods have been developed for the isolation of intact, infectious EHV-3 DNA from nucleocapsids and infected cells, and a variety of physicochemical, biochemical, and electron-microscopic methods have been applied to characterize EHV-3 DNA (Atherton *et al.*, 1982). CsCl analytical ultracentrifugation of EHV-3 DNA yielded a single species of DNA with a buoyant density of 1.727 g/cm^3, which corresponds to a G + C content of 67.9 moles%. Rate velocity centrifugation studies revealed that this DNA has a sedimentation coefficient of approximately 55.4 S, which equates to a molecular weight of approximately 93–100 md. In good agreement with this value, electron-microscopic measurements of the contour length of intact EHV-3 DNA molecules using simian virus 40 (SV40), EHV-1, and EHV-2 DNAs as markers indicated the size of the DNA to be 96–100 md. Last, restriction-enzyme digestion with a variety of enzymes yielded fragments that totaled to indicate a molecular weight of 95–96 md (Table I).

The key properties of EHV DNAs are summarized in Table I. One property of these DNAs shared with those of other members of the Herpetoviridae is their fragmentation after denaturation with alkali. Alkali treatment of EHV-1 (Dauenhauer, Henry, and O'Callaghan, unpublished), EHV-2 (Wharton *et al.*, 1981), or EHV-3 (Atherton *et al.*, 1982) DNAs resulted in the generation of a heterogeneous population of DNA fragments as determined by rate velocity sedimentation analyses in which T_4 DNA was employed as marker. Sedimentation patterns observed under these conditions as compared to T_4 DNA, which sedimented as a single, sharp band, indicate that these double-stranded viral DNAs contain a number of single-stranded regions or gaps. Recent findings that carefully extracted, nondenatured EHV DNA can be repaired *in vitro* and has sites for lambda exonuclease digestions suggest that gaps exist within the native molecule (Sullivan *et al.*, 1982). Whether these gaps occur at specific sites or in a random fashion is not known; however, the observation that variations occur in the alkaline density-gradient sedimentation profile of different DNA preparations of the same virus type is consistent with the idea that these nicks are not located at specific sites.

2. Isomeric Arrangement

Insight into the structure of EHV DNAs has been obtained recently by employing restriction-enzyme and electron-microscopic methods to analyze EHV-1 (Henry *et al.*, 1981, and Whalley *et al.*, 1981) and EHV-3 DNAs (Atherton, *et al.*, 1982). Restriction-enzyme and blot hybridization analyses revealed that EHV-1 DNA contains three terminal fragments as well as 0.5 M fragments that exhibit significant homology and are located at one terminus or at a fixed location within the molecule. These and additional findings indicated that this herpesvirus genome is a 92-md linear, double-stranded DNA molecule and is comprised of two segments designated as L (long) and S (short), which are approximately 71.6 and 20.4 md, respectively. The 0.5 M fragments contain inverted repeat sequences and are located at the ends of the S region and bracket the unique sequences of this region; thus, this arrangement allows the S region to invert relative to the L region, which lacks inverted sequences and is comprised only of long unique (U_L) sequences in a fixed orientation. Therefore, two structural arrangements or isomers of the EHV-1 genome exist, and DNA from infectious virions consists of two equimolar populations that differ solely in the relative orientation of the S region and are designated as prototype (P) and inverted (I). The restriction-enzyme maps and overall arrangements of these two EHV-1 DNA populations are presented in Fig. 2.

Confirmation of the two-isomer structure of the EHV-1 genome has been obtained by electron-microscopic analysis of self-annealed EHV-1 DNA molecules (Ruyechan *et al.*, 1982; Sheldrick, personal communication). Examination of EHV-1 DNA strands allowed to reanneal reveals structures containing a single-stranded loop (the U_S segment) contiguous to a double-stranded region (the inverted repeat sequences of the S region) that terminates in a single-stranded stretch (Fig. 3A and B). The double-stranded region of the molecule in Fig. 3B is shorter than that contained in the structure in Fig. 3A, and the single-stranded tail is quite short, indicating that the molecule contained single-stranded interruptions at these points. More than 70% of the molecules examined appear to have single-stranded breaks in the double-stranded or single-stranded regions or both, but molecules containing double loops or molecules with loops less than one or greater than two phage fd DNA contour lengths were not observed. The average molecular weight of the loops that correspond to the short unique (U_S) segment of the S region equated to 6.4 ± 1.1 md using DNAs of SV40 and phage fd as internal molecular-weight markers. Measurement of the S-region terminal repeat (TR) sequences of more than 60 molecules yielded a value of 6.4 ± 0.89 md, which is in good agreement with the estimate from restriction-enzyme analyses (Henry *et al.*, 1981).

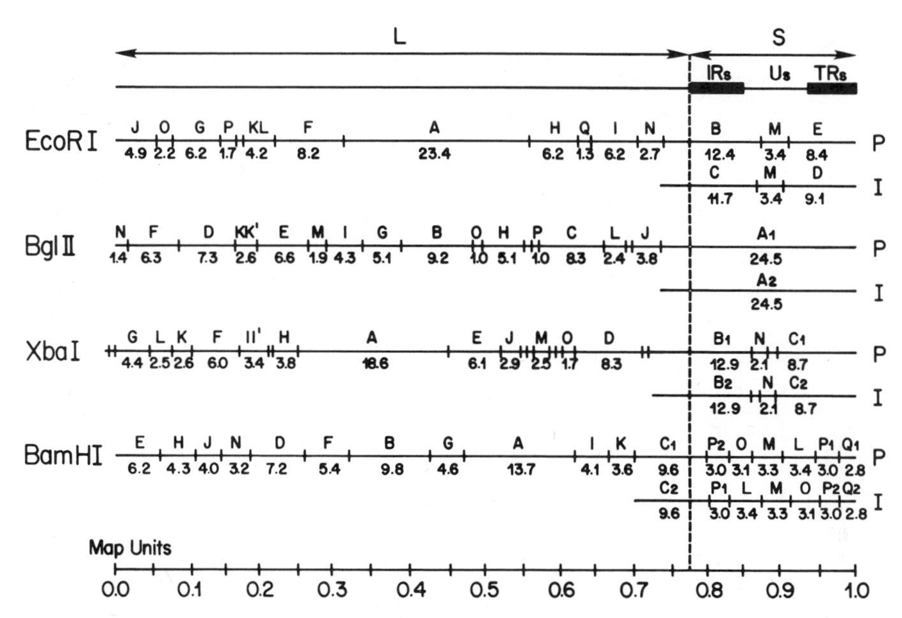

FIGURE 2. Restriction-enzyme maps of the genome of EHV-1. EHV-1 and EHV-3 DNAs have a similar overall genomic structure in which the S region can invert because its short unique (U_S) sequences are bracketed by inverted repeat sequences located at the free S-region terminus (TR_S) and the internal terminus (IR_S). The S region is covalently joined to the sequences of the L region, which exists in one fixed orientation relative to the S region. Two isomeric arrangements of the DNA molecule exist in equimolar proportion and are referred to as prototype (P) and inverted (I). Reproduced from Henry *et al.* (1981) with permission from Academic Press.

Recently, we have initiated restriction-enzyme and electron-microscopic studies to elucidate the structure of EHV-3 DNA. All findings indicate that this EHV DNA is a 96-md linear, double-stranded DNA that is composed of an L region of approximately 73–76 md covalently linked to a 20–23 md S region; the S region contains an approximate 5.7-md U_S segment bounded by 8-md inverted repeat sequences that allow the S region to invert relative to the fixed L region and the DNA to exist in two isomeric arrangements. These conclusions are based on restriction-enzyme and blot hybridization studies that revealed that: (1) four 0.5-M fragments were present; (2) all 0.5-M fragments shared significant homology; and (3) three terminal fragments were present, two of which were 0.5-M fragments. Thus, a set of DNA sequences located at one terminus is repeated within the genome, and two populations of molecules exist with regard to the orientation of these repeat sequences. In addition, electron-microscopic examination of self-reannealed EHV-3 DNA molecules revealed structures that contained a single-stranded loop (equivalent to a duplex molecular weight of 5.7 md) at one end of the molecule contiguous to a double-stranded region that terminated in a long single-stranded segment (Fig. 3C and D). These structures were identical in morphology

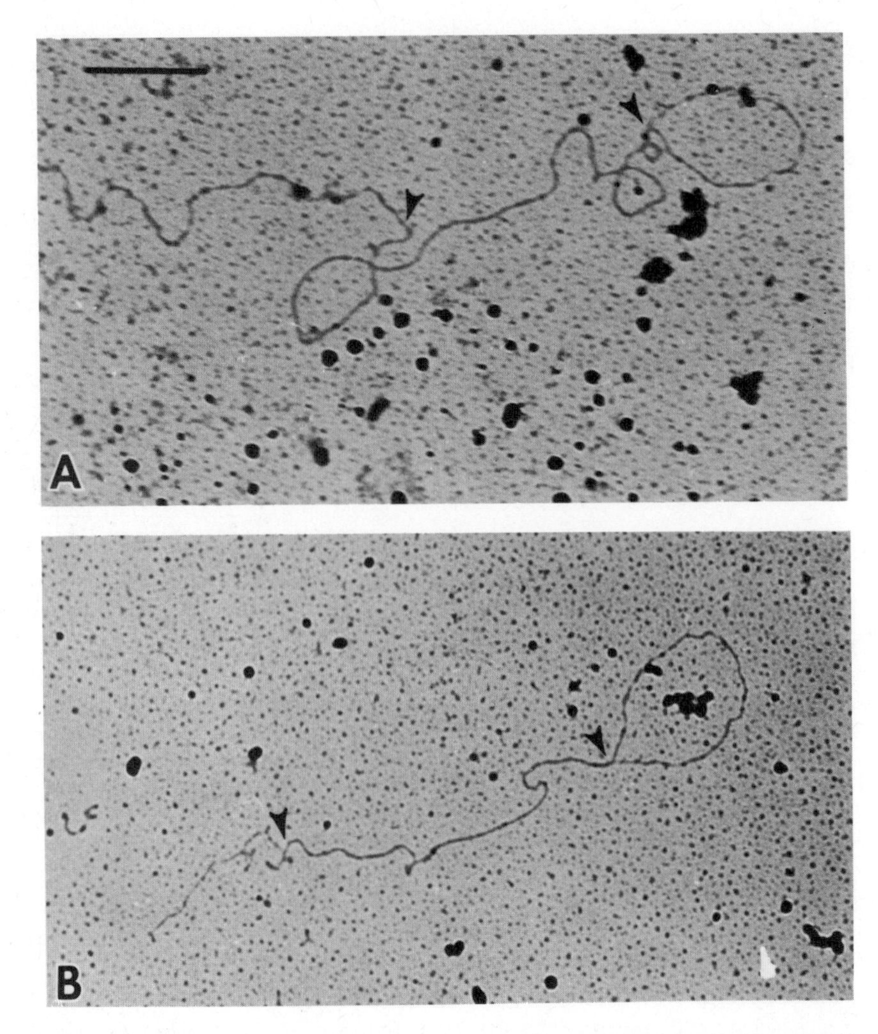

FIGURE 3. Electron micrographs of self-annealed EHV-1 (A, B) and EHV-3 (C, D) DNA molecules showing the S region. The arrows delineate the double-stranded segment corresponding to the inverted repeats; the short unique (U_S) sequences appear as single-stranded loops. Electron-microscopic findings support data of restriction-enzyme analyses and show that the S regions of EHV-1 and EHV-3 DNA molecules have inverted repeat sequences that can reanneal. Scale bar: 500 nm. For details, see the text, Ruyechan *et al.* (1982) and Atherton *et al.* (1982).

to those observed for EHV-1 DNA (Ruyechan *et al.*, 1982). Last, preliminary results of restriction-enzyme mapping with several enzymes confirm this model for EHV-3 DNA structure. Thus, EHV-1 and EHV-3 have an identical genomic structure and one that is similar to that reported for pseudorabies virus DNA (Stevely, 1977; Ben Porat *et al.*, 1979) and varicella–zoster DNA (Dumas *et al.*, 1981; Straus *et al.*, 1982).

Very little is known about the molecular anatomy of EHV-2 DNA.

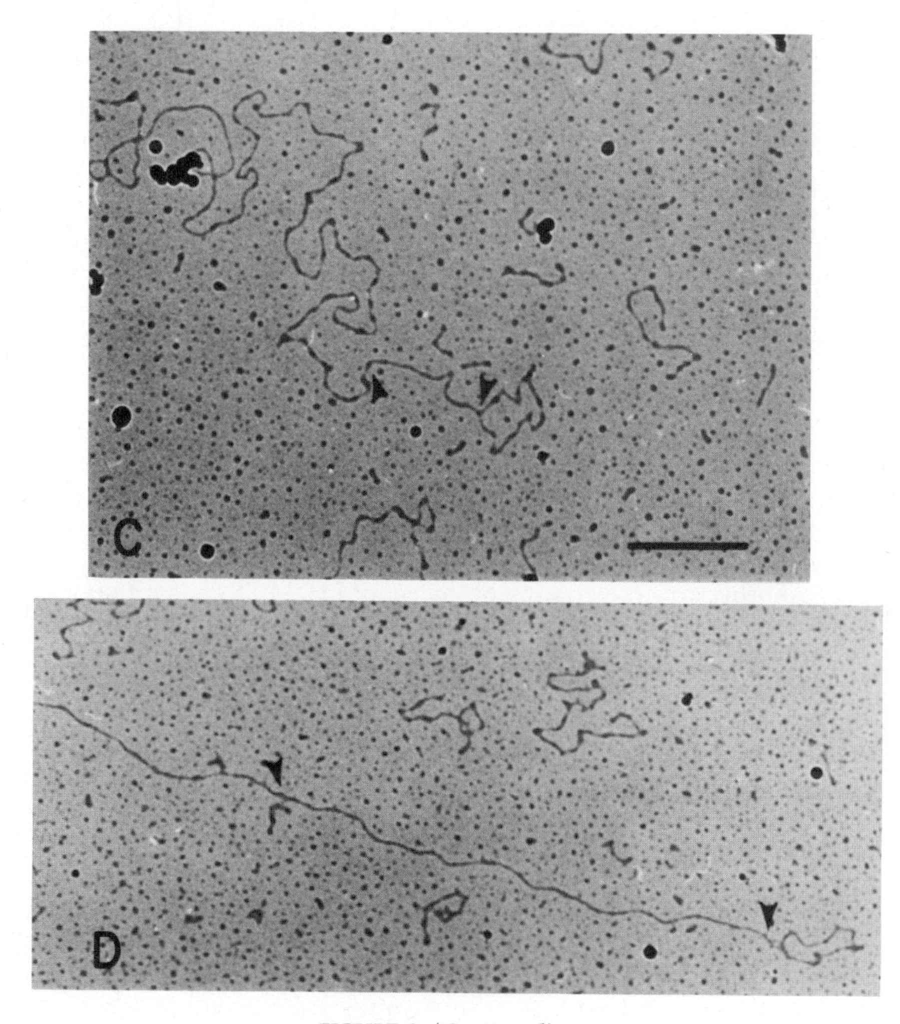

FIGURE 3. (*Continued*)

Restriction-enzyme digestions of this DNA yield complex profiles with regard to fragment number and molarity and indicate that several submolar fragments are present. Recently, electron microscopic examination of EHV-2 self-reannealed DNA molecules revealed loop-like structures similar to those observed for EHV-1 and EHV-3 DNAs. The presence of these single-stranded loops (equivalent to 5.8 md $\pm$ 1 md of 2X DNA) that terminate in a double-stranded stem (3-5 md) which is contiguous to a long, single-stranded linear region suggests that inverted repeat sequences are present and bracket a region of unique DNA sequences (Staczek, Ruyechan and O'Callaghan, in preparation). These preliminary findings suggest that the genome of this cytomegalovirus exists in more than one isomeric form.

3. Variability

As discussed above, there appears to be a significant level of variability in EHV-2 DNA. This variability may be similar to areas of heterogeneity detected in EHV-1 DNA and found to map at both the L and S termini and at map position 0.18 within the L region. It is possible that the variation in the size of restriction-enzyme fragments located at the L and S terminal fragments is due to imprecision in the mechanism for cleavage of the replicating DNA into unit size molecules, and thus small repetitive segments of DNA are excised or added in varying numbers. Such variability has been reported for other herpesvirus DNAs, especially HSV (G. Hayward, personal communication; see Spear and Roizman, 1980). The reason for the "split variable" region consisting of two or three very discrete versions of the same fragment differing in molecular weight by ± 0.3 md and located at 0.18 map position is unclear, since these fragments are a great distance from the joint region and either end of the molecule.

Another type of variation observed for EHV-1 DNA is intratypic variability reported to exist among different strains of this equine herpesvirus. Such variability is not uncommon in the Herpetoviridae family, and recently Spear and Roizman (1980) have reviewed the large amount of data showing the variation in restriction enzyme profiles of different isolates of the same HSV type. They note that variation may exist at each restriction enzyme site and that there may be more than 2^{19} different HSV strains in the human population. O'Callaghan *et al.* (1978) pointed out the significant variation in the restriction enzyme patterns of the genomes of the *in vivo* (hamster; EHV-1 ha) and *in vitro* (tissue culture; EHV-1 tc) passaged EHV-1 of the fetal Kentucky A strain. Recently, Allen *et al.* (1983a) observed the rapid development of variability in the restriction enzyme profiles of EHV-1 passaged in the hamster *in vivo* and in non-equine cells *in vitro*. Variability in the EHV-1 genome was found to develop to a lesser extent upon passage of this virus in its natural host, the horse, (Allen *et al.* 1983a) and to be detectable when the DNAs from 297 EHV-1 field isolates were analyzed (Allen *et al.* 1983b).

Sabine *et al.* (1981) were the first to demonstrate differences in the restriction-endonuclease profiles of DNAs of abortigenic (fetal) vs. respiratory (nonabortigenic) isolates of EHV-1 and suggested that differentiation of EHV-1 isolates as abortigenic or nonabortigenic subtypes may be possible by restriction-endonuclease analysis. Recently, these observations have been extended by Studdert *et al.* (1981) and Allen and Turtinen (1982), who showed that the restriction-enzyme profiles of DNAs of EHV-1 isolates from animals with respiratory disease were similar to each other, but different from the patterns for DNAs of fetal isolates. On the basis of these findings, Studdert *et al.* (1981) have proposed that respiratory viruses analyzed in their study be designated as protypic of equine herpesvirus type 4, a new equine herpesvirus type. Classification of respiratory strains of EHV-1 as a new virus type (type 4) on the basis of this single criterion may be premature, since Allen, Bryans, and coworkers (personal communication) have observed that EHV-1 respiratory isolates can cause abortigenic disease in pregnant mares. Until more data are collected, it seems prudent to refer to fetal and respiratory isolates as different strains or, perhaps, subtypes.

4. Cloning of Sequences

To provide quantities of viral DNA sequences necessary for detailed mapping of specific regions of the viral genome, biochemical analyses of EHV-transformed cells, elucidation of the amount and nature of DNA sequences shared among the three EHV genomes, and eventual determination of the gene products of specific DNA sequences, we have initiated experiments to clone EHV DNA fragments by recombinant DNA methods. Recently, we described the construction of recombinant plasmids containing EHV-1 DNA restriction fragments the total coding capacity of which exceeded 95% of the viral genome (Robinson *et al.*, 1981b). The experimental approach involved the insertion of *Bam*HI and *Eco*RI fragments of standard EHV-1 DNA at the appropiate restriction sites of the *E. coli* plasmid pBR322. Since sequences located at the L terminus are not repeated within the genome, as is the case for the inverted repeat sequences located at the S-region terminus (see above), fragments mapping at approximately 0.0–0.06 were cloned by covalently linking these sequences to oligonucleotide linkers that contained restriction-enzyme recognition sites for either *Eco*RI or *Bam*HI.

Bacteria carrying plasmids with the inserted EHV-1 *Eco*RI or *Bam*HI fragments were selected and screened for resistance to the appropiate antibiotic(s). Antibiotic-resistant transformants that demonstrated positive hybridization with the proper ^{32}P-labeled EHV-1 restriction fragments were picked and propagated to provide crude DNA lysates that were restricted and processed to provide the cloned EHV-1 fragments. To date, this library of EHV-1 *Bam*HI and *Eco*RI fragments has been employed in a variety of investigations of the structure and biology of EHV DNA. Initial experiment with these cloned fragments have confirmed the restriction-enzyme maps of the EHV-1 genome (Robinson *et al.*, 1981b), identified the origin of DNA sequences present in EHV-1 defective DNA (Robinson *et al.*, 1981b; O'Callaghan *et al.*, 1981a; Dauenhauer, Robinson, and O'Callaghan, in prep.), and revealed the nature and arrangement of viral DNA sequences in EHV-1-transformed cells (Robinson *et al.*, 1981a, b; Robinson and O'Callaghan, 1981, 1983). Experimental protocols similar to those used to clone EHV-1 DNA fragments are being employed to provide libraries of cloned fragments of EHV-2, EHV-3, and defective EHV-1 DNAs.

5. Virion Polypeptides

Improvements in the methods for the purification of herpes virions and subviral particles and in the techniques for separation of large numbers of proteins have permitted workers from several laboratories to ascertain the polypeptide composition of a variety of herpesviruses. Over the last ten years since Abodeely *et al.* (1971) first reported that 20 different proteins were present in EHV-1 virions as judged by staining of polyacrylamide gels after electrophoresis of solubilized proteins of the KyA-ha virions, our knowledge of the protein composition of these complex equine herpesviruses has been expanded. Using techniques with increased resolving power, we have attempted to define the glycosylated proteins, to separate proteins of similar size, and to assign the location of the proteins to specific subvirion components. These studies have revealed that EHV-1 and EHV-3 virions are comprised of apparently 30 polypeptides that vary greatly in nature, size, and relative amount (Perdue *et al.*, 1974, 1975, 1976; Kemp *et al.*, 1974; Allen and Bryans, 1976; O'Callaghan and Randall, 1976; O'Callaghan *et al.*, 1977a, 1978; Vernon *et al.*, 1978; Allen and Randall, 1979; Turtinen *et al.*, 1981; Turtinen and Allen, 1983).

The proteins present in EHV-1 virions of the Kentucky A strain passaged both *in vivo* in the hamster and in cell culture and of the Army 183 strain, a respiratory isolate that can cause abortigenic disease in pregnant mares (Bryans, 1980), are listed in Table II. The virion is comprised of at least 28 polypeptides that range in size from approximately 270,000 to 16,000 daltons. The majority of these proteins and all the glycoproteins are located in the envelope as judged by comparative analyses of nucleocapsids and enveloped virions and by isotopic surface labeling methods. The overall electrophoretic protein patterns of these EHV-1 strains were found to be similar, with the exception that VP 8 (175,000) was not detected in Army 183 or the tissue-culture (L-M cell)-adapted KyA strain of EHV-1. These results of the reports of O'Callaghan and co-workers (Perdue *et al.*, 1974, 1975, 1976; O'Callaghan and Randall, 1976; O'Callaghan *et al.*, 1977a) and of the more recent studies by Allen and co-workers (Turtinen *et al.*, 1981; Turtinen and Allen, 1983), who identified glycoproteins by measuring incorporation of [^{3}H] glucosamine, by enzymatically labeling surface glycoproteins *in vitro* with [^{3}H] sodium borohydride or UDP-[^{14}C]galactose, and by staining with Alcian Blue dye, indicate that approximately 14 glycoproteins are present in the EHV-1 envelope. Fluorography indicates that there are 8 major glycoproteins: VP 1 (260,000), 2 (260,000), 10 (138,000), 13 (93,000), 14 (87,000), 18 (60,000), 21 (50,000), and 22a (46,000). Proteins VP 8, 9b, 16, 17, 23, and 25 appear to be present in smaller amounts and were judged to be minor glycoproteins. Whether each of these 14 glycoproteins is in fact a unique protein species is not known at present, but the possibility that some of these proteins have precursor–product relationships and/or have an identical amino acid sequence but differ in electrophoretic mobility because of differences in carbohydrate content or structure seems likely as judged from recent findings of HSV glycoproteins (see Spear and Roizman, 1980; G.H. Cohen *et al.*, 1980). Interestingly, Turtinen and Allen (1982) found that all EHV-1 glycoproteins were solubilized to varying degrees by treatment of purified virions with the detergent NP-40. Glycoproteins 10 and 13 were extremely susceptible to this treatment, confirming the observation of Vernon *et al.* (1978), who proposed that VP 10 (138,000) may be a major component of the surface spikes.

Several proteins not present in purified nucleocapsids (see below) or on the outer surface of the virion envelope (as judged by surface labeling and susceptibility to detergent) have been classified tentatively as tegument proteins (see Table II). These proteins are VP 11 (122,000), 12 (110,000), 15 (73,000), 19 (58,000), 26A (18,000), and 27 (16,000). A 59,000-dalton protein, designated VP 19, was found to be associated with nucleoscapsids by Kemp *et al.* (1974) and Perdue *et al.* (1975), and it is possible that this major capsid protein masked the presence of a minor tegument protein in the analysis of the cell-culture EHV-1 virion (Table II).

A significant amount of information has been obtained on the structural protein composition of different species of EHV-1 nucleocapsids isolated from the nuclei of infected L-M cells (Perdue *et al.*, 1975, 1976) and hepatocyte nuclei of infected hamsters (Kemp, 1975; O'Callaghan *et al.*, 1977a, 1978). The multistep procedure of Perdue *et al.* (1975) used to isolate the capsid species involved clarification of nuclear extracts by low-speed centrifugation, centrifugation of this extract through 35% sucrose to remove membranous debris, two or more cycles of isopyknic banding in Renografin density gradients, and separation of capsid types by rate velocity centrifugation in sucrose gradients. The procedure yielded capsids of high purity as judged by the criteria of electron microscopy, removal of radiolabeled cellular macromolecules, and attainment of reproducible data for the protein composition of the capsid types. Three capsid species were obtained by this procedure: (1) L [(light) $\rho = 1.237$ g/cm^3] capsids, which lacked an internal core structure and were judged to be empty shells; (2) I [(intermediate) $\rho = 1.244$ g/cm^3], which were devoid of a mature core structure, but possessed an internal crosslike or "four-pointed star" structure; and (3) H [(heavy) $\rho = 1.258$ g/cm^3] capsids, which were rich in DNA and exhibited the morphology of core-containing nucleocapsids. All three capsid types have been described in cross sections of EHV-1-infected cells by several workers and are similar in ultrastructure to capsids described for other herpesviruses (see O'Callaghan and Randall, 1976; O'Callaghan *et al.*, 1978).

The capsid species were shown to differ not only in morphology, density, sedimentation properties, and DNA content, but also in structural protein and amino acid compositions (Perdue *et al.*, 1975, 1976; O'Callaghan *et al.* 1977a, 1978; O'Callaghan and Randall, 1976). H capsids were comprised of six major proteins with apparent molecular weights of 150,000, 59,000, 45,000, 37,000, 30,000, and 18,000; more recent analyses using improved systems of polyacrylamide gel electrophoresis (PAGE) have confirmed that EHV-1 capsids of the Army 183 and KyA-

TABLE II. Structural Proteins of Equine Herpesvirus Virions

EHV-1 virion proteins[a]				EHV-1 virion proteins[b]				EHV-3 virion proteins[c]	
Protein	Mol. wt. ($\times 10^3$)	Nature	Location	Protein	Mol. wt. ($\times 10^3$)	Nature	Location	Protein	Mol. wt. ($\times 10^3$)
1A	270–276	—	E	1	260	G	E	1	>220
1B	270–276	—	e	2	260	G	E	2	215
2	252–257	—	e	3	200	G	E	3	180
3	249–256	—	e	4	200	G	E	4	168
4	224–234	L	e	5	200	G	E	5	160
5	213–220	L	e	6	200	G	E	6	148(NC)
6	200–203	G	e	7	200	G	E	7	135
7	188–198	G	e	8	175	G	E	8	113
8	173–176	—	e	9A	150	—	NC	9	110
9A,I	147–151	—	NC	9B	150	G	E	10	105
9B	147–151	GL	e	10	138	G	E	11	95
m2	130	—	—	11	122	—	T	12	85
10A	127–130	GL	E	12	110	—	T	13	80
10B	127–130	—	E	13	93	G	E	14	72
11	118–120	—	E	14	87	G	E	15	65
12	108–111	GL	e	15	73	—	T	16	62
m4	110	—	—	16	65	G	E	17	60(NC)
13	92–97	GL	E	17	62	G	E	18	58
14	81–87	GL	E	18	60	G	E	19	56

15	73–75	—	e	19	58	—	T	20	53(NC)
16A	67–75	—	E	20	53	—	NC	21	50
16B	67–75	—	E	21	50	G	E	22	48(NC)
17	63–75	G	E	22A	46	G	E	23	44
18	60–65	G	E	22B	44	—	NC	24	39
19,II	59	—	NC	23	36	G	E	25	34
20	52–57	GL	E	24A	35	—	NC	26	31(NC)
21	48–49	GL	e	24B	29	—	NC	27	25
22A,III	45–46	—	NC	25	24	G	E	28	23.5
22B	42	—	—	26A	18	—	T	29	23
23,IV	37–38	—	NC	26B	17	—	NC	30	21
24A	33–38	L	E	27	16	—	T	31	20.5
m8,IVa	30	—	NC					32	20
24B	29	GL	e					33	16(NC)
25	24–26	L	e					34	14
26,V	16–20	—	NC						
27	16	GL	e						

[a] Composite of data for the Kentucky A strain of EHV-1 propagated both in Syrian hamsters (KyA-ha) and in cell culture. Data are taken from O'Callaghan and Randall (1976), Perdue *et al.* (1974, 1975, 1976), and Kemp *et al.* (1974). (G) Glycoprotein as judged by incorporation of radiolabeled glucosamine or fucose or both; (L) lipoprotein as judged by incorporation of radiolabeled choline; (NC) nucleocapsid protein; (E) proteins located in envelope or tegument and judged to be present in major amount; (e) minor protein located in tegument or envelope. Proteins I, II, III, IV, IVa, and V were major nucleocapsid proteins; proteins m2, m4, and m8 were designated as possible minor nucleocapsid proteins.

[b] Composite of data for the A 183 strain and KyA-ha strains of EHV-1. Data are taken from Turtinen and Allen (1983). VP 8 was present only in the KyA-ha strain. (G) Glycoprotein as determined by incorporation of radiolabeled glucosamine, enzyme-catalyzed addition of UDP-[$^{14}C_3$]galactose, chemical reduction of oxidized galactose residues with [^{3}H]sodium borohydride, or staining with Alcian Blue dye; (E) envelope protein; (NC) nucleocapsid protein; (T) tegument protein, a protein present neither in the nucleocapsid nor on the outer surface of viral envelope.

[c] Composite of data taken from Allen and Randall (1979) for the prototype strain (1118) of EHV-3. (NC) Nucleocapsid protein.

ha strains are comprised of six major polypeptides (Table II); the sizes of these proteins were reported to be 150,000, 53,000, 44,000, 35,000, 29,000, and 17,000. Considering that different gel systems were employed to analyze the various EHV-1 strains, it is likely that the 53,000-dalton nucleocapsid protein (VP 20) detected in the Army 183 strain of EHV-1 capsids corresponds to the 58,000-dalton protein (VP 19, nucleocapsid protein II) present in the nucleocapsids of the KyA cell-culture strain of EHV-1. The I capsid preparations were found to contain these same six proteins; however, the 30,000-dalton protein was present only in small amounts. This nucleocapsid protein corresponds to virion protein m8 of the cell-culture strain or VP 24B (29,000) of the other two EHV-1 strains (see Table II) (Perdue *et al.*, 1975). Last, the empty (L) capsids were shown to be comprised of only four of these proteins: 150,000, 59,000, 37,000, and 18,000. The 150,000-dalton protein (VP 9A) accounted for approximately 65% of the total capsid protein, whereas proteins 59,000 and 37,000 accounted for approximately 10 and 12%, respectively (Perdue *et al.*, 1975; O'Callaghan *et al.*, 1977a, 1978; O'Callaghan and Randall, 1976). These findings suggest that the 45,000- and 30,000-dalton proteins are important components of the EHV-1 core. The fact that the 30,000-dalton protein is present in significant amounts only in the H capsids suggests that it may have an important role in the packaging and/or arrangement of the viral genome within the core structure.

The amino acid compositions of each of the three EHV-1 capsid species isolated from both infected L-M cell nuclei (Perdue *et al.*, 1976; O'Callaghan and Randall, 1976) and infected hamster liver cell nuclei (O'Callaghan *et al.*, 1977a) were found to be grossly similar to that of the virion (O'Callaghan *et al.*, 1972a), and the composition of each capsid species from each of the two host-cell systems was virtually identical. The amino acid compositions of L and I capsids were somewhat similar, but differed from that of the H capsids, which was high in the content of lysine, serine, and glutamic acid. The large glutamic acid content in these core-containing capsids indicates that one or both of the putative core proteins—45,000 and 30,000—are extremely rich in this amino acid. A large content of this amino acid has been reported for mature HSV capsids isolated from virions by detergent treatment (Dreesman *et al.*, 1972). It is possible that poly (glutamic acid) is present in the herpesviral core and functions in the collapse and packaging of the viral DNA, as has been shown to occur in the assembly of some DNA-containing bacteriophage (Laemmli, 1975).

There has been only one report of the structural proteins of EHV-3 virions. Allen and Randall (1979) found that purified EHV-3 virions contain a minimum of 34 electrophoretically distinct polypeptides that vary in size from greater than 220,000 to 14,000 daltons (Table II). Examination of nucleocapsids purified from the EHV-3-infected equine cells revealed that six major virion proteins were present in the nucleocapsid: VP 6 (148,000), 17 (60,000), 20 (53,000), 22 (48,000), 26 (31,000), and 33 (16,000).

These results were in general agreement with those reported previously for EHV-3 capsid proteins (Allen and Bryans, 1976), with the exception that nucleocapsid proteins II and III, reported to be approximately 59,000 and 48,000 in the initial study, were found to have molecular weights of approximately 60,000 and 53,000 in the improved gel system. Allen and Bryans (1976) isolated two major and one minor species of EHV-3 capsids from the nuclei of infected equine epithelial cells (Allen and Bryans, 1974) and found that the most rapidly sedimenting species (B capsids) contained 7-fold more DNA than the slowest-sedimenting capsids (A capsids).The B capsid was found to be comprised of the six major capsid proteins, and electron-microscopic studies showed it to be a core-containing nucleocapsid (O'Callaghan *et al.*, 1978). The protein composition of the A capsids, which were shown by electron microscopy to be an apparent empty capsid type that lacks any electron-dense internal structure, was found to differ from that of B capsids, since the 16,000–18,000 protein was not present in significant amounts and the 30,000 band appeared to be comprised of two proteins that were similar in size (Allen and Bryans, 1976). Thus, these initial studies of the proteins of EHV-3 virions indicate that DNA-containing capsids of both EHV-1 and EHV-3 are comprised of six major proteins and that a protein of approximately 150,000 is the major capsid component. Since EHV-1 and EHV-3 DNAs exhibit approximately 10% homology (see below), it is likely that several viral proteins are closely related. Whether these shared DNA sequences code for capsid or envelope proteins is but one of the many questions concerning the molecular anatomy of these two herpesviruses that remain to be answered. Last, it should be noted that virtually nothing is known about the structural proteins of EHV-2. Preliminary investigations of the proteins of EHV-2 nucleocapsids suggest that this cytomegalovirus differs markedly from EHV-1 in protein composition (O'Callaghan, unpublished).

IV. REPLICATION

A. Introduction

The replicative cycle for EHV-1 has been studied in some detail, in cell culture as well as in animals. Bracken and Randall (1957) and O'Callaghan *et al.* (1972b) have both described the rather remarkable cycle in the young Syrian hamster. The basic observation is that following the injection of 10^8 LD_{50} (per animal), all animals die at 12 $\pm$ 0.5 hr, with a marked viremia (10^9–10^{10} virions/ml blood), and with more than 95% of the hepatic parenchymal cells showing a typical intranuclear inclusion. This is, to our knowledge, the best-defined herpesviral animal model, and its unique features have allowed measurements normally reserved for "one-step" growth kinetics in cell cultures. Features of the replicative cycles of the three equine herpesviruses are summarized in Table III. The

TABLE III. Growth Characteristics of the Equine Herpesviruses[a]

Virus–host system	Time of maximum titer (postinfection)	Virus produced	Cell-associated virus
EHV-1–hamster	10–12 hr	$10^{8.5}$ LD_{50}/g liver	[b]
EHV-1–cell culture, mouse L-M cells	18–24 hr	200 PFU/cell	5%
EHV-2–cell culture, equine cells	96–120 hr	20 PFU/cell	100%
EHV-3–cell culture, equine cells	9–12 hr	200 PFU/cell	95%

[a] Data were estimated as follows: For EHV-1–hamster, replication was quantitated by collecting livers from three animals or blood from five animals at various times after intraperitoneal inoculation with 1×10^8 LD_{50} units of virus; the samples were weighed, pooled, and homogenized, and the LD_{50} titer of various dilutions of each sample was assayed by inoculation of at least four animals (O'Callaghan *et al.*, 1972b). EHV-1 replication in L-M cells was measured by plaque assay (O'Callaghan *et al.*, 1968b; Perdue *et al.*, 1974). EHV-2 replications in equine (KyED) cells was measured by plaque assay (Wharton *et al.*, 1981), as was EHV-3 replication in equine (ETCC) cells (Allen and Bryans, 1974).
[b] Not applicable.

cycle of EHV-1 in cell culture is similar to that in the hamster, although slightly slower. In contrast, EHV-3 (in cell culture) grows even more rapidly than EHV-1 in the animal, while EHV-2 (also in cell culture) grows much more slowly, behaving as a cytomegalovirus. It may further be seen from Table III that with EHV-2, essentially all progeny virus remains cell-associated, another feature characteristic of cytomegaloviruses.

B. Early Events: Morphological

Attachment, penetration, and uncoating of EHV-1 have been observed by electron microscopy of infected cell cultures. Viral attachment occurs quite rapidly, because by 5 min after the addition of virus, pseudopodia of the host cells (L-M cells) have already begun to surround the virions, and by 15 min, virions may be observed in cytoplasmic vacuoles. Uncoating apparently takes somewhat longer, as observed by electron microscopy, but biochemical evidence (transcription) indicates that uncoating has occurred effectively before 4 hr postinfection (see Section IV.C). Because enveloped virus was observed inside cytoplasmic vacuoles, it was concluded that EHV-1 gained entry by viropexis (Abodeely *et al.*, 1970). To our knowledge, no similar studies of the early morphological events have been made of EHV-2 or -3; it may be presumed that they would not differ greatly from EHV-1, with the possible exception of EHV-2, which is almost entirely cell-associated.

C. Transcription

Several lines of evidence suggest that transcription has begun by 4 hr postinfection (p.i) with EHV-1. First H.L. Huang *et al.* (1971) showed—

by filter hybridization and competition studies—that EHV-1 viral specific messenger RNA was already present (although in small quantity) by 4 hr p.i. of L-M cells. This RNA was also present at late times (12 hr p.i) as well as additional species of viral specific RNA not detected at 4 hr p.i. J.C. Cohen *et al.* (1975b) subsequently showed that transcription could best be explained on the basis of two abundancy classes with different ratios at different times postinfection. Finally, the effect of inhibition of DNA synthesis by 5-fluoro-2′-deoxyuridine (FUdR) has shed some light on transcription. O'Callaghan *et al.* (1968b) showed that in the presence of FUdR, the synthesis of capsids continued, despite the absence of progeny DNA synthesis. Two conclusions may be drawn from this: First, assembly of capsids may occur in the absence of progeny DNA. Second, sufficient sequences of the viral genome are transcribed from the parental genome to code for the capsid proteins. The second of these, however, must be modified by the findings of J.C. Cohen *et al.* (1977b), who showed that in L-M cells infected with EHV-1 in the presence of FUdR, transcription was altered so that by 4 hr p.i. sequences of viral specific RNA could be detected that otherwise could be found only at later times. The mechanism of this shift is not known, but it is tempting to suppose that because viral genome replication had been stopped, more of the parental genome was available for transcription and there was reiterated transcription of some portions of the genome that normally would be represented by undetectably low quantities of transcripts at early times.

D. DNA Synthesis

The idea that equine abortion virus (as EHV-1 was known in the 1950s) contained DNA was at least in the back of the minds of Randall and Bracken because of their biochemical attempts to demonstrate accumulation of DNA in the nuclei of hamsters infected with EHV-1 (Bracken and Randall, 1957). The main evidence they produced, however, was histochemical, and this showed that the intranuclear inclusions typical of EHV-1 were Feulgen-positive [i.e., contained DNA different from host DNA that had been clumped and marginated around the periphery of the nuclei (Randall and Bracken, 1957)]. Subsequently, Gentry *et al.* (1961, 1962) investigated the synthesis of this DNA by using [^{3}H]deoxythymidine ([^{3}H]-dT), autoradiography, and isolation of liver-cell nuclei from EHV-1-infected hamsters. The conclusive demonstration that EHV-1 contained DNA was made by Darlington and Randall (1963), who in addition showed that the viral DNA was of relatively high guanosine plus cytosine (G + C) content. Subsequently, Soehner *et al.* (1965) characterized the EHV-1 DNA, O'Callaghan *et al.* (1972b) described the kinetics of its synthesis, and Campbell *et al.* (1976) demonstrated by Cot analysis the cyclic appearance of a species of viral DNA that had a higher than usual G+C content while lacking some of the sequences found in standard virus. This was the first report of defective interfering herpesvirus in an animal system. Indeed, it must be emphasized that *all* the studies reported in this section were carried out with the Syrian hamster–EHV-1 system. Finally, inhibition of viral DNA synthesis in this system by antivirals has been reported from two laboratories: Lieberman *et al.* (1972) with arabinosyladenine and Aswell *et al.* (1977) with arabinosylthymine. This work is discussed in greater detail in Section VIII.A.

The kinetics of viral DNA synthesis have been worked out in cell culture both for EHV-1 (O'Callaghan *et al.*, 1968a,b) and for EHV-3 (Allen and Bryans, 1976). In both these studies, $[^3H]$-dT was given to cell cultures at various times postinfection, and pulse-labeled DNA was extracted. Viral DNA was then separated from host DNA by salt-elution from methylated albumin–kieselguhr columns or by preparative density-gradient centrifugation with CsCl. In either case, the identity of the DNA was confirmed by hybridization to purified viral DNA. Similar studies were carried out in the hamster by O'Callaghan *et al.* (1972b). The results of these studies (Table IV) show that the peak DNA synthetic rate for EHV-3 occurs at 4 hr p.i., much earlier than with EHV-1 in L-M cells (12 hr p.i.) or even with EHV-1 in hamsters (8 hr p.i.). An interesting dependence on the cell cycle was reported by Lawrence (1971a), who showed—by autoradiographic analyses of EHV-1-infected synchronized KB cells—that EHV-1 DNA synthesis could begin only in cells in the S phase. Although possibly true for the KB cell, EHV-1 replicates quite well in hepatic parenchymal cells that are resting and that do not enter S phase during the infection (O'Callaghan *et al.*, 1977a). This dependency does not appear to be universal with EHV-1.

E. New Enzymatic Activities following Infection with Equine Herpesviruses

1. Introduction

Changes in several enzymatic activities have been described following infection of cells with equine herpesviruses. In some cases, these have led to the characterization of entirely new enzymes (deoxythymidine kinase and DNA polymerase) that are now known to be coded for by the viral genome; at the other extreme is the protein kinase (of possible viral origin) located within the virion of EHV-1. In between are very interesting fish swimming in muddy water.

2. Viral-Coded Enzymes

Perhaps the best-characterized viral-coded enzyme—at least among the herpesviruses—is deoxythymidine kinase (dTK). While our understanding of other features of the equine herpesviruses, especially EHV-1, has generally paralleled (in a chronological sense) our understanding of

TABLE IV. Correlation of Time of Maximum DNA Synthesis with Time of Maximum Titer for EHV-1 and EHV-3[a]

Virus–host system	Time of maximum DNA synthesis (postinfection)	Time of maximum titer (postinfection)
EHV-1–hamster	8 hr	10–12 hr
EHV-1–cell culture, mouse L-M cells	12 hr	18–24 hr
EHV-3–cell culture, equine cells	4 hr	9–12 hr

[a] Data were estimated as follows: Infected cells or animals were pulse-labeled with [^{3}H]dT at different times during the infection cycle, and the amount of radioactivity incorporated into viral DNA was determined as described for the EHV-1–hamster system by O'Callaghan *et al.* (1972b), for the EHV-1–L-M cell system by O'Callaghan *et al.* (1968a,b), and for the EHV-3–equine (ETCC) cell culture system by Allen and Bryans (1976). Time of maximum virus titer as in Table III.

herpes simplex virus (HSV), there was a considerable lag between the first report that HSV specified a dTK (Kit and Dubbs, 1963a,b) and the first suggestion that EHV-1 also specified this activity (Gentry and Aswell, 1975). In fact, at this time there were already data from Kit's group (Kit *et al.*, 1975; Leung *et al.*, 1975) as well as from Subak-Sharpe's group (Jamieson *et al.*, 1974) that suggested that EHV-1 did not specify a dTK. The reasons for this discrepancy are not clear. In the case of Jamieson *et al.* (1974), the strains employed were British and may have been of respiratory origin as opposed to the virus used by Gentry and Aswell (1975), which is of fetal origin. Although there is a growing body of evidence (Allen *et al.*, 1983c) that there are two subsets of EHV-1 that differ considerably in a variety of ways, there are no data which suggest that the respiratory subset of EHV-1 lacks a dTK. In the case of Leung *et al.* (1975), dTK$^-$ cells were not used, and it has since been shown that the EHV-1 dTK cannot be separated from mammalian cytosol dTK by the procedures [(PAGE)] they employed (Allen *et al.*, 1979). The EHV-1 strain Leung and co-workers used (Ky-D) is of fetal origin. As indicated above, the earliest suggestion that EHV-1 might specify a dTK was made by Gentry and Aswell in 1975. This was based on the sensitivity of EHV-1 replication to arabinosylthymine (ara-T), an inhibitor of DNA synthesis that is phosphorylated by various herpesviral-coded dTKs but not significantly (with one or two exceptions) by mammalian cytosol dTKs. Since then, the dTKs of EHV-1 and 3 have both been characterized thoroughly (Allen *et al.*, 1978b,c, 1979; McGowan *et al.*, 1980a,b). The EHV-1 dTK was purified by dT affinity-column chromatography and shown to be EHV-1-specific by inhibition by antisera as well as studies in which dTK$^-$ 3T3 cells were biochemically transformed by UV-irradiated EHV-1. The new dTK in these transformed cells was identical to EHV-1 TK in its properties and different from the cytosol dTK of dTK$^+$ 3T3 cells (Allen *et al.*, 1978c). The evidence for viral specificity of the EHV-3 dTK rests mainly on inhibition by specific antisera and a set of properties that distinguish it from host cytosol dTK (EHV-3 does not replicate in any of the dTK$^-$ cells available in our laboratory and must be propagated in cells of equine origin). The main features of the dTKs of EHV-1 and EHV-3 are compared in Table V.

Except for the values for the K_m for dT and the effectiveness of dGTP as phosphate donor, the two viral enzymes resemble each other more than either resembles the corresponding host cytosol dTK. The two host dTKs shown also resemble each other. Basically, the viral dTKs bind ara-T much better than the host dTKs, but do not bind dTTP as well. The main difference between the EHV-1 and EHV-3 dTKS is in the ability of the former to use dGTP as a phosphate donor.

In other respects, both the viral enzymes appear to be dimeric, EHV-1 with a molecular weight of 85,000 (Allen *et al.*, 1979) for the dimer and EHV-3 with a molecular weight of 83,000. Neither dTK could phosphorylate dC, which appears to set them apart from the dTK of HSV, which can (Jamieson and Subak-Sharpe, 1974).

The other enzyme that has been characterized and that is almost certainly coded for by the EHV-1 genome is DNA polymerase (DNA pol) (Kemp *et al.*, 1975; J.C. Cohen *et al.*, 1975a; Allen *et al.*, 1977b). The enzyme was purified from liver-cell nuclei of EHV-1-infected hamsters, a rich source. Nuclei were prepared (Kemp *et al.*, 1975), sonicated, and the supernatant (containing the DNA pol activity) applied to a diethylaminoethyl (DEAE) column. The DNA pol, which did not adhere to the column, was collected, dialyzed, and applied to a different DEAE column in a different buffer. It was eluted and fractions containing activity in high salt (150 mM K_2SO_4) were pooled and applied in turn to DNA cellulose, phosphocellulose, and hydroxylapatite. The overall yield was 1–2%, representing a 500-fold purification; the resulting specific activity was 10,000 U/mg protein (Allen *et al.*, 1977b). DNA pol thus purified gave a single band in sodium dodecyl sulfate–PAGE. Several properties of the EHV-1 DNA pol are shown in Table VI.

Rabbits were immunized with the purified DNA pol; the resulting antisera inhibited the DNA pol by more than 90%. When DNA pol was

TABLE V. Comparison of the dTKs of EHV-1, EHV-3, and Equine and Murine Cytosol

Property	EHV-1[a]	EHV-3[b]	KyED[b]	3T3[a]
K_m dT	5 μM	1.4 μM	3 μM	5 μM
K_i ara-T	60 μM	34 μM	210 μM	180 μM[b]
K_i dTTP	150 μM	52 μM	2.6 μM	5 μM
dCTP as phosphate donor[c]	89	107	10	14
dGTP as phosphate donor[c]	65	9	5	16

[a] Data from Allen *et al.* (1979), except the K_i for ara-T.
[b] Data from McGowan (1980).
[c] Phosphorylation with ATP as donor was arbitrarily set at 100 for each of the different dTKs.

TABLE VI. Properties of the DNA Polymerase of EHV-1

Optimal salt concentration	150 mM
pH optimum	8–9
Time of appearance	4 hr postinfection
Molecular weight	160,000
Number of subunits	1
Sedimentation coefficient	7.4 S
Axial ratio	8
Preferred template	Activated DNA
Divalent cation required	Mg^{2+} (2–6 mM)
Inhibitors	*N*-ethyl maleimide, phosphonoacetic acid, zinc
Nuclease activity	None
Reducing agent required?	Yes (dithiothreitol)

prepared from EHV-1 mouse L-M cells, it was also inhibited by the antisera by more than 80%, in contrast to DNA pol prepared from equine cells infected with EHV-3, which was not inhibited (<5%) significantly. For these reasons, and because of the activity in high salt, we suggest that it has been shown that the EHV-1 DNA pol is viral-specified. While we also believe that EHV-3 specifies its own DNA pol, the level of proof on that point is not as strong, because of a lack of data. The principal feature distinguishing these DNA pols (as well as other herpesvirus-specified DNA pols) from eukaryotic DNA pol α is the increased activity in high salt (Weissbach *et al.*, 1973).

3. Viral-Induced Enzymes

In this category are two enzymes, ribonucleotide reductase and deoxyribonuclease (DNase), the sources of which are uncertain. Of these two, the DNase activity was found first, by Stock and Gentry (1969). This work had its origin in the observation of Randall and Walker (1963) that in L-M cells prelabeled with [^{14}C]-dT, the DNA was substantially degraded to acid-soluble material during infection with EHV-1. Later, Gentry *et al.* (1964) found that in this system, viral replication was not blocked by methotrexate (used to produce the thymineless state and hence to inhibit replication of DNA viruses), which implied that the synthesis of progeny viral DNA was made possible by significant quantities of dTMP resulting from the degradation of host DNA. It remained for Randall *et al.* (1965) to show that the viral stock used in these studies had become contaminated by *Mycoplasma hominis*, which was responsible for the production of DNase and the subsequent degradation of the host DNA. It is important to note that this type of contamination cannot be detected by routine monitoring of cell-culture stocks—it is the virus stock itself that must be cultured. Stock and Gentry (1969) then showed that the mycoplasmal DNase was a DNase I and was excreted into the medium, with a maximum level at about 20–24 hr p.i. There was also found, however, a DNase activity in L cells infected with EHV-1 alone (that is, without the contaminating *M. hominis*). This activity reached a peak at about 12 hr p.i. and was not excreted into the medium. It is possible that the latter DNase is viral-coded, because Francke *et al.* (1978) have described a mutant HSV-2 in which expression of DNase is temperature-sensitive. Somewhat later, Gentry *et al.* (1975) showed that infection of L cells with EHV-1 rendered them permeable to added DNase I, which suggested how the mycoplasmal DNase might get into the cells, and that the cellular DNA was accessible to the added DNase whether in infected cells or in uninfected cells made permeable by scraping off the plastic surface. The most interesting question arising from this series of papers is how EHV-1 can replicate in the presence of DNase without having its own DNA degraded. This question is still unresolved. The other enzyme in this category was first suspected when it was found by J.C. Cohen *et al.* (1975a) that EHV-1 replication was resistant to hydroxyurea, an inhibitor of ribonucleotide reductase (the principal endogenous source for deoxyribonucleotides to be used in the synthesis of DNA). More specifically, it was found that in hydroxyurea-treated cultures, only EHV-1 DNA was made, that there was no significant degradation of host DNA with subsequent reutilization of the deoxyribonucleotides thus produced, and that hydroxyurea did not inhibit either EHV-1 or host-cell DNA pol. Further, studies with the inhibitor of protein synthesis cycloheximide showed that a protein other than the viral DNA pol was required by 4 hr p.i. if viral DNA synthesis was to proceed in the presence of hydroxyurea. It was next shown by J.C. Cohen *et al.* (1977a) that a new ribonucleotide reductase activity did indeed appear in L cells by 4 hr p.i. with EHV-1. A 28% increase in CDP reductase activity was observed by 6 hr p.i., and in contrast to the activity seen in control cultures, that in infected cultures did not require ATP for activity *in vitro*, had a lower apparent K_m, and had reduced sensitivity to hydroxyurea added to the reaction mixtures. Subsequently, Allen *et al.* (1978a) showed that the replication of EHV-3 was also resistant to inhibition by hydroxyurea and that both EHV-1 and EHV-3 replicated effectively in the presence of dT at 2 mM, a concentration sufficient to stop DNA synthesis completely in uninfected cells. Both hydroxyurea and dTTP are known to inhibit ribonucleotide reductase (it is also of interest that because the EHV-1 and 3 dTKs are insensitive to negative feedback by dTTP, the intracellular level of dTTP might well be elevated in infected cells even more than the dTTP level in uninfected cells treated with dT at 2 mM). It should also be mentioned that an altered ribonucleotide reductase has been described in Epstein–Barr-virus-infected cells (Henry *et al.*, 1978) and that the reductase of HSV-1 has similar properties (Huszar and Bachetti, 1981). Although the evidence for a viral origin of the new ribonucleotide reductase activity is in some respects stronger than that for DNase, neither activity has thus far been manipulated genetically for any herpes-virus (except for the HSV-2 that is temperature-sensitive for producing DNase mentioned above), which is why these two enzymes are grouped together and are considered less certain to have originated with the virus. In contrast, both dTK and DNA pol can be mapped, dTK as dTK$^-$ mutants and DNA pol as phosphonoacetic-acid-resistant mutants.

4. Virion-Associated Protein Kinase

Although there is no evidence linking the protein kinase activity to the genetic capability of EHV-1, it is present in highly purified virions. For that reason, it is discussed separately. The demonstration by Randall *et al.* (1972) that virions of EHV-1 contained a protein kinase was the first such observation for a herpesvirus and the second such study with any animal virus [the first was a paper by Strand and August (1971) in which protein kinase activity was described in avian myeloblastosis, Rauscher leukemia, and vesicular stomatitis viruses—all enveloped—but not in poliovirus or adenovirus—both nonenveloped]. The EHV-1 enzyme was found with $\gamma[^{32}P]$-ATP as phosphate donor and viral proteins (purified virions) as acceptor. The activity required Mg^{2+}, was cAMP-independent, and was enhanced by added protamine or arginine-rich histone. It appeared that serine and threonine residues were the principal phosphate acceptors. When the virions labeled during the reaction were analyzed by PAGE, it was found that all 17 of the viral proteins recogized at that time were labeled. The origin and significance of the EHV-1 protein kinase are not clear; it is tempting to speculate that it is involved in the transformation of cells to a malignant state (discussed elsewhere in this chapter), but to speculate in any detail would be premature. In any case, the recent work of Lemaster and Roizman (1980) suggesting a viral-coded origin for HSV-1 protein kinase suggests a similar origin for the EHV-1 enzyme.

F. Protein Synthesis

Almost nothing is known about the synthesis of EHV proteins, and only very elementary investigations have been carried out with but one of the three EHV serotypes. Lawrence (1971b) found that infection of human KB cells with the KyA cell-culture strain of EHV-1 resulted in a decreased rate of total protein synthesis, beginning at 6 hr p.i. and continuing to decrease during the infection cycle. O'Callaghan *et al.* (1968b) found little difference in the rate of total protein synthesis in control and EHV-1-infected mouse L-M cells and suggested that infected cells undergo a smooth transition from the synthesis of cellular protein to that of viral protein. Studies on the synthesis of capsid proteins in EHV-1-infected LM cells revealed that the major capsid protein is produced in large quantities as compared to the amounts of other capsid polypeptides and that these proteins are rapidly transported to the nucleus for capsid assembly (Henry *et al.*, 1980). Recently, Yeo *et al.* (1981) have presented indirect evidence that a nonstructural DNA-binding protein is synthesized in EHV-1-infected cells. This protein is antigenically related to the major virus-specific DNA-binding protein present in HSV-1- and HSV-2-infected cells. Further, this HSV protein was purified and characterized by Powell *et al.* (1981). Additional experiments have confirmed this initial finding and indicate that this EHV-1 nonstructural protein is produced in EHV-1-transformed and tumor cells (Caughman, Dauenhauer, Powell, and O'Callaghan, unpublished). Studies concerning the properties and synthesis of viral-coded proteins that have known enzymatic activity have been described above.

In the case of EHV-3, Allen and Bryans (1976) developed an *in vitro* translation system comprised of components from equine cells and rabbit reticulocytes and showed that oligo(dT)–cellulose chromatography of polysomal RNA from EHV-3-infected cells yielded RNA preparations that could be translated in this system. It was found that translation of 12–32 S RNA species yielded polypeptides that comigrated with viral polypeptides in polyacryamide gels. Analysis of the translation products precipitated by antiserum to EHV-3 nucleocapsids confirmed that the major capsid proteins were synthesized, indicating that these viral proteins are the direct products of translation and are not derived from cleavage of larger precursor proteins.

The elegant studies of Honess and Roizman (1974, 1975) (for a review, see Spear and Roizman, 1980) have shown that HSV polypeptide synthesis is coordinately regulated and sequentially ordered in a cascade fashion. On the basis of the temporal order and requirements for their expression, the polypeptides fall into three major groups, designated as α, β, and γ. The α gene products are made first ("immediate early") and are thought to represent viral genes containing promotors recognized by host RNA polymerase as alpha messenger RNAs (mRNAs) are made and translocated into the cytoplasm in the presence of inhibitors of protein synthesis. Genes coding for the β polypeptides ("early proteins") require functional α proteins but not viral DNA synthesis for their expression. The γ gene products ("late proteins") are made in significant amounts only late in infection, and their synthesis is dependent on viral DNA synthesis. The findings that EHV-1 mRNA synthesis is temporally regulated (H.L. Huang *et al.*, 1971), that temporal classes of viral transcripts are made up of sequences that differ in molar concentration (J.C. Cohen *et al.*, 1975b), and that inhibition of viral DNA synthesis alters both the production of certain viral proteins (O'Callaghan *et al.*, 1968b) and transcript classes (J.C. Cohen *et al.*, 1977b) suggest that EHV-1 gene expression is also highly regulated and sequentially ordered. However, nothing is known about the mechanisms that regulate EHV gene expression, and very little is known about the nature of the viral gene products made at different times during the infection cycle.

G. Capsid Assembly and Envelopment

A large body of literature from electron-microscopic examination of herpesvirus-infected cells has revealed that the assembly of herpesvirus nucleocapsids occurs within the nucleus of the infected cell and that the principal site of capsid envelopment is the inner nuclear membrane (for reviews, see Spear and Roizman, 1980; O'Callaghan and Randall, 1976). Similar findings have been obtained for the maturation processes of the equine herpesviruses (O'Callaghan and Randall, 1976; O'Callaghan *et al.*, 1978). In thin sections of EHV-infected cells, nucleocapsids occur in a variety of morphological forms, and the establishment of the precursor–product relationship among these different particle types is difficult, if not impossible, to ascertain by electron-microscopic observations. Essentially, two general classes of capsid types are seen in EHV-infected cells. The first are capsids that are referred to as empty because they exhibit no internal ultrastructure and lack an electron-dense core. These capsids are deficient in viral DNA (Perdue *et al.*, 1976) and have been described for EHV-1 (Arhelger *et al.*, 1963; Darlington and James, 1966; O'Callaghan *et al.*, 1968b, 1977a, 1978; Abodeely *et al.*, 1970; Perdue *et al.*, 1975, 1976; O'Callaghan and Randall, 1976; Henry *et al.*, 1980), EHV-2 (Wharton *et al.*, 1981), and EHV-3 (Allen and Bryans, 1976; O'Callaghan *et al.*, 1978; Atherton *et al.*, 1982). The second major class of EHV nucleocapsids are those that contain some type of internal structure and include a variety of capsid types such as (1) DNA-rich, toroidal core-containing capsids referred to as H capsids for EHV-1 (Perdue *et al.*, 1975, 1976; O'Callaghan and Randall, 1976; O'Callaghan *et al.*, 1977a) and EHV-2 (Wharton *et al.*, 1981) and as B capsids for EHV-3 (Allen and Bryans, 1976; O'Callaghan *et al.*, 1978); (2) capsids with an electron-lucent, cross-shaped internal structure such as the I capsids of EHV-1 (Perdue *et al.*, 1975, 1976) and EHV-2 (Wharton *et al.*, 1981); and (3) capsids that appear as two concentric ring structures as well as other forms the morphology of which may reflect different angles of sectioning through a maturing capsid (see Perdue *et al.*, 1976).

In an attempt to elucidate the role of these capsid types in herpesvirus assembly, biochemical studies designed to follow the fate of L, I, and H capsid species were carried out in EHV-1-infected L-M cells (Perdue *et al.*, 1975, 1976; O'Callaghan and Randall, 1976). Pulse-labeling and pulse–chase experiments with radiolabeled amino acids indicated that radioactivity incorporated into L nucleocapsids during the early hours of infection remained associated with these empty capsids during the entire chase period, suggesting that these empty (L) capsids are not the chief precursor to I or H capsids or both. However, it was found that label could be chased into H capsids as the chase period was increased, and a concomitant decrease in the radioactivity of I capsids was observed. In addition, H-capsid label appeared to become associated with enveloped virions. These findings suggest that I capsids are the chief precursor to the core-containing H nucleocapsids and that L capsids are primarily defective by-products that accumulate during infection, possibly because of excessive synthesis of the four proteins that comprise the shell (Henry *et al.*, 1980). In support of this model, described elsewhere in more detail (see O'Callaghan and Randall, 1976), it was found that L capsids are very minor species in hepatocyte nuclei of EHV-1-infected hamsters (O'Callaghan *et al.*, 1977a). Furthermore, this model of herpesvirus assembly proposed that I capsids accept the viral DNA molecule and that the DNA is, or becomes, associated with the two major core proteins [45,000 and 30,000 (see Table II)] and is collapsed and folded into a toroidlike structure, thereby forming a mature (H) capsid. The possibilities that poly (glutamic acid) detected in H capsids as well as events of phosphorylation and dephosphorylation of certain viral proteins important in the maturation process may function in the collapse and packaging of the viral DNA molecule were also proposed (Perdue *et al.*, 1976; O'Callaghan and Randall, 1976).

Support for this model of herpesvirus capsid assembly has been obtained by Ladin *et al.* (1980), who used several DNA$^+$ temperature-sensitive (ts) mutants of pseudorabies virus to demonstrate that shiftdown to the permissive temperature allowed mutants defective in the capsid-maturation process to develop into mature capsids. Electron-microscopic experiments and measurements of capsid types by sedimentation analyses indicated that the number of empty capsid types present in the ts-mutant-infected cell decreased with a concomitant accumulation of full capsids, indicating that immature capsids are direct precursors to full capsids.

In general, two major mechanisms, have been described for the envelopment process of EHV capsids. One mechanism is a budding process in which mature, core-containing nucleocapsids, possibly capsids to which certain tegument proteins have been added, recognize modified areas of the inner lamella of the nuclear membrane and bud outward and thereby acquire the modified nuclear membrane as the virus envelope. This budding process has been described for all three EHV serotypes (Darlington and James, 1966; Darlington and Moss, 1969; O'Callaghan and Randall, 1976; O'Callaghan *et al.*, 1978; Wharton *et al.*, 1981). The second major mechanism involves the apparent proliferation of areas of the nuclear membrane that invaginate into the nucleoplasm and form convolutions of membrane structures that react with and thereby envelop mature nucleocapsids. This process has been observed in both EHV-1- and EHV-2-infected cells (O'Callaghan and Randall, 1976; O'Callaghan *et al.*,

1978; Wharton *et al.*, 1981). Interestingly, both mechanisms seem to occur in EHV-1- and EHV-2-infected cells.

H. Effect of Equine Herpesvirus Infection on Host Macromolecular Synthesis

The effect of EHV infection on host macromolecular synthesis has been studied only for EHV-1 and EHV-3. The most definitive study on the effect of EHV infection on cellular DNA synthesis was carried out in EHV-1-infected L-M cells (O'Callaghan *et al.*, 1968a,b). Measurement of the syntheses of cellular and viral DNAs was accomplished by separation of host and viral DNA by chromatography on methylated albumin–kieselguhr or by isopyknic centrifugation in CsCl. It was found that there is a marked and progressive inhibition of cellular DNA synthesis beginning at approximately 6 hr p.i. and continuing throughout the infection cycle. By the time of maximal EHV-1 DNA replication at 12 hr p.i., cellular DNA synthesis was inhibited by 95% as compared to that observed in uninfected cells. A drastic inhibition in cellular DNA synthesis was also observed in EHV-3-infected cells (Allen and Bryans, 1977). As early as 2 hr p.i., total DNA synthesis in EHV-3-infected equine cells was reduced to levels equating to only 10% of that observed in control-cell cultures. After this early time, total DNA synthesis, as measured by incorporation of radiolabeled thymidine, decreased to very low levels.

In contrast to the nearly complete inhibition of cellular DNA synthesis observed in EHV-1 and EHV-3 infection of log-phase cells in culture, cellular DNA synthesis in the resting cells of the hamster liver was virtually unaltered during EHV-1 KyA-ha infection (O'Callaghan *et al.*, 1972b). A low level of cellular DNA synthesis was observed in this organ, and it remained constant throughout EHV-1 infection. Since EHV-1 replication in the liver involves the parenchymal cells (see above), it is possible that this low level of host DNA synthesis can be attributed to DNA replication in nonparenchymal cells that have escaped infection. The replication of this herpesvirus in these resting cells, however, does indicate that a requirement for cell DNA synthesis does not exist for EHV-1 replication.

Studies on the mechanism of inhibition of host DNA synthesis in EHV-1-infected cultured mouse cells (J. C. Cohen *et al.*, 1975a) and in hepatocyte nuclei from infected hamsters (Kemp *et al.*, 1975) revealed that the pronounced inhibition of cellular DNA synthesis corresponds to a marked inhibition of host DNA pol activity. However, little is known about the mechanism responsible for this reduction in host DNA pol activity. No evidence was found that a selective degradation of cellular DNA synthesis occurs during EHV-1 infection.

Examination of RNA synthesis in EHV-1- (O'Callaghan *et al.*, 1968b;

Lawrence, 1971b; Kemp *et al.*, 1975) and EHV-3- (Allen and Bryans, 1977b) infected cells has revealed a marked suppression of cellular RNA synthesis during the infection cycle. The inhibition of total RNA synthesis began at 2 hr p.i., the time of immediate early transcription (J.C. Cohen *et al.*, 1977b), and progressed from 60% inhibition at this time to greater than 95% inhibition by 12 hr p.i., the time of maximal viral DNA replication. A comparable pattern of inhibition of total RNA synthesis was found in EHV-1 infection of mouse and human cells as well as in EHV-3 infection of equine cells.

Randall and co-workers (Szabocsik, 1971; Kemp *et al.*, 1975) showed that the inhibition of cellular RNA synthesis affected the synthesis of all size classes of both nuclear and cytoplasmic cellular RNAs. Both transfer RNA and host mRNA syntheses were markedly inhibited by 4 hr p.i., and 18 S and 28 S ribosomal RNA synthesis was completely blocked by 5 hr p.i. Examination of the RNA polymerase activities of EHV-1-infected hepatocyte nuclei revealed that RNA polymerase II was responsible for the production of viral mRNA synthesis *in vitro*, but that this enzyme activity became progressively inhibited during EHV-1 infection (Kemp *et al.*, 1975). In addition, most of the greatly reduced production of cellular mRNA species in this isolated nuclei system could be accounted for by the reduction in RNA polymerase II activity.

As discussed (in Section IV. F), little is known about protein synthesis in EHV-infected cells except that cellular protein synthesis is markedly and progressively inhibited during EHV-1 infection. No investigations on the effect of equine cytomegalovirus infection on host macromolecular synthesis have been reported.

V. DEFECTIVE INTERFERING PARTICLES AND SYSTEMS OF PERSISTENT INFECTION

Serial, high-multiplicity passage of animal viruses of virtually every family has been shown to result in the generation of defective viral particles that possess the capacity to inhibit the replication of the parental (homologous) standard virus (see A.S. Huang and Baltimore, 1977). These defective particles have been shown to be deletion mutants that are usually much less genetically complex than the standard virus and thus have a selective advantage in replication and become abundant during passage. Although these defective particles interfere with the replication of standard virus at the molecular level, hence the name defective interfering (DI) particles, they require standard virus as helper to provide replicative functions or structural gene products or both. In the last few years, a growing body of literature has suggested that DI particles may alter the outcome of natural viral infections, initiate persistent or slow viral infections or both, and possibly play a role in oncogenic transformation in the case of tumor viruses (see O'Callaghan *et al.*, 1981a; Dauenhauer *et al.*, 1982). In addition, DI particles offer the investigator simplified versions of the more complex standard virus for the study of viral gene function and possibly for use as vaccines or reagents in investigations of viral pathogenesis and host response to infection.

Since herpesvirus DI particles were first described for herpes simplex virus (HSV) in 1973 (Bronson *et al.*, 1974), several additional members of the Herpetoviridae family have been shown to be capable of generating DI particles (Frenkel, 1981). Several workers have recently reviewed the structure and replication of HSV DI particles (Frenkel *et al.*, 1980; Frenkel, 1981; Becker and Rabkin, 1981; Cuifo and Hayward 1981; Kaerner, 1981), and Frenkel (1981) has presented a very comprehensive discussion of all defective herpesviruses described to date. This section will address the generation and properties only of EHV DI particles and will review the systems of persistent infection established in hamster embryo cells by infection with EHV-1 DI particle preparations or by high multiplicity infection with equine cytomegalovirus (ECMV).

A. Generation of Equine Herpesvirus Type 1 Defective Interfering Particles *in Vivo*

To our knowledge, EHV-1 is the only herpesvirus shown to produce DI particles in an *in vivo* system (Campbell *et al.*, 1976). Prolonged passage of EHV-1ha in the Syrian hamster model at high multiplicity resulted in the generation of a virus population that contained viral DNA with densities of 1.724 g/cm^3 as well as 1.716 g/cm^3, the density of standard EHV-1 DNA (see Section III.B.I). The high-density variant was judged to be DI virus because: (1) it was generated in a cyclic fashion when passaged in the animal; (2) inoculation of hamsters with virus rich in the high-density variant DNA resulted in prolonged survival of the animal, whereas animals inoculated with standard virus succumb to a fatal hepatitis within 12–14 hr postinfection (O'Callaghan *et al.*, 1972b); and (3) a correlation existed between the increase in the proportion of virus containing the high-density variant DNA and the reduction in specific infectivity.

Biochemical studies employing the techniques of CsC1 analytical ultracentrifugation, thermal chromatography on hydroxylapatite, and DNA hybridization analyses indicated that the heavy DNA species was a double-stranded EHV-1 DNA molecule and was similar in molecular size to that of standard DNA, but contained sequences equating to only 50% of the infectious genome. Considerable differences in the distribution of sequences rich in G + C were found to exist between the standard and heavy viral DNA species, suggesting that major segments of A + T-rich sequences were deleted in the defective DNA molecules.

B. Biochemical and Biological Properties of Equine Herpesvirus Type 1 Defective Interfering Particles Generated in Cell Culture

Since experiments on the interaction of DI particles at the cellular and molecular levels are difficult to conduct in the intact animal, EHV-1 DI particles were generated in L-M cell suspension cultures by serial, high-multiplicity passage (O'Callaghan *et al.*, 1978, 1981a; Henry *et al.*, 1979, 1980; Dauenhauer *et al.*, 1982). The scheme of experiments to generate DI-particle preparations and to assay their biochemical and biological properties is presented in Fig. 4. Serial, high-multiplicity passage of EHV-1 resulted in the rapid generation of DI particles as evidenced by the cyclic curve obtained in infectious virus titer (Fig. 5). The amplitude of these cyclic fluctuations in infectious virus titer became pronounced after only a few passages, and as much as a 2000-fold decrease in titer was obtained. Particles in these serial passages possessed high levels of interfering activity that was specific for homologous (EHV-1) virus and that could be abolished by UV irradiation of the serially passaged virus prior to its use in interference assays (Henry *et al.*, 1979; Dauenhauer *et al.*, 1982). Thus, expression of defective DNA is required for the interference capacity of the DI particle.

CsCl isopyknic analysis of the DNA species present in purified viral particles obtained from these serial passages revealed that a high-density DNA was present and that this DNA had a density (1.724 g/cm^3) identical to that of the defective DNA produced in the *in vivo* model. Analyses of each of the serial passages showed that the 1.724 g/cm^3 DNA species is produced in a cyclic manner and that its production follows the cyclic fluctuations in infectious virus titer and the decrease in specific infectivity [particle/plaque-forming unit (PFU) ratio] (Table VII). Also, it was found that the level of interference activity of particles isolated from each serial passage corresponded to the relative increase in the amount of this DNA species, confirming that the heavy DNA was the genome of the DI particle (Henry *et al.*, 1979; Dauenhauer *et al.*, 1982).

Characterization of the DNA extracted from purified particles of the serial passages rich in the 1.725 g/cm^3 DNA revealed that it was EHV-1 DNA and was similar to standard viral DNA in overall size, but was genetically less complex as determined by DNA reassociation analyses. It was found that the DNA became less complex with passage, lacked major portions of the standard genome, and contained significant amounts of reiterated sequences (Henry *et al.*, 1979). Restriction-endonuclease digestion of DNA preparations rich in defective DNA, as judged by measurement of the amount of the heavy DNA species in the analytical ultracentrifuge, revealed that defective DNA differed markedly in structure from standard EHV-1 DNA (Fig. 6). The most obvious difference was that fragments not present in standard DNA were major components of the defective DNA. These fragments, such as the *Bgl*II supramolar A

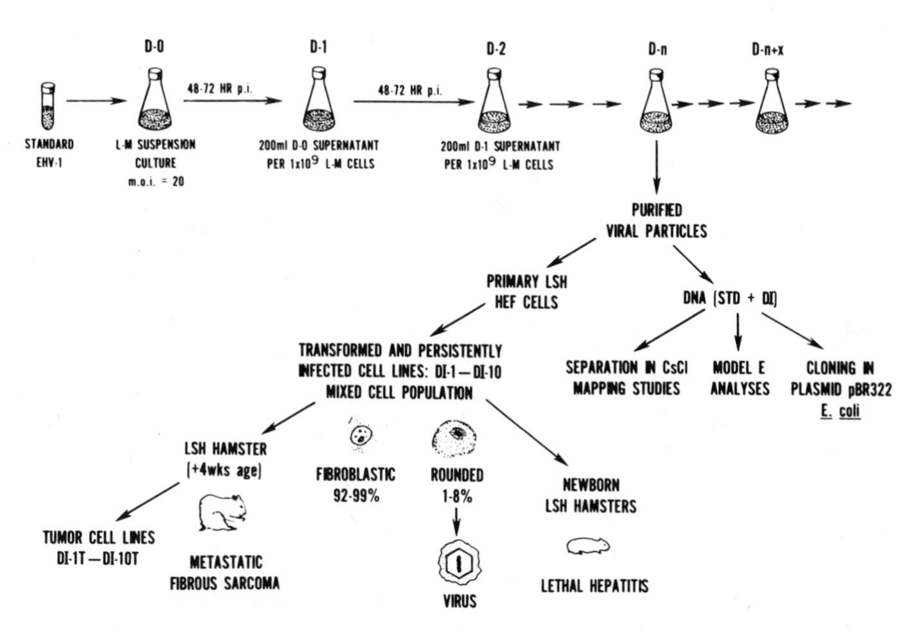

FIGURE 4. Scheme outlining the experimental approaches to generate EHV-1 DI particle preparations and to ascertain their biological and biochemical properties. *Top: Generation.* EHV-1 DI particles are generated by serial, high-multiplicity passage in L-M cell suspension culture by infecting 1×10^9 cells with 200 ml cell-free virus from previous passage. In initial infection, cells were infected at multiplicity of 20 PFU/cell with standard virus. *Bottom left: Biological properties.* Infection of permissive, primary LSH hamster embryo cells results in coestablishment of oncogenic transformation (major population of cells) and persistent infection (minor population of cells). Inoculation of these mixed cell populations, (DI cells) into adult hamsters results in production of fibrosarcomas that are transplantable and may be used to establish tumor-cell lines (DI-T cells). *Bottom right: Biochemical properties.* Viral particles purified from EHV-1 DI particle preparations may be used as a source to obtain purified defective DNA for (1) mapping by restriction-enzyme and electron-microscopic methods, (2) analyses by isopyknic banding in the Model E ultracentrifuge, and (3) cloning as recombinant DNA.

TABLE VII. Cyclic Appearance of Infectious Virus, Total Particles, and Heavy Viral DNA Species in Serial, High-Multiplicity Passages of EHV-1[a]

Virus passage	PFU/ml[b]	Particles/ml[b]	Particles/PFU	Composition of viral DNA (%)[c]	
				Standard (1.716 g/cm³)	Heavy (1.724 g/cm³)
Standard	9×10^8	15×10^9	16	100	0
D-1	2×10^7	11.9×10^8	60	95	5
D-2	4×10^7	47.2×10^8	118	79	21
D-3	1×10^8	47.6×10^8	48	70	30
D-4	9×10^7	32.3×10^8	36	65	35
D-5	5×10^7	15.4×10^8	31	30	70
D-6	1×10^6	18.1×10^8	1810	37	63
D-7	5×10^6	32.5×10^8	650	65	35
D-8	1.5×10^7	55.3×10^8	369	46	54
D-9	2.5×10^6	4.6×10^8	184	26	74
D-10	4×10^5	2.8×10^8	700	1	99
D-11	2×10^7	9.6×10^8	48	29	71
D-12	2×10^6	5.3×10^8	265	14	86
D-13	2×10^5	7×10^7	350	1	99
D-14	3×10^7	16.6×10^8	55	38	62
D-15	1.5×10^6	8.2×10^8	547	1	99
D-16	2×10^4	2×10^6	100	1	99

[a] L-M cells were initially infected with standard virus at a multiplicity of 20 PFU/cell, and the infection was allowed to proceed for 48 hr. Subsequent passages were made with cell-free supernatant at a multiplicity of 200 ml of cell-free supernatant per 1×10^9 cells as outlined in Fig. 4. Data presented here are from a series of serial passages distinct from the series presented in Fig. 5.

[b] Samples of virus supernatant were assayed for infectious virus by plaque assay and for total particles by a particle-counting method described by O'Callaghan *et al.* (1968b) and Henry *et al.* (1979).

[c] Total viral particles were purified from supernatants of each passage by the method of Perdue *et al.* (1974). Particles were lysed directly in ultracentrifuge cell, and the viral DNA and reference DNA (*M. luteus*; $\rho = 1.731$ g/cm³) were subjected to CsCl isopyknic banding in the Model E analytical ultracentrifuge. The percentage of each DNA species was measured by scanning the areas under each absorbance peak with a densitometer (Henry *et al.*, 1981).

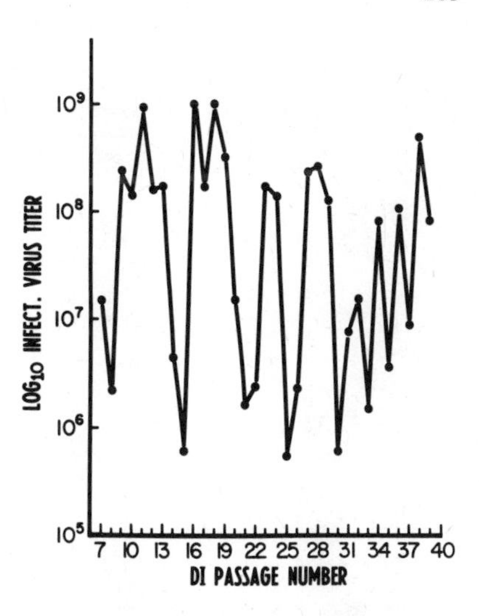

FIGURE 5. Cyclic fluctuations in infectious virus titer obtained by serial, high-multiplicity passage of EHV-1 in L-M cell suspension culture. In the initial passage, 1×10^9 L-M cells were infected with 20 PFU/cell of standard EHV-1 and were incubated for 48 hr at 37°C. For subsequent passages, 1×10^9 cells were infected with 200 ml cell-free supernatant of the previous virus passage.

fragment (Fig. 6, arrow), were present in amounts that were proportional to the content of defective DNA and could account, on the basis of radioactivity present in each fragment, for the overwhelming bulk of the defective DNA. In addition, it was found that fragments containing suequences mapping to the short (S) region were not present, suggesting that S-region sequences were either deleted or grossly altered in the defective genome.

To ascertain the nature of the DNA sequences present in the defective genome, the heavy DNA species was separated from standard DNA by subjecting DNA extracted from purified particles rich in defective DNA to CsCl isopyknic banding in a vertical rotor system (Fig. 7). This method achieved excellent separation of the two DNA species, and preparations of defective DNA entirely free of standard DNA, as judged by analytical ultracentrifugation analysis (Fig. 8), were obtained by this method. The nature of DNA sequences comprising the defective DNA species was ascertained by restriction-enzyme digestion analyses, hybridization of ^{32}P-labeled defective DNA to restriction-enzyme fragments of standard EHV-1 DNA by Southern blot hybridization methods (Southern, 1975; O'Callaghan *et al.*, 1981a; Robinson *et al.*, 1981b), and partial denaturation–electron-microscopic techniques (Dauenhauer, Ruyechan, and O'Callaghan, in prep.). Restriction-enzyme digestion patterns of EHV-1 defective DNA with any of several different enzymes gave results that confirmed those presented in Fig. 6. Digestion of defective DNA with *Eco*RI, *Xba*I, *Bgl*II, *Bcl*I and *Sma*I yielded a large fragment(s) of approximately 25–30 md as well as a large population of very small fragments that migrated with mobilities indicative of a size of 2–0.1 md. Digestion with any of several other enzymes (e.g., *Pst*I, *Hin*cI, *Hae*I, *Hae*II, *Sst*I, *Sst*II, *Xho*I, *Msp*I, *Hpa*II, *Hind*III, *Sal*I) yielded only a smear of unresolvable fragments with sizes of approximately 2.5 md to less than 0.1 md (Dauen-

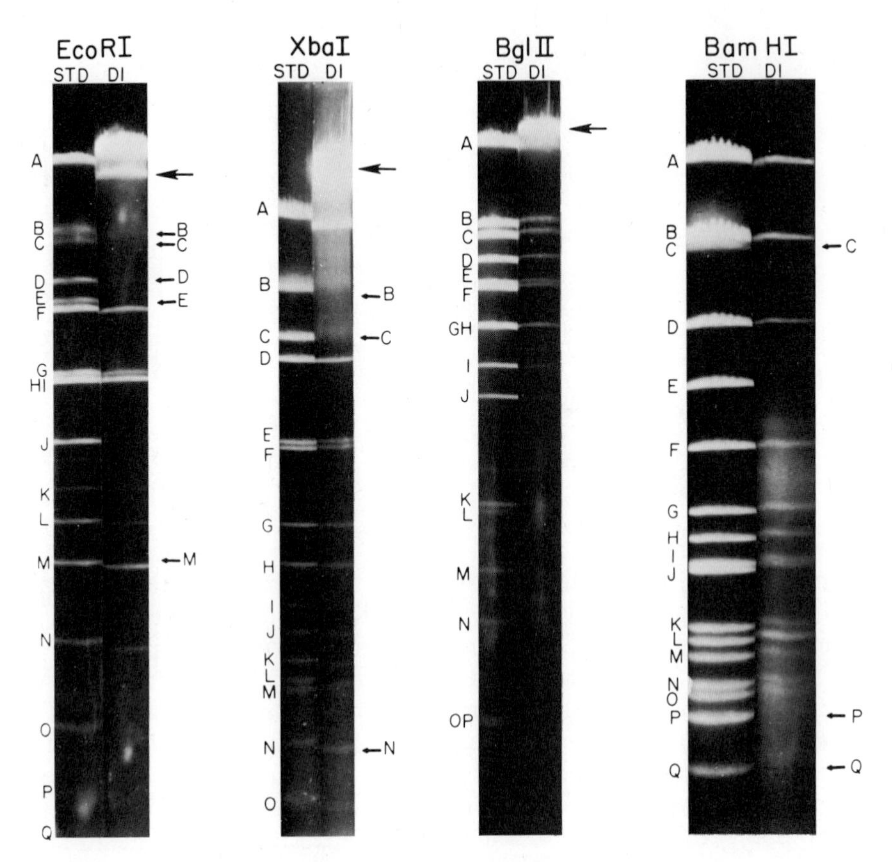

FIGURE 6. Restriction-endonuclease analysis of DNA from high-multiplicity, serially passaged EHV-1 enriched for DI particles. DNA obtained from purified virus rich in particles (80–90%) containing the high-density DI genome species (ρ = 1.724 g/cm^3 vs. 1.716 g/cm^3 for standard EHV-1) as judged by analytical ultracentrifugation was cleaved with the four restriction enzymes shown at top and compared with similar profiles of standard DNA (shown at left in each pair). Electrophoresis of fragments in cylindrical agarose gels of DI-rich DNA cleaved with *Bgl*II reveals the generation of an additional high-molecular-weight fragment (see arrows) shown to contain the bulk of the DI genome when the amount of radioactivity in slices of these gels was measured. The disappearance of 0.5 M *Eco*RI fragments B and E and C and D in DI DNA indicated a loss of cleavage sites external to the unique S region (see Fig. 2). Also observed was the retention of the S-region *Eco*RI fragment M (1.0 M) in DI DNA and the evolution of a new fragment (see arrow), possibly orginating from inverted repeat sequences joined in a head-to-tail arrangement. Similarly, cleavage with *XbaI* yielded an extra-high-molecular-weight band (see arrows) with a concomitant loss of S-region flanking sequence B and C. As with *Eco*RI, the *XBa*I 1.0 M S-region fragment N was not lost. In the case of *Bam*HI cleavage, the 1.0 M *Bam*HI-C fragment, end fragment Q, and penultimate fragment P were also lost from the S region.

hauer *et al.*, 1981; Baumann, Dauenhauer, Staczek, and O'Callaghan, in prep.). No fragments similar in size to those obtained from digestion of standard DNA were obtained. Blot hybridization analyses using defective DNA as probe showed that radiolabeled defective DNA hybridized only to standard EHV-1 DNA fragments that were located in the S region (Fig. 9). The most intense homology was observed with the fragments that contain the inverted repeat sequences. Findings to date indicate that if U_S sequences are present in the DI DNA, they comprise only a small proportion of this DNA as compared to that contributed by the inverted repeat sequences. The very weak homology to fragments mapping at the L-region terminus is thought to be due to the limited homology between terminal sequences of EHV-1 DNA that allow the molecule to circularize as proposed by Sheldrick (see Spear and Roïzman, 1980). Partial denaturation mapping studies comparing the anatomy of standard and defective DNAs in the electron microscope confirm that defective DNA molecules are similar in size to that of standard DNA and are comprised of a large repeat unit of approximately 30 md that in turn is comprised of several small repeat units. Although the exact arrangement of sequences comprising the defective DNA remains to be elucidated, it is evident that this EHV-1 DI particle population is a Class I defective herpesvirus (Frenkel, 1981), since its genome contains reiterations of sequences arising

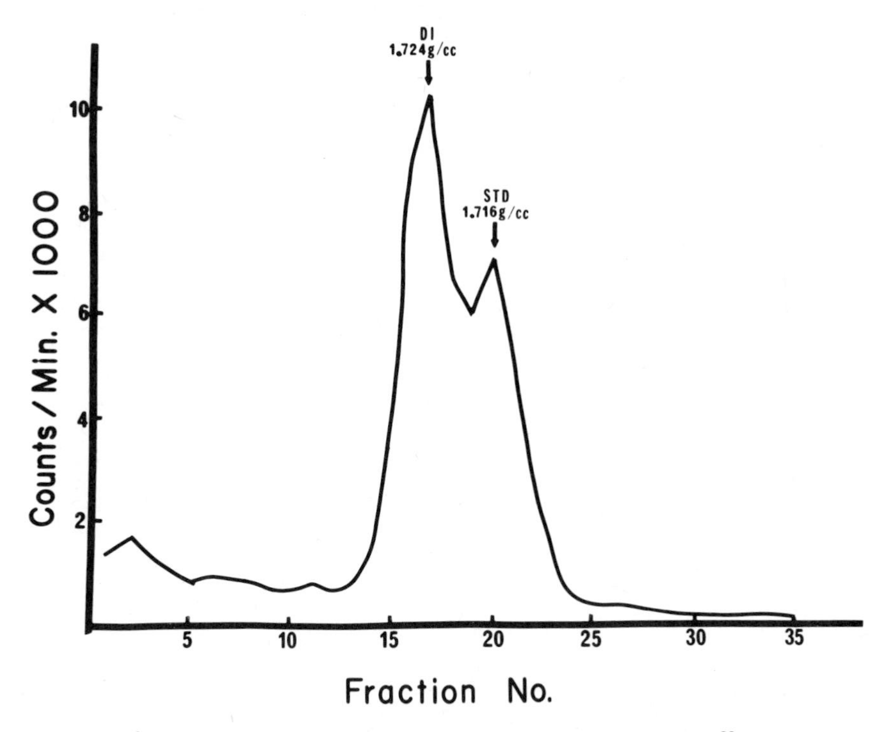

FIGURE 7. Purification of EHV-1 defective DNA. Radioactivity profile of ^{32}P-labeled DNA in a mixed preparation of EHV-1 DI particles and standard virus following CsCl isopycnic banding in the VTi65 vertical rotor at 35,000 rpm for 20 hr. Gradient fractions were collected by drops after bottom puncture, and each fraction was counted by the Cherenkov method. Radioactivity peaks correspond to the DI (ρ = 1.724 g/cm^3) and standard (ρ = 1.716 g/cm^3) DNA species present in the preparation. Quantitative separation of defective DNA and standard DNA was verified by analytical ultracentrifugation studies (see Fig. 8).

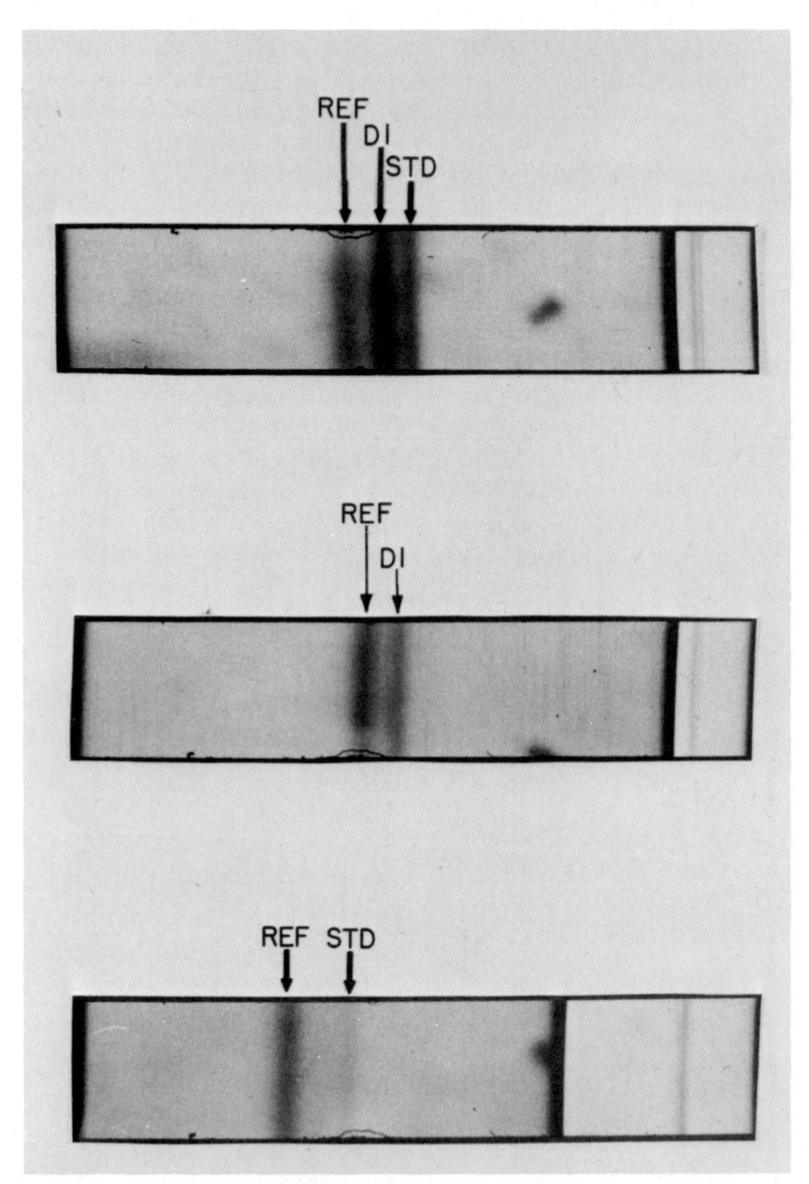

FIGURE 8. Model E analytical ultracentrifuge photographs demonstrating quantitative separation of DI ($\rho = 1.724$ g/cm^3) and standard EHV-1 ($\rho = 1.716$g/cm^3) DNAs by isopycnic banding in a vertical rotor (see Fig. 7). Each DNA sample was mixed with reference (REF) DNA from *M. luteus* ($\rho = 1.731$ g/cm^3) and CsCl to an average density of 1.70 g/cm^3. The samples were centrifuged at 44,770 rpm in an AnF rotor, and the DNA UV absorbance was photographed at 24 hr. *Top:* Analysis of DNA sample containing defective and standard EHV-1 DNAs. *Middle:* Model E analysis of DNA in the heavy peak obtained by isopycnic banding of DNA from EHV-1 DI particle preparations in a vertical rotor (see Fig. 7). *Bottom:* Model E analysis of DNA in the light peak obtained by isopycnic banding of DNA from EHV-1 DI particle preparations in a vertical rotor (see Fig. 7).

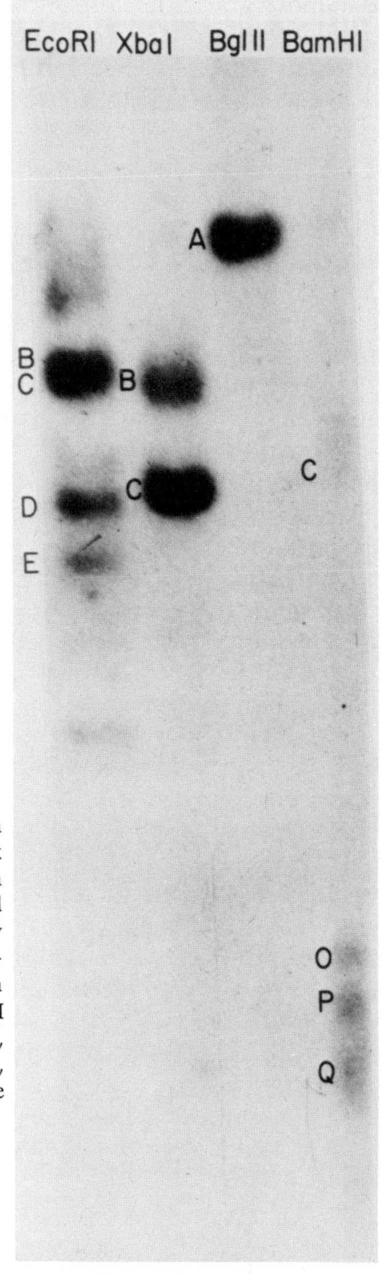

FIGURE 9. Identification of DNA sequences in EHV-1 defective DNA. Hybridization of ^{32}P-nick translated *Bgl*II supramolar A fragment with Southern blots of standard EHV-1 DNA digested with the restriction enzymes listed. Essentially identical results were obtained with ^{32}P-labeled defective DNA purified by the method outlined in Fig. 7. Hybridization of these probe DNAs to *Eco*RI fragments B, C, D, and E, *Xba*I fragments B and C, the *Bgl*II A fragment, and *Bam*HI fragments C, O, P, and Q confirmed the S-region origin of defective DNA.

exclusively from either end (inverted repeats) of the S region of the parental DNA (O'Callaghan *et al.*, 1981a; Robinson *et al.*, 1981b; Dauenhauer, Staczek, and O'Callaghan, in prep.). At present, experiments to ascertain the exact arrangement of the S-region sequences within this defective DNA are continuing. Our recent success in cloning the *Bgl*II supramolar fragment unique to defective DNA by ligating this fragment to *Bam*HI sites of the plasmid pBR322 should provide quantities of DNA sufficient for these investigations (Robinson *et al.*, 1981b).

Frenkel (1981) has reviewed the evidence from investigations of HSV and pseudorabies defective viruses showing that overproduction of certain polypeptides in cells infected with DI preparations results from abundant transcription of gene templates present in the reiterated defective genome and that expression of DNA sequences of the defective repeat unit is regulated by mechanisms operating in standard virus infection. Certain HSV defective populations have been shown to have the capacity to interfere with the transcription of helper standard virus and thereby effect significant decreases in the synthesis of total viral polypeptides (Frenkel *et al.*, 1981). Initial studies of viral protein synthesis and capsid production in cells infected with EHV-1 preparations enriched for Class I defective particles have shown that viral structural polypeptide synthesis is greatly reduced and that the reduction in total viral particles may be explained in terms of these alterations (Henry *et al.*, 1980). The synthesis of the major capsid protein (VP 9A, 150,000 daltons) was markedly inhibited, and the amount of each of the three capsid types—L, I, and H—was reduced as compared to the levels of each in standard virus infection. Interestingly, the 46,000 nucleocapsid protein, which is a major component of the EHV-1 core structure and is present in large amounts in immature (I species) capsids, was overproduced in cells infected with defective virus preparations as compared to the amount of this protein made in standard virus infection. The overproduction of this capsid component concomitant with the reduction in the synthesis of the four polypeptides that comprise the capsid resulted in the favored assembly of I capsids in these cells. In addition to the overproduction of this structural protein, a large nonstructural polypeptide of approximately 175,000 daltons was also produced in quantities much greater than that observed in standard virus infection. Although it would be premature to attempt to interpret these observations until the sequences coding for these EHV-1 polypeptides are mapped, it is obvious that major alterations in the programming of EHV-1 gene expression occur in permissive cells infected with defective virus. Some of these alterations may modify cytocidal functions and allow viral genes with limited expression in the replication cycle of standard virus to be dominantly expressed in the replication of defective virus.

C. Coestablishment of Persistent Infection and Oncogenic Transformation by Equine Herpesvirus Type 1 Defective Interfering Particle Preparations

The observation that infection of L-M cells with EHV-1 DI particle preparations caused alterations in the biochemical events of virus replication prompted an investigation of the biological responses elicited by infection of primary mammalian cultures with EHV-1 defective virus preparations. Although infection of primary hamster embryo (HE) cell cultures with standard EHV-1 caused marked cytopathological response accompanied by production of infectious virus and eventual lysis of the cells (O'Callaghan *et al.*, 1978; Robinson *et al.*, 1980a), it was found that infection of these primary HE cultures with viable EHV-1 preparations containing DI particles generated by either a few or a large number of serial high-multiplicity passages in L-M cells resulted in the coestablishment of persistent infection and oncogenic transformation (Vance *et al.*, 1978; Robinson *et al.*, 1980c; O'Callaghan *et al.*, 1981a; Dauenhauer *et al.*, 1981, 1982). Numerous hamster cell lines were established using either passage D-7 of the initial series of serial high-multiplicity passages of EHV-1 or passage X-16, which was the 23rd passage of a second series of high-multiplicity passages used to generate EHV-1 Class I defective particles. Both DI particle preparations used in these biological studies were shown to be comprised of +65% defective virus as judged by measurement of the amount of the high-density DNA. In both series of experiments, many cell lines were independently established from the large number of foci of cells that survived the infection with these viable EHV-1 DI particle preparations. Ten cell lines selected at random have been investigated to date: (1) DI-1 to DI-4 are representative of cell lines established by infection of HE cells with DI particle passage D-7 and (2) DI-5 to DI-10 are representative of cell lines established by infection with DI particle preparation D-16.

Investigations to date have shown that all DI cell lines established by infection with either series of EHV-1 defective particles share many, but not all, key properties (Robinson *et al.*, 1980b,c, 1981a,b; O'Callaghan *et al.*, 1981a,b; Robinson and O'Callaghan, 1981; Dauenhauer *et al.*, 1982; Dauenhauer, Staczek, Robinson, and O'Callaghan, in prep.). Some of these properties are listed in Table VIII, and a summary of the characteristics of these cells is presented below.

DI cell cultures are comprised of a mixed cell population. Investigations using the techniques of light microscopy, scanning electron microscopy, and transmission electron microscopy have shown that all cell lines are comprised of two basic cell types. The major population is a large, spindlelike, fibroblastic cell type that has the morphology of oncogenically transformed cells. Documentation for this is provided below. The second cell type, which was present in much lesser amounts, exhibited a rounded, cytopathic appearance similar to that of EHV-1-infected HE cells during the early stages of productive infection. The ratio of fibroblastic to rounded cell type varied from approximately 7 : 1 to 20 : 1, but the ratio of these cell types in each cell line has remained constant throughout their continuing life span.

Additional evidence for two distinct cell types has been obtained by indirect immunofluorescent-antibody and cell-sorting studies. Staining with antibody to EHV-1 structural or nonstructural proteins, or both, by

TABLE VIII. Biological Properties of EHV-1 Persistently Infected Hamster Embryo Cells

Cell[a]	Morphology[b]	Supernatant virus titer (PFU/ml)[c]	Cells persistently infected (%)[d]		Cells expressing viral antigens (%)[e]	Resistance to superinfection[f]		Inoculation of adult hamsters[g]
			IC	EM		EHV-1	VSV	
Normal HE	Normal fibroblastic	0	0	0	0	No	No	No pathology
DI-1	Mixed	1.1×10^5	3.9	2.6	+75	Yes	No	Fibrosarcoma
DI-2	Mixed	4.4×10^4	4.6	3.2	+75	Yes	No	Fibrosarcoma
DI-3	Mixed	3.3×10^4	8.1	7.0	+75	Yes	No	Fibrosarcoma
DI-4	Mixed	5.3×10^3	4.3	9.0	+75	Yes	No	Fibrosarcoma
DI-5	Mixed	1×10^6	3–7	5	+75	Yes	No	Fibrosarcoma
DI-6	Mixed	2×10^6	3–10	2	+75	Yes	No	Fibrosarcoma
DI-7	Mixed	3×10^4	5–17	5	+75	Yes	No	Fibrosarcoma
DI-8	Mixed	1×10^3	4–11	2	+75	Yes	No	Fibrosarcoma
DI-9	Mixed	5×10^3	7–8	2	+75	Yes	No	Fibrosarcoma
DI-10	Mixed	1×10^3	13–20	7	+75	Yes	No	Fibrosarcoma

[a] DI-1 to DI-4 cell lines were established by infecting primary HE cells with passage D-16 of EHV-1 high-multiplicity, serially passaged virus (70% defective DNA). DI-5 to DI-10 cells were established by infecting HE cells with passage D-7 (65% defective DNA) virus of another serial, high-multiplicity passaged virus. Surviving cells or foci from separate cultures were picked and propagated as described by Robinson *et al.* (1980c) and Dauenhauer *et al.* (1982).

[b] The major cell population exhibited the morphology of EHV-1-transformed cells; the minor cell population had a rounded cell morphology typical of EHV-1 productively infected cells at early stages of the development of cytopathic response.

[c] Typical infectious virus titer (PFU/ml) of 25 cm^2 confluent cell culture containing 8 ml medium. Passages assayed varied from 50 to 80 of these immortal cell lines.

[d] In infectious-center (IC) assays, cells were plated onto L-M cell monolayers, overlaid, and incubated for 5 days prior to counting plaques. In electron-microscopic (EM) assays, numerous thin sections of several hundred cells were examined for virions, subviral particles, and morphological features of virus replication.

[e] Cells on coverslips were fixed with methanol or acetone and reacted with either hamster antisera to EHV-1 virions or infected-cell extracts. After rinsing, the cells were reacted with fluorescein-conjugated rabbit anti-hamster immunoglobulin, dried, and examined in a fluorescent microscope. At least 1000 cells were scored. All controls (normal HE cells reacted with antisera to EHV-1 and DI cells reacted with normal hamster serum) failed to demonstrate fluorescence.

[f] Monolayer cultures were challenged with standard EHV-1 or vesicular stomatitis virus (VSV) at a multiplicity of infection of 1–50 PFU/cell. Cells were monitored for virus cytopathology, and samples taken at 2 and 24 hr postinfection were assayed for infectious virus. All VSV titers increased by a factor of at least 1×10^6 (ratio of 24 hr to 2 hr titer); EHV-1 titers increased by a factor of 5 or less.

[g] LSH Syrian hamsters, 4 weeks old, were inoculated subcutaneously with 1×10^5 to 2.5×10^6 cells. Tumors identified as malignant fibrosarcomas developed within 6 weeks or less in 100% of the animals. At least 20 animals were tested for each cell type.

immunofluorescent techniques revealed that the rounded cell population stained much more intensely than the major fibroblastic cell type with antibody to EHV-1 structural proteins and exhibited surface fluorescence in the entire cell membrane. In contrast, fluorescence in the major cell type was more localized and more intense in assays using antisera to nonstructural viral proteins. Some of the DI cell cultures have been analyzed by cytofluorography, and these investigations comfirm that the population is comprised of two cell types on the basis of several parameters such as size, density, and DNA content.

DI cell cultures are persistently infected. All cell lines established by infection with EHV-1 DI particle preparations contain a persistently infected cell population that continues to release infectious EHV-1. The amount of virus produced varies from approximately 1×10^3 to 2×10^6, but significant fluctuations in virus titer have been observed for each cell line in the +36 months since they were first established. Infectious-center assays and electron-microscopic studies revealed that the minor cell population is the virus producer. The infectious virus produced is EHV-1 that is identical to the standard input virus as judged by the restriction-enzyme profiles of the DNA of the infectious virus, growth characteristics and host range in cell culture, and capacity to induce a fulminant hepatitis when inoculated intraperitoneally into newborn hamsters.

Although the amount of infectious virus released by each producer cell has been calculated to be in the range of 20 PFU/cell or less, DNA hybridization analyses indicate that the persistently infected cells harbor as many as 250–400 EHV-1 genome copies per cell. Nothing is known about the state(s) in which these viral genomes are maintained in the producer cells.

The virus-producer cell population is maintained as the minor cell population. All of EHV-1 persistently infected cell cultures have survived repeated passage for more than 3 years. The producer population has been maintained as the minor cell population, and subjection of the DI cell lines to a variety of experimental conditions has failed to alter the state of persistency or the relative number of producer cells in each culture. These conditions included incubation of the cells at different temperatures, cocultivation of the cells with permissive L-M cells, treatment of the cultures for brief periods of time with iododeoxyuridine or hydroxyurea, and superinfection with standard EHV-1. Interestingly, both cell types were resistant to superinfection with homologous virus, but were susceptible to superinfection with heterologous viruses such as vesicular stomatitis virus or mengo virus.

The major cell population is comprised of EHV-1 oncogenically transformed cells. In addition to the establishment of EHV-1 persistency, infection of permissive HE cells with defective virus preparations resulted in coestablishment of oncogenic transformation. Experiments cataloging the biological, biochemical, and immunological properties of the DI-1 to DI-10 cell lines have shown that the major cell population is comprised of EHV-1 oncogenically transformed cells. All DI cell lines exhibit the biological properties of virus-transformed cells: altered morphology, indefinite passage in culture (immortality), increased saturation density, shorter generation time, reduced requirement for serum, colony formation at high efficiency in soft agar or methylcellulose, resistance to superinfection with homologous virus, and abnormal karyotype. In addition, the essential biological property—oncogenicity in the animal—has been demonstrated, since inoculation of adult syngeneic animals with any of the DI cell lines results in the rapid development of fibrosarcomas at the site of inoculation. These tumors are highly malignant, and metastasis to major organs such as lung, kidney, and heart has been documented. The tumors are transplantable and can be readily established as tumor cell lines *in vitro* (DI-T cells).

That the nonproducing cell population constitutes oncogenically transformed cells has been shown by the demonstration that cell lines established from single nonproducer cells are oncogenic in the hamster. To date, efforts to clone the EHV-1 producer cells and establish cell lines comprised only of producer cells have been unsuccessful. It remains to be determined whether this failure is due to the inability of virus-producer cells to survive in culture in the absence of the other cell population or whether it is a consequence of a complex process in which producer cells are derived from nonproducer cells at a rate equal to that at which persistently infected cells succumb to infection. Indeed, these questions address the mechanism of persistency *per se*, and little is known of the biochemical and biological events of this altered outcome of virus infection. It should be noted that all tumor-cell lines derived from DI-cell-induced tumors are virus nonproducers and all attempts to detect or induce infectious virus or virus particle production in these tumor cells have been unsuccessful. In addition, as will be discussed below, cell lines established from clones of DI nonproducer cells as well as DI tumor cells contain integrated EHV-1 DNA sequences that represent only specific portions of the viral genome and produce EHV-1 polypeptides detectable by immunological methods. Last, since at least 75% (a conservative estimate) of the cells in the DI-1 to DI-10 cell lines express viral antigens, it is apparent that most of the nonproducer transformed cells in the persistently infected cultures harbor viral genetic information.

Several lines of evidence support the conclusion that EHV-1 DI particles initiate the persistent state of infection observed for the DI cell cultures. First, the presence of DI particles is required for the establishment of the persistency, and infection of these permissive cells with virus preparations free of defective virus resulted in cell lysis. Second, all EHV-1 persistently infected cell cultures are resistant to superinfection with homologous virus. This observation is consistent with those of other systems of persistent infection established by defective viruses (A.S. Huang, 1973; A.S. Huang and Baltimore, 1977; Roux and Holland, 1979, 1980; Meinkoth and Kennedy, 1980; Andzhaparidze *et al.*, 1981; Stollar, 1979; De and Nayak, 1980). Third, preparations of defective EHV-1 used to initiate the persistency have extremely high levels of interference capacity and can inhibit production of standard EHV-1 by greater than 1000-fold. Therefore, it has been postulated that these Class I DI particles alter the events of the replicative cycle of standard virus such that its cytocidal functions are suppressed or modified and thereby the persistent state of infection can result (Robinson *et al.*, 1980c; O'Callaghan *et al.*, 1981a). Evidence that viral DNA sequences mapping at approximately 0.30–0.40 map unit of the L region are associated with EHV-1 oncogenesis and maintenance of the oncogenic state is discussed in Section VI. Therefore, it has been proposed that these DI particles with DNA comprised only of S-region DNA sequences do not initiate the oncogenic transformation *per se*, but temper cytocidal functions and thereby allow standard virus DNA sequences associated with oncogenesis to be dominantly expressed (Robinson *et al.*, 1980c, 1981b; O'Callaghan *et al.*, 1981a,b). Such a model for the indirect role of defective herpesviruses in oncogenic transformation offers one possible explanation as to how cells permissive for highly lytic members of the Herpetoviridae family escape destruction and become oncogenically transformed.

The capacity of EHV-1 to establish persistent infection of the Jijoye human lymphoblastoid cell line derived from a Burkitt lymphoma has been documented (Roumillat *et al.*, 1979). Less than 15% of the total cell population of these Epstein–Barr virus (EBV)-immortalized cells expressed EHV-1 antigen, but this subpopulation contained 80% of the cells expressing EBV capsid antigen. It was found that infectious EHV-1 was produced in these cultures and that as EHV-1-infected cells succumbed due to lysis, additional cells were recruited to replicate EHV-1. The authors proposed that cell lysis may be due to activation of latent EBV in the EHV-1-superinfected cells and that this system may offer a model to study persistent infection as well as activation of latent virus. The question of whether the preparations of the KyD strain of EHV-1 used to establish this persistency contain DI particles was not addressed by these workers.

D. Mechanism for Maintenance of Equine Herpesvirus Type 1 Persistent Infection

Three mediators have been reported to be involved in the establishment and maintenance of persistent viral infections of animal cells. The isolation of temperature-sensitive (ts) mutants from persistently infected cells and also the establishment of persistence by ts virus preparations have implicated their possible role in persistency for Newcastle disease virus (Preble and Youngner, 1973a,b, 1975), reovirus (see Ahmed *et al.*, 1980), vesicular stomatitis virus [(VSV) Wagner, 1974; Stanners and Goldberg, 1975], and other viruses (see Dal Canto and Rabinowitz, 1981). In long-term persistence in BHK_{21} cells established using VSV virions and DI particles, Holland *et al.* (1976) found that ts and small plaque mutants evolved that contributed to the stability of persistence. For many years, a role for interferon in establishment and/or maintenance of persistent infections has also been proposed by several investigators working with a variety of viruses (Henle *et al.*, 1959; Glasgow and Habel, 1962; Enzmann, 1973; Nishiyama, 1977). Recently, Sekellick and Marcus (1978, 1979) have proposed that induction of interferon may occur by DI particles or by ts mutants and that these dual factors may play a synergistic role in virus persistence. The involvement of DI particles as a mediator of persistent infection has been demonstrated in model systems using numerous animal viruses (see Reichmann and Schnitzlein, 1979; Andzhaparidze *et al.*, 1981; De and Nayak, 1980; Stollar, 1979).

Although it is known that many herpesviruses cause persistent infections in man and animals and that severe disease may result or be associated with these infections in the case of certain viruses such as EBV, HSV, and human cytomegalovirus, very little is known about the exact events at the molecular and cellular levels responsible for maintenance of the persistency. Recently, factors responsible for maintaining the state of persistent infection have been examined in the EHV-1 persistently infected cell-culture system (Dauenhauer *et al.*, 1982). Initial experiments were conducted to ascertain whether the DI-1 to DI-4 cells produced DI particles that may function in maintaining the state of virus persistency as reported for other animal viruses.

Three experimental approaches were employed to assay for the continued production of DI particles by these persistently infected cells (Table IX). Initially, interference assays using supernatant fluids of DI cell lines were performed in L-M cells, and the results revealed that virus released from three of the DI cell lines tested could inhibit standard virus replication by 21–81%. Since the conditions of the assay were quite stringent—2 ml DI cell supernatant were employed to inhibit 2×10^8 cells infected at a multiplicity of 2 PFU/cell—it is evident that the virus released from these persistently infected cells has significant interference capacity. Next, the DNA present in the particles released by the DI-1 to DI-4 cells was examined for defective DNA by restriction-endonuclease digestion. DNA extracted from particles purified from 3 liters of supernatant fluid of each cell type was found to contain the large *Bgl*II supramolar A fragment unique to the Class I defective genome (see Fig. 6). This fragment that comprises the bulk of the defective DNA was a major component of the DNA of viral particles being produced by the DI-1 to DI-4 cells. Thus, these data confirm the presence of DI particles in the virus being released by these persistently infected cell lines.

Last, CsCl isopyknic analyses of these viral DNA preparations in the

TABLE IX. Detection of DI Particles in Virus Released From EHV-1 Persistently Infected–Transformed Cell Lines DI-1 to DI-4

Cell line	Interference with standard EHV-1 (%)[a]	Presence of 35 md *Bgl*II Fragment[b]	Presence of 1.724 g/cm³ DNA[c]
DI-1	62	+	+
DI-2	81	+	+
DI-3	—	+	+
DI-4	21	+	+

[a] Interference assays involved treatment of 1×10^8 L-M cells with medium (control) or 2 ml supernatant from the DI-1 to DI-4 cells. After a 1-hr attachment period, the cells were collected, washed, and challenged with standard virus at a multiplicity of 2 PFU/cell. Total virus produced by 48 hr postinfection was assayed by plaque titration. The interference assay data indicate the percentages of inhibition of standard EHV-1 replication obtained at 48 hr postinfection in L-M cells pretreated with DI-cell supernatants prior to standard virus challenge as compared to virus titers of L-M cells mock-infected prior to challenge with standard virus.

[b] Viral DNA extracted from particles released from each of the four DI cell lines was digested with *Bgl*II, and the resultant fragments were analyzed by agarose gel electrophoresis. All these viral DNAs were found to contain a 35-md fragment unique to DI-rich DNA preparations digested with *Bgl*II (see Fig. 6).

[c] Density analyses of DNAs contained in viral particles relesed from each DI cell line were performed in the Model E analytical ultracentrifuge. Virus released from all four cell lines was found to contain both standard EHV-1 DNA (ρ = 1.716 g/cm³) and a small amount (<10%) of the defective DNA species (ρ = 1.724 g/cm³).

analytical ultracentrifuge revealed the presence of the 1.724 g/cm³ species of defective DNA in all four DNA preparations (Table IX). Quantitation of the percentage of DI particle DNA in the released virus by measuring the relative amounts of each density species by laser scanning techniques indicated that a significant amount of the defective DNA was present in all four samples and that it comprised approximately 3–7% of total DNA.

Since the production of significant levels of interferon or ts mutants or both has been shown to function in concert with DI particles in some virus systems to maintain the persistent state, DI-1 to DI-4 cells were assayed for these possible mediators of persistency (Table X). As judged by employment of the VSV–HE cell assay system, which is quite sensitive in detecting the antiviral activity of interferon, the DI-1 to DI-4 cell supernatants did not produce significant quantities of interferon. The production of EHV-1 ts mutants by these persistently infected cells was also ruled out (Table X). Several series of experiments were conducted in which the DI cells *per se* or samples of virus released from these cells were cultivated at various temperatures and the virus titer was assayed at these same temperatures. Results from all these experiments failed to indicate the presence of EHV-1 ts mutants in the DI cell lines. At all temperatures, virus released from the DI cells replicated in hamster cells to titers similar to those of standard virus. Also, cultivation of the DI cells at various temperatures did not result in major differences in the level of virus production as compared to titers of standard virus produced in HE cells at these same temperatures. Therefore, all findings to date

TABLE X. Mechanism of Maintenance of EHV-1 Persistent Infection: Failure to Detect Interferon and EHV-1 ts Mutants in DI-1 to DI-4 Cell Lines

Interferon assay[a]: Assay for antiviral activity in VSV–hamster cell system

VSV titer (PFU/ml)	
Untreated	Experimental
6.9×10^5	Hamster interferon-treated: 4.2×10^2
6.9×10^5	DI-1 supernatant-treated: 6.5×10^5
6.9×10^5	DI-2 supernatant-treated: 7.8×10^5
6.9×10^5	DI-3 supernatant-treated: 3.1×10^5
6.9×10^5	DI-4 supernatant-treated: 9.5×10^5

ts Mutant assay[b]: Effect of temperature on EHV-1 production

	Temperature of virus growth and titration		
Source of virus	33°C	37°C	39°C
Standard EHV-1	6.3×10^4	5.1×10^5	4.3×10^4
DI-1 cells	4.3×10^5	2.5×10^5	2.8×10^5
DI-2 cells	3.5×10^5	4.9×10^5	4.1×10^5
DI-3 cells	1.9×10^4	8.2×10^4	1.9×10^4
DI-4 cells	7.8×10^4	2.0×10^5	7.6×10^3

[a] HE cell cultures were incubated with either clarified (80,000*g*, 1 hr) supernatants of normal HE cells, or hamster interferon, or clarified supernatants from DI-1 to DI-4 cell lines for 15 hr at 37°C. Cultures were washed once and were challenged with VSV at a multiplicity of 1 PFU/cell. Maximal VSV titers obtained after incubation for 24–48 hr are shown.

[b] Primary HE cell cultures were infected at an input multiplicity of 0.1 PFU/cell with either standard EHV-1 (produced in primary HE cell cultures) or EHV-1 released from DI-1 to DI-4 persistently infected cell lines. Cultures were incubated at the temperatures indicated, and maximal virus titer was determined by plaque assays conducted at the same temperature.

indicate that the persistency established by infection with defective virus has been maintained and that these cells continue to produce significant quantities of EHV-1 DI particles that may function in the maintenance of the persistent infection.

The failure to detect a subpopulation of DI particles in the DI-5 to DI-10 cells described by Robinson *et al.* (1980c) could have been influenced by one or more of three factors: (1) The DI-1 to DI-4 persistently infected cell lines were established by infection with DI particles that had been propagated for 23 consecutive passages, whereas those established previously were transformed by DI particles generated within only 7 serial passages (Robinson *et al.*, 1980c). Our earlier findings (Henry *et al.*, 1979) suggested that with continued passage, EHV-1 DI particles became "more defective"; that is (a) the magnitude of the cyclic fluctuations in infectious virus titer increased with continued passage; (b) the number of particles required to cause a specific reduction in virus titer decreased with continued passage; and (c) the genomes of the DI particles became genetically less complex with continued passage. Thus, multiply passaged DI particle preparations may contain particles the genome of which has evolved to a stable genetic structure that has a selective advantage for replication and thus for survival and continued production in persistently infected cells. (2) The second factor that may account for the lack of DI particle production in the DI-5 to DI-10 cell lines is that those cultures arose from selected dense foci that grew rapidly after infection. These cells may not have contained DI particles, or, possibly, they lost the capacity to produce DI particles in the selection process, whereas DI-1 to DI-4 persistently infected–transformed cells were propagated from the total primary hamster cell culture and these conditions may have allowed the selection of cells that produce DI particles. (3) Last, it is possible that the DI-5 to DI-10 cells produce small quantities of defective DNA that is not packaged into particles and released from these cells. All assays for defective particles in the DI-5 to DI-10 cells have analyzed particles released into the supernatant fluids of these cultures. Experiments to examine the intracellular viral DNA for the defective DNA species must be undertaken before defective virus can be ruled out as a mediator in maintaining the persistent infection in these cells. It should be noted that preliminary experiments failed to detect significant levels of interferon or EHV-1 ts mutants in the DI-5 to DI-10 cultures.

E. Coestablishment of Persistent Infection and Oncogenic Transformation by High-Multiplicity Infection with Equine Cytomegalovirus

HE cell cultures infected with ECMV (EHV-2) at multiplicities of 1–10 *PFU*/cell do not exhibit cytopathic effects or support ECM replication at levels detectable by plaque assay in permissive equine cells (O'Callaghan *et al.*, 1978; Wharton *et al.*, 1981). However, infection of primary HE cell cultures with ECMV at very high multiplicities of infection (1500 particles/cell) resulted in the altered cell responses of persistent infection and oncogenic transformation (O'Callaghan *et al.*, 1981a; Staczek, Wharton, Dauenhauer, and O'Callaghan, in prep.). Foci of HE cells resulting from ECMV infection under these conditions were picked and used to establish cell lines. Three of these cell lines have been studied, and some of their key properties are listed in Table XI. Cell lines EC-1, EC-2, and EC-3 display all the biological properties of oncogenically transformed cells: immortality, growth to high saturation density, reduced serum requirement, short generation time, and capacity to form colonies in soft agar. The cells were shown to be highly oncogenic in newborn or adult syngeneic animals, and tumors developed at the site of inoculation within 4 weeks or less. The tumors metastasized to several major organs and have been shown to be transplantable. Tumor-cell lines, designated as EC-1T, -2T, and -3T, have been established, and all have been shown to be virus nonproducers. Numerous attempts to detect or induce virus in the tumor-cell lines by a variety of techniques have been unsuccessful.

TABLE XI. Biological and Biochemical Properties of Persistently Infected and Oncogenically Transformed Hamster Embryo Cells Established by High-Multiplicity Infection with ECMV[a]

Cell	Immortality[b]	ECMV (PFU/ml)[c]	Cells persistently infected (%)[d]		Resistant to superinfection?[e]		Oncogenic in hamster[f]	ECMV DNA copies/cell[g]	Cells expressing ECMV antigens (%)[h]
			IC	EM	ECMV	HSV-1			
Normal HE	No	0	0	0	No	No	No	No	0
EC-1	Yes	0	0	0	Yes	No	Yes	<0.05	+75%
EC-2	Yes	$>1 \times 10^4$	2%	0.5%	Yes	No	Yes	0.5	+75%
EC-3	Yes	$>1 \times 10^4$	2–5%	0.8%	Yes	No	Yes	15–49	+75%
EC-1T	Yes	0	0	0	Yes	No	Yes	<0.05	+75%
EC-2T	Yes	0	0	0	Yes	No	Yes	<0.05	+75%
EC-3T	Yes	0	0	0	Yes	No	Yes	<0.05	+75%

[a] Primary HE cell cultures were infected with ECMV at an input multiplicity of 1500 particles/cell. Foci of transformed cells were picked and maintained in culture.
[b] Control HE cells did not survive passage after 8–12 subcultivations; all EC and EC-T cell lines continue to replicate after more than 2 years in culture.
[c] ECMV was assayed in equine cells as described by Wharton *et al.* (1981).
[d] Infectious-center (IC) assays were carried on in permissive equine cells. In electron-microscopic (EM) assays, numerous thin sections containing several thousand cells were examined.
[e] Cells at various passages were challenged with 1–20 PFU/cell of ECMV or HSV-1. "Yes" indicates failure of cells to exhibit alterations in ECMV titer, cellular morphology, and cell growth; "No" indicates marked cytopathology and eventual cell lysis.
[f] Cells, 1×10^5 to 2.5×10^6, were injected subcutaneously. "Yes" indicates development of metastatic fibrosarcomas within 6 weeks.
[g] DNA extracted from HE (control), EC, or EC-T cells was allowed to reanneal in the presence of ^{32}P-labeled ECMV DNA ($2–6 \times 10^8$ cpm/μg) at molar ratios of 50–500 (cell/viral DNA). Kinetics of duplex formation, measured by hydroxylapatite chromatography, were analyzed by linear regression and compared to reconstruction data to determine ECMV equivalents/cell.
[h] Immunofluorescence was assayed by the indirect method using antisera to ECMV-infected cell extracts and fluoroscein-isothiocyanate-conjugated anti-IgG.

Two of the transformed cell lines, EC-2 and EC-3, have been shown to contain a population of persistently infected cells, although only a small portion (approximately 2%) of the total cells release infectious ECMV or contain particles with herpesvirus morphology (Table XI). The released virus has been shown to be ECMV by several criteria, such as host range, growth properties, and restriction-enzyme digestion of the viral DNA. Titers of supernatant virus fluctuate in these two cell lines, but titers range from 1×10^4 to 1×10^5 PFU/ml, which equates to approximately 1–20 PFU of released virus per producer cell. The two cell lines have continued to produce virus since their establishment, and the percentage of producer cells has not varied during the life-span of the cultures. Thus, these HE cell cultures are persistently infected with cytomegalovirus and a steady state of virus production without significant cytopathology has developed. Interestingly, these cells are resistant to superinfection with either ECMV or EHV-1, but not to superinfection with HSV-1 or VSV. Superinfection with homologous virus (ECMV) or EHV-1, the genome of which has less than 5% homology with ECMV (see below), does not alter the state of persistent infection. The mechamism of this resistance to ECMV and EHV-1 is not known, but it is not at the level of early events, since these equine herpesviruses enter the cells and are uncoated as judged by electron-microscopic examination.

The mechanism(s) for initiation and maintenance of this ECMV persistency in hamster cells is not known. Although the role of ECMV DI particles cannot be ruled out with certitude, virus preparations used to establish the EC-cell systems were comprised of virions that had been plaque-purified, passaged only at very low (<0.01 PFU/cell) multiplicities of infection in equine cell types, and shown to contain a single density species of viral DNA (1.7165 g/cm^3). Obviously, the nonpermissiveness of hamster cells for this cytomegalovirus of limited host range (Wharton *et al.*, 1981) may be overcome, at least in a small portion of these cells, by high-multiplicity infection, whereas abortive replication of ECMV occurs in the majority of the cells, resulting in oncogenic transformation.

DNA reassociation analyses of the two ECMV persistently infected cell lines revealed that EC-2 and EC-3 cell cultures contain copies of the viral genome in amounts of 0.5 and 15–49 copies per cell, respectively. When these values are corrected for the number of producer cells, it can be estimated that there are at least several hundred genome copies per EC-3 producer cell and more than 20 genome copies per EC-2 producer cell. Repeated hybridization analyses of the DNA extracted from the nonproducer, transformed EC-1 cell line and from all tumor-cell lines have indicated that a small portion of the ECMV genome is present in small amounts in these cells. In support of these findings, recent studies using the indirect immunofluorescent-antibody assay have detected ECMV antigens in the overwhelming portion of all cells of the EC and EC tumor-cell lines, indicating that ECMV DNA sequences present in these cells are expressed. In addition, small levels of antibody to ECMV proteins have been detected very recently in the sera of tumor-bearing animals, confirming that these ECMV-transformed cells possess viral antigens (Staczek, Dauenhauer, and O'Callaghan, in prep.).

VI. ONCOGENIC POTENTIAL OF EQUINE HERPESVIRUSES

As discussed in the preceding sections, investigations in our laboratories and those of our colleagues have shown that equine herpesviruses offer model systems to examine the biochemical and biological parameters of herpesvirus cytocidal infection, abortive infection, persistent infection, and biochemical transformation. Since the initial report that a herpesvirus is the causative agent of a naturally occurring lymphoproliferative disease in chickens known as Marek's disease (Churchhill and Biggs, 1967), a growing body of literature has indicated that herpesviruses may play a role in several neoplastic diseases of man and animals and that investigation of herpesvirus-transformed cells may provide basic information of the molecular events associated with and necessary for oncogenesis. In recent years, we have demonstrated that all three equine herpesviruses possess the potential to oncogenically transform mammalian embryo cells and that EHV tumor models developed in our laboratory are suitable to investigate basic questions concerning the role of viral genes and gene products in the oncogenic process (O'Callaghan *et al.*, 1977b, 1981,b; Allen *et al.*, 1977c, 1978d; Robinson *et al.*, 1980a–c, 1981a,b; Atherton *et al.*, 1980, 1981; Robinson and O'Callaghan, 1981, 1983; Dauenhauer *et al.*, 1982; Robinson and O'Callaghan, in prep.; Staczek, Wharton, Dauenhauer, and O'Callaghan, in prep.). Since these three herpesviruses share DNA sequences that account for less than 10% of the total genome, it should be possible to ascertain whether DNA sequences associated with oncogenic transformation are type-unique or type-specific and whether the interactions of EHV genes and gene products responsible for oncogenic transformation are part of a common pathway for these three herpesviruses that differ in biological and biochemical properties.

A. Properties of Equine Herpesvirus Type 1 Oncogenically Transformed Cells

Three models of EHV-1 oncogenically transformed and tumor cells, each resulting from a different type of virus–cell interaction, have been established and investigated in our laboratory (Table XII):

1. Oncogenically transformed hamster embryo (HE) cell lines were established by infection of permissive, primary LSH HE cells with UV-irradiated standard EHV-1. These cell lines, designated LSEH cells, are oncogenic in the LSH hamster, and the highly malignant fibrosarcomas

TABLE XII. Biological and Biochemical Properties of Primary Hamster and Mouse Embryo Cells Oncogenically Transformed by EHV-1

Oncogenic transformation conditions	Embryo cell type	Cell line or tumor tissue	Cells producing EHV-1 (%)[a]	Resistant to EHV-1 superinfection[b]	Oncogenic in animal[c]	EHV-1 genome (%)[d]	Genome equivalents per cell[d]	Cells expressing EHV-1 antigens (%)[e]
Infection with UV-irradiated standard EHV-1	Hamster	LSEH-3	0	Yes	Yes	3.8	1.0	+75
	Hamster	LSEH-4	0	Yes	Yes	20.9	2.2	+75
	Hamster	LSEH-8	0	Yes	Yes	3.3	0.3	+75
	Hamster	LSEH-3T	0	Yes	Yes	2.9	1.2	+75
	Hamster	LSEH-4T	0	Yes	Yes	5.1	2.5	+75
	Hamster	LSEH-8T	0	Yes	Yes	2.2	3.0	+75
	Hamster	Tumor tissue 4T	—	—	Yes	12.2	4.0	+ + + +
	Hamster	Tumor tissue 8T	—	—	Yes	2.2	3.0	+ + + +
Infection with standard virus or transfection with EHV-1 DNA	Mouse	BALB/c ME	0	Yes	Yes	8–15	1–4	+40
	Mouse	BALB/c ME-T	0	Yes	Yes	8–15	1–4	+40
Infection with viable EHV-1 virus containing defective interfering (DI) particles	Hamster	DI-5	3–7%	Yes	Yes	100	25.3	+75
	Hamster	DI-8	4–11%	Yes	Yes	100	26.6	+75
	Hamster	DI-10	13–20%	Yes	Yes	100	43.7	+75
	Hamster	DI-5T	0	Yes	Yes	42.2	2.2	+75
	Hamster	DI-7T	0	Yes	Yes	44.2	2.2	+75
	Hamster	DI-8T	0	Yes	Yes	42.6	3.1	+75
	Hamster	DI-10T	0	Yes	Yes	56.7	3.8	+75

[a] The percentage of cells producing virus was measured by infectious-center assay and by electron-microscopic examination of thin sections in which 1000 or more cells were scored.

[b] All cells failed to exhibit cyopathology, to succumb to cell lysis, and/or to produce virus (or produce larger quantities of virus in the case of the persistently infected cell lines) after challenge with 1–20 PFU/cell of EHV-1.

[c] Fibrosarcomas developed in 100% of animals within 4–6 weeks at the site of inoculation of 1×10^5 to 2.5×10^6 cells.

[d] DNA-reassociation hybridization analyses were carred out by reannealing denatured cell DNA in the presence of ^{32}P or ^{125}I (alpha dNTP)-labeled EHV-1 DNA (2–8×10^8 cpm/μg) at molar ratios of 50–500 (cell/viral DNA). Calculations of "EHV-1 genome (%)" and "Genome equivalents per cell" were made as described by Robinson *et al.* (1980a,c).

[e] Determined by indirect immunofluorescence assays using antisera to EHV-1 structural or nonstructural proteins or both. (+ + + +) Denotes that virtually all cells in sections of tumor tissues exhibited viral antigen.

induced by inoculation of these cells are transplantable and have been used to establish a number of tumor-cell lines (LSEH-T cells) (O'Callaghan *et al.*, 1977b, 1981a,b; Robinson *et al.*, 1980a,b, 1981a,b; Robinson and O'Callaghan, 1981, 1983).

2. Oncogenically transformed mouse embryo cell lines were established either by infection of nonpermissive, primary BALB/c mouse embryo cell cultures with viable standard EHV-1 or by transfection of these cells with EHV-1 DNA preparations shown to contain infectious DNA as judged by its transfectivity for permissive equine cells. These cell lines (ME cells) are oncogenic in BALB/c mice, and the metastatic fibrosarcomas that developed are transplantable and have been employed to establish tumor cell lines (ME-T cells) (Allen *et al.*, 1977c, 1978d, Baumann and O'Callaghan, unpublished).

3. Additional series of oncogenically transformed HE cell lines were established by infection of primary HE cultures with viable EHV-1 preparations that had been passaged serially at high multiplicities of infection and were shown to be comprised of standard virus and DI particles (+65%). These cell lines, designated as DI-1 to DI-4 and as DI-5 to DI-10, were comprised, respectively, of a minor population of persistently infected cells and a major population of oncogenically transformed cells (see Table V). All the DI cell lines were oncogenic in adult hamsters, and the resulting metastatic sarcomas were transplantable. Tumor-cell lines, designated as DI-T cells, have been established for all cell lines. In addition, numerous oncogenically transformed cell lines have been established by cloning individual nonproducer cells from the mixed cell population of the DI cell lines (Vance *et al.*, 1983; Robinson *et al.*, 1980b,c, 1981a,b; O'Callaghan *et al.*, 1981a,b; Dauenhauer *et al.*, 1982).

All these EHV-1-transformed cell lines have been shown to exhibit the biological properties typical of virus-transformed cells: immortality, growth to high saturation density, reduced requirement for serum for normal growth, short generation time, ability to form colonies in soft agar or methylcellulose, resistance to superinfection with homologous virus, aneuploid karyotypes, and oncogenicity in syngeneic animals. In the case of the DI cell lines that contain a population of persistently infected cells, cell lines established from the nonproducer, transformed cell population by cloning methods exhibited all of the aforementioned biological properties of virus-transformed cells. In all cases, the tumor-cell lines derived from tumor tissues in the hamster (LSEH-T cell lines and DI-T cell lines) or mouse (ME-T cell lines) were shown to exhibit these same key properties and to cause metastatic fibrosarcomas when inoculated in the appropriate animal.

The role of EHV-1 in the initiation and maintenance of the oncogenic transformed state was supported by the demonstration that all transformed and tumor-cell lines of the three tumor systems harbor EHV-1

DNA sequences and continue to do so after repeated passage in cell culture or in the animal (tumorigenesis) (Table XII). In all cases, DNA hybridization analyses conducted under stringent reannealing conditions that allow the detection of viral sequences only 2.8 kilobases (kb) in size revealed the presence of subgenomic EHV-1 DNA sequences. LSEH transformed and tumor cells as well as these tumor tissues harbored only a few copies of viral sequences equating to 2.2–20.9% of the whole genome. Interestingly, the LSEH-4 transformed cells contained 20.9% of the genome in amounts that averaged approximately 2.2 genome copies/cell; however, the complexity of EHV-1 DNA sequences was less in the tumor tissues (12.2%) and tumor-cell lines (5.1%), but these viral sequences were present in larger amounts. These findings would indicate that some viral sequences are lost with passage of these cells while other sequences are retained and increase in amount. This interpretation of the data was substantiated in later studies (see below).

As expected, the DI cell lines that are comprised of a subpopulation of producer, persistently infected cells contained copies of the entire EHV-1 genome in large amounts. However, analyses of the transformed cells established by cloning techniques and of the nonproducer tumor cell lines showed that the oncogenically transformed cell type contained a few copies per cell of viral sequences equating to approximately 40–60% (Table XII). Preliminary hybridization investigations of these cells passaged for more than 30 subcultivations after establishment from tumor tissues indicate that EHV-1 DNA sequences are retained but that the complexity of these EHV-1 sequences decreases, suggesting that certain sequences are lost or become reduced to amounts not detectable by these hybridization assays.

Subgenomic amounts (8–15%) of EHV-1 DNA in low copy numbers (1–4 copies/cell)were detected by Cot analysis in BALB/c mouse embryo cells transformed by EHV-1 (virus or DNA) and in progeny tumor-cell lines. Interestingly, the amount and the complexity of viral sequences retained in several independently transformed mouse embryo cell lines did not vary significantly, and no obvious correlation could be found between the method of transformation—infection or transfection—and the amount of viral DNA harbored. Biological assays attempting to detect virus or to induce virus production by a variety of means have failed, supporting the hybridization findings that only subgenomic DNA sequences are present in these cells.

Recent immunological, biological, and biochemical investigations of the EHV-1-transformed and tumor cells of all three systems have provided evidence that viral DNA sequences harbored by these cells are expressed. Three independent lines of evidence permit this conclusion:

1. *Viral antigens can be detected by immunofluorescence.* Viral antigens were detected in all EHV-1-transformed and tumor-cell lines by

indirect immunofluorescence using antisera to EHV-1 structural or nonstructural proteins or both. Viral proteins were found on the surface of the transformed and tumor cells of all three models; in the two hamster cell transformation systems (LSEH, LSEH-T, and DI-T cells), viral antigens were also detected in the cytoplasm, especially in the perinuclear region. Positive immunofluorescence was obtained with rabbit antisera to EHV-1-infected cell extracts, hyperimmune anti-EHV-1 mouse or hamster sera, and tumor-bearer sera of hamsters inoculated with LSEH, LSEH-T, or DI-T cells. Absorption of these antisera with normal cells, chemically transformed cells, and/or SV40-transformed cells removed antibody that reacted weakly with normal mouse or hamster cells, but did not reduce the intense fluorescence obtained for the transformed and tumor cells.

Very recently, we (Caughman, Dauenhauer, and O'Callaghan, unpublished) have observed that EHV-1-transformed and tumor hamster cells, but not normal HE cells, react with an antiserum containing antibodies to the herpes simplex virus 2 (HSV-2)-infected-cell-specific protein 11/12 (ICSP 11/12) (provided by Dr. K.L. Powell). This HSV-2 nonstructural protein has been shown to be a major DNA-binding protein that is normally present in nuclei of HSV-2-infected cells, but is expressed in the cytoplasm of cells of certain human cancer tissues (see Kaufman *et al.*, 1981). In the case of the EHV-1-transformed cells, the immunofluorescence indicates that the protein(s) reacting with this antiserum is located in the cytoplasm. These findings, considered in light of the observation (Yeo *et al.*, 1981; K.L. Powell, personal communication) that EHV-1 codes for a protein that shares antigenic determinants with the HSV-2 ICSP 11/12 protein, suggest a possible role of these related proteins in herpesvirus transformation.

2. *Tumor-bearer serum contains anti-EHV-1 antibody.* Assay of the sera of hamsters bearing sarcomas induced by inoculation of LSEH, LSEH-T, and DI-T cells—all being comprised only of nonproducer, transformed cells—has revealed that very significant levels of EHV-1 neutralizing antibody are present. ND_{50} titers ranged from 1:4 to greater than 1:100 in the case of hamsters inoculated with the three cell types (Robinson *et al.*, 1980a,c).

3. *Polypeptides of transformed cells are immunoprecipitated by anti-EHV-1 specific sera.* Preliminary results from experiments in which solubilized proteins, labeled *in vivo* with [^{35}S]-methionine, of LSEH and LSEH-T cells were immunoprecipitated with antisera specific to EHV-1 polypeptides revealed that the immunoprecipitates contained proteins with electrophoretic mobilities identical to those of several EHV-1 coded polypeptides. Immunoprecipitates of transformed cell extracts incubated with normal hamster sera and immunoprecipitates of normal cell extracts incubated with anti-EHV-1 sera failed to display these polypeptides when analyzed by PAGE (Caughman and O'Callaghan, unpublished).

B. Identification of Equine Herpesvirus Type 1 DNA Sequences Integrated in LSEH Transformed and Tumor Hamster Embryo Cell Lines

To ascertain the identity and arrangement of EHV-1 DNA sequences present in the viral-transformed and tumor-cell lines, blot hybridization analyses using a library of cloned EHV-1 restriction-enzyme fragments as probes were undertaken to identify viral sequences in host DNA fragments generated by digestion with *Eco*RI, *Bgl*II, *Xba*I and *Bam*HI (Robinson *et al.*, 1981b; Robinson and O'Callaghan, 1981, 1983; O'Callaghan *et al.*, 1981a,b). Selection of the probes (Table XIII) and the enzymes for this experimental approach was based on the following lines of reasoning: (1) These cloned restriction fragments represent highly purified viral sequences of known map positions. (2) The coding capacities of several of the fragments overlap and thus allow more definitive mapping of viral sequences within the cell. (3) The fragments encompass DNA sequences from the entire EHV-1 genome and allowed detection of all possible viral sequences. (4) The fragment sizes are within the limits demarcated by reconstruction experiments for detection of a 2-md sequence of EHV-1

TABLE XIII. Cloned EHV-1 Restriction Fragments Used in Blot Hybridizations to Detect Viral Sequences Present in EHV-1-Transformed and Tumor Cells

Fragment[a]	Mol. wt. ($\times 10^6$)	Kilobases	Map units
1. *Bam*HI-E[b]	6.2	9.3	0.00–0.06
2. *Bam*HI-H[b]	4.3	6.5	0.07–0.11
3. *Bgl*II-D[b]	7.3	11.0	0.08–0.16
4. *Bam*HI-N[b]	3.2	4.8	0.15–0.19
5. *Bam*HI-D[b]	7.2	10.8	0.19–0.27
6. *Eco*RI-F[b]	8.2	12.3	0.21–0.30
7. *Bam*HI-F[b]	5.4	8.1	0.27–0.32
8. *Bam*HI-B[b]	9.8	14.7	0.32–0.43
9. *Bgl*II-B[b]	9.2	13.8	0.38–0.48
10. *Bam*HI-A[b]	13.7	20.6	0.48–0.63
11. *Xba*I-J[c]	2.9	4.4	0.52–0.55
12. *Bgl*II-C[b]	8.3	12.5	0.57–0.66
13. *Bgl*II-L[b]	2.4	3.6	0.66–0.68
14. *Bam*HI-K[b]	3.6	5.4	0.67–0.71
15. *Bgl*II-J[b]	3.8	5.7	0.69–0.73
16. *Eco*RI-B[b]	12.4	18.6	0.73–0.87
17. *Eco*RI-M[b]	3.4	5.1	0.86–0.90
18. *Eco*RI-D[c]	9.1	13.7	0.90–1.00

[a] All restriction fragments were ^{32}P-labeled by nick translation and denatured at 110°C for 10 min in a 0.3 N NaOH solution followed by neutralization prior to use in hybridizations.
[b] EHV-1 DNA restriction fragments were cloned into the plasmid pBR322 and propagated in *Escherichia coli* strains HB101 and C600SF8 as described in Robinson *et al.* (1981b).
[c] The source of these fragments was from purified bands resolved by electrophoresis in agarose gels (Henry *et al.*, 1981).

DNA present in less than one copy per cell; DNA-reassociation analyses revealed that all EHV-1 cells contained DNA sequences in amounts and of a complexity above these limits (see Table XII). (5) Restriction-enzyme maps are available for the four enzymes used in this study—*Eco*RI, *Bgl*II, *Xba*I, and *Bam*HI. Therefore, digestion of cellular DNAs with each of these four enzymes and subsequent analyses of these blotted fragments for viral sequences with the library of 18 EHV-1 DNA fragments listed in Table XIII are instructive in determining the structural arrangment of EHV-1 sequences within these cells.

In the last year, blot hybridization analyses have been carried out for several different passages of the LSEH-3, -4, and -8 transformed cells, LSEH-3T, -4T, and -8T tumor cells, and some of the DI-T (tumor) cell lines. In all these experiments, blots of these cellular DNAs generated by digestion with *Eco*RI, *Bgl*II, and *Bam*HI were analyzed with all 18 probes (Table XIII) under conditions that permitted detection of a 2-md fragment of viral DNA present in less than one copy per cell (Robinson and O'Callaghan, 1981, 1983; O'Callaghan *et al.*, 1981a). Presentation of the enormous amount of data obtained from these investigations is beyond the scope of this chapter, and only the key findings will be summarized here:

1. All LSEH transformed cells and LSEH-T tumor cells contained limited portions of the EHV-1 genome, confirming results of the Cot hybridization analyses.

2. EHV-1 DNA sequences present in these cells were integrated into the hamster cell genome as colinear tracts of subgenomic viral sequences. That the viral DNA was integrated was indicated by the findings that in the cellular digestions, DNA fragments larger than the probe DNAs contained viral sequences, that the size and number of these large DNA fragments containing viral sequences varied for the four different enzymes used to restrict cell DNAs and varied for the different cell lines, that the size and number of fragments containing viral sequences detected by several of the probes were in excess of those predicted by cleavage of only viral DNA, and, last, that some bands containing viral sequences mapping at the terminus of a viral probe had a molecular weight unpredicted by digestion of only viral DNA.

3. LSEH-8 and LSEH-3 transformed cells contained EHV-1 sequences located only within map position 0.30–0.40 of the L region, since only the *Bam*HI-F (0.27–0.32) and *Bam*HI-B (0.32–0.43) probes hybridized to cellular DNA fragments. The patterns of these viral sequences in the LSEH-8 cells as revealed by digestion with the four restriction enzymes indicated that sequences mapping from approximately 0.32 to 0.38 are conjoined with two different sets of host flanking sequences, and thus at least two sites for integration exist in these cells. In the case of LSEH-3 cells, a similar situation existed, since integrated EHV-1 sequences also mapping at approximately 0.33–0.37 were detected; however, the two

sites of integration were different from those in the LSEH-8 cell. The pattern of integration of EHV-1 DNA sequences in the LSEH-4 transformed cells was quite complex when DNAs of cells at different passage levels were analyzed. DNA sequences mapping within the 0.30–0.40 portion of the viral genome were found in all LSEH-4 cells (Fig. 10); however, these cells harbored EHV-1 sequences detectable with several probes originating from other locations within the L region as well as from the S

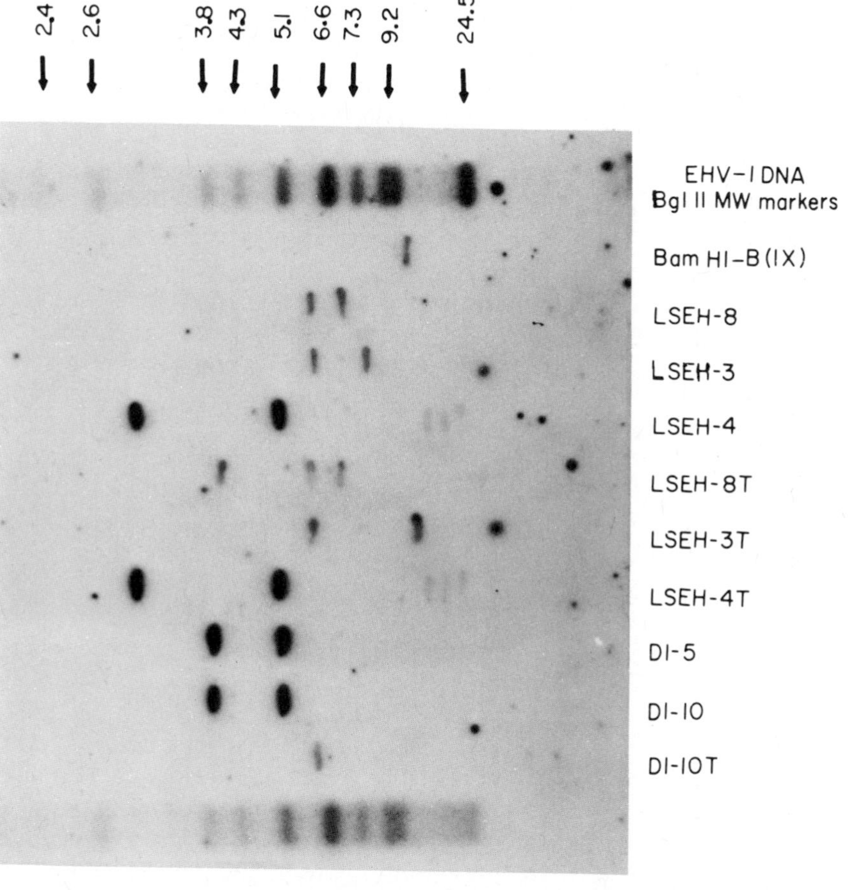

FIGURE 10. Integration of EHV-1 DNA sequences mapping within the *Bam*HI-B fragment (0.32–0.43) in all virus-transformed and progeny tumor-cell lines. EHV-1 DNA (marker) or cell DNAs (20 ng) were digested with *Bgl*II. The *Bam*HI-B cloned fragment (48.8 pg) was mixed with 20 μg hamster embryo DNA as control. Fragments were fractionated in 0.7% agarose gels and transferred to nitrocellulose filters by Southern blotting (Southern, 1975). *Bgl*II fragments hybridizing to the [^{32}P]-*Bam*HI-B fragment were detected by autoradiography as described by Robinson *et al.* (1981b). EHV-1-transformed (LSEH) and tumor (LSEH-T) cells were established by infection of HE cells with UV-irradiated virus (see Table XII); DI (transformed) and DI-T (tumor) cells were established by infection with viable defective interfering (DI) particle preparations (see Tables VIII and XII).

region of the EHV-1 chromosome. The blot hybridization data indicated that integrated copies of viral sequences originating from the 0.3 portion of the EHV-1 genome were present as three separate insertion units: two of these were organized as a single copy of the 0.26–0.42 sequence, and the third insertion was comprised of multiple copies of this sequence. Digestion of LSEH-4 cell DNA with enzymes that would cut within the 0.26–0.42 sequence generated EHV-1 fragments of a size indicating that these multiple copies in the third insertion site were arranged as colinear, tandem repeats. Also scans of the autoradiographs with a soft laser densitometer supported the interpretation that multiple copies were integrated within one DNA band, since the intensity of this fragment(s) generated by enzymes that would not cleave within the tandem repeat viral sequence was severalfold greater than the intensity of fragments containing the single insertion.

In addition to the 0.26–0.42 sequence, EHV-1 DNA sequences mapping at approximately 0.07–0.11 in the L region and at 0.79–0.82 in the S region of the EHV-1 genome were also detected as separate insertions of integrated sequences in LSEH-4 cells passaged for 25 subcultivations or fewer. When DNAs from highly passaged LSEH-4 cells were examined, only sequences mapping at 0.26–0.42 were detected (Fig. 11). These results indicate that excision of some viral sequences occurred with passage while amplification of other sequences took place, confirming the findings from DNA-reassociation studies discussed above (see Table XII). However, it is possible that a selection process(es) occurred and that the "deleted sequences" remain present in very small amounts in a subpopulation of the cells and/or that the amplified 0.26–0.42 sequence was present in some cells in early passages and these cells increased in proportion with passage.

4. Thus, all LSEH transformed cells examined to date harbor EHV-1 sequences mapping at approximately 0.32–0.38, and this sequence common to all cells is retained during prolonged passage (passage 125) (O'Callaghan *et al.*, 1981b; Robinson and O'Callaghan, 1981, 1983).

5. The hybridization analyses of the tumor-cell lines derived from the LSEH transformed cells revealed that the EHV-1 sequence mapping at approximately 0.3–0.4 is retained in an integrated state during passage of the transformed cells in the animal (tumorigenesis) and during passage of the tumor cell lines in culture. Briefly, it was found that this EHV-1 sequence increased markedly in amount, as judged by the intensities of the autoradiographs of the cell fragments harboring the viral DNA. The patterns of the cellular DNA bands harboring this sequence in the LSEH-8T cells indicated that tandem duplication of the integrated sequence had occurred within both integration sites and that some viral sequences located at the 0.31–0.33 position were deleted within both copies of each repeat. Thus, only the *Bam*HI-B probe (.032–0.43) hybridized to the cellular DNA bands of these tumor cells. In the LSEH-3T tumor cells, it was found that one of the two 0.33–0.37 sequences integrated in the

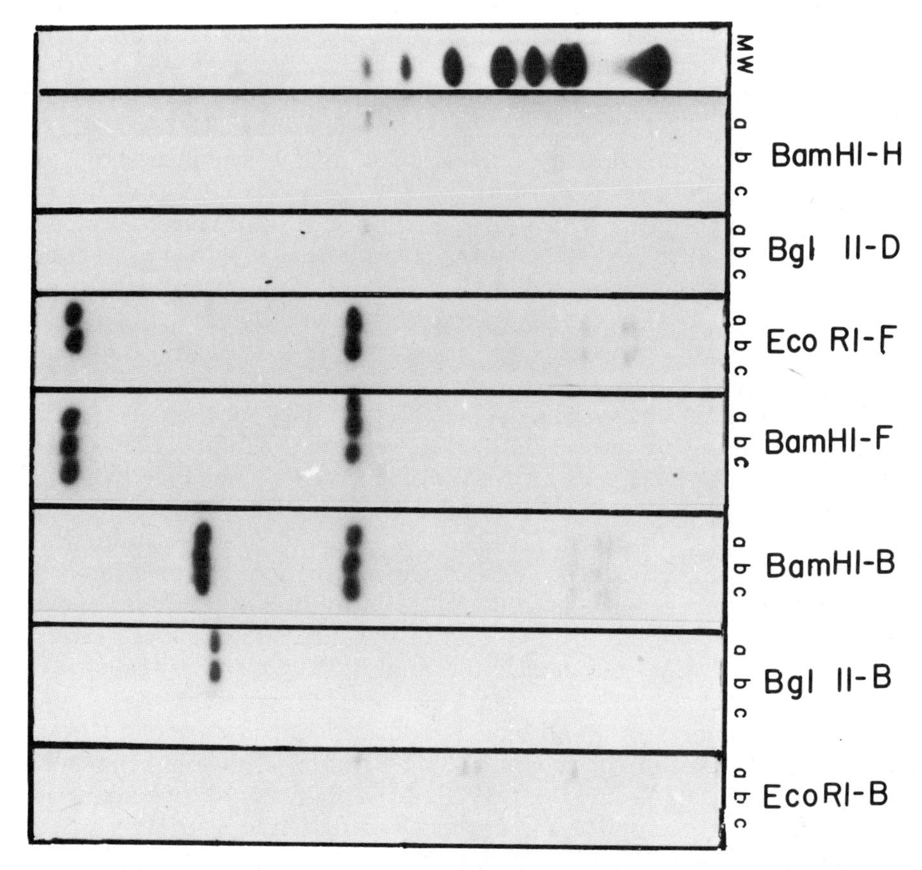

FIGURE 11. Retention of selective EHV-1 sequences as determined by blot hybridization of *Bgl*II digests of DNA from LSEH-4 transformed cells at different passages with ^{32}P-labeled cloned restriction fragments of EHV-1 DNA. DNAs from LSEH-4 cells at three levels of passage were analyzed: (a) passage 25; (b) passage 60–66; and (c) passage 125. The figure is a composite of separate hybridizations.

parental transformed cells had undergone a tandem duplication in a colinear fashion. In the case of the LSEH-4T cells, the copies of the 0.26–0.42 sequence that had become amplified with passage *in vitro* were found to have remained stably integrated during tumorigenesis, and the data indicated that additional amplification of these sequences had occurred during tumorigenesis.

The results of these analyses of the LSEH and LSEH-T cells clearly demonstrate that multiple sites, perhaps at random, exist for the integration of EHV-1 subgenomic sequences and that all transformed and tumor cells possess one or more copies of viral DNA sequences mapping colinearly from approximately 0.32–0.38 in the viral genome. This EHV-1 sequence common to all cells, termed the "consensus sequence," was found to be retained in an integrated state in all these transformed and tumor-cell lines (see Fig. 10) and to be amplified at some sites to a copy number severalfold greater than that found in the early passages of the transformed cells. Only DNA sequences mapping at this position within the L region of the EHV-1 genome were retained in these LSEH and LSEH-T hamster cells transformed by infection with UV-irradiated standard EHV-1.

That the consensus sequence was retained after prolonged passage of the cells in culture or in the animal as tumors was shown by experiments similar to that presented in Fig. 11. Blot hybridization analyses of *Bgl*II-restricted DNAs from LSEH-4 transformed cells at different passage levels were carried out with several probes of ^{32}-P-labeled EHV-1 cloned fragments that contained EHV-1 sequences detected in very early passages of these cells. As shown, viral sequences detectable by the *Bam*HI-F probe (0.27–0.32), *Bam*HI-B probe (0.32–0.43), and *Bgl*II-B probe (0.38–0.48) were retained in these cells at passage levels of 25 (lane a), 66 (lane b), and 125 (lane c). However on prolonged passage of the cells, sequences mapping within 0.07–0.11, 0.08–0.16, 0.21–0.30, 0.38–0.48, and 0.69–0.73 were lost or were reduced to levels below detection. The mechanisms for deletion of these sequences, for amplification of the consensus sequence, and for deletion of small portions of integrated viral sequences (possibly by a process responsible for amplification of other sequences) remain to be elucidated.

Results obtained from all blot hybridization analyses of the DNAs of these LSEH transformed cells and LSEH-T tumor cells using 4 enzymes to restrict cell DNAs and 18 EHV-1 restriction-enzyme fragments as probes have been compiled and organized as tentative restriction maps that show the identity and arrangement of the consensus sequence (Fig. 12).

C. Blot Hybridization Analyses of Defective Interfering Tumor Cells

Recently, experiments have been initiated to investigate EHV-1 DNA sequences harbored in the HE cells transformed by infection with EHV-1 preparations enriched for defective interfering (DI) particles. The experimental approach is similar to that employed for analyses of EHV-1 sequences in the LSEH tumor model. Initial results reveal that EHV-1 specific DNA sequences are detectable with many of the 18 probes (Fig. 13), confirming the DNA-reassociation findings that sequences equating to more than 40% of the EHV-1 genome are retained in these cells. For example, viral sequences mapping from approximately 0.30 to 0.45, 0.57 to 0.68 and 0.73 to 1.0 were detected in DI-5T cell DNA restricted with *Eco*RI. Since there are no *Eco*RI sites between 0.30 and 0.45 in the EHV-1 genome [see EHV-1 map (Fig. 2)], the 19.8- and 18.1-md fragments detected with the *Bam*HI-F, *Bam*HI-B, and BglII-B probes may represent the

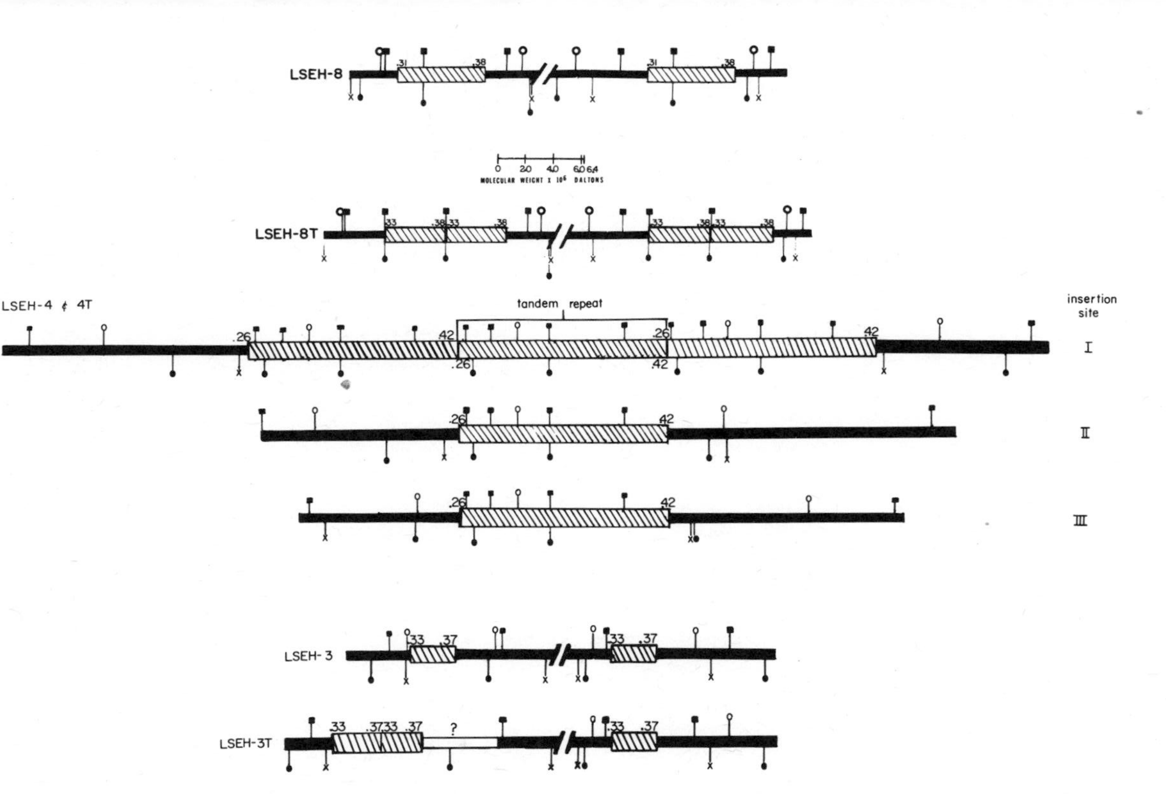

FIGURE 12. Restriction cleavage maps of integrated EHV-1 DNA sequences and flanking cellular DNA sequences of EHV-1 transformed and tumor HE cells. Maps were constructed from blot hybridization analyses of LSEH and LSEH-T cells DNAs digested with four different restriction enzymes and reacted with 18 ^{32}P-labeled restriction-enzyme fragments as probes (see Table XIII). Heavy solid lines represent cellular DNA. Restriction-enzyme sites are represented by the following symbols: (○) *Eco*RI; (■) *Bgl*II; (×) *Xba*I; (●) *Bam*HI.

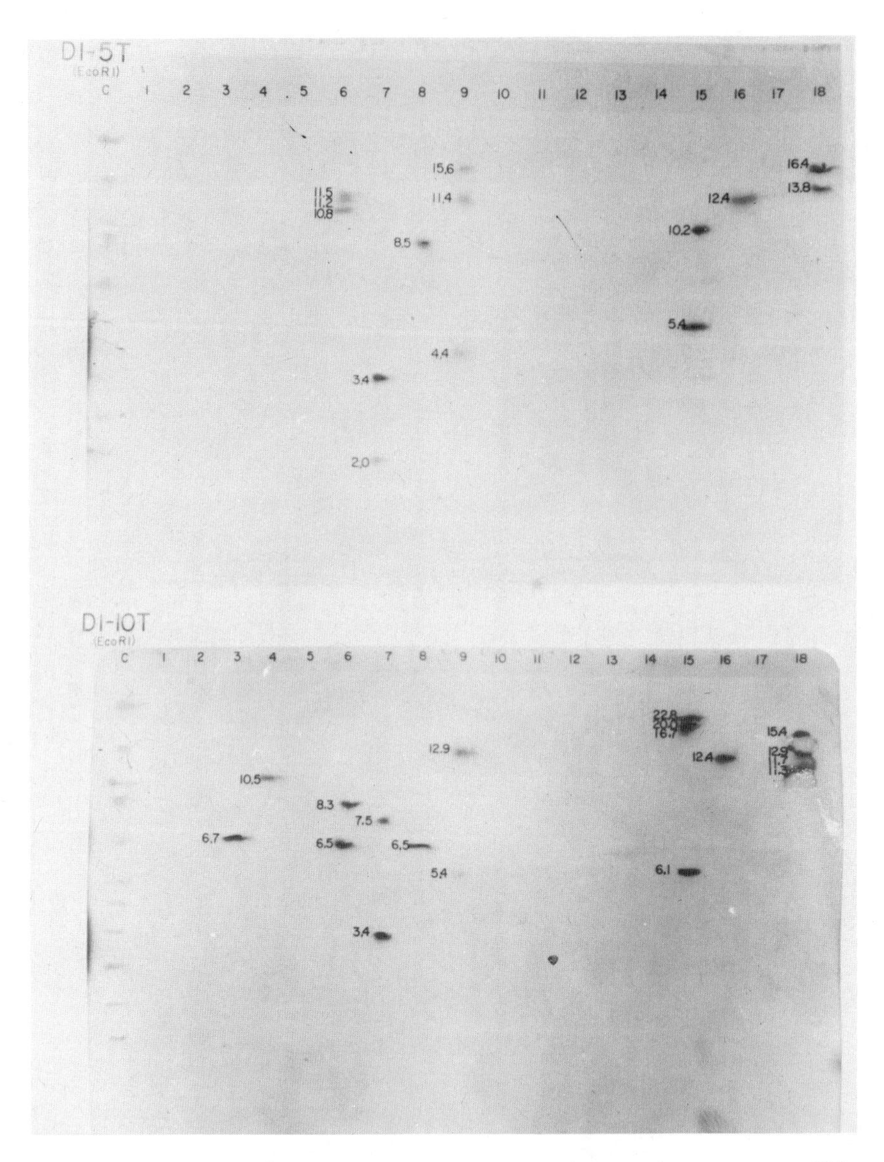

FIGURE 13. Blot hybridization analyses of *Eco*RI digests of DI-5 and DI-10 tumor-cell line DNAs with ^{32}P-labeled cloned restriction-enzyme fragments of EHV-1 DNA listed in Table XIII. Methods were similar to those described in the Fig. 10 caption.

termini of separate viral insertions. Of the three fragments that hybridized to the *Bgl*II-C probe (0.57–0.66), only the 1.3-md fragment was predicted from *Eco*RI digestion. A 4.2-md fragment was detected with both the *Bgl*II-L and *Bam*HI-K fragments. The results from these and additional hybridizations suggest that a viral sequence colinear to sequences in the genome at 0.57–0.68 may be arranged between sequences mapping at 0.45

and 0.73 in both viral insertions. Results from other blot hybridizations in which DI-T cell DNAs were restricted with other enzymes showed that cellular DNA bands were of a number and size that were consistent with the integration of these L-region viral sequences that encompassed the 0.32–0.38 consensus sequence. In all DI-T cells examined to date, sequences mapping within the 0.3–0.4 locus have been detected (see Fig. 10).

The results from hybridizations with probes derived from the EHV-1 S region revealed that these sequences were present in high-molecular-weight bands of cellular DNA, suggesting integration of S-region DNA sequences. The presence of S-region fragments in the DI-T cells is not suprising if one considers that the transforming virus preparation in this tumor model was comprised of approximately 70% defective DNA, which is composed of reiterated copies of S-region sequences, primarily those located in the inverted repeats. Last, it must be remembered that the finding that these sequences appear to be covalently linked to those of cellular DNA does not prove that this integration occurs within high-molecular-weight, chromosomal cellular DNA.

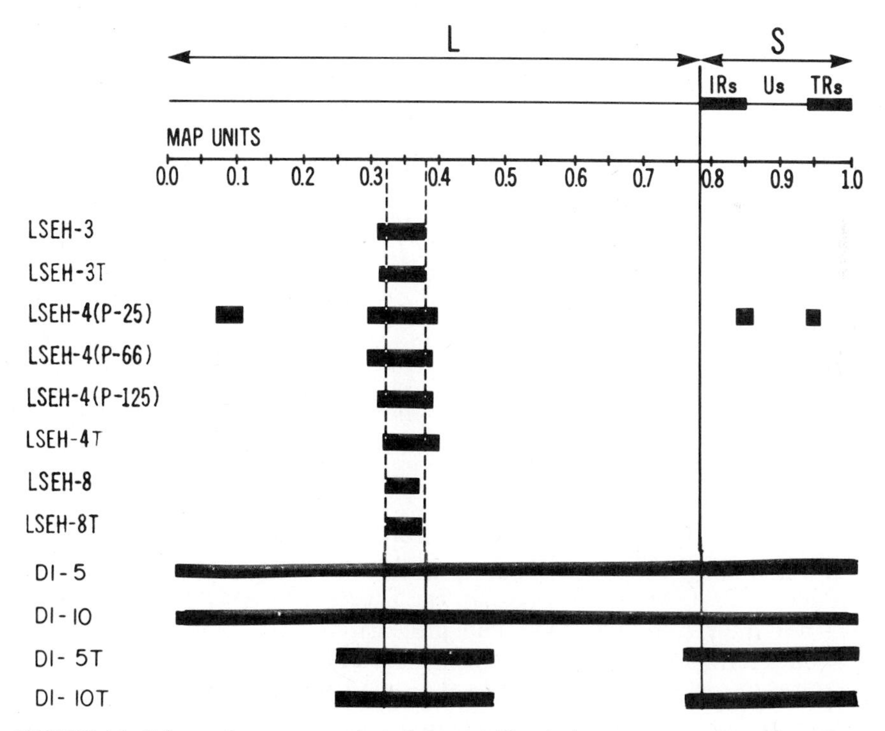

FIGURE 14. Schematic representation of the specific viral sequences contained in EHV-1-transformed and tumor cells examined to date. The viral sequence mapping at approximately 0.32–0.38 was present in all cell DNAs and has been designated as "consensus sequence." DI-5 and DI-10 cell lines contain a population of virus producer, persistently infected cells and therefore contain all EHV-1 DNA sequences (see Table VIII).

D. Construction and Mapping of Recombinant Phages Containing LSEH-4 Conjoint Sequences

Results of these biochemical investigations of EHV-1 oncogenically transformed cells suggest that multiple sites—perhaps a large number of random sites—exist within the hamster-cell genome for integration of viral subgenomic fragments. Direct evidence for the integration of EHV-1 sequences could be obtained by cloning conjoint fragments containing viral and cellular sequences as recombinant DNA molecules. Acquisition of a number of conjoint fragments from various integration sites of different cell lines could be used in nucleotide-sequencing experiments to ascertain whether integration sites share common features, in transfection experiments to determine whether such fragments have transforming potential, and eventually in biochemical experiments to define the gene products encoded by these viral genes. Recently, we have initiated experiments of this type and have constructed recombinant phages that contained covalently linked viral and host sequences isolated from the LSEH-4 transformed cell line (Robinson *et al.*, 1981a; Robinson, Tucker, and O'Callaghan, in prep.).

A schematic outline of the cloning protocol used to construct lambda recombinant phages containing conjoint sequences from EHV-1-transformed cell DNA is presented in Fig. 15. The LSEH-4 transformed cell line was chosen because of its multiple viral insertions and the differences detected between viral sequence arrangements in it and its progeny tumor cell line (Robinson and O'Callaghan, 1981, 1983). Partial *Eco*RI digests of LSEH-4 DNA that provided a distribution of possible cellular sequences were sedimented in sucrose gradients, and DNA molecules in fractions shown to contain 10–20 kb fragments were ligated to the 31-kb *Eco*RI arms of Charon 4A. Methods for the ligation of these DNAs, the *in vitro* packaging of the ligation products into infectious lambda particles, and propagation of phage in *E. coli* strain K802 were essentially those of Blattner *et al.* (1978). The genomic libraries, preserved by plaque amplification and storage at 4°C, were prepared for screening of EHV-1 sequences by plating phages on megaplates (300,000 PFU per 42 × 32 cm cafeteria tray), incubating overnight at 37°C, and transferring plaques to nitrocellulose sheets, followed by denaturation, neutralization, and baking of DNA in the plaques to the filters. Following initial screening with the [^{32}P]-*Xba*I-A probe (0.30–0.42), plaques within a 4-mm circumference of positive autoradiographic spots were picked, propagated, and replated. Subsequent screenings were done until 90% of the selected plaques were positive for the *Xba*I-A probe. Clones found to contain conjoint sequences were isolated from these sets of plaque-purified phages by screening with [^{32}P]-*Bam*HI-D (0.26–0.32), *Eco*RI-F (0.21–0.30), and *Bgl*II-I (0.27–0.32) probes, which were selected because they share homology with viral sequences thought to be located at the termini of the viral insertions within the cell DNA (see Fig. 12).

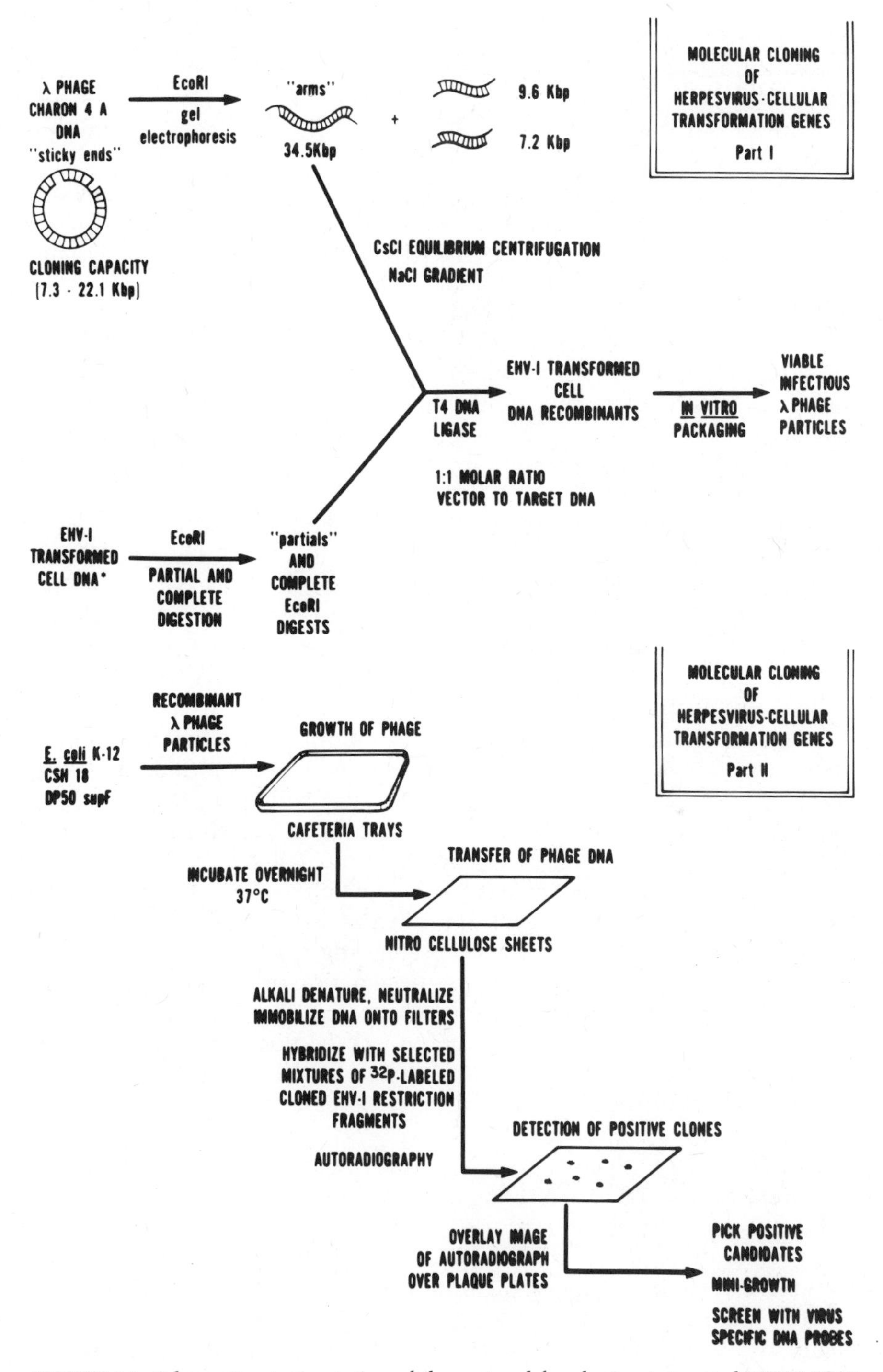

FIGURE 15. Schematic representation of the protocol for cloning integrated EHV-1 DNA sequences from LSEH-4 tranformed cells.

From the screening of more than one million recombinant phages, five clones (designated as LS4-121, -063, -414, -464, and -708) were isolated and shown to contain conjoints of LSEH-4 cellular–EHV-1 DNA sequences. Physical restriction-enzyme maps of the five isolated conjoint sequences were obtained to discriminate host from viral sequences and to ascertain from which viral insertions in this cell line these clones were derived. Since target DNA insertions were introduced at the *Eco*RI sites of the Charon 4A DNA, primary *Eco*RI digestion of these recombinant phage DNAs released the target DNAs, which could be purified by CsC1 equilibrium ultracentrifugation and resolved by agarose-gel electrophoresis. Restriction-enzyme digestion of these DNAs, typified by the restriction-enzyme digestion profiles presented in Fig. 16 and blot hybridization of the restriction-enzyme fragments in these gels with [^{32}P]-*Bam*HI-D (0.19–0.27) and [^{32}P]-*Bam*HI-B (0.32–0.43) fragments as probes were conducted to identify the origin of the fragments. Control experiments employing [^{32}P]-Charon 4A DNA as probe confirmed the absence of phage DNA sequences in the target DNAs liberated from these clones. Several series of experiments of this type in which several EHV-1 DNAs were employed as probes have allowed us to construct models depicting the arrangement of viral and host sequences in these cloned fragments containing the conjoint sequences (Fig. 17).

The results confirm the covalent linkage of hamster and viral sequences within the LSEH-4 transformed cells and the arrangements of these viral sequences as repeated copies within some insertion sites. These results indicated that clones LS4-121 and LS4-063 originated from opposing termini of the same integration site, which contained only a single copy of the aforementioned EHV-1 transformation consensus sequence (integration site 1). Clones LS4-414 and LS4-464 were derived from the opposing termini of integration site 2, which was comprised of multiple copies of the viral consensus sequence. Clone LS4-708 was comprised of viral consensus sequences originating solely from the internal viral sequences (0.30–0.40). However, designation of this clone to any of the three integration sites of viral consensus sequences was not possible by these methods. Further experiments will be done to map these sequences more finely and to ascertain whether they have transforming potential when introduced into hamster embryo cells by transfection techniques.

E. Oncogenic Transformation of Hamster Embryo Cells by Equine Cytomegalovirus

As discussed in Section V, high-multiplicity infection of primary LSH HE cells with equine cytomegalovirus (ECMV) resulted in the coestablishment of persistent infection and oncogenic transformation. In addition to the virus nonproducer EC-1 cell line (see Table XI), more than 50 ECMV-transformed, nonproducer cell lines have been established by

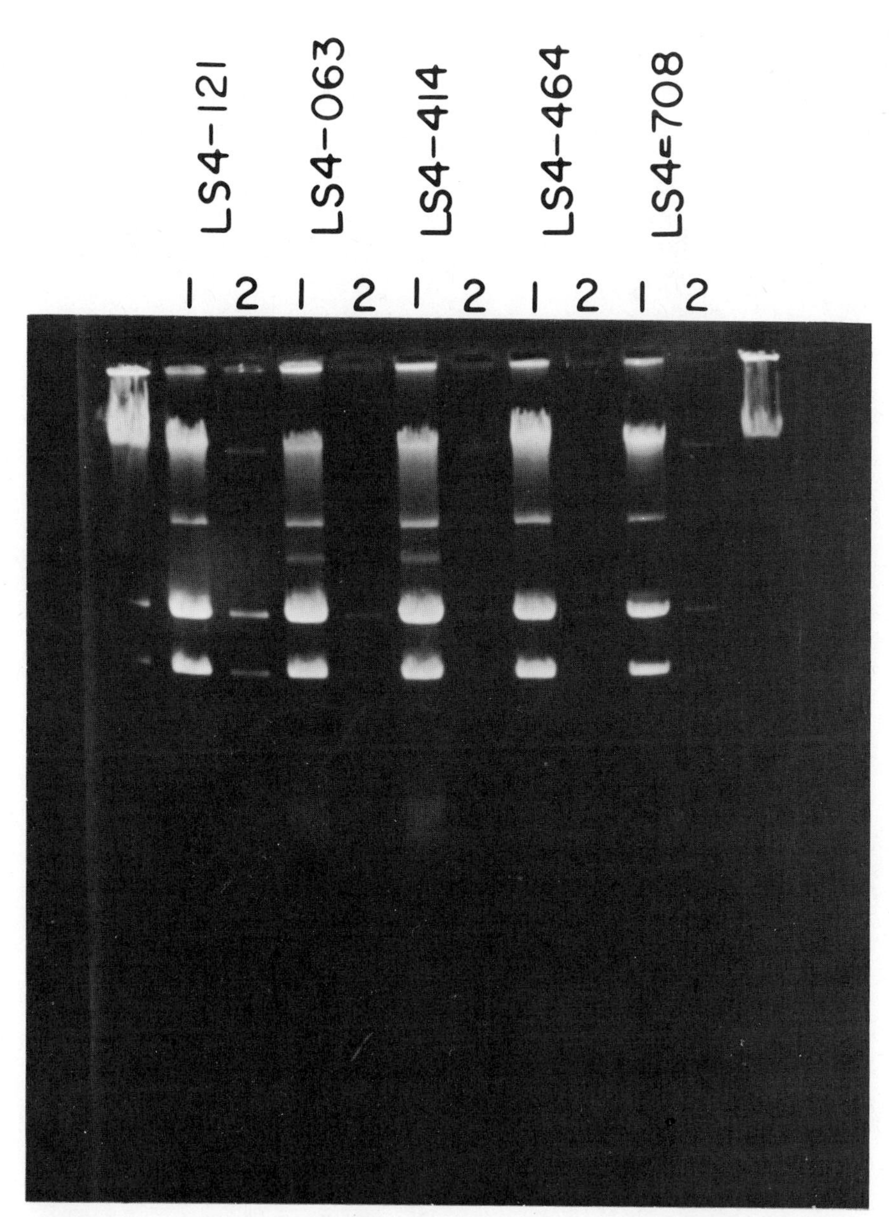

FIGURE 16. Typical restriction-enzyme digestion analysis of recombinant phage DNAs containing EHV-1 specific DNA sequences. DNAs were digested with *Bgl*II (1) or *Eco*RI (2), fractionated in 0.8% agarose gels, stained with ethidium bromide, and visualized by UV illumination. Lanes at the end contain undigested Charon 4A DNA and served as molecular-weight markers. The presence of viral sequences was shown by preparing Southern blots of these gels that were then subjected to hybridization with ^{32}P-labeled EHV-1 *Bam*HI-B and *Bam*HI-F fragments.

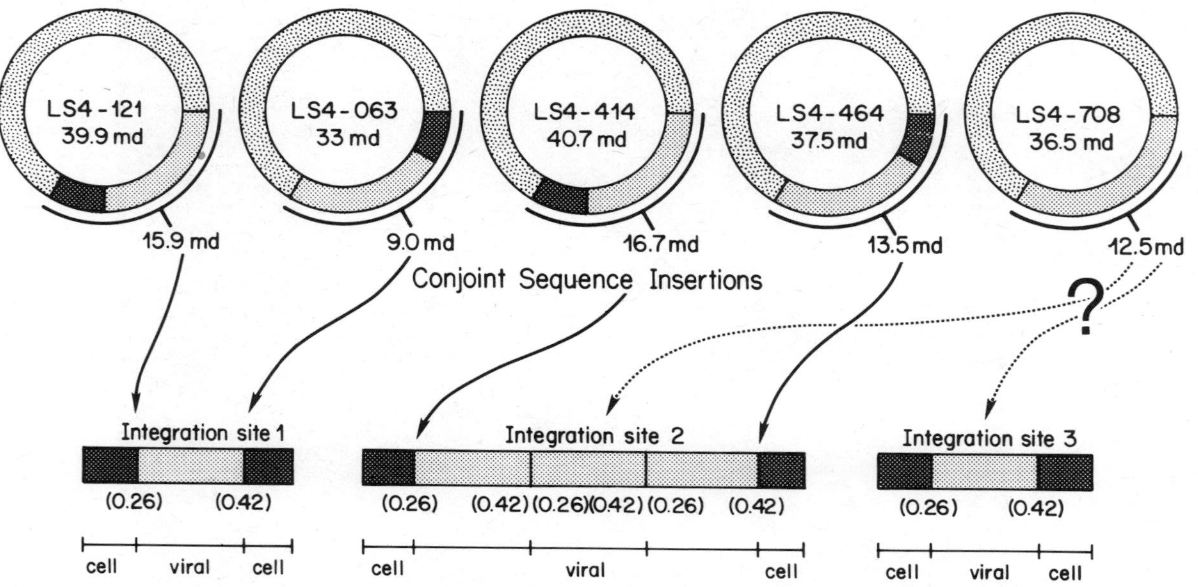

FIGURE 17. Models showing the arrangements of EHV-1 and LSEH-4 conjoint sequences cloned as recombinant DNAs in λ Charon 4A bacteriophage. Screening of more than one million phage by the *in situ* plaque hybridization method yielded five clones that were amplified, rescreened, and grown in large quantities. DNAs from these clones, designated as LS4-121, LS4-063, LS4-414, LS4-464, and LS4-708, were analyzed by restriction-enzyme mapping with several enzymes and by blot hybridization analyses using specific EHV-1 sequences as probes. Results from these experiments provided data to construct the models presented here.

cloning single nonproducer cells from the EC-2 and EC-3 cell cultures (Staczek and O'Callaghan, unpublished). All these ECMV-transformed cells examined to date as well as their progeny tumor-cell lines have been shown to be highly oncogenic in the LSH hamster, to be resistant to ECMV and EHV-1 superinfection, to contain small amounts of a limited portion (<0.05%) of the ECMV genome, and to express ECMV-specific proteins. Although the identity and arrangement of ECMV DNA sequences within these oncogenically transformed and tumor hamster cells are not known, recent "dot hybridization" assays have confirmed the presence of ECMV DNA sequences in these cells after prolonged passage in culture and in the animal as tumor tissues. Therefore, the EC and EC-T cell lines seem to offer an excellent model to examine the biochemical and biological parameters of ECMV oncogenesis. As clones of ECMV specific fragments become available by recombinant DNA techniques used to clone EHV-1 sequences (Robinson *et al.*, 1981b) and the mapping of the ECMV genome is completed, it should be possible to utilize these cloned fragments to elucidate the identity and arrangement of ECMV DNA sequences in these transformed cell lines and tumor tissues. If specific ECMV sequences were found to be integrated within these transformed hamster cells, as is the case in the EHV-1 hamster tumor model, it should be possible to ascertain whether these viral sequences are related to those of the EHV-1 genome and whether sites for insertion of herpesviral DNA are identical in LSH hamster cells transformed by these related herpesviruses.

F. Oncogenic Transformation of Hamster Embryo Cells by Equine Herpesvirus Type 3

Last, the oncogenic potential of the third member of the EHV triad, EHV-3 (equine coital exanthema virus), has been demonstrated in our laboratory (Atherton *et al.*, 1980, 1981; Sullivan, Atherton, Staczek, Dauenhauer, and O'Callaghan, in prep.). Infection of nonpermissive, primary LSH HE cells resulted in abortive infection that led to the development of foci containing cells with altered morphology and growth characteristics. Five independently derived hamster cell lines resulting from abortive infection with this equine veneral disease (EVD) virus have been characterized and designated as EVD-1, -2, -3, -4, and -5. All cell lines exhibit the key biological properties of transformed cells: immortality, altered morphology, growth to high saturation, aneuploid karyotype, ability to form colonies in soft agar or methylcellulose, and oncogenicity in the syngeneic animal. Tumors resulting from inoculation of hamsters with these cells developed rapidly, metastasized to several major organs, and were transplantable (Atherton *et al.*, 1980, 1981). Recently, these EVD-transformed and tumor-cell lines have been shown to express EHV-3-specific proteins by indirect immunfluorescent-antibody assays, indicating that EHV-3 DNA sequences are harbored and expressed in these cells after prolonged passage in cell culture and in the animal (Caughman, Sullivan, and O'Callaghan, unpublished). Also, as was found in the case of the EHV-1-transformed and hamster tumor-cell lines, these EHV-3-transformed cells contain cytoplasmic antigen(s) that reacts with antisera to the HSV ICSP 11/12 polypeptide. Repeated attempts to detect infectious virus or to induce virus or particle formation, or both, in these cells by a variety of treatments (cocultivation with permissive equine cells, growth at various temperatures, iododeoxyuridine–treatment) have failed. At this writing, experiments are in progress to assay the EHV-3-transformed and tumor-cell lines for specific viral sequences.

G. Concluding Remarks Concerning Equine Herpesvirus Oncogenesis

Results of biochemical investigations of cells transformed by adenoviruses, papovaviruses, and retroviruses have supported the hypothesis that oncogenic transformation by animal viruses involves the transmission of integrated specific viral DNA sequences, or DNA copies of RNA sequences in the case of RNA tumor viruses, and the ultimate expression of these viral genes to produce one or more proteins that initiate and maintain the transformed state. In the case of oncogenic transformation by herpesviruses, several laboratories have proposed that specific regions of the HSV-1 or HSV-2 genome have transforming potential. Sequences associated with HSV-1 morphological transformation have been reported to map at approximately 0.31–0.42 (Camacho and Spear, 1978; Reyes *et al.*, 1979), while those of HSV-2 have been found to be located at position 0.58–0.62 in most recent investigations (Galloway and McDougall, 1981; Reyes *et al.* 1979). The findings that all EHV-1-transformed and tumor-cell lines of both EHV-1 tumor models in the hamster contain DNA sequences mapping at approximately 0.32–0.38 (see Fig. 14) and that these EHV-1- as well as EHV-3-transformed and tumor cells express one or more proteins that react with antisera to an HSV-1 protein encoded by sequences within a putative morphological transforming region suggest that viral sequences associated with transformation may be shared by these herpesviruses.

Identification of specific EHV-1 sequences in EHV-1 oncogenically transformed hamster cells and the successful cloning of a fragment containing conjoint sequences of viral and cell DNAs indicate the validity of blot hybridization analyses and recombinant DNA methodologies to study herpesvirus transformation. Application of these experimental approaches to the EHV-1 mouse tumor model and to the hamster tumor models established for ECMV and EHV-3 may allow us to address basic questions concerning the nature of herpesvirus DNA sequences associated with oncogenesis and to define the arrangement of these viral genes within the genomes of virus-transformed cells of different genera.

VII. EVOLUTIONARY RELATIONSHIPS

A. Among the Equine Herpesviruses

The three known equine herpesviruses, EHV-1, -2, and -3, are serologically distinct (Allen *et al.*, 1977a; Allen and Randall, 1979; Wharton *et al.*, 1981). As discussed elsewhere, there is a further possibility that those viruses previously classified as EHV-1 may in fact be further distributed in two subclasses. DNA–DNA hybridization, as well as analysis by Southern blots, has provided a substantial amount of information about the genetic relatedness of these viruses. Allen *et al.* (1977a) were the first to explore the relatedness of EHV-1 and EHV-3 by DNA–DNA reassociation kinetics. By these procedures, as well as hybridization of RNA from EHV-3-infected cells to filter-immobilized DNA of EHV-1, it was concluded that genomic homology between EHV-1 and EHV-3 amounted to only 2–5%. Similar studies (Wharton *et al.*, 1981) suggested that EHV-2 and EHV-1 also shared little genomic homology. More recently, Staczek and Atherton (1981) and Staczek *et al.*, (1983) have made a comparative study of the extent of homology among all three of these viruses. A summary of the relationships is shown in Table XIV.

These data suggest that EHV-1 and EHV-3 are somewhat more closely related than either is to EHV-2, which seems reasonable, since EHV-2 has the characteristics of a cytomegalovirus. Very recently, Allen and Turtinen (1982) reported that DNAs from an EHV-1 respiratory strain and an EHV-1 abortigenic strain exhibit only 17% homology; these data support the suggestion to classify EHV-1 respiratory strains as EHV-4.

Staczek *et al.* (1983) have further analyzed the locations of genomic homology between EHV-1 and EHV-3 by hybridization of various restricted fragments. In general, although areas of homology appear in sev-

TABLE XIV. Percent Homology of EHV-1, EHV-2, and EHV-3 Genomes[a]

	Restricted DNA		
[^{32}P]-DNA	EHV-1	EHV-2	EHV-3
EHV-1	100	2.0	8.9
EHV-2	2.9	100	1.0
EHV-3	7.8	2.0	100

[a] Data from Staczek and Atherton (1981) and Staczek and O'Callaghan (in prep.).

eral spots in the U_L and U_S regions, there is consistent homology in the joint region when genomic [^{32}P]-EHV-3 is hybridized to various fragments of EHV-1 as well as when the reciprocal analyses are made. Further, Hay and Bartoski (personal communication) and Hayward (personal communication) have both observed homology between the joint regions of EHV-1 and HSV-1, which suggests that this structure (and by implication the ability to isomerize) may be rather ancient and important features of these herpesviruses. In contrast, Rand and Ben-Porat (1980) have done similar analyses between pseudorabies (PsR) virus and HSV-1 and 2; they report that most of the homology is distributed through the U_L regions, with less in the U_S and practically none in the extreme ends of the inverted repeat of PsR virus.

B. Between the Equine and Other Herpesviruses

Aside from the unpublished work of Hay and Bartoski and of Hayward (mentioned above), little direct comparison of DNA–DNA homology has been made between any of the equine and nonequine herpesviruses, although Ludwig *et al.* (1971) did report that labeled RNA from HSV-1-infected cells failed to show any homology to EHV-3 DNA. Perhaps the best data on the relatedness of EHV-1 and nonequine herpesviruses have been published by Yeo *et al.* (1981) in a study of cross-reacting antigens of HSV-1 and 2, bovine mammillitis virus (BMV) PsR virus, and EHV-1. These viruses were clearly separable into two groups, HSV-1 and -2, and BMV; and EHV-1 and PSR virus. The members of the first group shared numerous antigens with each other, but few with EHV-1 and only one with PsR virus. PsR virus and EHV-1 shared few antigens, although more than either did with the first group. Perhaps the most interesting finding of all was that all five viruses shared the DNA-binding infected-cell-specific protein 11 12, which suggests that this structure may be older and more important than others. Unfortunately, there are no data on relationships between the cytomegaloviruses of equines and other mammalian species.

C. Origins of the Herpesviruses

It has been suggested (Reanney, 1981), on the basis of evidence for gene splicing, that the herpesviruses originated from eukaryotes. Such an origin, initially perhaps a temporary colonization, would have been followed by losses of various functions until a state of obligate intracellular parasitism obtained. Of the various functions that might be attributed to a precursor of herpesviruses, one has apparently been retained not only by most of the herpesviruses but also by other large DNA viruses—the ability to salvage thymine compounds. This ability—whether reflected in deoxythymidine kinase (dTK) (Animal cells, mitochondria of animal origin, bacteria, large DNA viruses), nucleoside phosphotransferase (plants), or thymine-7-hydroxylase (yeasts)—seems ubiquitous in life, and its occurrence in the herpesviruses suggests their common origin with cellular forms (Gentry *et al.*, 1982). There is an important exception within the herpesviruses: human and mouse cytomegaloviruses do not appear to specify a dTK (Estes and Huang, 1977; Miller *et al.*, 1977; Müller and Hudson, 1977; Zavada *et al.*, 1976). One can speculate that this function has simply been deleted in these viruses during the course of evolution. If this has indeed happened, it would suggest further that at least the cytomegaloviruses have origins that predate the main events in mammalian speciation (if the lack of a dTK proves to be general). It could, of course, be argued that the viral-coded DNA polymerases [which resemble eukaryotic more than prokaryotic enzymes, at least in their inhibition by aphidicolin (Pedrali-Noy and Spadari, 1980)] are a better indicator of a cellular ancestry than is dTK, but this does not distinguish between DNA viruses and RNA viruses, most of which also code for one or more polymerases. Finally, there remains at least one herpesvirus the minimum age of which can be told with more confidence, because the evidence is good that it predates the biological isolation of Australia, which occurred during the late Cretaceous, more than 75 million years ago (Dodson, 1960). It is the recently reported Parma wallaby herpesvirus, widespread in Australian marsupials, which grows well only in marsupial cells (it has apparently not been tested in cells of New World marsupials) and poorly in BHK or not at all in eight other eutherian cell lines (Finnie *et al.*, 1976; Webber and Whalley, 1978; Whalley and Webber, 1979). It should prove a most interesting candidate for further study of the evolution of herpesviruses.

VIII. ANTIVIRAL PROSPECTS

A. Predictive Value of the Equine Herpesvirus Type 1–Hamster Model

Although the cell-culture systems with EHV-1 and EHV-3 do respond to antiviral agents in the same way as herpes simplex virus (HSV), their utility in predicting antiviral activity in human herpetic infections is not unique, at least when considered as an individual system. With EHV-1, however, the cell-culture system becomes an important screen preliminary to animal studies with the EHV-1–hamster model, which does offer unique advantages as a screen for antiviral activity with human herpesviruses. The main advantage with the EHV-1–hamster model is in the quality of the quantitation. As discussed above, the animals die with great regularity, depending on the dose of virus employed; this may be as early as 12 ± 0.5 hr postinjection (with a dose of 10^8 LD_{50} per animal) to 72 ± 6 hr (with a dose of 10 LD_{50}). This allows two measurements of antiviral activity, average time of death as well as survival. A typical plot of an antiviral trial [with arabinosylthymine (ara-T)] is shown in Fig. 18 (Aswell *et al.*, 1977). A second advantage is that the infectious process involves the liver, and although human hepatitis caused by HSV is rare, it does occur (Conner *et al.*, 1979). In any case, the EHV-1 model involves a simple but rigorous test of any antiherpesviral drug. A final advantage that applies to a few situations derives from the high levels of deoxycytidine deaminase present in hamster liver. In this respect, hamsters resemble humans, who also have high levels of this enzyme, but not mice or rats, which have rather low levels (Aswell and Gentry, 1977). This feature becomes important in the case of 5-methyl arabinosylcytosine (5 Me-araC)which is apparently not phosphorylated by either the host dTK or dCK or the viral dTK. It is, however deaminated to ara-T by deoxycytidine deaminase and shows activity typical of ara-T in those systems that can deaminate it (Aswell and Gentry, 1977).

Several compounds with good antiviral activity against HSV in cell culture as well as animals have been tested in the EHV-1 animal system, with good agreement. These include arabinosyl adenine (Lieberman *et al.*, 1972), phosphonoacetic acid (Aswell *et al.*, 1977), and arabinosylthymine (Aswell *et al.*, 1977; Gentry *et al.*, 1978, 1979, 1981). 5-Methyl ara-C has also been tested in this system and appears to be at least as active as ara-T if not more so (Gentry *et al.*, 1981).

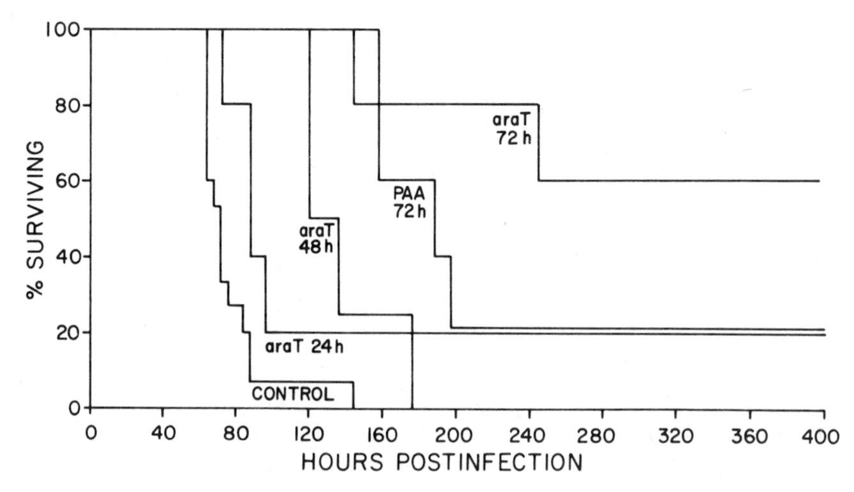

FIGURE 18. Antiherpes activity of ara-T *in vivo*. Each 25–30 g Syrian hamster received 10 LD_{50} of EHV-1 in 1 ml by intraperitoneal injection at 0 hr. ara-T (5 mg/animal per injection, 5 mg/ml of water) or phosphonoacetate [(PAA) 10 mg/animal per injection, 10 mg/ml of water] was given intraperitoneally at 4-hr intervals, beginning at 4 hr postinoculation with EHV-1. The drugs were continued for the indicated intervals to a total of 6 injections per 24 hr, 12 per 48 hr, and 18 per 72 hr. The control group contained 15 animals; each of the other groups contained 5 except for the one receiving ara-T for 48 hr, which contained 4 (Aswell *et al.*, 1977).

B. Control of Equine Herpesviral Disease by Chemotherapy

Although we are not aware of any clinical trials of antiviral compounds in equine herpesviral disease, there is no inherent reason that those that are effective against EHV-1 and EHV-3 in cell culture and EHV-1 in the hamster should not also be effective in the horse. In addition to the compounds named above, 5-ethyldeoxyuridine and acycloguanosine may also be effective against EHV-1 as well as EHV-3. While EHV-1 is a serious economic problem—the first foal sired by triple crown winner Secretariat was lost to an EHV-1 infection—the use of any antiherpetic in horses would be much less than any similar drug in humans, and because of the size differences the individual dose would be much larger. These factors suggest that veterinary usage of any antiherpetic drug will follow human use; it will probably be economical to use only drugs that have already been developed and placed on the market for systemic use in humans. What the future holds in this direction cannot be predicted with certainty.

ACKNOWLEDGMENTS. We thank Steven Dauenhauer for preparing the photographs, Linda Devine, Marcie Rehfeldt, Jeanette Mize, and Katie Carey for secretarial assistance, and Drs. John Bryans and George Allen for providing preprints and unpublished findings. We acknowledge with gratitude all our former and present postdoctoral and doctoral students who have contributed to our investigations of the equine herpesviruses. Support for these investigations was provided by Research Grants AI 02 032, S-507-RR-05386, and DE-05089 from the National Institutes of Health, PCM-78-22700 from the National Science Foundation, CH-74 and an Institutional Grant from the American Cancer Society, and funds from the Grayson Foundation. Last, we thank our editor, Bernard Roizman, for his patience and understanding.

REFERENCES

Abodeely, R.A., Lawson, L.A., and Randall, C.C., 1970, Morphology and entry of enveloped and deenveloped equine abortion (herpes) virus., *J. Virol.* **5:**513.

Abodeely, R.A., Palmer, E., Lawson, L.A., and Randall, C.C., 1971, The proteins of enveloped and deenveloped equine abortion (herpes) virus and the separated envelope, *Virology* **44:**146.

Ahmed, R., Chakraborty, P.R., Graham, A.F., Ramig, R.F., and Fields, B.N., 1980, Genetic variation during persistent reovirus infection: Presence of extragenically suppressed temperature-sensitive lesion in wild-type virus isolated from persistently infected L cells, *J. Virol.* **34:**383.

Allen, G.P., and Bryans, J.T., 1974, Studies of an established equine cell line derived from a transitional cell carcinoma, *Am. J. Vet. Res.* **35:**1153.

Allen, G.P., and Bryans, J.T., 1976, Cell-free synthesis of equine herpesvirus type 3 nucleocapsid polypeptides, *Virology* **69:**751.

Allen, G.P., and Bryans, J.T., 1977, Replication of equine herpesvirus type 3: Kinetics of infectious particle formation and virus nucleic acid synthesis, *J. Gen. Virol.* **34:**421.

Allen, G.P., and Randall, C.C., 1978, Biological properties of equine herpesvirus, type 1 DNA: Transfectivity and transforming capacity, *Infect. Immun.* **22:**34.

Allen, G.P., and Randall, C.C., 1979, Proteins of equine herpesvirus type 3. 1. Polypeptides of the purified virion, *Virology* **92:**252.

Allen, G.P., and Turtinen, L.W., 1982, Assessment of the base homology between the two subtypes of equine herpesvirus 1. *J. Virol.* **44:**249.

Allen, G.P., O'Callaghan, D.J., and Randall, C.C., 1977a, Genetic relatedness of equine herpesviruses type 1 and 3, *J. Virol.* **24:**761.

Allen, G.P., O'Callaghan, D.J., and Randall, C.C., 1977b, Purification and characterization of equine herpesvirus-induced DNA polymerase, *Virology* **76:**395.

Allen, G.P., O'Callaghan, D.J., and Randall, C.C., 1977c, Oncogenic transformation of nonpermissive murine cells by intact equine herpesvirus type 1 DNA, in: Proceedings International Symposium on Oncogenesis and Herpesviruses, p. 77, Cambridge, Massachusetts.

Allen, G.P., Cohen, J.C., Randall, C.C., and O'Callaghan, D.J., 1978a, Replication of equine herpesvirus type 1 and type 3: Resistance to hydroxyurea and thymidine, *Intervirology* **9:**276.

Allen, G.P., McGowan, J.J., Bryans, J.T., Randall, C.C., and Gentry, G.A., 1978b, Induction of deoxythymidine kinase activity in several mammalian cell lines after infection with six different strains of equine herpesvirus type 1 (EHV-1), *Virology* **90:**351.

Allen, G.P., McGowan, J.J., Gentry, G.A., and Randall, C.C., 1978c, Biochemical transformation of deoxythymidine kinase-deficient mouse cells with UV-irradiated equine herpesvirus type 1, *J. Virol.* **28:**361.

Allen, G.P., O'Callaghan, D.J., and Randall, C.C., 1978d, Oncogenic transformation of nonpermissive murine cells by viable equine herpesvirus type 1 (EHV-1) and EHV-1 DNA, in: *Oncogenesis and Herpesviruses III*, Part 1, *DNA of Herpesviruses, Viral Antigens, Cell–Virus Interaction* (G. de Thé and F. Rapp, eds.), pp. 509–516, International Agency for Research on Cancer, Lyon, France.

Allen, G.P., McGowan,J.J., Randall, C.C., Mancini, W., Cheng, Y.-C., and Gentry, G.A., 1979, Purification and characterization of deoxythymidine kinase (dTK) induced in dTK-3T3 mouse cells by equine herpesvirus type 1 (EHV-1), *Virology* **92:**367.

Allen, G.P., Yeargan, M.R., and Bryans, J.T., 1983a, Alterations in the genome of equine herpesvirus 1 after *in vitro* and *in vivo* virus passage. *Infect. Immun.* (in press).

Allen, G.P., Yeargan, M.R., Turtinen, L.W. and Bryans, J.T., 1983b, Rapid emergence of a new dominant field strain of equine abortion virus (equine herpesvirus 1). *Science* (in press).

Allen, G.P., Yeargan, M.R., Turtinen, L.W., Bryans, J.T., and McCollum, W.H., 1983c, Molecular epidemiologic studies of equine herpesvirus 1 infections by restriction endonuclease fingerprinting of viral DNA. *Amer. J. Vet. Res.*

Anderson, K., and Goodpasture, E.W., 1942, Infection of newborn Syrian hamsters with the virus of mare abortion, *Am. J. Pathol.* **18:**555.

Andzhaparidze, O.G., Bogomolova, N.N., Boriskin, Y.S., Bektemirova, M.S., and Drynov, I.D., 1981, Comparative study of rabies virus persistence in human and hamster cell lines, *J. Virol.* **37:**1.

Arhelger, R.B., Darlington, R.W., and Randall, C.C., 1963, An electron microscopic study of equine abortion virus infection in hamster liver, *Am. J. Pathol.* **42:**703.

Aswell, J.F., and Gentry, G.A., 1977, Cell-dependent antiherpesviral activity of 5-methylarabinosylcytosine, an intracellular araT donor, *Ann. N. Y. Acad. Sci.* **284:**342.

Aswell, J.F., Allen G.P., Jamieson, A.T., Campbell, D.E., and Gentry, G.A., 1977, Antiviral activity of arabinosylthymine in herpesviral replication: Mechanism of action *in vivo* and *in vitro*, *Antimicrob. Agents Chemother.* **12:**243.

Atherton, S.S., Robinson, R. A., and O'Callaghan, D.J., 1980, Equine herpesvirus type 3

(EHV-3): Genomic structure and transforming potential, Fifth International Herpesvirus Workshop, Cold Spring Harbor Laboratory, p. 53.

Atherton, S.S., Robinson, R.A., and O'Callaghan, D.J., 1981, Genomic structure and oncogenic potential of equine coital exanthema virus (equine herpesvirus type 3, EHV-3), *Abstr. Annu. Meeting Am. Soc. Microbiol.*, Dallas, Texas, p. 226.

Atherton, S.S., Sullivan, D.C., Dauenhauer, S.A., Ruyechan, W.T., and O'Callaghan, D.J., 1982, Properties of the genome of equine herpesvirus type 3 (submitted).

Bagust, T.J., Pascoe, R.R., and Harden, T.J., 1972, Studies on equine herpesviruses. 3. The incidence in Queensland of three different equine herpesvirus infections, *Aust. Vet. J.* **48:**47.

Becker, Y. and Rabkin, S., 1981, Biosynthesis of defective HSV DNA, in: *Developements in Molecular Virology*, Vol. I, *Herpesvirus DNA* (Y. Becker, ed.), pp. 185–195, Nijhoff, The Hague.

Ben Porat, T., Rixon, F.J., and Blankenship M., 1979, Analysis of the structure of the genome of pseudorabies virus, *Virology* **95:**285.

Blattner, F.R., Blechl, A.E., Thompson, K.D., Richards, J.E., Slightom, J.L., Tucker, P.W., and Smithies, O., 1978, Human and mouse DNA. I. Cloning of fetal gammaglobin and mouse alphaglobin DNA: Preparation and screening of shotgun collections, *Science* **202:**1279.

Bracken, E.C., and Randall, C.C., 1957, Studies on hepatitis in hamsters infected with equine abortion virus. II. Changes in protein, nucleic acid, and weight of isolated hepatic nuclei, *Am. J. Pathol.* **33:**1009.

Bronson, D.L., Dreesman, G.R., Biswal, N., and Benyesh-Melnick M., 1974, Defective virions of herpes simplex virus, *Intervirology* **1:**141.

Bryans, J.T., 1968, The herpesvirus in disease of the horse, Proceedings of the 14th Annual Convention of the American Association of Equine Practitioners, Philadelphia, pp. 119–125.

Bryans, J.T., 1969, On immunity to disease caused by equine herpesvirus 1, *J. Am. Vet. Med. Assoc.* **156:**294.

Bryans, J.T., 1978, Immunization of pregnant mares with an inactivated equine herpesvirus 1 vaccine, *Proceedings of the 4th International Conference of Equine Infectious Diseases*, Lyon, 1976 (J. Bryans and H. Gerber, eds.), pp. 83–92, Veterinary Publications, Princeton, New Jersey.

Bryans, J. T., 1980, Herpesvirus diseases affecting reproduction in the horse, *Vet. Clin. North Am. : Large Anim. Pract.* **2:**303.

Bryans, J.T., 1981, Application of management procedures and prophylactic immunization to the control of equine rhinopneumonitis, Proceedings of the 26th Annual Convention of the American Association of Equine Practitioners, Anaheim, California 1980, pp. 259–272.

Bryans, J.T., and Allen, G.P., 1973, *In vitro* and *in vivo* studies of equine coital exanthema, in: *Proceedings of the 3rd International Conference on Equine Infectious Disease* (J. Bryans and H. Gerber, eds.), pp. 322–336, S. Karger, Basel.

Bryans, J.T., and Allen, G.P., 1982, Application of a chemically inactivated, adjuvanted vaccine to control abortigenic infection of mares by equine herpesvirus 1, Proceeding of the 17th International Congress of the International Association of Biological Standardization, Lyon, France, 1981, S. Karger, Basel (in press).

Bryans, J.T., and Prickett, M.E., 1970, A consideration of the pathogenesis of abortigenic disease caused by equine herpesvirus 1, in *Proceedings of the 2nd International Conference of Equine Infectous Diseases*, Paris, 1969 (J. Bryans and H. Gerber, eds.), pp. 34–40, S. Karger, Basel.

Bryans, J.T., Swerczek, T.W., Darlington, R.W., and Crowe, M.W., 1977, Neonatal foal disease associated with prenatal infection by equine herpesvirus 1, *J. Equine Med. Surg.* **1:**20.

Burki, F., Pickler, L., and Sibalin, M., 1973, Klassifizierung des Isolates 53/69 aus equinem

Coitalexanthem als eigenständiges equine Herpesvirus, *Zentralbl. Bakteriol. Parasitenkd. Infektionskr. Hyg. Abt. 1: Orig. Reihe A* **225:**438.

Burrows, R., 1968, Some observations on the viral aetiology of upper respiratory disease of British horses, 1965–1967, *Bull. Off. Int. Epizootol.* **70:**Rep. No. 306.

Burrows, R., 1970, The general virology of the herpesvirus group, *Proceeding of the 2nd International Conference of Equine Infectious Diseases,* Paris, 1969 (J. Bryans and H. Gerber, eds.), pp. 1–12, S. Karger, Basel.

Burrows, R., 1973, Discussion, in: *Proceedings of the 3rd International Conference of Equine Infectious Diseases,* Paris, 1972 (J. Bryans and H. Gerber, eds.), p. 340, S. Karger, Basel.

Burrows, R., and Goodridge, D., 1973, *In vivo* and *in vitro* studies of equine rhinopneumonitis virus strains, in: *Proceedings of the 3rd International Conference of Equine Infectious Diseases,* Paris, 1972 (J. Bryans and H. Gerber, eds.), pp. 306–321, S. Karger, Basel.

Camacho, A., and Spear, P.G., 1978, Transformation of hamster embryo fibroblasts by a specific fragment of the herpes simplex virus genome, *Cell* **15:**993.

Campbell, D.E., Kemp, M.C., Perdue, M.L., Randall, C.C., and Gentry, G.A., 1976, Equine herpesvirus *in vivo*: Cyclic production of a DNA density variant, *Virology* **69:**737.

Charlton, R.M., Mitchell, D., Girard, A., and Corner, A.H., 1976, Meningoencephalitis in horses associated with equine herpesvirus I infection, *Vet. Pathol.* **13:**59.

Churchill, A.E., and Biggs, P.M., 1967, Agent of Marek's disease in tissue culture, *Nature (London)* **215:**528.

Cohen, G.H., Long, D., and Eisenberg, R.J., 1980, Synthesis and processing of glycoproteins gD and gC of herpes simplex virus type 1, *J. Virol.* **36:**429.

Cohen, J.C., Perdue, M.L., Randall, C.C., and O'Callaghan, D.J., 1975a, Replication of equine herpesvirus type 1: Resistance to hydroxyurea, *Virology* **67:**56.

Cohen, J.C., Randall, C.C., and O'Callaghan, D.J., 1975b, Transcription of equine herpesvirus type 1: Evidence for classes of transcripts differing in abundance, *Virology* **68:**561.

Cohen, J.C., Henry, B.E., Randall, C.C., and O'Callaghan, D.J., 1977a, Ribonucleotide reductase activity in hydroxyurea-resistant herpesvirus replication, *Proc. Soc. Expt. Biol. Med.* **155:**395.

Cohen, J.C., Perdue, M.L., Randall, C.C., and O'Callaghan, D.J., 1977b, Herpesvirus transcription: Altered regulation induced by FUdR, *Virology* **76:**621.

Conner, R.W., Lorts, G., and Gilbert, D.N., 1979, Lethal herpes simplex virus type 1 hepatitis in a normal adult, *Gastroenterology* **76:**590.

Cuifo, D.M., and Hayward, G.S., 1981, Tandem repeat defective DNA from the L segment of the HSV genome in: *Developments in Molecular Virology,* Vol. I, *Herpesvirus DNA* (Y. Becker ed.),pp. 107–128, Nijhoff, The Hague.

Dal Canto, M.C., and Rabinowitz, S.G., 1981, Murine central nervous system infection by a viral temperature-sensitive mutant, *Am. J. Pathol.* **102:**412.

Dalsgaard, H., 1970, Enzootic paresis in horses in relation to outbreaks of rhinopneumonitis (virus abortion), *Medlemsbl. Dan. Drylaegeforen.* **53:**71.

Darlington, R.W., 1978, The role of equine macrophages in resistance or susceptilility to infection by equine herpesvirus 1, in: *Proceedings of the 4th International Conference of Equine Infectious Diseases,* Lyon, 1976 (J. Bryan and H. Gerber, eds.), pp. 129–139, Grayson Foundation, Lexington, Kentucky.

Darlington, R.W., and James, C., 1966, Biological and morphological aspects of the growth of equine abortion virus, *J. Bacteriol.* **92:**250.

Darlington, R.W., and Moss, L.H., 1968, Herpesvirus envelopment, *J. Virol.* **2:**48.

Darlington, R.W., and Moss, L.H., 1969, The envelope of herpesvirus, *Prog. Med. Virol.* **11:**16.

Darlington, R.W., and Randall, C.C., 1963, The nucleic acid content of equine abortion virus, *Virology* **19:**322.

Dauenhauer, S.A., Robinson, R.A., and O'Callaghan, D.J., 1981, Structure of the DNA of equine herpesvirus type 1 (EHV-1) defective interfering (DI) particles that establish persistent infection and oncogenic transformation, *Fed. Proc. Fed. Am. Soc. Exp. Biol.* **40:**1827.

Dauenhauer, S.A., Robinson, R.A., and O'Callaghan, D.J., 1982, Chronic production of defective interfering particles by hamster embryo cultures of herpesvirus persistently infected and oncogenically transformed cells, *J. Gen. Virol.* (in press).

De, B.K., and Nayak, D.P., 1980, Defective interfering influenza viruses and host cells: Establishment and maintenance of persistent influenza virus infection in MDBK and HeLa cells, *J. Virol.* **36:**847.

Dimock, W.W., 1940, The diagnosis of virus abortion in mares, *J. Am. Vet. Med. Assoc.* **96:**665.

Dimock, W.W., and Edwards, P.R., 1932, Infections of fetuses and foals, Kentucky Agriculture Experiment Station, Bulletin 333, Lexington.

Dimock,W.W., and Edwards, P.R., 1933, Is there a filterable virus of abortion in mares?, Kentucky Agriculture Experiment Station, Bulletin 333, Supplement, Lexington.

Dimock, W.W., and Edwards, P.R., 1936, The differential diagnosis of equine abortion with special reference to a hitherto undescribed form of epizootic abortion in mares, *Cornell Vet.* **26:**231.

Dimock, W.W., Edwards, P.R., and Bruner, D.W., 1942, Equine virus abortion, Kentucky Agriculture Experiment Station, Bulletin 426, Lexington.

Dinter, Z., and Klingeborn, B., 1976, Serological study of an outbreak of paresis due to equid herpesvirus I (EHV-I),*Vet. Rec.* **99:**10.

Dodson, E.O., 1960, *Evolution: Process and Product*, pp. 63 and 173, Reinhold, New York.

Doll, E.R., 1953, Intrauterine and intrafetal inoculations with equine abortion virus in pregnant mares, *Cornell Vet.* **43:**112.

Doll, E.R., 1961, Immunization against viral rhinopneumonitis of horses with live virus propagated in hamsters, *J. Am. Vet. Med. Assoc.* **139:**1324.

Doll, E.R., 1962, A planned infection program for controlling abortion, *Blood Horse* **83:**1214.

Doll, E.R., and Bryans, J.T., 1962a, Development of complement-fixing and virus neutralizing antibodies in viral pneumonitis of horses, *Am. J. Vet. Res.* **23:**843.

Doll, E.R., and Bryans, J.T., 1962b, Incubation periods for abortion in equine viral rhinopneumonitis, *J. Am. Vet. Med. Assoc.* **141:**351.

Doll, E.R., and Bryans, J.T., 1963a, Epizootiology of equine viral rhinopneumonitis, *J Am. Vet. Med. Assoc.* **142:**31.

Doll, E.R., and Bryans, J.T., 1963b, Immunization of young horses against viral rhinopneumonitis, *Cornell Vet.* **53:**24.

Doll, E.R., and Bryans, J.T., 1963c, A planned infection program for immunizing mares against viral rhinopneumonitis, *Cornell Vet.* **53:**249.

Doll, E.R., and Kintner, J.H., 1954, A comparative study of the equine abortion and equine influenza viruses, *Cornell Vet.* **44:**355.

Doll, E.R., and Wallace, M.E., 1954, Cultivation of the equine abortion and equine influenza viruses on the chorioallantoic membrane of chicken embryos, *Cornell Vet.* **44:**453.

Doll, E.R., Richards, M.G., and Wallace, M.E., 1953, Adaptation of the equine abortion virus to suckling Syrian hamsters, *Cornell Vet.* **43:**551.

Doll, E.R., Richards, M.G., and Wallace, M.E., 1954a, Cultivation of the equine influenza virus in suckling Syrian hamsters: Its similarity to the equine abortion virus, *Cornell Vet.* **44:**133.

Doll, E.R., Wallace, M.E., and Richards, M.G., 1954b, Thermal, hematological and serological responses of weanling horses following inoculation with equine abortion virus: Its similarity to equine influenza, *Cornell Vet.* **44:**181.

Doll, E.R., Crowe, M.E., Bryans, J.T., and McCollum, W.H., 1955, Infection immunity in equine virus abortion, *Cornell Vet.* **45:**387.

Doll, E.R., Bryans, J.T., McCollum, W.H., and Crowe, M.E.W., 1957, Isolation of a filterable agent causing arteritis of horses and abortion by mares: Its differentiation from the equine abortion (influenza) virus, *Cornell Vet.* **57:**3.

Dreesman, G.R., Suriano, J.R., Swartz, S.K., and McCombs, R.M., 1972, Characterization of the herpes virion. I. Purification and amino acid composition of nucleocapsids, *Virology* **50:**528.

Dumas, A.M., Geelen, J.L., Westrate, M.W., Wertheim, P., and Van Der Noordaa, J., 1981, *Xba*I, *Pst*I, and *Bgl*III restriction enzyme maps of the two orientations of the varicella–zoster genome, *J. Virol.* **39:**390.

Enzmann, P.J., 1973, Induction of an interferon-like substance in persistently infected *Aedes albopictus* cells, *Arch. Gesamte Virusforsch.* **40:**382.

Erasmus, B.J., 1966, The activation of herpesvirus infections of the respiratory tract in horses, in: *Proceedings of the 1st International Conference of Equine Infectious Diseases*, Stresa (J. Bryan, ed.), pp. 117–121, Grayson Foundation, Lexington, Kentucky.

Erasmus, B.J. 1970, Equine cytomegaloviruses, in: *Proceedings of the 2nd International Conference of Equine Infectious Diseases*, Paris, 1969 (J. Bryans and H. Gerber, eds.), pp. 46–55, S. Karger, Basel.

Estes, J.E., and Huang, E.S., 1977, Stimulation of cellular thymidine kinase by human cytomegalovirus, *J. Virol.* **24:**13.

Finnie, E.P., Littlejohns, I.R., and Acland, H.M., 1976, Mortalities in Parma wallabies (*Macropus parma*) associated with a probable herpesvirus, *Aust. Vet. J.* **52:**293.

Francke, B., Moss, H., Timbury, M.C., and Hay, J., 1978, Alkaline DNase activity in cells infected with a temperature-sensitive mutant of herpes simplex virus type 2, *J. Virol.* **26:**209.

Frenkel, N., 1981, Defective interfering herpesviruses, in: *The Human Herpesviruses: An Interdisciplinary Perspective* (A.J. Nahmias, W.R. Dowdle, and R.F. Schinazi, eds.), pp. 91–120, Elsevier, New York.

Frenkel, N., Locker, H., and Vlazny, D.A., 1980, Studies of defective herpes simplex viruses, *Ann. N.Y. Acad. Sci.* **354:**347.

Frenkel, N., Locker, H., and Vlazny, D., 1981, Structure and expression of class I and class II defective interfering HSV genomes, in: *Developments in Molecular Virology*, Vol. I, *Herpesvirus DNA* (Y. Becker, ed.), pp. 149–184, Nijhoff, The Hague.

Furlong, D., Swift, J., and Roizman, B., 1972, Arrangement of herpesvirus deoxyribonucleic acid in the core, *J. Virol.* **10:**1071.

Galloway, D.A., and McDougall, J.R., 1981, Transformation of rodent cells by a cloned DNA fragment of herpes simplex virus type 2, *J. Virol.* **38:**749.

Gentry, G.A., and Aswell, J.F., 1975, Inhibition of herpes simplex virus replication by araT, *Virology* **65:**294.

Gentry, G.A., and Randall, C.C., 1960, Studies on hepatitis in hamsters infected with equine abortion virus. III. Nuclear protein changes. A histochemical study. *Amer. J. Pathol.* **37:**433.

Gentry, G.A., Randall, C.C., and Darlington, R.W., 1960, Composition of RNA and DNA of citric acid-isolated liver nuclei from hamsters infected with equine abortion virus (EAV), *Virology* **11:**773.

Gentry, G.A., Randall, C.C., Walker, B.M., and Rawson, J.E., 1961, Labelling of liver nuclei with tritiated thymidine in hamsters infected with equine abortion virus, *Biochim. Biophys. Acta* **47:**212.

Gentry, G.A., Walker, B.M., and Randall, C.C., 1962, Correlated autoradiographic and biochemical study of DNA labelling in equine abortion virus hepatitis, *Virology* **16:**460.

Gentry, G.A., Lawson, L.A., and Randall, C.C., 1964, Replication of a DNA virus in thymidine-deficient mammalian cells, *J. Bacteriol.* **88:**1324.

Gentry, G.A., Evans, S., and Campbell, D.E., 1975, Herpesvirus-induced availability of host DNA to exogenous endonuclease, *Acta Virol.* **19:**401.

Gentry, G.A., Aswell, J.F., Allen, G.P., and Campbell, D.E., 1978, Arabinosylthymine, *in vivo* effectiveness in a systemic herpesvirus infection, in: *Proceedings of the 3rd International Symposium on Herpesviruses and Oncogenesis*, Cambridge, Massachusetts, IARC Sci. Publ. No. 24, pp. 1007–1012.

Gentry, G.A., McGowan, J., Barnett, J., Nevins, R., and Allen, G., 1979, Arabinosylthymine, a selective inhibitor of herpesvirus replication: Current status of *in vivo* studies, *Adv. Ophthalmol.* **38:**164.

Gentry, G.A., Nevins, R., McGowan, J.J., Barnett, J.M., and Holton, R., 1981, Factors in the design of pyrimidine nucleoside analogs as antiherpesviral drugs, in: *Proceedings of the 2nd International Symposium on Antiviral Chemotherapy*, Hamburg, 1980 (K. Gauri, ed.), pp. 61–74, Academic Press, New York.

Girard, A., Greig, A.S., and Mitchell, D., 1968, A virus associated with vulvitis and balantitis in the horse: A preliminary report, *Can. J. Comp. Med.* **32:**603.

Glasgow, L.A., and Habel, K., 1962, The role of interferon in vaccinia virus infection of mouse embryo tissue culture, *J. Exp. Med.* **115:**503.

Gutekunst, D.E., Malmquist, W.A., and Becvar, C.S., 1978, Antigenic relatedness of equine herpes virus type 1 and 3, *Arch. Virol.* **56:**33.

Henle, W., Henle, G., Deinhardt, F., and Bergs, V.V., 1959, Studies on persistent infections in tissue cultures. IV. Evidence for the production of an interferon in MCN cells by myxoviruses, *J. Exp. Med.* **110:**525.

Henry, B.E., Glaser, R., Hewetson, J., and O'Callaghan, D.J., 1978, Expression of altered ribonucleotide reductase activity associated with the replication of the Epstein–Barr virus, *Virology* **89:**262.

Henry, B.E., Newcomb, W.W., and O'Callaghan, D.J., 1979, Biological and biochemical properties of defective interfering particles of equine herpesvirus type 1, *Virology* **92:**495.

Henry, B.E., Newcomb, W.W., and O'Callaghan, D.J., 1980, Alterations in viral protein synthesis and capsid production in infection with DI particles of herpesviruses, *J. Gen. Virol.* **47:**343.

Henry, B.E., Robinson, R.A., Dauenhauer, S.A., Atherton, S.S., Hayward, G.S., and O'Callaghan, D.J., 1981, Structure of the genome of equine herpesvirus type 1, *Virology* **115:**97.

Holland, J.J., Villarreal, L.P., Welsh, R.M., Oldstone, M.B.A., Kohne, D., Lazzarini, R., and Scolnick, E., 1976, Long-term persistent vesicular stomatitis virus and rabies virus infection of cells *in vitro*, *J. Gen. Virol.* **33:**193.

Honess, R.W., and Roizman, B., 1974, Regulation of herpesvirus macromolecular synthesis. I. Cascade regulation of the synthesis of three groups of viral proteins. *J. Virol.* **14:**8.

Honess, R.W., and Roizman, B., 1975, Regulation of herpesvirus macromolecular synthesis: Sequential transition of polypeptide synthesis requires functional viral polypeptides. *Proc. Natl. Acad. Sci. USA,* **72:**1276.

Huang, A.S., 1973, Defective interfering viruses, *Annu. Rev. Microbiol.* **27:**101.

Huang, A.S., and Baltimore, D., 1977, Defective interfering animal viruses, in: *Comprehensive Virology*, Vol. 10 (H. Fraenkel-Conrat and R.R. Wagner, eds.), pp. 73–116, Plenum Press, New York.

Huang, H.L., Szabocsik, J.M., Randall, C.C., and Gentry, G.A., 1971, Equine abortion (herpes) virus-specific RNA, *Virology* **45:**381.

Huszar, D., and Bachetti, S., 1981, Partial purification and characterization of the ribonucleotide reductase induced by herpes simplex virus infection of mammalian cells, *J. Virol.* **37:**580.

Jackson, T., and Kendrick, J.W., 1971, Paralysis of horses associated with equine herpesvirus 1 infection, *J. Am. Vet. Med. Assoc.* **158:**1351.

Jackson, T.A., Osburn, B.I., Cordy, D.R., and Kendrick, J.W., 1977, Equine herpesvirus infection of horses: Studies on the experimentally induced neurologic disease, *Am. J. Vet. Res.* **38:**709.

Jamieson, A.T., and Subak-Sharpe, J.H., 1974, Biochemical studies on the herpes simplex virus-specified deoxypyrimidine kinase activity, *J. Gen. Virol.* **24:**481.

Jamieson, A.T., Gentry, G.A., and Subak-Sharpe, J.H., 1974, Induction of both thymidine and deoxycytidine kinase activity by herpes viruses, *J. Gen. Virol.* **24:**465.

Jeleff, W., 1957, Beitrag zur fetalen Histopathologie des Virusabortes der Strete mit besonder Berücksichtigung der Differentialdiagnose, *Arch. Exp. Vet. Med.* **2:**906.

Jones, T.C., Sleiser, C.A., Mauer, F.D., Hale, M.W., and Roby, T.D., 1948, Transmission and immunization studies on equine influenza, *Am. J. Vet. Res.* **9**:243.

Kaerner, H.C., 1981, Structure and physical mapping of different classes of defective HSV-1 Ang DNA, in: *Developments in Molecular Virology*, Vol. I, *Herpesvirus DNA* (Y. Becker, ed.), pp. 129–148. Nijhoff, The Hague.

Karpas, A., 1966, Characterization of a new herpes-like virus isolated from foal kidney, *Ann. Inst. Pasteur* **110**:688.

Kawakami, Y., and Shimuzu, T., 1978, Combined immunizing effects of live and inactivated equine herpesviruses in horses, in: *Proc. 4th Int. Conf. Equine Infectious Diseases*, Lyon, 1976 (J. Bryans and H. Gerber, eds.), pp. 78–82, Veterinary Publications, Princeton, New Jersey.

Kawakami, Y., Kaji, T., Sugimura, K., Ishitoni, R., Shimuzu, T., and Matumoto, M., 1959, Histopathological study of aborted fetuses naturally infected with equine abortion virus with some epidemiological findings, *Jpn. J. Exp. Med.* **29**:635.

Kawakami, Y., Kaji, T., Ishitoni, R., Shimizu, T., and Matumoto, M., 1962, Etiologic study of an outbreak of acute respiratory disease among colts due to equine rhinopneumonitis virus, *Jpn. J. Exp. Med.* **32**:211.

Kaufman, R.H., Dreesman, G.R., Burek, J., Korhonen, M.O., Matson, D.O., Melnick, J.L., Powell, R.L., Purifoy, D.J., Courtney, R.J., and Adam, E., 1981, Herpesvirus-induced antigens in squamous cell carcinoma *in situ* of the vulva, *N. Engl. J. Med.* **305**:483.

Kemp, M.C., 1975, Equine herpes virus type 1: Structural proteins and viral-induced alterations of RNA and DNA polymerase activities, Doctoral dissertation, University of Mississippi Medical Center, Jackson.

Kemp, M.C., Perdue, M.L., Rogers, H.W., O'Callaghan, D.J., and Randall, C.C., 1974, Structural polypeptides of the hamster strain of equine herpes virus type 1: Products associated with purification, *Virology* **61**:361.

Kemp, M.C., Cohen, J.C., O'Callaghan, D.J., and Randall, C.C., 1975, Equine herpesvirus-induced alterations in nuclear RNA and DNA polymerase activities, *Virology* **68**:467.

Kit, S., and Dubbs, D.R., 1963a, Acquisition of thymidine kinase activity by herpes simplex infected mouse fibroblast cells, *Biochem. Biophys. Res. Commun.* **11**:55.

Kit, S., and Dubbs, D.R., 1963b, Non-functional thymidine kinase cistron in bromodeoxyuridine resistant strains of herpes simplex virus, *Biochem. Biophys. Res. Commun.* **13**:500.

Kit, S., Leung, W.C., Jorgensen, G.N., Trkula, D., and Dubbs, D.R., 1975, Viral-induced thymidine kinase isozymes, *Prog. Med. Virol.* **21**:13.

Kress, F., 1941, Virus abortus bei der Stute, *Wien. Tieraerztl. Wochenschr.* **28**:320.

Kress, F., 1944, Uber virusbedingten Abortus bei Pferdestuten in der Donaugauen, *Z. Infektionskr. Haustiere* **60**:180.

Krogsrud, J., and Onstad, O., 1971, Equine coital exanthema: Isolation of a virus and transmission experiments, *Acta Vet. Scand.* **12**:1.

Ladin, B.F., Blankenship, M.L., and Ben-Porat, T., 1980, Replication of herpes-virus DNA. V. Maturation of concatemeric DNA of pseudorabies virus to genome length is related to capsid formation, *J. Virol.* **33**:1151.

Laemmli, U.K., 1975, Characterization of DNA condensates induced by poly(ethylene oxide) and polylysine, *Proc. Natl. Acad. Sci. U.S.A.* **72**:4288.

Lawrence, W.C., 1971a, Evidence for a relationship between equine abortion (herpes) virus deoxyribonucleic acid synthesis and the S phase of the KB cell mitotic cycle, *J. Virol.* **7**:736.

Lawrence, W.C., 1971b, Nucleic acid and protein synthesis in KB cells infected with equine abortion virus (equine herpesvirus type 1), *Am. J. Vet. Res.* **32**:41.

Lemaster, S., and Roizman, B., 1980, Herpes simplex virus phosphoproteins. II. Characterization of the virion protein kinase and of the polypeptides phosphorylated in the virion, *J. Virol.* **35**:798.

Leung, W.C., Dubbs, D.R., Trkula, D., and Kit, S., 1975, Mitochondrial and herpesvirus-specific deoxypyrimidine kinases, *J. Virol.* **16**:486.

Lieberman, M., Pascale, A., Schafer, T.W., and Came, P.E., 1972, Effect of antiviral agents in equine abortion virus-infected hamsters, *Antimicrob. Agents Chemother.* **1:**143.
Little, P.B., and Thorsen, J., 1974, Virus involvement in equine paresis, *Vet. Rec.* **95:**575.
Ludwig, H., Biswal, N., Bryans, J.T., and McCombs, R.M., 1971, Some properties of the DNA from a new equine herpesvirus, *Virology* **45:**534.
Manninger, R., 1949, Studies on infectious abortion in mares due to a filterable virus, *Acta Vet. Hung.* **1:**1.
Manniger, R., and Csontos, J., 1941, Virusabortus der Stuten, *Dtsch. Tieraerztl. Wochenschr.* **49:**103.
Mathews, R.E.F., 1979, Classification and nomenclature of viruses, *Intervirology* **12:**166.
Matumoto, M., Ishizaki, R., and Shimizu, T., 1965, Serological survey of equine rhinopneumonitis virus infection among horses in various countries, *Arch. Virusforsch.* **15:**609.
Mayr, A., Bohm, H.O., Brill, J., and Woyciechowska, S., 1965, Charakterisierung eines Stutenabortvirus aus Polen and Vergleich mit bekannten rhinopneumonitis Virus–Stammen des Pferdes, *Arch. Virusforsch.* **17:**216.
McCollum, W.H., 1966, Equine rhinopneumonitis, in: *Proceedings of the 1st International Conference on Equine Infectious Diseases*, Stresa, 1966 (J. Bryans, ed.), pp. 94–100, Grayson Foundation, Lexington, Kentucky.
McCollum, W.H., Doll, E.R., Wilson, J.C., and Johnson, C.B., 1962, Isolation and propagation of equine rhinopneumonitis virus in primary monolayer kidney cell cultures of domestic animals, *Cornell Vet.* **52:**164.
McGowan, J.J., 1980, Purification and characterization of cellular and herpesviral-induced deoxythymidine kinases, Doctoral dissertation, University of Mississippi Medical Center, Jackson.
McGowan, J.J., Allen, G.P., and Gentry, G.A., 1979, Purification and biochemical characterization of deoxythymidine kinase of deoxythymidine kinase-deficient mouse 3T3 cells biochemically transformed by equine herpesvirus type 1, *Infect. Immun.* **25:**610.
McGowan, J.J., Allen, G.P., Barnett, J.M., and Gentry, G.A., 1980a, Biochemical characterization of equine herpesvirus type 3-induced deoxythymidine kinase purified from lytically infected horse embryo dermal fibroblasts, *J. Virol.* **34:**474.
McGowan, J.J., Allen, G.P., Barnett, J.M., and Gentry, G.A., 1980b, Deoxythymidine kinase metabolism in equine herpesvirus type 3 infected horse embryo dermal fibroblasts, *Virology* **101:**516.
Meinkoth, J., and Kennedy, S.I.T., 1980, Semliki Forest virus persistence in mouse L929 cells, *Virology* **100:**141.
Miller, R., Iltis, J.P., and Rapp, F., 1977, Differential effect of arabinofuranosylthymine on the replication of human herpesvirus, *J. Virol.* **23:**679.
Miller, W.C., 1966, Personal communication in Burrows, R., and Goodridge, D., 1973, *In vivo* and *in vitro* studies of equine rhinopnuemonitis virus strains, in: *Proceedings of the 3rd International Conference on Equine Infectious Diseases*, Paris, 1972 (J. Bryans and H. Gerber, eds.), p. 306, S. Karger, Basel.
Müller, M.T., and Hudson, J.B., 1977, Thymidine kinase activity in mouse 3T3 cells infected by murine cytomegalovirus (MCV), *Virology* **80:**430.
Mumford, J.A., and Thompson, S.R., 1978, Serological methods for identification of slowly-growing herpesviruses isolated from the respiratory tract of horses, in: *Proceedings of the 4th International Conference on Equine Infectious Disease*, Lyon, 1976 (J. Bryans and H. Gerger, eds.), pp. 49–52, Veterinary Publications, Princeton, New Jersey.
Nazerian, K., 1974, DNA configuration in the core of Marek's disease virus, *J. Virol.* **13:**1148.
Nishiyama, Y., 1977, Studies of L cells persistently infected with VSV: Factors involved in the regulation of persistent infection, *J. Gen. Virol.* **35:**265.
O'Callaghan, D.J., and Randall, C.C., 1976, Molecular anatomy of herpesviruses: Recent studies, *Prog. Med. Virol.* **22:**152.
O'Callaghan, D.J., Cheevers, W.P., Gentry, G.A., and Randall, C.C., 1968a, Kinetics of viral

and cellular DNA synthesis in equine abortion (herpes) virus infection of L-M cells, *Virology* **36:**104.

O'Callaghan, D.J., Hyde, J.M., Gentry, G.A., and Randall, C.C., 1968b, Kinetics of viral deoxyribonucleic acid, protein, and infectious particle production and alterations in host macromolecular synthesis in equine abortion (herpes) virus-infected cells, *J. Virol.* **2:**793.

O'Callaghan, D.J., Rogers, H.W., and Randall, C.C., 1972a, Amino acid composition of equine abortion (herpes) virus, *Virology* **47:**842.

O'Callaghan, R., Randall, C.C., and Gentry, G.A., 1972b, Herpesvirus replication *in vivo*, *Virology* **49:**784.

O'Callaghan, D.J., Kemp, M.C., and Randall, C.C., 1977a, Properties of nucleocapsid species isolated from an *in vivo* herpesvirus infection, *J. Gen. Virology* **37:**585.

O'Callaghan, D.J., Randall, C.C., Duff, R., Allen, G., Jackson, J., and Blasecki, J.W., 1977b, Oncogenic transformation of inbred hamster cells by equine herpesvirus type 1, in: Proceedings of the 3rd International Symposium on Oncogenesis and Herpesviruses, p. 177, Cambridge, Massachusetts.

O'Callaghan, D.J., Allen, G. P., and Randall, C.C., 1978, Structure and replication of equine herpesvirus, in: *Equine Infectious Diseases IV* (J.T. Bryans and H. Gerber, eds.), pp. 1–32, Veterinary Publications, Princeton, New Jersey.

O'Callaghan, D.J., Henry, B.E., Wharton, J.H., Dauenhauer, S.A., Vance, R.B., Staczek, J., and Robinson, R.A., 1981a, Equine herpesvirus: Biochemical studies on genomic structure, DI particles, oncogenic transformation, and persistent infection, in: *Developments in Molecular Virology*, Vol. I, *Herpesvirus DNA* (Y. Becker, ed.), pp. 387–418, Nijhoff, The Hague.

O'Callaghan, D.J., Robinson, R.A., and Dauenhauer, S., 1981b, Herpesviral DNA sequences associated with oncogenesis and persistent infection, pp. 328, Fifth International Congress of Virology, Strasbourg, France.

Palmer, E.L., Martin, M.L., and Gray, G.W., 1975, The ultrastructure of disrupted herpesvirus nucleocapsids, *Virology* **65:**260.

Pascoe, E.R., Spradbrow, P.B., and Bagust, T.J., 1968, Equine coital exanthema, *Aust. Vet. J.* **44:**485.

Pedrali-Noy, G., and Spadari, S., 1980, Mechanism of inhibition of herpes simplex virus and vaccinia virus DNA polymerases by aphidicolin, a highly specific inhibitor of DNA replication in eucaryotes, *J. Virol.* **36:**457.

Perdue, M.L., 1975, Studies of the molecular anatomy of equine herpesvirus type-1 (EHV-1), Doctoral dissertation, University of Mississippi Medical Center, Jackson.

Perdue, M.L., Kemp, M.C., Randall, C.C., and O'Callaghan, D.J., 1974, Studies of the molecular anatomy of the L-M cell strain of equine herpesvirus type 1 proteins of the nucleocapsid and intact virion, *Virol.* **59:**201.

Perdue, M.L., Cohen, J.C., Kemp, M.C., Randall, C.C., and O'Callaghan, D.J., 1975, Characterization of three species of nucleocapsids of equine herpes-virus type 1 (EHV-1), *Virology* **64:**187.

Perdue, M.L., Cohen, J.C., Randall, C.C., and O'Callaghan, D.J., 1976, Biochemical studies of the maturation of herpesvirus nucleocapsid species, *Virology* **74:**194.

Petzoldt, K., 1970, Equine–coital exanthem, *Berl. Muench. Tieraerztl. Wochenschr.* **83:**93.

Petzoldt, K., Luttman, U., Pohlenz, J., and Teichert, U., 1972, Virological studies on the central nervous systems of equine fetuses and the lesions in mares with central nervous symptoms after abortion caused by equine herpesvirus 1, *Schweiz. Arch. Tierheilkd.* **114:**129.

Plummer, G., and Waterson, A.P., 1963, Equine herpes viruses, *Virology* **19:**412.

Plummer, G., Bowling, C.P., and Goodheart, C.R., 1969a, Comparison of four horse herpesviruses, *J. Virol.* **4:**738.

Plummer, G., Goodheart, C.R., Henson, D., and Bowling, C.P., 1969b, A comparative study of the DNA density and behavior in tissue cultures of fourteen different herpesviruses, *Virology* **39:**134.

Plummer, G., Goodheart, C.R., and Studdert, M.J., 1973, Equine herpesviruses, antigenic relationships and deoxyribonucleic acid densities, *Infect. Immun.* **8**:621.

Powell, K.L., Littler, E., and Purifoy, D.J.M., 1981, Nonstructural proteins of herpes simplex virus. II. Major virus-specific DNA-binding protein, *J. Virol.* **39**:894.

Preble, O.T., and Youngner, J.S., 1973a, Temperature-sensitive defect of mutants isolated from L cells persistently infected with Newcastle disease virus, *J. Virol.* **12**:472.

Preble, O.T., and Youngner, J.S., 1973b, Selection of temperature-sensitive mutants during persistent infection: Role in maintenance of persistent Newcastle disease virus infections of L cells, *J. Virol.* **12**:481.

Preble, O.T., and Youngner, J.S., 1975, Temperature-sensitive viruses and the etiology of chronic and inapparent infections, *J. Infect. Dis.* **131**:467.

Prickett, M.E., 1970a, The pathology of disease caused by equine herpesvirus 1, in: *Proceedings of the 2nd International Conference of Equine Infectious Diseases*, Paris, 1969 (J. Bryans and H. Gerber, eds.), pp. 24–33, S. Karger, Basel.

Prickett, M.E., 1970b, Discussion and short communication, in: *Proceedings of the 2nd International Conference of Equine Infectious Diseases*, Paris, 1969 (J. Bryans and H. Gerber, eds.), p. 60, S. Karger, Basel.

Rand, T.H., and Ben-Porat, T., 1980, Distribution of sequences homologous to the DNA of herpes simplex virus, type 1 and 2, in the genome of pseudorabies virus, *Intervirology* **13**:48.

Randall, C.C., 1945a, Propagation of equine abortion virus in tissue of other animals, *Fed. Proc. Fed. Am. Soc. Exp. Biol.* **13**:605.

Randall, C.C., 1954b, Propagation *in vitro* of equine abortion virus in human epithelial cells (strain Hela, Gey carcinoma of cervix), *Am. J. Pathol.* **30**:659.

Randall, C.C., and Bracken, E.C., 1957, Studies on hepatitis in hamsters infected with equine abortion virus. I. Sequential development of inclusions and the growth cycle, *Am. J. Pathol.* **33**:709.

Randall, C.C., and Lawson, L.A., 1962, Adaptation of equine abortion virus to Earle's L-cells in serum-free medium with plaque formation, *Proc. Soc. Exp. Biol. Med.* **110**:487.

Randall, C.C., and Walker, B.M., 1963, Degradation of deoxyribonucleic acid and alteration of nucleic acid metabolism in suspension cultures of L-M cells infected with equine abortion virus, *J. Bacteriol.* **86**:138.

Randall, C.C., Ryden, F.W., Doll, E.R., and Schell, F.S., 1953, Cultivation of equine abortion virus in fetal horse tissue *in vitro, Am. J. Pathol.* **29**:139.

Randall, C.C., Stevens, W.C., and Bracken, E.C., 1954a, Viremia in hamsters inoculated with equine abortion virus, *Am. J. Pathol.* **30**:654.

Randall, C.C., Turner, D., and Doll, E.R., 1954b, The cultivation of equine abortion virus in cat tissue *in vitro, Am. J. Pathol.* **30**:1049.

Randall, C.C., Todd, W.M., Gentry, G.A., and Bracken, E.C., 1959, Relationships of virus multiplication to the development of morphological and biochemical changes in the cell, *Ann. N.Y. Acad. Sci.* **81**:38.

Randall, C.C., Gafford, L.G., Gentry, G.A., and Lawson, L.A., 1965, Lability of host-cell DNA in growing cell cultures due to Mycoplasma, *Science* **149**:1098.

Randall, C.C., Rogers, H.W., Downer, D.N., and Gentry, G.A., 1972, Protein kinase activity in equine herpesvirus, *J. Virol.* **9**:216.

Reanney, D.C., 1981, Evolutionary virology, a molecular overview, in: *The Human Herpesviruses: An Interdisciplinary Perspective* (A.J. Nahmias, W.R. Dowdle, and R.F. Schinazi, eds.), p. 519, Elsevier, New York.

Reichmann, M.E., and Schnitzlein, W.M., 1979, Defective interfering particles of rhabdoviruses, in: *Current Topics in Microbiology and Immunology*, Vol. 86 (W. Arber, S. Falkow, W. Henle, P.H. Hofschneider, J.H. Humphrey, J. Klein, P. Koldovský, H. Koprowski, O. Maaløe, F. Melchers, R. Rott, H.G. Schweiger, L. Syrucek, and P.K. Vogt, eds.), pp. 123–168, Springer–Verlag, Berlin and Heidleberg.

Reyes, G.R., LaFemina, R., Hayward, S.D., and Hayward, G.S., 1979, Morphological transformation by DNA fragments of human herpes viruses: Evidence for two distinct transforming regions in HSV-1 and HSV-2 and lack of correlation with biochemical transfer of the thymidine kinase gene, *Cold Spring Harbor Symp. Quant. Biol.* **44:**629.

Robinson, R.A., and O'Callaghan, D.J., 1981, The organization of integrated herpesvirus DNA sequences in equine herpesvirus type 1 transformed and tumor cells, in: *Developments in Molecular Virology*, Vol. I, *Herpesvirus DNA* (Y. Becker, ed.), pp. 419–436, Nijoff, The Hague.

Robinson, R.A., and O'Callaghan, D.J., 1983, A specific viral DNA sequence is stably integrated in herpesvirus oncogenically-transformed cells. *Cell* (in press).

Robinson, R.A., Henry, B.E., Duff, R.G., and O'Callaghan, D.J., 1980a, Oncogenic transformation by equine herpesviruses (EHV). I. Properties of hamster embryo cells transformed by UV-irradiated EHV-1, *Virology* **101:**335.

Robinson, R.A., Tucker, P.W., and O'Callaghan, D.J., 1980b, Mapping and molecular cloning of host genomic and herpesvirus oncogenic DNA sequences, Fifth International Herpesvirus Workshop, p. 120, Cold Spring Harbor Laboratory, Cold Spring Harbor, New York.

Robinson, R.A., Vance, R.B., and O'Callaghan, D.J., 1980c, Oncogenic transformation by equine herpesvirus (EHV). II. Co-establishment of persistent infection and oncogenic transformation of hamster embryo cells by equine herpesvirus type 1 preparations enriched for EHV-1 defective interfering particles, *J. Virol.* **36:**204.

Robinson, R.A., Dauenhauer, S.A., Baumann, R., and O'Callaghan, D.J., 1981a, Structure and arrangement of the genome of EHV-1 DI particles and of integrated herpesvirus sequences in transformed and tumor cells, International Workshop on Herpesvirus, p. 221, Bologna, Italy.

Robinson, R.A., Tucker, P.W., Dauenhauer, S.A., and O'Callaghan, D.J., 1981b, Molecular cloning of equine herpesvirus type 1 DNA: Analysis of standard and defective viral genomes and viral sequences in oncogenically transformed cells, *Proc. Natl. Acad. Sci. U.S.A.* **78:**6684.

Roizman, B., 1973, Provisional labels for herpesviruses, *J. Gen. Virol.* **20:**417.

Roizman, B., and Furlong, D., 1974, The replication of herpesvirus, in: *Comprehensive Virology*, Vol. 3 (H. Fraenkel–Conrat and R. R. Wagner eds.), pp. 229–403, Plenum Press, New York.

Roumillat, L.F., Feorino, P.M., and Lukert, P.D., 1979, Persistent infection of a human lymphoblastoid cell line with equine herpesvirus type 1, *Infect. Immun.* **24:**539.

Roux, L., and Holland, J.J., 1979, Role of defective interfering particles of Sendai virus in persistent infections, *Virology* **93:**91.

Roux, L., and Holland, J.J., 1980, Viral genome synthesis in BHK21 cells persistently infected with Sendai virus, *Virology* **100:**53.

Russell, W.C., and Crawford, L.V., 1964, Properties of the nucleic acids from some herpes group viruses, *Virology* **22:**288.

Ruyechan, W.T., Dauenhauer, S.A., and O'Callaghan, D.J., 1982, Electron microscopic study of equine herpesvirus type 1 DNA, *J. Virol.* **42:**297.

Sabine, M., Robertson, G.R., and Whalley, J.M., 1981, Differentiation of subtypes of equine herpesvirus 1 by restriction endonuclease analyses, *Aust. Vet. J.* **57:**148.

Salyi, J., 1942, Beitrag zur Pathohistologie des Virusabortus der Stuten, *Arch. Wiss. Prakt. Tierheilkd.* **77:**244.

Saxegaard, F., 1966, Isolation and identification of equine rhinopneumonitis virus 1 (equine abortion virus) from cases of abortion and paralysis, *Nord. Vet. Med.* **18:**501.

Sekellick, M.J., and Marcus, P.I., 1978, Persistent infection. I. Interferon-inducing defective-interfering particles as mediators of cell sparing: Possible role in persistent infection by vesicular stomatitis virus, *Virology* **85:** 175.

Sekellick, M.J., and Marcus, P.I., 1979, Persistent infection. II. Interferon-inducing temperature-sensitive mutants as mediators of cell sparing: Possible role in persistent infection by vesicular stomatitis virus, *Virology* **95:**36.

Shimizu, T., Ishizaki, R., Kono, Y., and Ishii, S., 1957, Propagation of equine abortion virus in horse kidney culture, *Jpn. J. Exp. Med.* **27:**175.

Shimuzu, T., Ishizaki, R., Ishii, S., Kawakami, Y., Sugimura, K., and Matumoto, M., 1959, Isolation of equine abortion virus in horse kidney cell culture, *Jpn. J. Exp. Med.* **29:**643.
Soehner, R.L., Gentry, G.A., and Randall, C.C., 1965, Some physiocochemical properties of equine abortion virus nucleic acid, *Virology* **26:**394.
Southern, E., 1975, Detection of specific sequences among DNA fragments separated by gel electrophoresis, *J. Mol. Biol.* **98:**503.
Sovinova, O., Tumova, M.B., Pouska, F., and Nemec, J., 1958, Isolation of a virus causing respiratory disease in horses, *Acta Virol.* **2:**52.
Spear, P.G., and Roizman, B., 1980, Herpes simplex viruses, in: *Molecular Biology of Tumor Viruses*, Part 2, *DNA Tumor Viruses* (J. Tooze, ed.), pp. 615–745, Cold Spring Harbor Laboratory, New York.
Staczek, J., and Atherton, S., 1981, Genomic structure and oncogenic potential of equine cytomegalovirus and equine coital exanthema virus, in: *Proceedings of the International Workshop on Herpesviruses*, Bologna, 1981, p. 227, Esculapio, Pub. Co., Bologna.
Staczek, J., Wharton, J.H., Dauenhauer, S.A., and O'Callaghan, D.J., 1983, Oncogenic transformation by equine herpesviruses (EHV): Co-establishment of persistent infection and oncogenic transformation of hamster embryo cells by equine cytomegalovirus (submitted).
Stanners, C.P., and Goldberg, V.J., 1975, On the mechanism of neurotropism of vesicular stomatitis virus in newborn hamsters: Studies with temperature-sensitive mutants, *J. Gen. Virol.* **29:**281.
Stevely, W.S., 1977, Inverted repetition in the chromosome of pseudorabies virus, *J. Virol.* **22:**232.
Stock, D.A., and Gentry, G.A., 1969, Mycoplasmal deoxyribonuclease activity in virus-infected L-cell cultures, *J. Virol.* **3:**313.
Stollar, V., 1979, Defective interfering particles of togaviruses, in: *Current Topics in Microbiology and Immunology*, Vol. 86 (W. Arber, S. Falkow, W. Henle, P.H. Hofschneider, J.H. Humphrey, J. Klein, P. Koldovský, H. Koprowski, O. Maaløe, F. Melchers, R. Rott, H.G. Schweiger, L. Syrucek, and P.K. Vogt, eds.), pp. 35–66,Springer-Verlag, Berlin and Heidelberg.
Strand, M., and August, J.T., 1971, Protein kinase and phosphate acceptor proteins in Rauscher murine leukemia virus, *Nature (London) New Biol.* **233:**137.
Straus, S.E., Owens, J., Ruyechan, W.T., Takiff, H.E., Casey, T.A., Vande Woude, G.F., and Hay, J., 1982, Molecular cloning and physical mapping of varicella zoster virus DNA, *Proc. Natl. Acad. Sci. U.S.A.* **79:**993.
Studdert, M.J., 1974, Comparative aspects of equine herpesvirus, *Cornell Vet.* **64:**94.
Studdert, M.J., and Blackney, M., 1979, Equine herpesviruses: On the differentiation of respiratory from foetal strains of type 1. *Aust. Vet. J.* **55:**488.
Studdert, M.J., Simpson, T., and Roizman, B., 1981, Differentiation of respiratory and abortigenic isolates of equine herpesvirus 1 by restriction endonucleases, *Science* **214:**562.
Sullivan, D., Atherton, S., Ruyechan, W., Dauenhauer, S., and O'Callaghan, D., 1982, Physical structure and properties of the equine herpesvirus type 3 genome. Proceedings of the Sixth Cold Spring Harbor Meeting on Herpesviruses, p. 181, Cold Spring Harbor Laboratory, New York.
Szabocsik, J.M., 1971, Effects of equine abortion (herpes) virus on RNA metabolism in L-M cells and the progress of viral induced changes in metabolically inhibited cells, Doctoral dissertation, University of Mississippi Medical Center, Jackson.
Turner, A.J., Studdert, M.J., and Peterson, J.E., 1970, Equine herpesviruses. 2. Persistence of equine herpesviruses in expermentally infected horses and experimental induction of abortion, *Aust. Vet. J.* **46:**90.
Turtinen, L.W., and Allen, G.P., 1983, Identification of the envelope glycoproteins of equine herpesvirus type 1, *J. Gen. Virol.* (in press).
Turtinen, L.W., Allen, G.P., Darlington, R.W., and Bryans, J.T., 1981, Serologic and molecular comparisons of several equine herpesvirus type 1 strains, *Am. J. Vet. Res.* **42:**2099.

Vance, R.B., Robinson, R.A., Henry, B.E., and O'Callaghan, D.J., 1978, Herpesvirus oncogenesis and persistent infection, *Clin. Res.* **26:**777.

Vernon, S.K., Lawrence, W.C., and Cohen, G.H., 1974, Morphological components of herpesvirus. I. Intercapsomeric fibrils and the geometry of the capsid, *Intervirology* **4:**237.

Vernon, S.K., Lawrence, W.C., Long, C.A., Cohen, G.H., and Rubin, B.A., 1978, Herpesvirus vaccine development: Study of virus morphological components, in: *New Trends and Developments in Vaccines* (A. Voller and H. Freidman, eds.), pp. 179–210, University Park Press, Baltimore.

Virat, J., Guillon, J.C., and Pouret, E., 1972, Etude d'un virus isolé chez une jument atleinte d'herpes genital, *Bull. Acad. Vet. Fr.* **45:**57.

Wagner, R.R., 1974, Pathogenicity and immunogenicity for mice of temperature-sensitive mutants of vesicular stomatitis virus, *Infect. Immun.* **10:**309.

Webber, C.E., and Whalley, J.M., 1978, Widespread occurrence in Australian marsupials of neutralizing antibodies to a herpesvirus from a Parma wallaby, *Aust. J. Exp. Biol. Med. Sci.* **56:**351.

Weissbach, A., Hong, S.L., Aucker, J., and Muller, R., 1973, Characterization of herpes simplex virus-induced deoxyribonucleic acid polymerase, *J. Biol. Chem.* **248:**6270.

Westerfield, C., and Dimock, W.W., 1946, The pathology of equine virus abortion, *J. Am. Vet. Med. Assoc.* **109:**101.

Whalley, J.M., and Webber, C.E., 1979, Characteristics of Parma wallaby herpes-virus grown in maruspial cells, *J. Gen. Virol.* **45:**423.

Whalley, J.M., Robertson, G.R., and Davison, A.J., 1981, Analysis of the genome of equine herpesvirus type 1: Arrangement of cleavage sites for restriction endonuclease EcoRI, Bgl II, and BamHI, *J. Gen. Virol.* **57:**307.

Wharton, J.H., Henry, B.E., and O'Callaghan, D.J., 1981, Equine cytomelgalovirus: Cultural characteristics and properties of viral DNA, *Virology* **109:**106.

Yeo, J., Killington, R.A., Watson, D.H., and Powell, K.L., 1981, Studies on cross-reactive antigens in the herpesvirus viruses. *Virology* **108:**256.

Zavada, V., Erban, V., Rezacora, D., and Vonka, V., 1976, Thymidine kinase in cytomegalovirus infected cells, *Arch. Virol.* **52:**333.

CHAPTER 6

Biology and Properties of Fish and Reptilian Herpesviruses

KEN WOLF

I. INTRODUCTION

This chapter updates and adds to information contained in earlier reviews of herpesvirus of reptiles and fishes (Wolf, 1973, 1979; Lunger and Clark, 1978; Clark and Lunger, 1981). This review includes at least 11, and possibly as many as 14, herpesviruses (Table I). The uncertainty of number stems from the fact that some of the agents have yet to be isolated, and therefore their antigenic, serological, and other critical relationships are unknown. Furthermore, the rather high degree of host specificity exhibited by most herpesviruses suggests the possibility that the two freshwater turtles (both family Testudinidae) could well be afflicted with a single virus. The same is true for the agent(s) seen in the venom of the three snakes of the family Elapidae; two of the snakes are members of the genus *Naja*, and all three share a common geographic area.

One generalization to be drawn from the collected reports is that in certain of their biological and biophysical properties, the herpesviruses of lower vertebrates share common characteristics. As an example, the representatives of poikilotherm vertebrate agents have temperature optima and ranges that conform to the preferred temperature or near-optimal environmental temperatures of their hosts.

Herpesvirus–host relationships of the fish and reptile agents also conform to patterns recognized among the herpesviruses of homeotherms. The channel catfish virus, *Herpesvirus salmonis*, and *Oncorhynchus masou* virus are agents of virulent disease and result in mortality of young

KEN WOLF • National Fish Health Research Laboratory, U.S. Fish and Wildlife Service, Kearneysville, West Virginia 25430.

TABLE I. Herpesviruses of the Classes Reptilia and Teleostomi (Bony Fishes)—1981

Agent	Scientific name of host(s)	Role or present status of virus
Class Reptilia		
Painted turtle herpesvirus	*Chrysemys picta*	Electron micrographs, diseased host
Pacific pond turtle herpesvirus	*Clemmys marmorata*	Electron micrographs, diseased hosts
Green sea turtle herpesvirus	*Chelonia mydas*	Etiological agent of gray patch disease
Iguana herpesvirus	*Iguana iguana*	Isolated, some evidence for virulence
Green lizard herpesvirus	*Lacerta viridis*	Electron micrographs, in papillomas
Elapid snake herpesvirus(es)	*Bungarus fasciatus*	Electron micrographs, role unknown
	Naja naja	Electron micrographs, role unknown
	Naja naja kaouthia	Electon micrographs, role unknown
Class Teleostomi		
Oncorhynchus masou virus (OMV)	*Oncorhynchus masou, O. kisutch*	Isolated, virulent, epithelial neoplasms
Herpesvirus salmonis	*Salmo gairdneri, O. nerka*	Isolated, experimentally virulent
Carp pox herpesvirus (*Herpesvirus epithelioma*)	*Cyprinus carpio* *C. carpio* × *Carassius auratus*	Electron micrographs, experimental transmission, epidermal hyperplasia
Channel catfish herpesvirus [Channel catfish virus (CCV)]	*Ictalurus punctatus*	Etiological agent of virulent disease
Turbot herpesvirus (*Herpesvirus scophthalmi*)	*Scophthalmus maximus*	Electron micrographs, cell hypertrophy
Walleye herpesvirus	*Stizostedion vitreum*	Isolated, epidermal hyperplasia

hosts. Although the story of salmonid herpesviruses is fragmentary, it is known that adult fish harbor the viruses, and that feature ensures generation-to-generation transmission or continuity of the pathogen. The gray patch disease of green sea turtles similarly fits the group of agents that are virulent for the young but incapable of evoking disease in older animals. Moreover, the skin, which is the target tissue of the gray patch agent, is similarly afflicted in other herpesvirus infections.

In some instances, neoplasms result from homeotherm herpesvirus infections. The agents associated with carp pox, walleye epidermal hyperplasia, and possibly the green iguana infection plausibly seem to play a causal role in neoplasia. It would not be surprising to find that these three viruses are highly virulent for the very young and that neoplasms are sequelae that occur only among some survivors.

Virulence for young hosts and neoplasm induction among survivors focus attention on a common research need in working with animals other than so-called domestic species and the need for young specific-pathogen-free animals for experimental infections. Some hatchery populations of fishes—notably some species of trout—have health histories that document freedom from specific pathogens; therefore, suitable test trout are available for research. At the other extreme, herpesviruses occur in such feral animals as turbot, iguanas, and cobras. How or where, for example, does one obtain young cobras, or more specifically, young cobras that do not harbor latent or interfering virus?

Just as most of the herpesviruses of homeotherms come from domesticated birds and mammals rather than from feral species, the herpesviruses of reptiles and fishes come from zoo specimens or from species under aquaculture. The agents themselves, like their hosts, originated in the wild but were recognized and came under scrutiny among animals in captivity—especially when they caused disease and mortality.

Knotty problems exist among the poikilotherm vertebrate viruses; several have yet to be isolated, e.g., carp pox agent and agent(s) from elapid snakes. The easiest to solve undoubtedly is the problem of cell toxicity of venom in which snake herpesvirus is found. Conceivably, one could infect embryonate snake eggs or hatchlings, produce a systemic infection, and derive a virus suspension free of lytic venom. In the two agents yet to be isolated from adult fishes, culture of neural tissue has not been reported for the channel catfish virus. The role of carp pox agent in young fish has yet to be determined; if it proves to be virulent, isolation of the virus from the young might be simple. Alternatively, and in either case, the virus may respond to temperature manipulation, as does the agent of Lucké adenocarcinoma of frogs.

Among viruses that have been isolated, those of fish and reptiles induce syncytia in susceptible cell cultures. Generally, too, the infected cells show basophilia, margination of chromatin, and intranuclear inclusions—all characteristics of homeotherm herpesviruses.

In this review, the word fish is used in the restricted sense and means only that class of vertebrates known as teleosts, or bony fishes. Still lower classes of vertebrates, loosely termed fishes, have yet to yield evidence of herpesvirus infections: class Chondrichthyes (sharks, skates, and rays) and class Agnatha (lampreys and hagfishes). Because herpesvirus has been visualized in mollusks, it is reasonable to expect that additional agents will eventually be found in animals such as sharks or lampreys. The search for such agents is now handicapped by the fact that cell lines have yet to be developed from either Chondrichthyes or Agnatha (Wolf and Mann, 1980).

Some of the original literature on the agents covered in this review contains qualifying terms such as herpeslike or herpesviruslike, but for convenience and simplicity, all are referred to here as herpesvirus.

II. REPTILIAN HERPESVIRUSES

A. Painted Turtle Herpesvirus

The painted turtle (*Chrysemys picta*) is a common chelonian member of the family Testudinidae; it is found in North American freshwater habitats from the Atlantic coast to eastern Washington State. A herpesvirus has been seen in affected tissues of a diseased adult male specimen from the Toronto, Ontario, Zoo (Cox *et al.*, 1980). Presumably, the turtle came from the local area.

The turtle had an abscessed swelling on the side of its head. After surgical excision, the site was topically disinfected with an iodophore and the animal injected with chloramphenicol and ascorbic acid. Nevertheless, the animal died 6 days later and was necropsied the same day. Gross findings included pulmonary edema, a friable and discolored liver, congested spleen, pale kidneys, and necrotic lesions on the plastron. Liver, lung, intestine, spleen, and kidney tissues were fixed for histopathological examination and electron microscopy.

Major pathological changes were found in the liver and lung. Many small foci of coagulation necrosis were present in the liver. The inflammatory response was minor, but nearby hepatocytes contained large eosinophilic intranuclear inclusions, and the nuclei additionally had marginated chromatin (Fig. 1). Bronchi showed accumulations of mononuclear and granulocytic cells, the latter often invading the mucosa. Lumens were usually filled with granulocytes, erythrocytes, and sloughed epithelial cells. Epithelial cells of the infundibula were metaplastic, vacuolated, and degenerate. Many cells displayed eosinophilic intranuclear inclusions (Fig. 2). Accumulations of inclusion-bearing epithelial cells, granulocytes, and fibrin were present in pulmonary infundibula. Occasional degenerated renal tubules contained mineralized debris, and small focal accumulations of granulocytes and mononuclear cells were present in interstitial tissue. Renal pathology, however, was judged to have resulted from inadequate water intake.

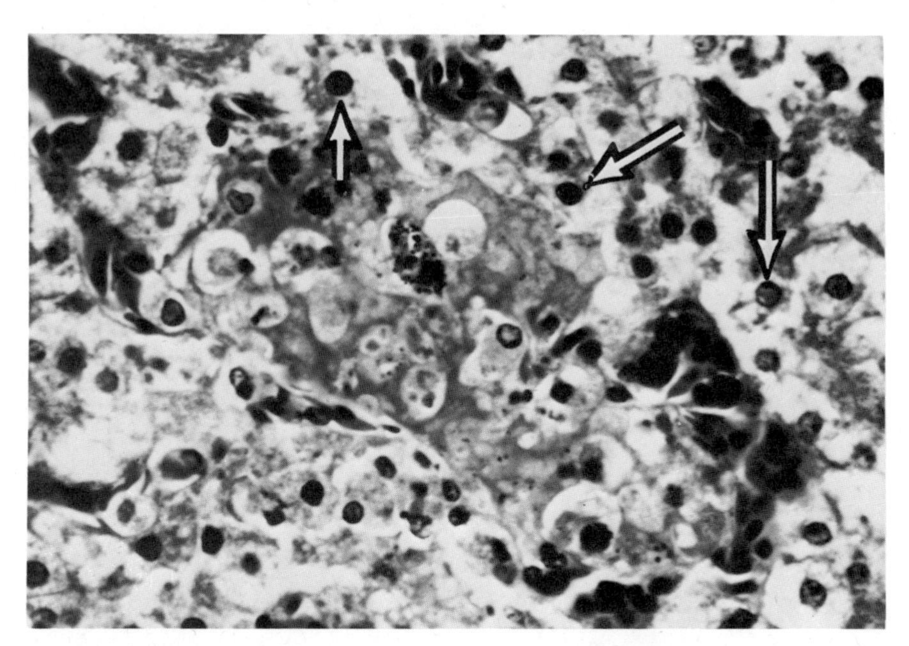

FIGURE 1. Necrotic focus in liver of painted turtle. Nearby hepatocytes contain dense intranuclear inclusions (arrows). Hematoxylin–eosin. ×560. From Cox *et al.* (1980). Reprinted with the permission for the Wildlife Disease Association.

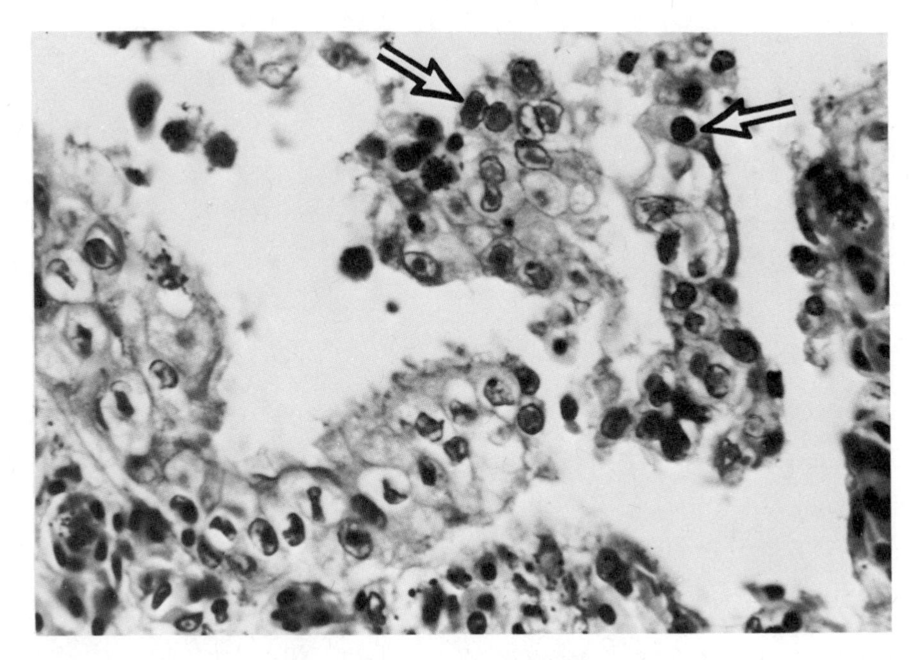

FIGURE 2. Section of painted turtle lung showing lumen with sloughed epithelial cells some of which have prominent intranuclear inclusions (arrows). Hemotoxylin–eosin. ×560. From Cox *et al. (1980). Reprinted with the permission of the Wildlife Disease Association.*

Evidence of the herpesvirus was obtained by electron microscopy. Hepatocytes and bronchial epithelial cells both showed numerous virions in cell nuclei (Fig. 3). Most of the particles were naked capsids measuring 85–115 nm, but some enveloped virions were also evident.

The authors noted that only fixed material was available, and accordingly neither animal inoculation nor isolation attempts could be carried out. They noted further, however, that the disease of the painted turtle resembled that of the Pacific pond turtle described by Frye *et al.* (1977).

B. Pacific Pond Turtle Herpesvirus

The Pacific pond turtle (*Clemmys marmorata*) is a common chelonian member of the family Testudinidae; it occupies fresh water and at

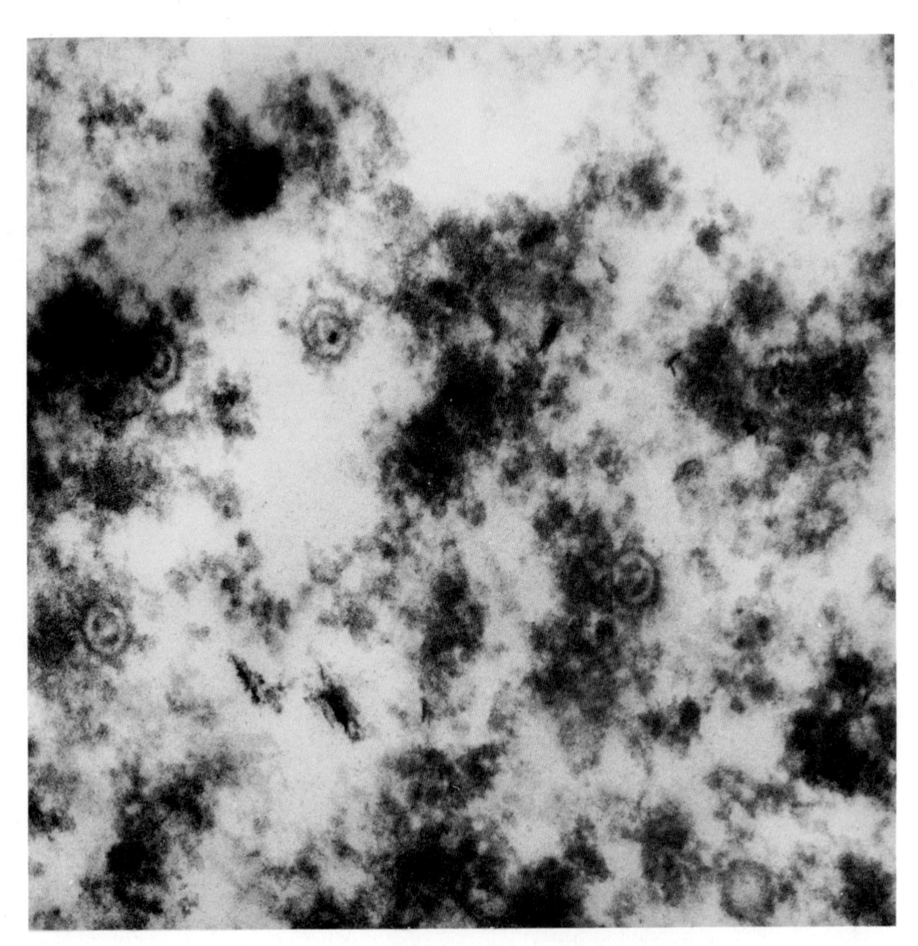

FIGURE 3. Herpesvirus particles in thin section of hepatocyte from a painted turtle. ×18,400. From Cox *et al.* (1980). Reprinted with the permission of the Wildlife Disease Association.

times brackish waters along the Pacific slope of the United States from San Diego to Puget Sound. Herpesvirus was seen in tissue sections of a diseased adult specimen that was necropsied after death (Frye *et al.*, 1977).

The case involved two wild specimens. The first was an adult female submitted for examination in a lethargic and anorexic condition and showing muscular weakness and feeble response to tactile stimulation. The skin of the neck and limbs, and the plastron, bore numerous petechiae and small ecchymotic hemorrhages. The turtle died within a few hours after admission.

Several weeks later, the owner submitted a companion adult male with identical signs, which also died shortly after admission. An acute or peracute disease course was indicated. Radiological, hematological, and histopathological examinations were carried out. Radiological examination showed no abnormalities, but blood films showed moderate lymphocytosis, a reduced eosinophil count, and vacuolated erythrocytes.

Internally, the liver and spleen of both turtles were slightly swollen; the kidneys were pale, but other organs appeared to be normal. Histological sections stained with hematoxylin and eosin showed acute hepatic necrosis. Many of the hepatocytes and some spleen and renal-tubule cells showed intranuclear inclusions and margination of chromatin. Small lymphocytes occurred in aggregates in liver, kidneys, and spleen, and the spleen was moderately hyperplastic.

Electron microscopy of thin sections of liver and spleen showed intranuclear icosahedra 100 nm in diameter. Some capsids were empty, but others bore electron-dense nucleoids. Enveloped particles (140 nm) occurred both extracellularly and in the cytoplasm. Bundles of dense fibers were present extracellularly in the cytoplasm and occasionally in cell nuclei. Although the point was not discussed by the authors, their micrograph of an intranuclear bundle resembled material found in cells infected with channel catfish virus or *Herpesvirus salmonis*.

The occurrence of similar disease signs and associated herpesvirus in two different freshwater turtles in the same family logically leads one to ask whether a single virus was involved. An opposing viewpoint is tenable, however, because the natural ranges of the two turtles do not overlap. One must then counter with questions regarding natural host–pathogen relationships and species composition of turtle populations kept in zoos or other captivity.

These matters may be clarified when new cases are brought to light, viral isolations are made, and definitive serological tests are carried out.

C. Green Sea Turtle Herpesvirus (Gray Patch Disease)

The green sea turtle (*Chelonia mydas*) is a large marine reptile native to the Americas and a member of the family Chelonidae. Its range extends along the Atlantic from New England to northern South America and in Pacific waters from southern California through Lower California. *Che-*

lonia mydas is one of the few turtles raised commercially for human food.

Herpesvirus involvement in what is termed gray patch disease first came to light with the report of Rebell *et al.* (1975), who described recurrent epizootics among hatchlings under intensive aquaculture at a commercial facility in the British West Indies. Each of nine successive hatches of young turtles underwent outbreaks of skin lesions involving 90–100% of the respective populations, and mortality of 5–20% was associated with the condition.

Other than the skin lesions from which the disease takes its name, it is difficult to define precisely the limits of the infection. With little doubt, the herpesvirus infection, or gray patch disease, is age-dependent and is very probably influenced by crowding, handling, organic pollution, fluctuating temperature, and hygiene—factors inherent in the commercial venture (Haines, 1978).

The virus is most virulent among animals 2–6 weeks old. Young victims do not develop skin lesions, possibly because the incubation period is so short. The sole finding is impaction of feces in the lower digestive tract. Under husbandry, mortality in this youngest age group may reach 20%, but experimental infections lead to 100% mortality, without skin lesions but again with impacted feces. Behavior changes were not mentioned.

Clinically obvious gray patch disease typically occurs when turtles are 2–4 months old. The condition occurs as either of two kinds of skin lesions. The lesser lesions are benign, sharply circumscribed, nonspreading papules on the neck and flippers; they eventually regress. The more severe lesions consist of spreading gray patches with superficial epidermal necrosis. Gray patches enlarge at a rate of about 5 mm per week; when most severe, the entire epidermis may be affected and the animals die (Haines, 1978). Elevated temperatures (about 30°C) seem to exacerbate the disease; mortality is higher in summer than in winter.

After the turtles reach 1 year of age, they no longer contract gray patch disease, nor can the condition be induced experimentally. Haines (1978) postulated that the animals, presumably survivors of earlier epizootics, have become immune. It would be interesting to determine whether previously unexposed yearlings are similarly resistant.

Histologically, the papules are small and sharply defined and show marked epidermal hyperkeratosis and acanthosis. The upper epidermal layers have slighlty basophilic nuclear inclusions surrounded by a clear halo, and chromatin is clumped at the periphery of the inclusions. The upper epidermis is infiltrated by eosinophils and the dermis by eosinophils and mononucleated cells.

The histopathological picture of spreading lesions is similar to that of papules, but more extensive. Keratosis is marked, and deeper layers of keratin show residual nuclei with inclusions and clumps of invading gram-negative rods. The upper keratin layer shows gram-positive cocci.

Epidermis beneath the keratin is acanthotic, and hyperplastic tissue is papillomatous. Eosinophils infiltrate the junction of keratin with epidermis. Late stages of the lesion show replacement of the epidermis by a necrotic crust, fibrin, and inflammatory cells. Neither polykaryocytes nor syncytia are mentioned.

Biopsies and skin scrapings examined by electron microscopy show abundant enveloped viral particles that measure 160–180 nm. Capsids measure 105–120 nm.

Experimental transmission was first effected by scratching flippers of 6– to 8-week-old hatchlings and inoculating the scratch lines with lesion material. Clinical signs appeared in 2–3 weeks among all those inoculated, and histology and electron microscopy confirmed the presence of virus (Rebell *et al.*, 1975). According to Haines (1978), turtles that are 2–4 months old can be experimentally infected by the scratch method. Clinical gray patch develops, but the animals recover.

Haines and Kleese (1977) investigated the effects of controlled temperature on the course of gray patch disease arising spontaneously among 3-week-old hatchlings under conditions that obviated other stress factors. Work was carried out at a location remote from the aquaculture facility. Experimental animals were free of gray patch disease at the start of the study, and the authors made a point of noting that hatchlings were from eggs that had been taken in the wild. Four different temperature regimens were used, and naturally occurring gray patch disease broke out among all groups. Controls held at constant 25°C first showed clinical signs at day 20, and the condition reached a peak incidence of about 90% at 32 days. The group subjected to a gradual temperature elevation to 30°C and then a reduction to 25°C had an incubation time of only 8 days; among them, the rate of disease increase was greater than that in controls, but at day 32 the final percentage was about the same. Temperature effects were most pronounced among the groups that were raised from 25 to 30°C and either held there or abruptly moved to 30°C for 4 days and then returned to 25°C. Turtles at constant 30°C showed an 8-day incubation time, but the incidence increased much more rapidly than in the controls; however, the final percentage was similar. Incubation was shortest—only 2 days—in animals subjected to sudden transfer to 30°C, then held there 4 days and returned to 25°C. The rate of disease appearance was the most rapid and the final incidence was highest, about 96%.

Severity of lesions was graded; the most severe disease occurred among turtles held at constant 30°C or subjected to sudden temperature changes. The trials were held over a period of 7 weeks, and it is worth noting that under near-ideal conditions, gray patch disease disappeared spontaneously 4–6 weeks after the trials were terminated.

Haines (1978) postulated that reduction of temperature may be one means of lowering mortality and morbidity from gray patch disease. One could add that mitigation of stress factors could also have a favorable effect. Haines (1978) also reported that successful vaccination was achieved. Virus was recovered from gray patch lesions, partly purified, inactivated, and injected intramuscularly into young turtles. Hatchlings so immunized proved to be protected against active virus administered by scratch inoculation.

Considerable effort has been spent in attempts to isolate the gray patch agent. Haines (1978) mentioned that mammalian, fish, and reptilian cell lines were tried, but that the only cells to show consistent cytopathic effects were derived from green sea turtle skin. Culture-grown virus was infective for turtles that were inoculated by the scratch method. Koment and Haines (1977) found that the virus replicated rapidly at 25°C and that it was distinguishable from channel catfish virus and herpes simplex; they found also that its effects *in vitro* were charactertistic of herpesviruses.

Circumstantial evidence suggests that the gray patch agent is carried by survivors that reach sexual maturity. Rebell *et al.* (1975) commented briefly on the "*prevalence of skin lesions in turtles of all ages*" (italics added). Perhaps skin lesions that persist in the adult harbor the virus and provide generation-to-generation transmission. One wonders, too, whether virus is transmitted vertically, perhaps infecting some embryos *in ovo*. However the virus is harbored or carried, the general pattern of virulence for the young and development of resistance with age is similar to that of channel catfish virus and the salmon herpesvirus *Oncorhynchus masou* virus.

D. Iguana Herpesvirus

The green iguana (*Iguana iguana*) is a New World herbivorous and terrestrial lizard of the family Iguanidae.

As part of a program of developing reptilian cell lines for virological application, Clark and Karzon (1972) prepared primary cultures from several tissues of an apparently normal adult male green iguana from a community lizard and turtle cage at the Buffalo, New York, Zoo. Cultures were incubated at 23 or 30°C, but cell outgrowth occurred only at 30°C, beginning within 2–10 days. From the 13th to the 30th day, syncytia became evident sequentially in cultures of liver, spleen, kidney, and heart; that development was followed by lysis and cell death. When infective culture fluid was inoculated onto the TH-1 cell line of box turtle (*Terrapene carolina*) heart origin, syncytia were induced. Similar primary cultures were prepared from kidney and lung tissues of a second adult male iguana from the same community cage; they too developed syncytia and degenerated. Although other tissues were cultured, viral effects did not become evident; in fact, heart tissues gave rise to the permanent iguana cell line IgH2, which subsequently proved susceptible to the iguana virus. In their efforts to detect virus by natural transmission, Clark and Karzon (1972) held an immature green iguana in the zoo's community cage for 2 months, but viral infection could not be demonstrated in the animal.

In vitro, the iguana virus grows at 23–36°C, but optimal temperature is 30°C. Syncytia are produced at 36°C, but no virus is released. Maximal titer at 23–30°C is about 10^7 plaque-forming units/ml; most infectivity is cell-associated, only 1% being released. The virus is ether-sensitive, inhibited by 5-bromodeoxyuridine, and substantially retained on filter membranes of 220-nm mean porosity. Thermal-inactivation data for virus in culture medium with 10% serum shows stability at 4°C (and presumably at lower temperatures as well) and one-hit kinetics; at 37°C, the half-life is about 13 hr. The virus is replicated by several cell lines of iguana origin, as well as by several lines from the box turtle. Syncytia, but otherwise nontransferable cytopathic effect (CPE), were produced in cell lines from other reptiles, the gekko, caiman, and python. No CPE occurred in cell lines from the sidenecked turtle, Russell's viper, fish, amphibian, chick embryo, or mammals. In all, 13 different vertebrate cell cultures or lines proved to be completely refractory.

In addition to syncytia, susceptible cell cultures showed prominent eosinophilic intranuclear inclusions and margination of chromatin.

Representative vertebrates of several classes were inoculated with iguana virus, but the agent was recovered only from reptiles (4/20), the greatest recovery being from young iguanas. Embryonate chicken eggs, white mice, slider turtles, three species of snakes, spectacled caiman, green anoles (an iguanid lizard), leopard frog, and American toad all proved resistant or refractory.

Twelve young iguanas were inoculated with virus by several different routes and held at 23 or 30°C. Seven animals died during the 15 days of observation, but no pattern of clinical signs was evident. Virus could not be recovered, and histopathological examination was not carried out. When the survivors were sacrificed, virus at low titer was recovered from the spleen of one. In contrast, culture of liver and kidney cells from four of the five survivors gave rise to syncytia within 1–4 weeks. In subsequent work with other vertebrates, culture of tissues from a single Tokay gecko and one of two box turtles revealed iguana virus.

The ultrastructural details of iguana virus were described and profusely illustrated by Zeigel and Clark (1972) (Fig. 4). Capsids of about 115 nm were found in the nucleus and more specifically in intranuclear inclusions. Capsomere number was estimated to be 162. Enveloped forms, 225 ± 75 nm, were found in cytoplasm and in extracellular spaces. A peculiarity of infected-cell cultures was the presence of clusters of 35-nm angular bodies within the nucleus, and in some instances encapsidation of the angular bodies was evident (Fig. 4).

Clark and Karzon (1972), who tested the iguana virus with antisera against one amphibian, one snake, one avian, and seven mammalian herpesviruses, found no neutralizing activity even at dilutions as low as 1 : 4. The iguana virus is therefore considered to be antigenically distinctive,

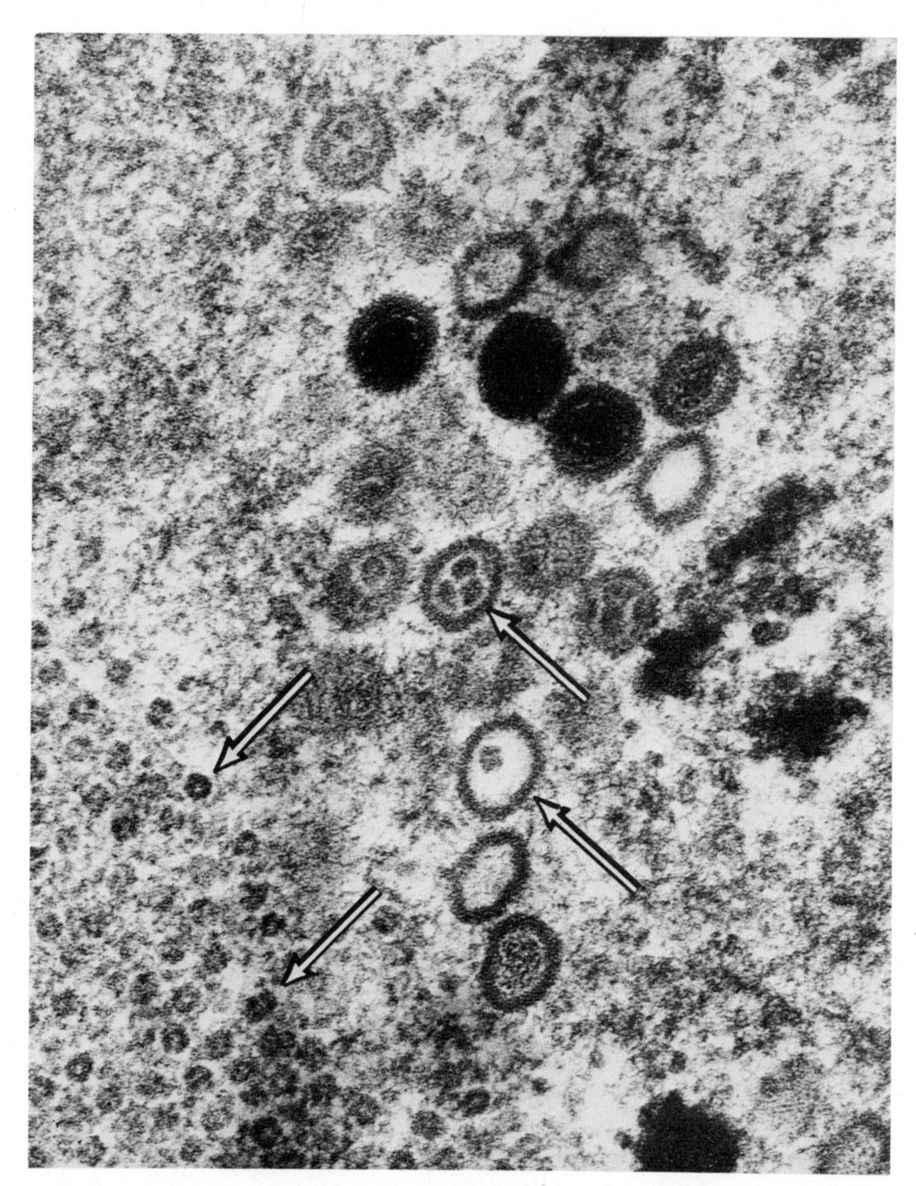

FIGURE 4. Thin section of nucleus of TH1 cell infected with the iguana herpesvirus. Empty and full capsids are in evidence. Several capsids enclose distinctive 35-nm bodies (center) and lie adjacent to a cluster of similarly shaped bodies (left). ×140,000. From Zeigel and Clark (1972). Reprinted with the permission of the American Society of Microbiology.

but its role in possible disease processes came to light during diagnostic work with animals from another population in California.

Frye *et al.* (1977) reported on findings in California involving seven iguanas with clinical signs of severe lymphocytosis, anorexia, lethargy, and loss of the normal brilliant green coloration. A single specimen had hemorrhaged, and its blood was devoid of thrombocytes. Another had enlarged kidneys. Histologically, all the lizards had severe splenic lymphoid hyperplasia, and most had significant histiocytic lymphoid infiltration of liver, spleen, myocardium, and bone marrow. Although organs were not specified, the authors stated that herpesvirus was found in the cells of two animals and, again without details, stated that host tissue cultures were infected. Morphologically, the virus was said to be similar to that of the Pacific pond turtle.

E. Green Lizard Herpesvirus

The green lizard (*Lacerta viridis*) is an Old World terrestrial reptile of the family Lacertidae.

Specimen lizards, some 30–40, were captured in Italy and taken to France, where they were held in a community terrarium with another reptile. Cutaneous plaquelike lesions were found on some of the lizards. Papillomatous in appearance, the 2–25 lesions per animal were 2–20 mm in diameter and usually persisted through hibernation; some, however, were shed at molt. The condition was significantly more common in females (70%) than in males, which additionally had only a few lesions. Moreover, the location of lesions differed with sex. Females showed lesions at the base of the tail and trunk, and in males the head and cervical regions were involved. The authors attributed the different locations to results of nuptial activities; biting aggression is directed to the head and neck of competing males, and females are seized by the trunk or base of the tail for copulation (Raynaud and Adrian, 1976). Thus, wounds are inflicted and infection is initiated.

Histologically, the lesions were benign papillomas consisting of hyperkeratinized tissue. Pigment cells occurred at depth, and the epidermis was hypertrophied. In some cells, chromatin had accumulated along the nuclear membrane. Atypical mitoses were not observed; neither was the muscle infiltrated by epidermis, nor were metastases found.

Evidence was found for three different viruses in papillomas from three lizards examined by electron microscopy. A single animal contained what was considered a papovavirus. Two lizards showed two kinds of virions, one considered reovirus and the other conforming to the herpesvirus group; however, neither experimental transmission nor isolation was reported.

F. Elapid Snake Herpesvirus

The banded krait (*Bungarus fasciatus*), Indian cobra (*Naja naja*), and Siamese cobra (*Naja naja kaouthia*) are poisonous terrestrial snakes of the family Elapidae and collectively are native to eastern Asia and India.

The snakes are widely exported to zoological gardens for display and to herpetaria and research laboratories. Herpesvirus has been found in the venom or venom glands of all three species, but no biological isolations have been made. Considering the relatedness of the three snakes and their geographic origin, it is conservatively prudent to discuss the associated herpesvirus as a single entity. Isolation and serological identification will provide definitive answers.

Herpesvirus of elapid snakes came to light with the work of Padgett and Levine (1966), who used reconstituted lyophilized crude venoms from the Indian cobra and banded krait in attempting to determine structural details of Rauscher leukemia virus. Their results showed herpesviruslike details previously unknown for, and inconsistent with, the Rauscher agent. Monroe *et al.* (1968) saw reason to question the findings and carried out a direct investigation of the two venoms themselves. Electron microscopy of thin sections and of negatively stained preparations showed that the venoms contained herpesvirus particles and considerable cellular debris. Clearly, the herpesvirus was a contaminant of the venoms and not structurally altered Rauscher virus.

Monroe *et al.* (1968) found both enveloped and nonenveloped particles and aggregates of virions numbering up to a hundred. Capsid size was 100–125 nm. Hollow capsomeres were visualized; they had lengths

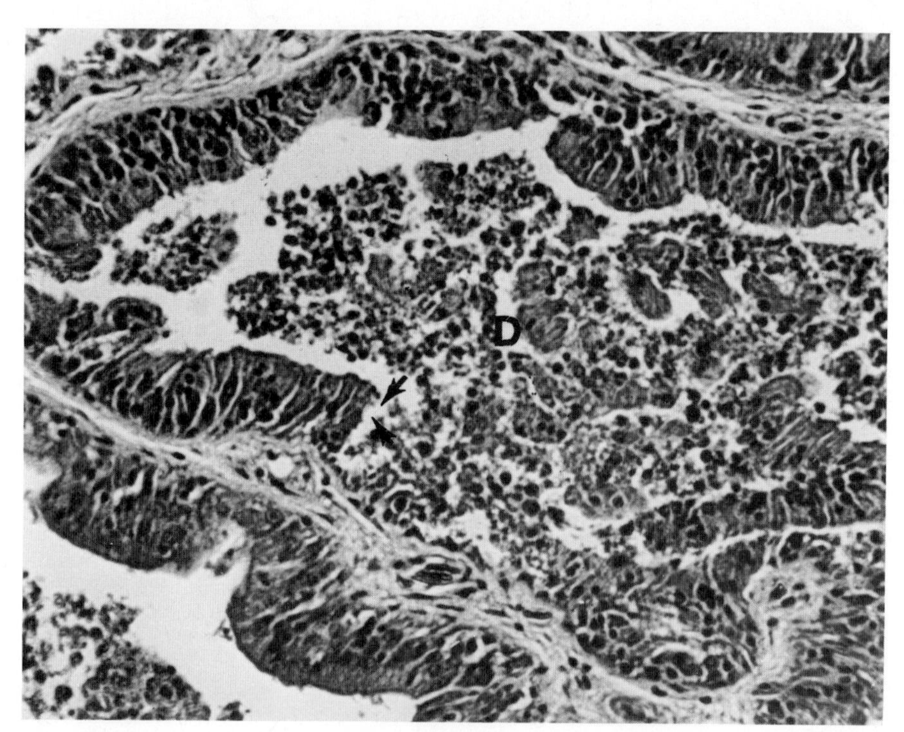

FIGURE 5. Section of Siamese cobra kidney showing patches of necrotic tubular epithelium and debris (D) in the lumen. Hematoxylin–eosin. ×125. From Simpson *et al.* (1979). Reprinted with the permission of the American Veterinary Medical Association.

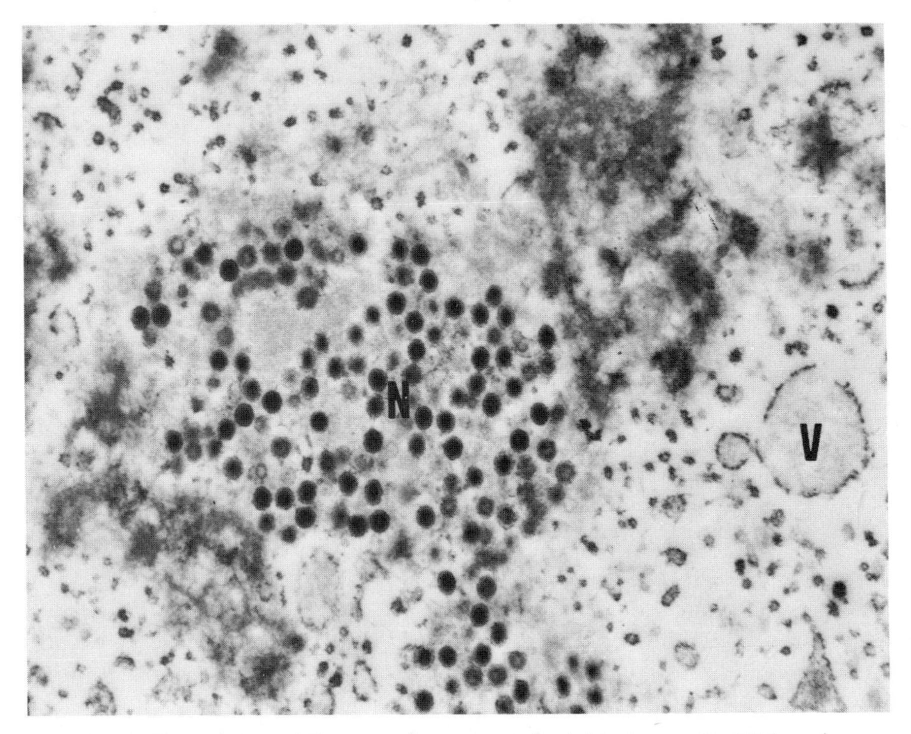

FIGURE 6. Thin section of Siamese cobra venom gland. Nucleocapsids (N) have been released from a cell nucleus into the lumen of a gland. Venom (V) occurs in coated vesicles. ×25,000. From Simpson *et al.* (1979). Reprinted with the permission of the American Veterinary Medical Association.

of 12 nm and diameters of about 8 nm. Capsomere number, however, could not be determined.

Thus far, there was no indication of effects that the virus had on the host snakes.

Simpson *et al.* (1979) added to the list of snakes with herpesvirus and in a sense provided the first clue to effects on the host. The Siamese cobra was the newly recognized source animal, and the reason for the discovery was that 20% of a shipment of 400 snakes from Thailand produced low-grade venoms that were abnormally thick. The condition was due to cellular contamination.

Simpson's group excised the venom glands of two specimens and prepared them for light and electron microscopy. The glands were enlarged, and lumens of some tubules contained appreciable cell debris. In limited areas, squamous epithelium was necrotic (Fig. 5). Microvilli were absent from surfaces of degenerated cells, and mononuclear infiltration was present in subepithelium of affected glands.

Electron microscopy of attached epithelial cells showed margination of nuclear chromatin and naked intranuclear capsids of 95–115 nm. Capsids were also found in necrotic cells (Fig. 6), and cell-free enveloped particles, 190–210 nm, were present in lumens of some venom glands.

Virus isolation attempted on viper heart cell cultures was not successful. The failure was attributed to cellulytic enzymes present in venom gland homogenate; inoculated cell sheets sloughed within 12 hr.

It is tantalizing to consider experimental transmission to young elapids, possible systemic infection, and possible isolation from internal-organ homogenates that would not be toxic to cell cultures.

III. TELEOST FISH HERPESVIRUSES

A. *Oncorhynchus masou* Virus

Members of the genus *Oncorhynchus* are Pacific salmon native to the higher latitudes of the northern hemisphere. They are mainly anadromous, but some small populations have become isolated and landlocked. An agent *Oncorhynchus masou* virus (OMV) was isolated from a landlocked population of *Oncorhynchus masou*, or masu salmon, a species native to Japan.

Initially during 1978, but again in the following year, Kimura *et al.* (1981a–c) isolated OMV during the course of a virological survey of propagated salmonids on Hokkaido. The source material was ovarian fluid from landlocked masu salmon that were of normal appearance but the population history of which showed a pattern of low survival among the progeny fry.

OMV has been sufficiently characterized to be firmly placed in the herpesvirus group (Kimura *et al.*, 1981a). Cultures of RTG-2, CHSE-214, KO-6, and other salmonid cell lines are susceptible; they show an initial rounding that is followed by syncytium formation and a terminal lysis. Small eosinophilic inclusions are present in the cytoplasm of infected cells. Commonly used nonsalmonid fish cell lines are refractory to OMV. The agent is both ether- and acid-labile (pH 3.0) and does not hemagglutinate human O+ cells. Replication is inhibited by levels of 50 μg/ml 5-iodo-2-deoxyuridine (IUdR), by phosphonacetate, and by acycloguanosine. Growth range is from 5 to 18°C, 15°C being optimal, and 25°C being nonpermissive. A one-step growth curve requires about 9 days, and maximal titer is about 10^6 median tissue-culture infectious doses ($TCID_{50}$)/ml. The agent's infectivity holds for at least 6 months at −80°C, but at −2°C, 99.9% of the infectivity is lost in 17 days. At temperatures of 15°C or higher, loss of infectivity is complete in 17 days.

Serologically, OMV is distinctive; it is neutralized by homologous antiserum but not by antiserum against any of the other salmonid viruses; infectious pancreatic necrosis, infectious hematopoietic necrosis, Egtved virus (viral hemorrhagic septicemia), or *Herpesvirus salmonis*.

Electron microscopy of infected cells shows hexagonal intranuclear capsids with a diameter of 115 nm. Particles in the process of budding are also found, and completely enveloped virions measure 200 × 240 nm. The calculated number of capsomeres is 162.

Successful experimental infection of young salmonids is readily accomplished by simple immersion in a suspension of OMV. There is evidence, however, that, as in the channel catfish herpesvirus, nonspecific resistance develops with age. Kimura *et al.* (1981a,c) exposed 100 3-month-old and 50 5-month-old chum salmon (*O. keta*) to 100 $TCID_{50}$/ml OMV for 1 hr at 10°C and held the fish for observation at 10–15°C. Mortality began among the younger fry at 11–12 days and among the older fry at about 20 days. During the next 50 days, 60% of the younger fry and 35% of the older fry died. Less than 5% of each control group died during the same period. Similar infection by immersion, followed by intraperitoneal injection of 200 $TCID_{50}$ virus, was tried with 8-month-old fish, but there was no evidence of susceptibility. Fish-to-fish transmission was effected by holding 5-month-old fry with fry that had been infected by immersion; the resulting mortality resembled that described for the original infection by immersion.

Coho salmon (*O. kisutch*), kokanee (*O. nerka*), and rainbow trout (*Salmo gairdneri*) proved to be susceptible to experimental inoculation with OMV.

Clinically, fry infected with OMV were inappetent and exophthalmic; they had petechiae on body surfaces and especially beneath the lower jaw. Internally, livers were mottled with white, and in advanced cases the whole liver was pearly white. The digestive tract was devoid of food, and in some fish, the spleen was swollen. The kidneys, prime targets for bacterial and viral pathogens of fish, appeared to be normal.

Histopathological examination revealed numerous foci of severe hepatocyte necrosis and syncytium formation in the liver (Kimura *et al.*, 1981a). Cardiac muscle was edematous, and partial necrosis was evident in some spleen sections. Pancreatic and renal tissues were essentially normal. Although specifics were not given, Kimura *et al.* (1981a) stated that necrosis was evident in all vital organs of month-old fish.

Electron microscopy of liver from infected fish showed intranuclear virus particles, some empty and some with nucleoids. Enveloped particles were found, but only infrequently.

The most interesting feature of OMV was the development of neoplasms among survivors of experimental infection (Kimura *et al.*, 1981b,c). Beginning at about 4 months and persisting at least 8 months postinfection, 60% of the 5-month-old chum salmon survivors developed tumors of differentiated epithelium. The growths occurred mostly about the head, on the mouth, under the gill covers, on the eyes, and sometimes on the caudal fin. One of the 52 survivors had a tumor in the kidney. Control fish held under the same conditions showed no tumors; accordingly, OMV is presumed to play a role in neoplasm induction.

Structurally, the neoplasms were papillomas of abnormally proliferating squamous epithelium, supported by a stroma of fine connective tissue (Kimura *et al.*, 1981b). Mitoses were abundant, but necrosis was absent in all but the single renal tumor; that growth displaced normal kidney tissues. Electron microscopy showed variability in nuclear size, but viral particles were not evident.

When virological examination was carried out on the neoplasms (liver, kidneys, heart, and spleen) of 11 tumor-bearing fish, serologically identified OMV was isolated from one of the tumors. Another tumor that was cultured showed cell outgrowth during the first 4 days; cytopathic effect (CPE) then appeared, and virus was recovered.

Kimura *et al.* (1981b) made a point of noting that tumor induction among chum salmon surviving experimental OMV infection was confirmed in additional work carried out during the year following their original observation. Time for tumor induction again was about 4 months. Moreover, coho salmon that survived experimental infection also developed tumors.

Virus-neutralizing activity against OMV was found in the serum from some of the normal and tumor-bearing survivors. Titers against 100 $TCID_{50}$ virus were 1:160 from those with neoplasms and 1:320 from those without tumors. Sera from uninoculated controls had titers of less than 1:20.

Many intriguing questions present themselves. First and foremost, are epithelial neoplasms evident on any of the spawning adult population? Second, is poor survival of progeny fry attributable to OMV? Do adults possess serum neutralizing activity against OMV, and if so, is there a correlation with presence or absence of the virus? One wonders too whether this particular herpesvirus is incorporated in the genome of the neoplastic cells. The existence of OMV in Japan and the single instance of *H. salmonis* in America lead to the question of whether they are one and the same agent. Kimura *et al.* (1981a) found that OMV antiserum did not neutralize *H. salmonis*. Early results of work with the two agents at the National Fish Health Research Laboratory have shown that the CPE of the two is distinctly different.

B. *Herpesvirus salmonis* (Salmonid Herpesvirus Disease)

Kokanee, also known in Japan as *himemasu*, are landlocked forms of *Oncorhynchus nerka*, a Pacific salmon. The rainbow trout is native to North America but widely introduced elsewhere in both hemispheres. The rainbow trout and kokanee are members of the family Salmonidae, and both fishes are highly regarded for sport and food. In North America, the annual production of rainbow trout exceeds 22 million kg, most of which is marketed as a fresh or frozen or otherwise processed product. In the United States, an agent subsequently named *Herpesvirus salmonis* (on deposit with the American Type Culture Collection) has been isolated from rainbow trout and a similar virus from *O. nerka* in Japan.

Investigations of excessive postspawning mortality among fish in a hatchery population of rainbow trout in the Pacific Northwest of the United States led to isolation of a virus in 1971. Identification was sought, but the agent was lost before a determination could be made. Additional isolations were made over the next several years, but it was not until 1975 that confirmation of the viral nature of the agent was obtained, the reason being that for reliable replication, incubation had to be at 10°C or less—lower than temperatures normally used for replicating viruses of coldwater fishes (Wolf *et al.*, 1978). In the meantime, Sano (1976) investigated annual epizootics among landlocked *O. nerka* fry that resulted in 80% mortality and had a history of occurrence since 1970. Sano also incubated cultures at 10°C, and concluded that the Nerka virus in Towanda Lake, Akita and Aomori Prefectures (NeVTA), was herpeslike, but that it differed from *H. salmonis*. Discussions by the American and Japanese investigators and comparisons of cell-culture results and histopathological changes led to the conclusion that *H. salmonis* and NeVTA were closely related and possibly identical (Wolf *et al.*, 1975). Serum neutralization tests have not been carried out; *H. salmonis* is poorly antigenic. Accordingly, and in view of the report by Kimura *et al.* (1981a), the possibility must be considered that NeVTA virus is the same as OMV and not *H. salmonis*.

Experimental inoculation of fry or small fingerling rainbow trout with *H. salmonis* and subsequent maintenance of the fish at 6–9°C resulted in a subacute to chronic viscerotropic disease with marked edema and necrosis and a heretofore unknown induction of syncytia in fish pancreatic acinar tissue (Wolf and Smith, 1981).

Behavior of young rainbow trout is characterized by anorexia and lethargy, with some victims showing loss of motor stability but maintaining a capability of brief bursts of erratic swimming when alarmed.

The first observable external sign is the presence of thick whitish mucoid casts trailing from the vent. Later, many of the fish have a distended abdomen and darken appreciably (Fig. 7). Marked exophthalmia is common, and hemorrhages may occur in the orbits and at the base of fins. The gills are pale.

Gross internal findings show overall pallor of the visceral mass. The body cavity contains slight to abundant amounts of ascitic fluid that may or may not be gelatinous. The anterior digestive tract is empty, but if any food is present, it is in the hind gut or rectal area. Viscera are flaccid; the liver may be mottled or hemorrhagic. Kidneys are pale but not obviously swollen. Limited hematological examination showed that peripheral blood contains blast cells and many immature red blood cells, but no cellular debris.

Major histopathological changes occur throughout the viscera, notably in the kidneys, respiratory organs, heart, digestive tract, and liver. Syncytia are produced in pancreatic acinar tissue (Fig. 8) and are considered pathognomonic for *H. salmonis* infection (Wolf and Smith, 1981).

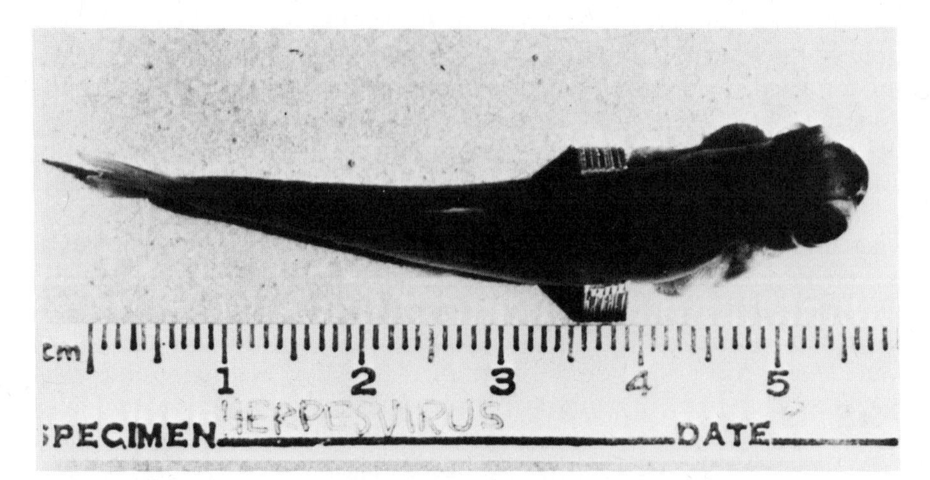

FIGURE 7. Rainbow trout fry that died as a result of experimental infection with *H. salmonis.* Notable features are marked melanism, distended abdomen, and severe exophthalmia. Courtesy of H. M. Stuckey.

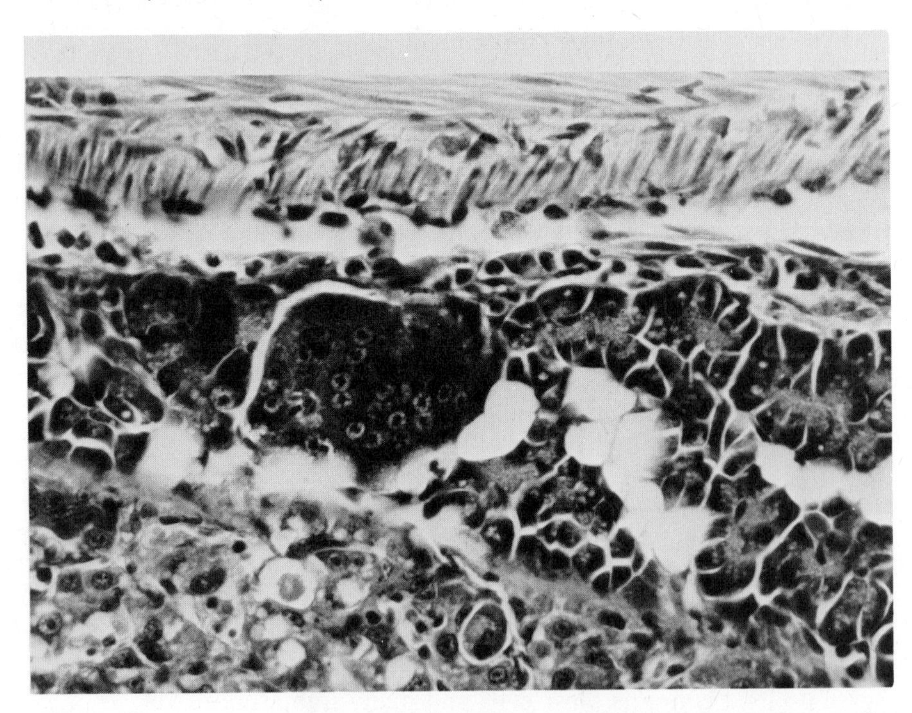

FIGURE 8. Syncytia in pancreatic acinar cells (left center) of rainbow trout fry parenterally infected with *H. salmonis*. Normal pancreatic tissue is at right center. Liver tissue at lower left is necrotic. Hematoxyin–eosin. ×200. From Wolf and Smith (1981). Reprinted with the permission of Blackwell Scientific Publications Ltd.

Kidneys show some hyperplasia of hematopoietic tissue, and generalized edema. Slight to moderate necrosis is evident in renal tubules, and some have serous material in the lumen. Glomeruli in general appear normal. Viral infectivity is highest in kidney tissues. Edema is widespread in the liver, and some specimens have areas of necrosis and hemorrhage or congestion (Fig. 8). Focal vacuolation is found in some livers. Cardiac tissue is grossly edematous and necrotic, and there is marked leukocytic infiltration involving both lymphocytes and polymorphonuclear cells. Gill epithelium shows edema and hypertrophy. Some lamellar epithelium separates from underlying connective tissue (Fig. 9), and a few specimens have gill hemorrhages. Pseudobranchs are typically edematous, and cell nuclei are hypertrophied and some have margination of chromatin (Fig. 10). Localized necrosis of pseudobranchs is slight to moderate. Edema of stomach tissue occurs in about half the specimens. The anterior intestine is mostly normal, but the hind gut or rectum has areas of marked necrosis, foci of infiltrating leukocytes in the submucosa, and frank sloughing of the mucosa into the lumen (Fig. 11). Sloughed mucosa undoubtedly constitutes the bulk of the fecal pseudocasts seen externally. Lesser changes are found in the eyes, spleen, brain, and ovaries of some specimens.

Herpesvirus salmonis in infected-cell cultures has a capsid size of 90–95 nm (Fig. 12), and enveloped forms are about 150 nm in diameter (Fig. 13). Nuclear replication and envelopment in the cytoplasm are typ-

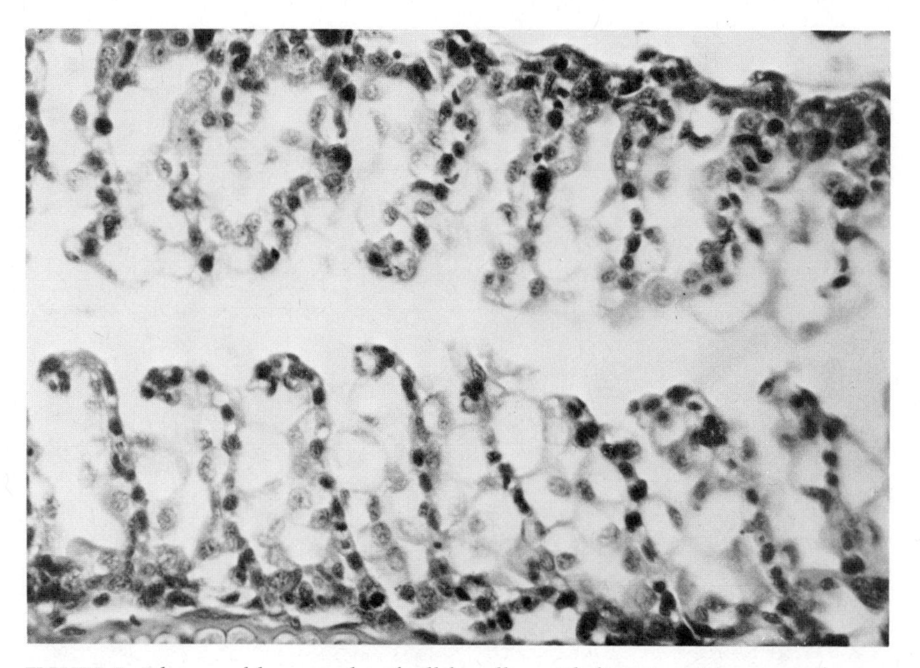

FIGURE 9. Edema and hypertrophy of gill lamellar epithelium in rainbow trout fry parenterally infected with *H. salmonis*. Hematoxylin–eosin. ×200. From Wolf and Smith (1981). Reprinted with the permission of Blackwell Scientific Publications Ltd.

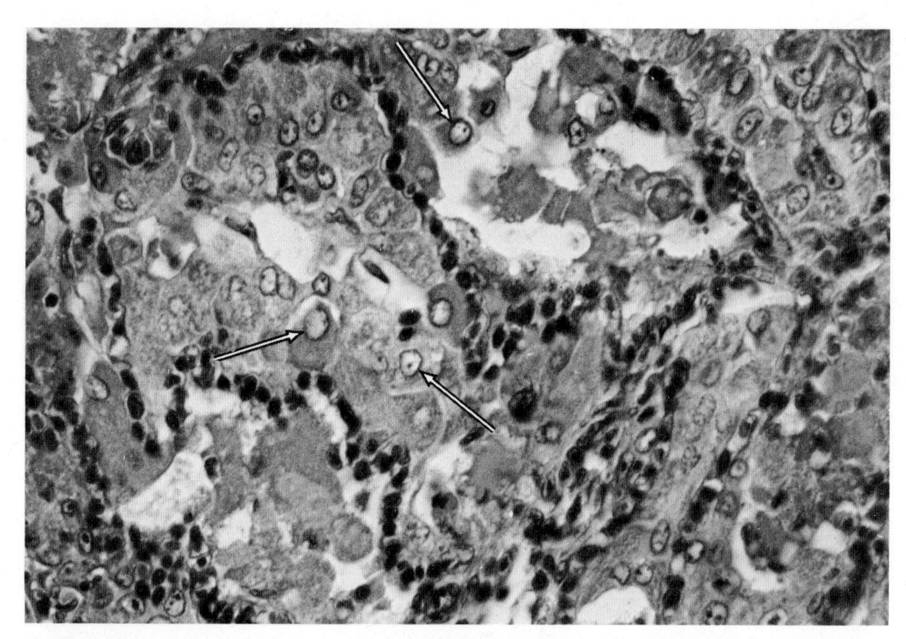

FIGURE 10. Edema, necrosis, and cellular hypertrophy in pseudobranch of rainbow trout fry parenterally infected with *H. salmonis.* Margination of chromatin (arrows) is common in hypertrophied cell nuclei. Hematoxylin–eosin. ×200. From Wolf and Smith (1981). Reprinted with permission of Blackwell Scientific Publications Ltd.

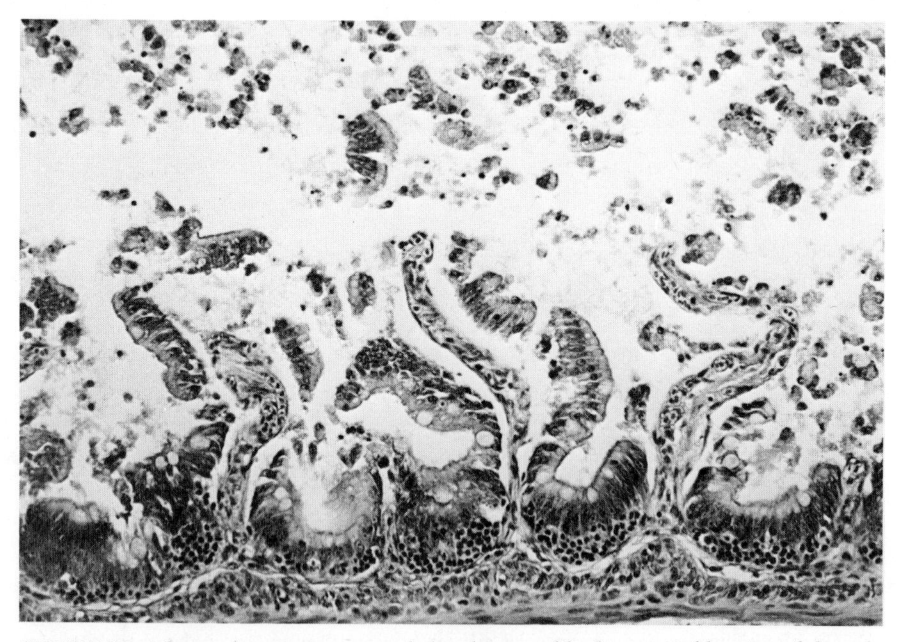

FIGURE 11. Advanced necrosis, mucosal sloughing, and leukocytic infiltration of the submucosa of rainbow trout fry parenterally infected with *H. salmonis.* Hematoxylin–eosin. ×80. From Wolf and Smith (1981). Reprinted with the permission of Blackwell Scientific Publications Ltd.

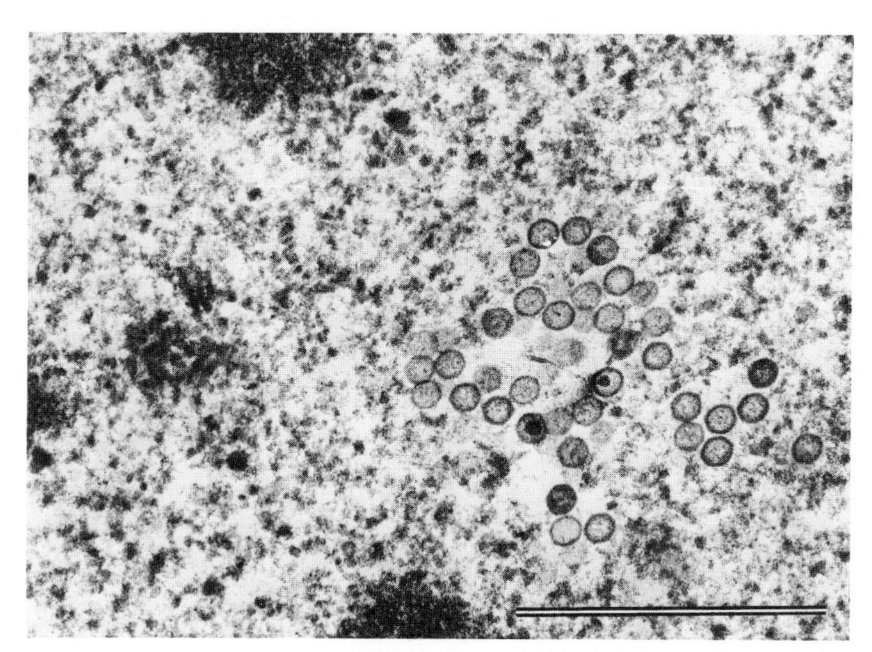

FIGURE 12. Portion of nucleus of RTG-2 cell infected with *H. salmonis*. Empty and full capsids in various stages of maturation. Scale bar: 1000 nm. From Wolf *et al.* (1978). Reprinted with the permission of the American Society of Microbiology.

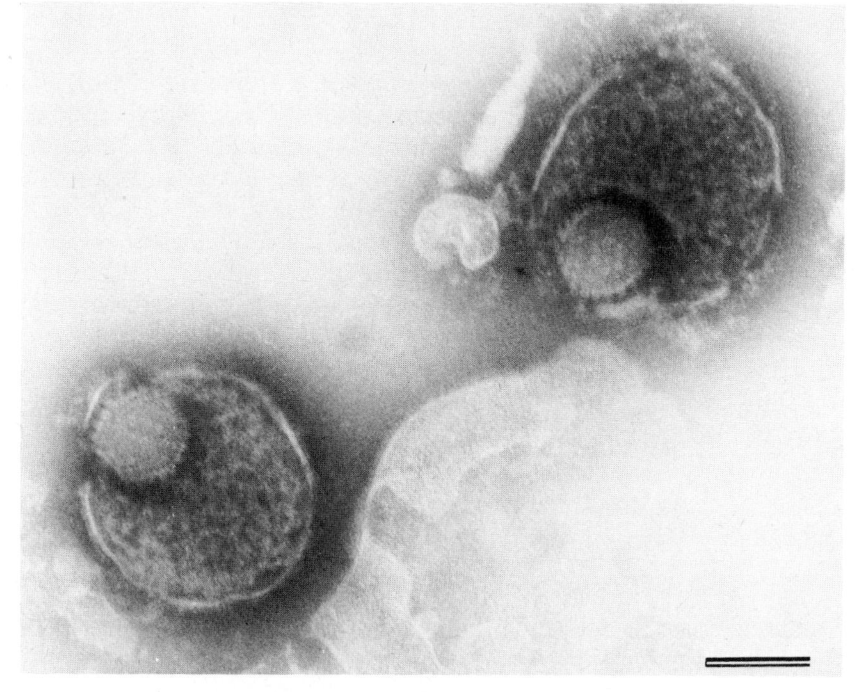

FIGURE 13. Negatively stained enveloped particles of *H. salmonis* Scale bar: 100 nm. From Wolf *et al.* (1978). Reprinted with the permission of the American Society of Microbiology.

ical of the group. Electron-dense lamellar or fibrillar material occurs in the nucleus and, though less abundant, resembles that seen in channel catfish virus infections (Fig. 14). Viral DNA has a buoyant density of 1.709 and a guanosine plus cytosine (G+C) value of 50%. As expected, the agent is ether- and acid-labile and without hemagglutinating capacity for human O+, rabbit, or rainbow trout erythrocytes. Replication occurs at 5–10°C, but is erratic at 15°C and does not occur at higher temperatures. The virus induces fusion of trout cells incubated at 0°C, but no replication occurs. The cryophilic nature of *H. salmonis* is also evident in its adsorption; other trout and salmon viruses are usually adsorbed for 30–60 min at 15°C, but *H. salmonis* must be adsorbed for at least 120 min at 10°C to achieve the full complement of plaques (Fig. 15).

The agent is replicated by the salmonid cells RTG-2, RTF-1, KF-1, and CHSE-214, but not by cell lines from nonsalmonid fish species. A one-step growth curve in RTG-2 cells at 10°C requires more than 4 days for completion (Fig. 16). At 24 hr, new virus appears in concert with formation of the earliest syncytia. Exponential growth takes place from 24 to 60 hr, and cell-associated virus exceeds released virus by a factor of about 10. Peak titers are achieved at 4–5 days, released virus attaining

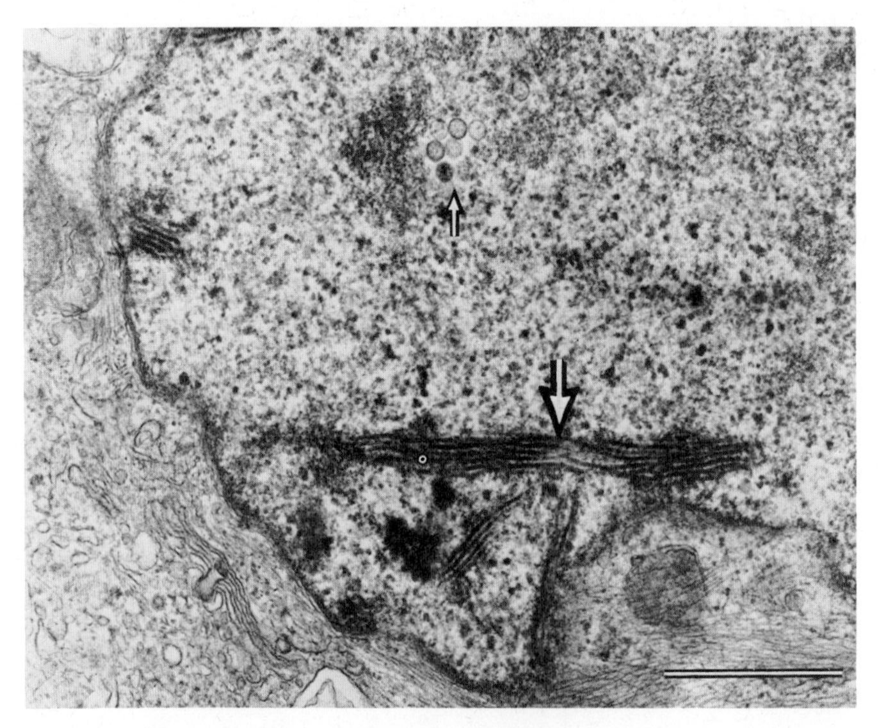

FIGURE 14. Portion of nucleus of RTG-2 cell infected with *H. salmonis*. Developing particles are indicated by small arrow and electron-dense fibrillar or lamellar material by large arrow. Scale bar: 1000 nm. From Wolf *et al.* (1978). Reprinted with the permission of the American Society of Microbiology.

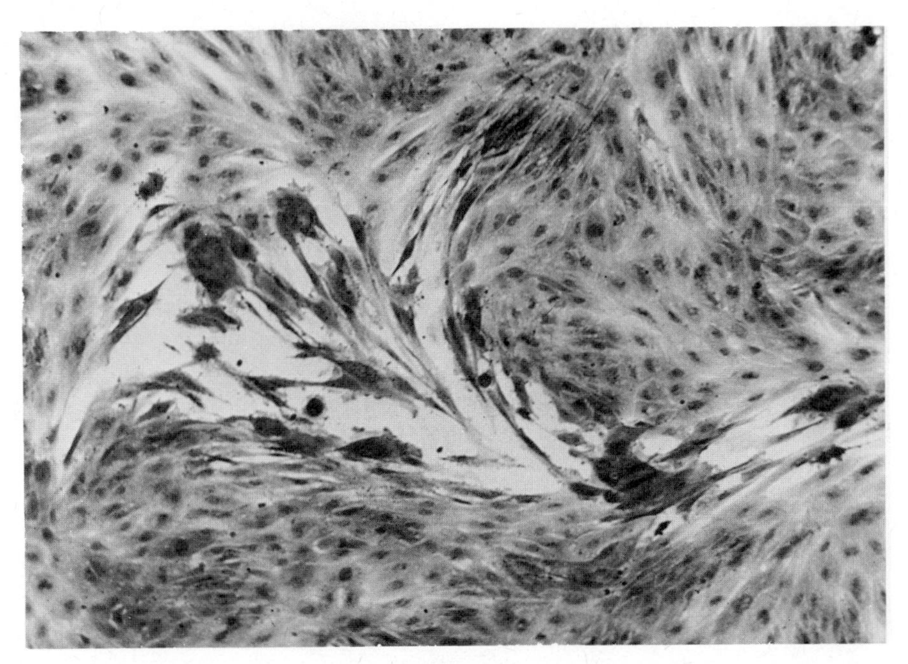

FIGURE 15. Plaque of *H. salmonis* in RTG-2 cells. Note that viral effects preferentially follow linear patterns within cell sheet. Crystal violet stain. From Wolf *et al.* (1978). Reprinted with the permission of the American Society of Microbiology.

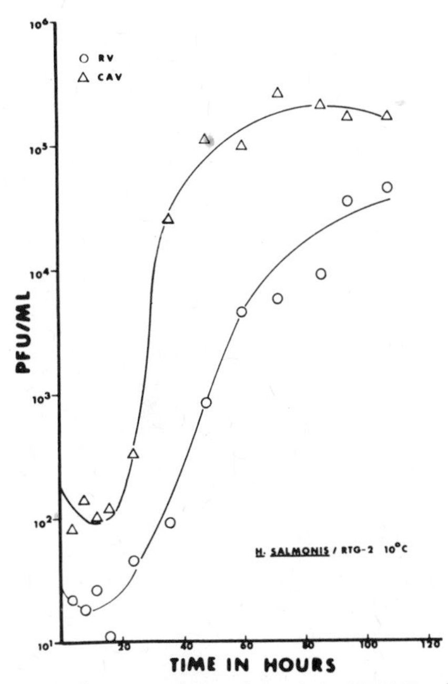

FIGURE 16. One-step growth curve of *H. salmonis* in RTG-2 cells at 10°C. (RV) Released virus; (CAV) cell-associated virus. From Wolf *et al.* (1978). Reprinted with the permission of the American Society of Microbiology.

about 5×10^4 plaque-forming units (PFU)/ml and cell-associated virus about 3×10^5 PFU/ml.

Morphological changes in infected-cell cultures begin with some rounding and increase in refractility (Fig. 17). Syncytium formation follows and lysis ensues. However, the lytic phase is not complete, and intact cell aggregates persist for days or even weeks. Stained cell sheets show basophilia, margination of nuclear chromatin, and Cowdry Type A intranuclear inclusions (Fig. 18).

Three cycles of freezing and thawing do not degrade infectivity of *H. salmonis*, but if the pH is 7.6 or greater, the agent is somewhat labile at −20 or even −80°C. Virus held at 4°C in culture medium without serum loses more than 99% of its infectivity within 3 weeks. Preferred storage is in infected-cell cultures in medium containing at least 10% serum, and at −80°C. Under these conditions, only 10% infectivity was lost in 18 months. Effects of lyophilization are not known.

Rainbow trout less than 6 months old and kept at 6–9°C are vulnerable to *H. salmonis* administered by injection, and among several trials the cumulative mortality ranged from 50 to 100%. Yearling rainbow trout are wholly resistant to *H. salmonis*. Under similar conditions, young

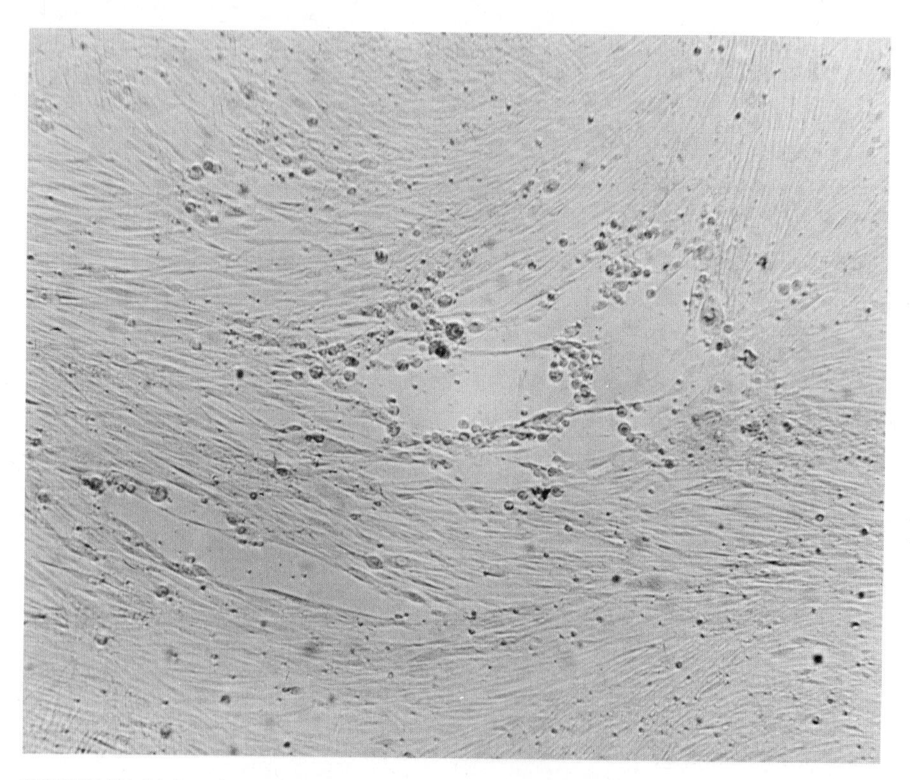

FIGURE 17. Living sheet of RTG-2 cells infected with *H. salmonis* showing characteristic foci of rounded cells and early evidence of fusion.

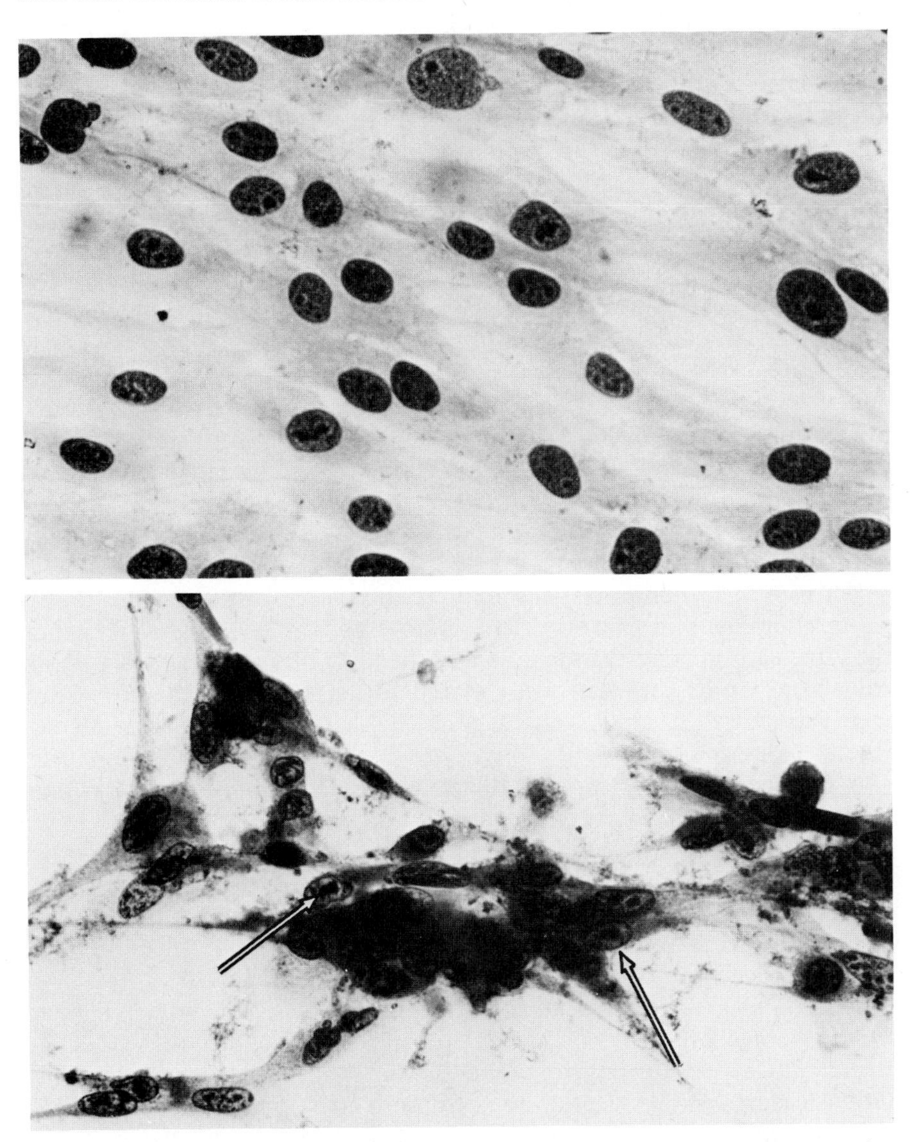

FIGURE 18. RTG-2 cells infected with *H. salmonis* showing marked basophilia, Cowdry Type A intranuclear inclusions, and fusion. *Top:* Uninoculated control. *Bottom:* Inoculated. Arrows indicate inclusions. From Wolf *et al.* (1978). Reprinted with the permission of the American Society of Microbiology.

Atlantic salmon (*Salmo salar*), brook trout (*Salvelinus fontinalis*), brown trout (*Salmo trutta*), and yearling *O. nerka* were refractory. In light of the Japanese virus NeVTA and its origin, it is noteworthy that *H. salmonis* could not be recovered from the American *O. nerka* specimens at termination of the work. After injection of virus, the first deaths occurred among rainbow trout at 25–33 days, and the disease course continued for several weeks. Virus was recoverable by direct inoculation of cell cultures with filtrates from infected tissues. Interestingly enough, transmission did not ensue when young susceptible trout were kept with moribund and dead victims of the infection.

Herpesvirus salmonis was apparently isolated in several different years from captive brood stock at a single hatchery. The brood stock was eventually disposed of, and the virus has not appeared again. Morevoer, virological examinations, during a 2-year period, of fish in all facilities in the contiguous states that received fertile eggs from the source hatchery yielded no incidence of *H. salmonis*. Such a single source for a fish virus has not been previously recognized. Failure of the virus to be transported with eggs from infected females might be due to elution during incubation, to the use of iodophores as egg disinfectants, and possibly also to the presence of temperatures at recipient facilities that were higher than tolerated by the virus. It is also conceivable that failure to show fish-to-fish transmission experimentally involves as yet unknown environmental factors.

It is perhaps somewhat academic to discuss control measures for *H. salmonis* infections, because there is now no way of knowing whether the virus caused or even contributed to the mortality of the adult rainbow trout from which the virus was isolated. Because yearling rainbow trout were refractory, extrapolation suggests that adults would also be refractory.

The peculiar low-temperature response of *H. salmonis* indicates that if an epizootic were to occur, elevation of the water temperature to 18°C, or whatever the fish could tolerate above 15°C, could greatly reduce losses from the virus.

Uncertainty as to whether the Japanese isolant NeVTA is the same as *H. salmonis* dictates that similarities and differences of the two agents be compared.

The NeVTA agent has an average diameter of 80 nm, and enveloped forms average 230 nm. Maximum titer in RTG-2 cells is $10^{4.5}$ $TCID_{50}$/ml, and infectivity during storage at 4°C was completely lost at 6 months. Other features of the virus were not described by Sano (1976). Cell cultures stained with hematoxylin and eosin showed syncytia and spherical eosinophilic inclusions in the cytoplasm, but neither nuclear inclusions nor margination of chromatin was mentioned. The greatest differences occur in histopathological changes; however, it must be remembered that two different host species are involved. In *O. nerka* having a natural infection, there is granular degeneration of skeletal muscle, leakage of proteinaceous material into Bowman's capsule, and vacuolation of pancreatic acinar cells—none of which was found in rainbow trout experimentally infected with *H. salmonis*. Both species, however, showed swelling and loss of gill epithelium. Whereas syncytia were found in the pancreas of fish infected by *H. salmonis*, Sano (1976) found syncytia in renal interstitial cells.

Further investigation will be required to resolve the question of identity of the two viruses.

C. Carp Pox Herpesvirus (*Herpesvirus epithelioma*)

The common carp *(Cyprinus carpio)* is a freshwater cyprinid originally native to China, where for millenia it was propagated for food and ornament. The carp was introduced to Europe during the 13th century and has since been widely grown as a food fish. The United States imported carp during the late 1800s, it being the expectation (which was never realized) that the New World, too, would regard carp highly as a food fish. Instead, the carp in America is generally disdained; it has adapted to feral life in a wide range of waters and is considered by most to be a coarse or trash fish.

In Europe, the condition known as carp pox or epithelioma papulosum has been recognized and described since the Middle Ages; in Western literature, it is the oldest known disease condition of fish. In the absence of recognized parasites and microbial pathogens, the etiology was uncertain, but the possibility of viral etiology was broached at the turn of the century when Lowenthal (1907) reported intranuclear inclusions and related them in inclusions of viral diseases then recognized. Subsequently, other workers postulated a viral etiology, but it remained for Schubert (1964) to provide the first concrete evidence in the form of micrographs of indisputable virions.

Carp pox is a benign, nodular, transiently chronic hyperplasia of epithelium usually occurring in multiple foci on skin and fins and occasionally on eyes and gills. The lesions are adequately vascularized so that they do not become necrotic. They do not occur internally; neither do they metastasize. Their appearance is smooth, waxy to glassy, and variously pigmented (Fig. 19). Most often, their color is similar to that of background skin, at times with a pinkish hue from vascularization. The lesions are up to several centimeters in diameter and one to several millimeters thick.

The effect of carp pox on the host is usually considered to be negligible, but some observers have attributed a slight retardation of growth to presence of the condition. The behavior of affected fish is normal, but when the lesions slough, pigmented scars follow, and in Europe, marketability of the product declines.

Histologically, the lesions consist of epidermal cells in the usual moderately stratified layers, but in places the cells are elongate and radiate outward to form papillae of highly proliferated cells. Mitoses are common, and the speed of proliferation often results in mucous cells occurring at some depth between papillae, literally being overgrown. Some of the neoplastic cells bear intranuclear inclusions, but that aspect is not universally present.

FIGURE 19. Common carp with multiple dermal lesions of "carp pox," circumstantially attributed to a herpesvirus. Courtesy of N.N. Fijan. From Wolf (1973). Reprinted with the permission of Academic Press.

Schubert (1964, 1966) published many illustrations of the virus, but only his 1964 report gives virion dimensions. Nuclear particles measured about 110 nm and polymorphic nucleoids about 50 nm. Enveloped virions (140–150 nm) were found only in cytoplasm.

Sonstegard and Sonstegard (1978) reported substantial breakthroughs in carp pox. The apparent enigma of why carp pox had not been found in the United States was explained when they reported its presence in the Laurentian Great Lakes. The disease had been here, but had not been recognized. More important, they described transmission by cohabitation or by the application of lesion material to abraded skin. In carp held at 10°C, 2 months' incubation was required. Hybrids of carp and goldfish (*Carassius auratus*) were susceptible, but the goldfish itself was refractory. The Sonstegards attempted to isolate the herpesvirus, using five different fish cell lines. No transformation was observed in BB (ictalurid), BF-2 (centrarchid), RTG-2 (salmonid), or FHM and CAR (cyprinid) cells. Failure to detect change in the cyprinid cell lines is particularly interesting; though the FHM is epithelial, the CAR is fibroblastic but of goldfish origin.

Carp pox has been clinically and histologically diagnosed in other cyprinids and even in fish of other families. Now that experimental transmission of the infection has been described, the matter of host range needs to be more accurately established. Alternatively, additional electron microscopy of lesions from other species of fish could be enlightening.

D. Channel Catfish Herpesvirus (Channel Catfish Virus)

The channel catfish (*Ictalurus punctatus*) is a member of the family Ictaluridae and native to North America, where it occurs widely in the warmer temperate latitudes. This esteemed food fish has been introduced elsewhere in the world because it grows rapidly and is well adapted to aquaculture. Over 45 million kg are raised annually in the United States, the greatest weight of any fish cultured in America.

Farm-raised channel catfish were the source of the first herpesvirus to be isolated from a fish. Relevant reviews of the catfish herpesvirus were given in reports by McCraren (1972), Wolf (1973), and Plumb (1977).

During the 1960s, the farm propagation of channel catfish underwent a marked expansion, and with it an intensification of its husbandry. As could well be expected, health problems were aggravated; notable among them were epizootics involving young-of-the-year that ran an acute course and resulted in high mortality. Fijan (1968) isolated an agent that has been unequivocally characterized as a herpesvirus and shown to be the etiological agent of the epizootics. Channel catfish virus (CCV) is available from the American Type Culture Collection as VR-665.

Under natural conditions, CCV causes a peracute hemorrhagic and systemic infection only among fry and fingerling *I. punctatus*. Incubation time can be as short as 72 hr. Blue catfish (*I. furcatus*) and walking catfish (*Clarias batrachus*, a close relative of the ictalurids) can be infected by injection. The degree of host specificity of CCV is exceptionally high.

CCV disease has a sudden onset and typically occurs at summer temperatures of 25°C or higher, particularly following stress of handling or low dissolved-oxygen levels. Victim fish show distress and move to pond margins. Many affected fish are lethargic, but some show a terminal agonal and convulsive spiral swimming. Food is refused.

Externally, victim fry are exophthalmic, and abdomens are distended. Hemorrhages are common along the ventral side, under the mouth, and in the base of fins, and surround a distended vent. Gills are pale and often hemorrhagic.

Internally, the digestive tract contains no food, kidneys are pale and enlarged, and the body cavity contains a pale yellow, sometimes reddish, fluid. Liver and spleen may be pale and enlarged, and viscera in general may show petechiation or hemorrhages.

Histopathological changes have been described in experimentally infected fish by Wolf *et al.* (1972) and in both naturally infected and experimentally infected fish by Major *et al.* (1975). Plumb *et al.* (1974) reported on both light and electron microscopy of tissues of experimentally infected specimens. Key features of the disease are hemorrhage, edema, and marked necrosis of kidneys, liver, and digestive tract. As in many fish viral diseases, the kidneys are particularly vulnerable. Hematopoietic and excretory tissues are edematous and show necrosis (Fig. 20). Infiltration of renal hematopoietic tissue occurs within a few hours after infection, and tubules show damage within 18 hr. Coincident with peak virus titer, kidney changes are severe by 96 hr. Liver shows necrosis, vascular congestion, hemorrhage, and obvious but regional edema (Fig. 21). In some cases, hepatocytes have eosinophilic cytoplasmic inclusions. Histopathological changes in liver occur later than in kidneys, and virus titers peak about a day later. The digestive tract has focal areas of macrophage concentration and shows evidence of edema. The mucosa and submucosa of some victims are necrotic and may slough into the lumen. Consistent with hemorrhages seen externally, extravasation of blood is found between muscle bundles.

CCV has been well characterized. Wolf and Darlington (1971) described many of the *in vitro* properties of the agent. Numerous studies of various biological features and host–pathogen relationships have been carried out during the past decade, the most relevant of which were cited

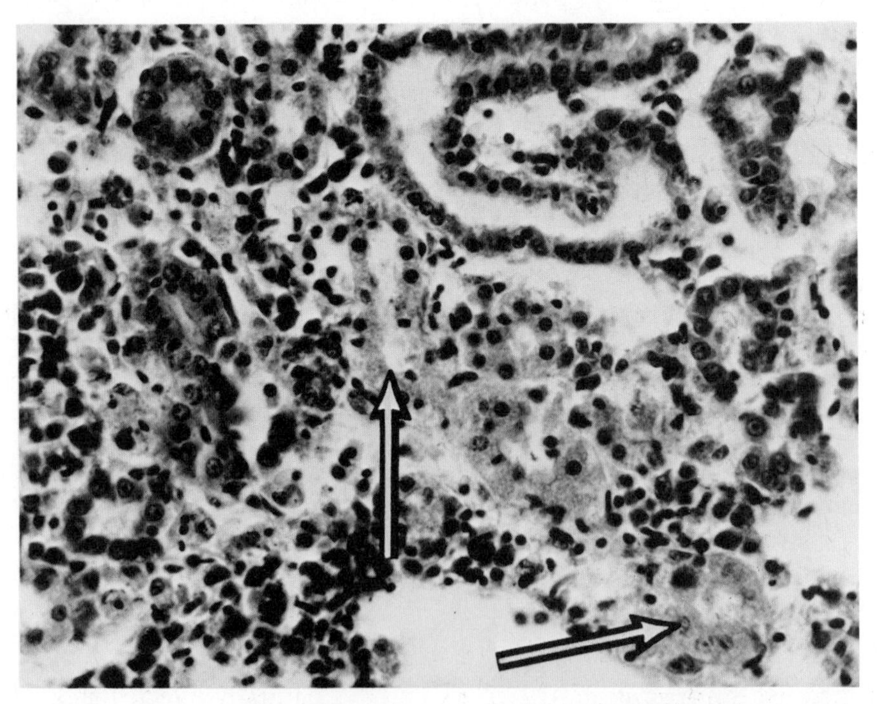

FIGURE 20. Channel catfish kidney showing generalized edema and tubule necrosis (arrows) resulting from experimental infection with CCV. Hematoxylin–eosin. From Wolf *et al.* (1972). Reprinted with the permission of the Journal of the Fisheries Research Board of Canada.

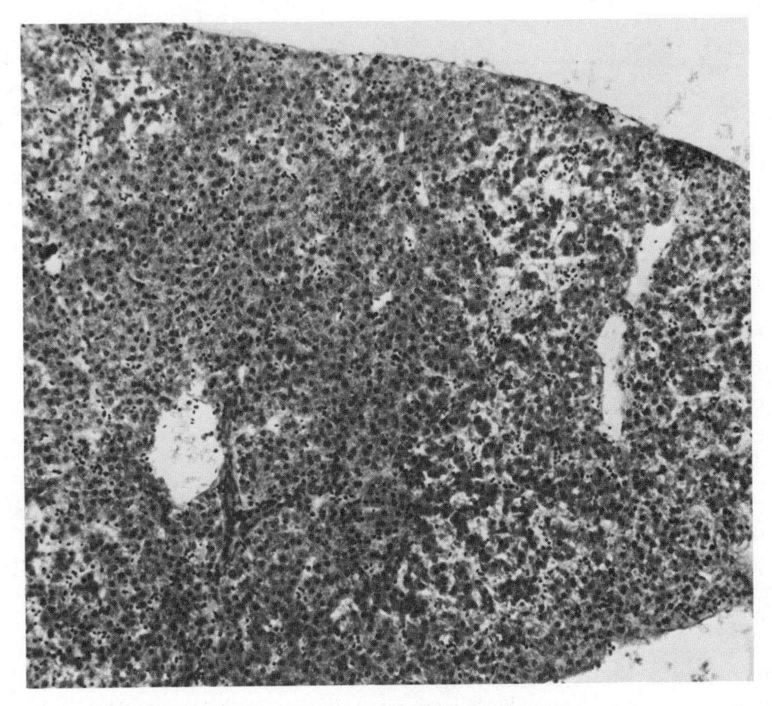

FIGURE 21. Channel catfish liver with regional edema, vascular congestion, and areas of necrosis following experimental infection with CCV. Hematoxylin–eosin.

by Plumb (1977). Goodheart and Plummer (1975) were the first to determine the G+C content, and Chousterman *et al.* (1979) confirmed the molecular weight and used restriction endonucleases to map the nucleotide sequences of the viral DNA. Thus far, CCV is the most extensively and intensively studied herpesvirus of a lower vertebrate.

The size of the CCV capsid is about 100 nm, and the nucleoid 40–50 nm (Fig. 22). Capsomere number is 162, and negatively stained enveloped forms measure 175–200 nm (Fig. 23). Buoyant density in CsCl is 1.715 g/cm^3, corresponding to a G+C value of 56% (Goodheart and Plummer, 1975). The molecular weight of the DNA genome is 86×10^6. Terminal ends of the molecule consist of a redundant sequence of 12×10^6 (Chousterman *et al.*, 1979).

When CCV is dried on concrete, glass, or netting, infectivity does not persist longer than 48 hr. Virus survives in pond water at 25°C no longer than 2 days, and is inactivated even more rapidly when in contact with pond soil. Infectivity is readily preserved by freezing in culture medium containing serum, but the pH should be in the physiological range; stability is greater at −80 than at −20°C. Infectivity in carcasses persists for several months at −20°C, but longer storage at that temperature is not recommended. The common practice of freezing and thawing tissue to yield virus degrades CCV. Three freeze–thaw cycles are each accom-

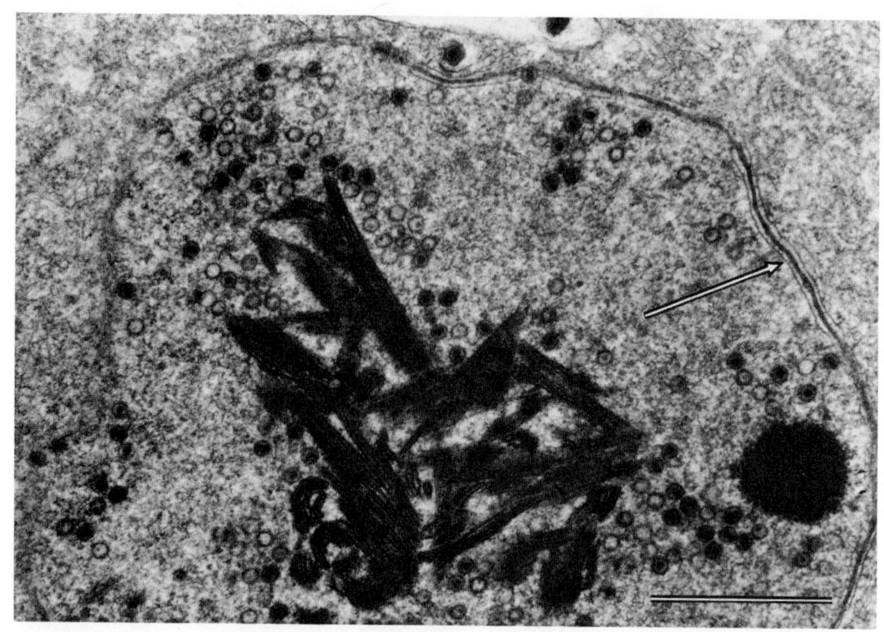

FIGURE 22. BB cell 10 hr after infection with CCV. Empty and full capsids occur in the nucleus together with electron-dense lamellar or fibrillar structures. In places, the nuclear membrane has been disrupted (lower left), and in others it has been reduplicated (arrow). Scale bar: 1 μm. From Wolf and Darlington (1971). Reprinted with the permission of the American Society of Microbiology.

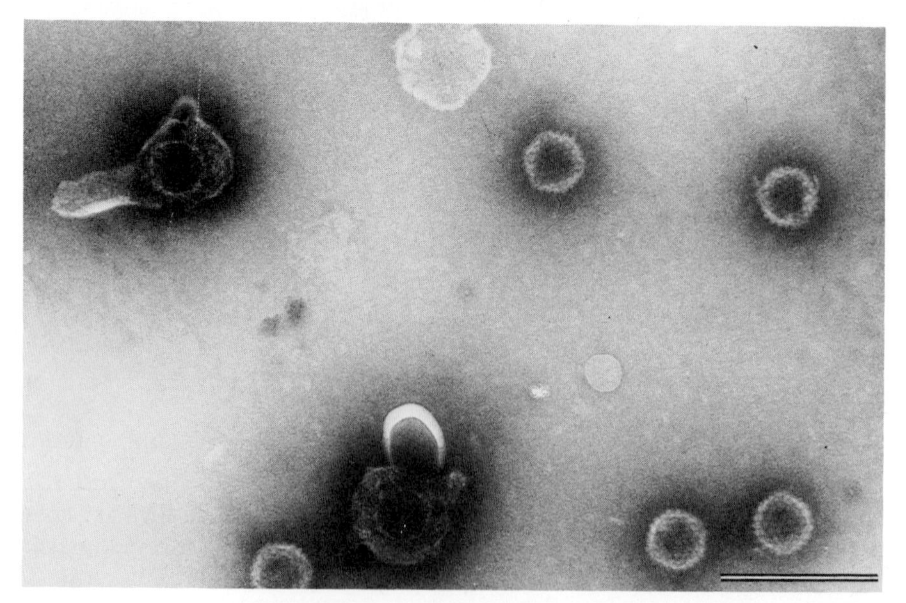

FIGURE 23. Negatively stained CCV particles. Scale bar: 0.25 μm. From Wolf and Darlington (1971). Reprinted with the permission of the Amrican Society of Microbiology.

panied by a decline in titer, and at the final assay, only 20% of the original infectivity remains.

Representative mammalian, avian, amphibian, and fish cell lines have been tested, but CCV has been replicated only by ictalurid and clariid fish cell lines. Most work has been done with the ictalurid lines BB and CCO (Bowser and Plumb, 1980). Temperature range for replication extends from 10 to 33°C, but yields are generally poor at 10–16°C (Fig. 24). Optimal growth occurs at 25–30°C, at which maximal yield is about 10^7 PFU/ml (Figs. 25 and 26). At 33°C, a one-step growth curve peaks at about 12 hr, but that temperature is higher than tolerated for sustained growth of the BB cell line. Regardless of temperature, the amounts of released virus and cell-associated virus are equal.

In vitro, CCV first causes cells to round somewhat and then induces syncytia and, ultimately, necrosis (Figs. 27 and 28). Polykaryocytes can become massive, but they ultimately undergo contraction and a terminal lysis ensues (Fig. 29). Stained cell cultures show a generalized basophilia, and classic Cowdry Type A intranuclear inclusions (Fig. 30). Electron microscopy of infected cells, whether in culture or in diseased tissues, shows the classic sequence of viral replication. Virions are first found as naked capsids in the nucleus and as enveloped forms in the cytoplasm or extracellulary. A noteworthy finding is the presence of large masses of electron-dense lamellae in the nucleus of infected cells (see Fig. 22). The masses of lamellae are variously postulated to be paramyelin or precursors to capsomeres or capsids.

Identification of CCV is most commonly accomplished with serum neutralization tests, but fluorescent-antibody techniques have also been used (Bowser, 1976).

Fish-to-fish transmission of CCV is readily demonstrated by holding susceptibles with fish undergoing an active infection or in effluent from

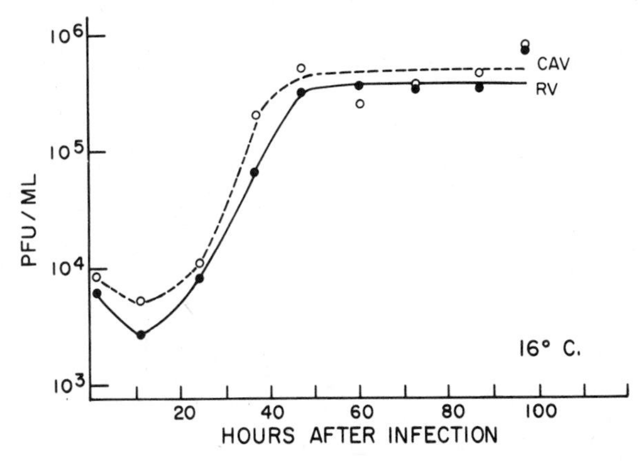

FIGURE 24. Representative growth curve of CCV in BB cells at 16°C in which virus yield exceeds input by a factor of only about 50. (RV) Released virus; (CAV) cell-associated virus. From Wolf (1973). Reprinted with the permission of Academic Press.

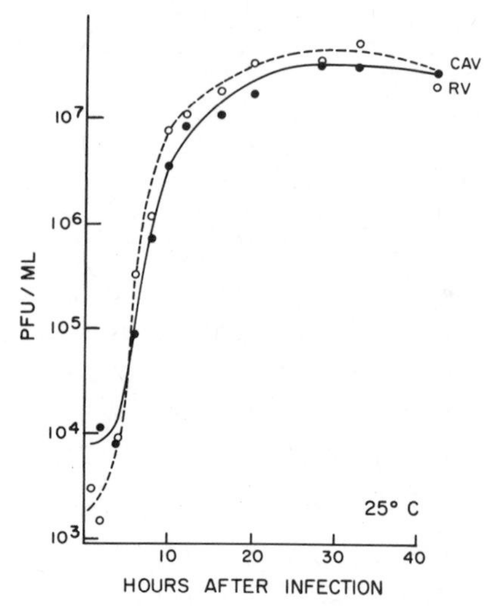

FIGURE 25. Representative growth curve of CCV in BB cells at optimal temperature of 25°C in which virus yield is several thousand times greater than input. (RV) Released virus; (CAV) cell-associated virus. From Wolf (1973). Reprinted with the permission of Academic Press.

them. Experimental infections are commonly initiated by intramuscular or intraperitoneal injection, by feeding, or by brushing infective material on the gills. Simple immersion in a viral suspension is also effective and has the advantage of being the most natural and least traumatic.

A wealth of circumstantial evidence has been accumulated in support of vertical transmission; it is the means that best explains the regular occurrence of CCV in the progeny of certain populations of captive brood

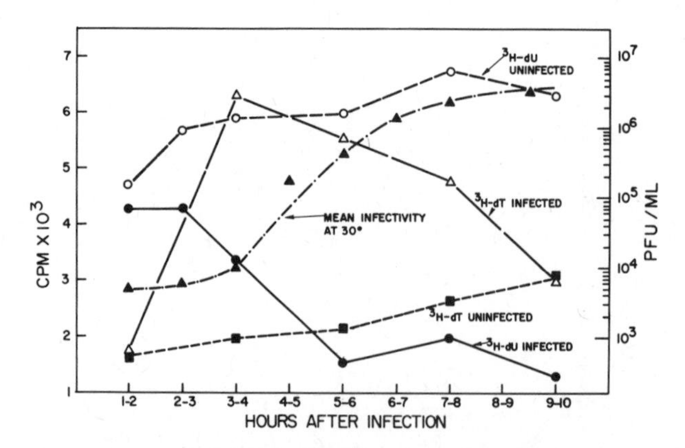

FIGURE 26. A comparison of [^{3}H]deoxyuridine (^{3}H-dU) and [^{3}H]deoxythymidine (^{3}H-dT) uptake at 30°C by infected and uninfected BB cells. An average one-step growth curve is included for reference purposes. From Wolf (1973). Reprinted with the permission of Academic Press.

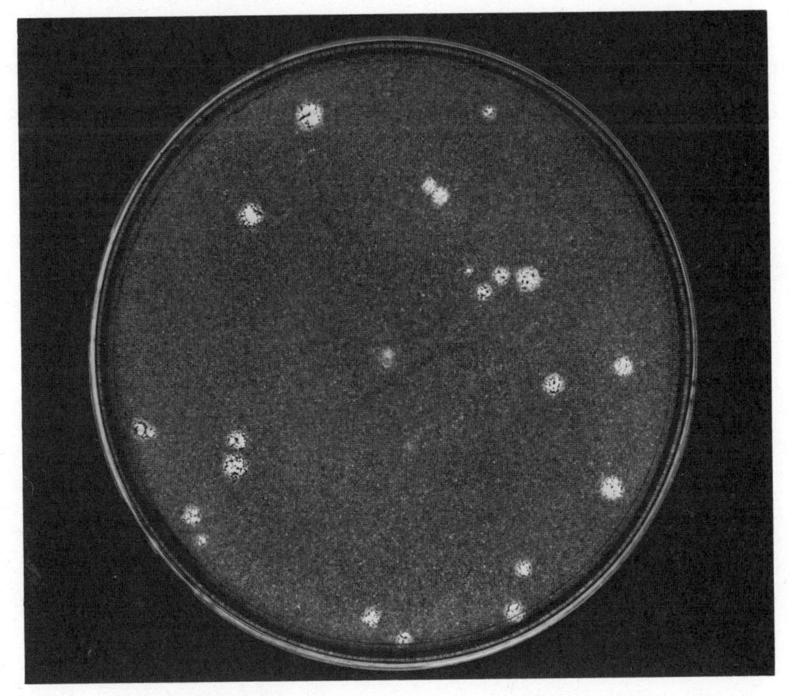

FIGURE 27. CCV plaques on BB cells consisting, in gross appearance, of peripheral syncytia around central areas of necrosis.

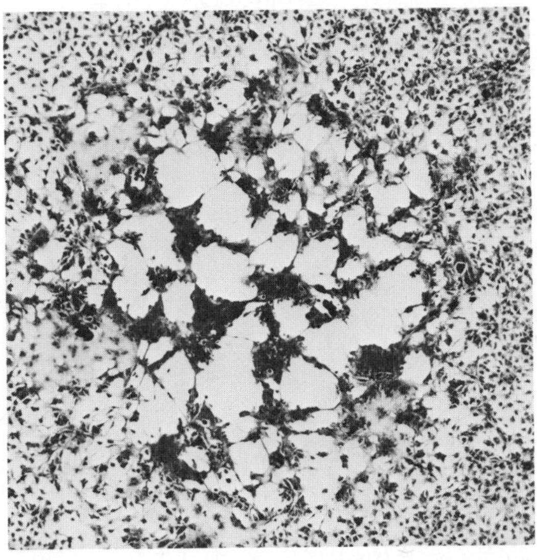

FIGURE 28. Close-up of CCV plaque in BB cells showing peripheral pyknosis and areas of fused cells. The plaque center shows advanced contraction and necrosis. From Wolf and Darlington (1971). Reprinted with the permission of the American Society of Microbiology.

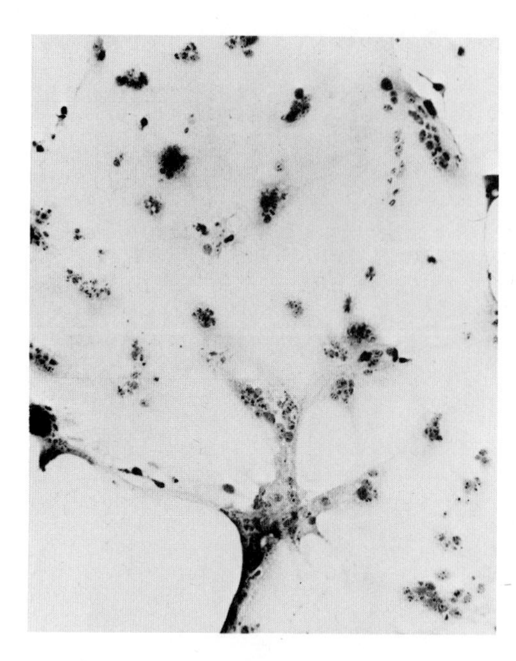

FIGURE 29. Large syncytium in BB cell culture induced by CCV. From Wolf and Darlington (1971). Reprinted with the permission of the American Society of Microbiology.

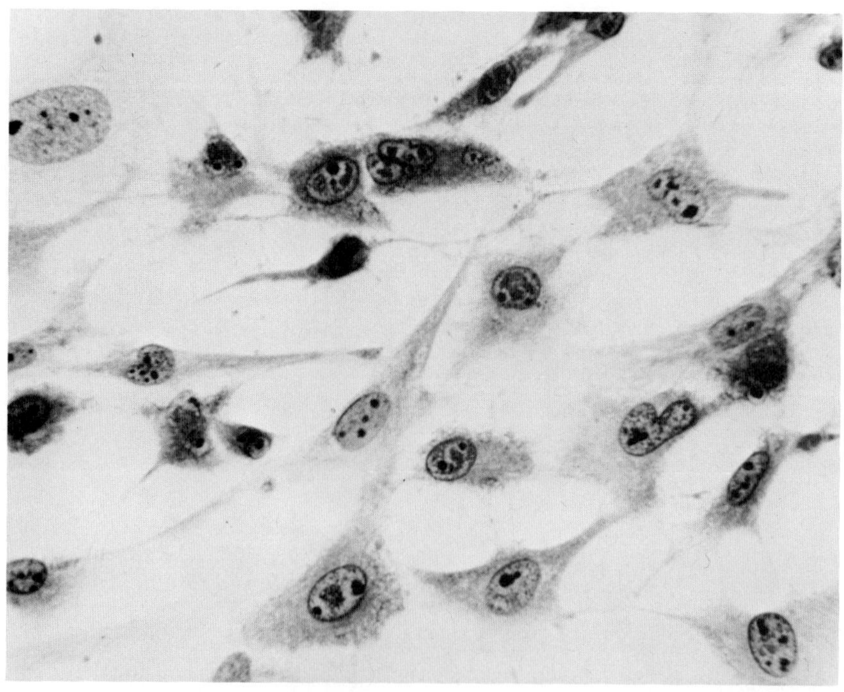

FIGURE 30. Margination of chromatin and Cowdry Type A intranuclear inclusions caused by CCV in BB cells. From Wolf (1973). Reprinted with the permission of Academic Press.

stock. However, the numerous attempts to isolate, or otherwise even to demonstrate, the virus in adult fish have thus far failed. No method of isolation, direct culture of tissues, or cocultivation has been productive. Notably lacking have been efforts to culture neural tissue or to induce the release of CCV by temperature or hormonal manipulation. It seems certain that adults must harbor the virus and that its release is related to the seasonal reproduction cycle or temperature change or both.

Indirect evidence supports the postulated source of CCV. Plumb (1977) assayed sera of 71 adults from a population with a history of progeny with CCV. By conventional standards, 67 (94%) showed a neutralization index equal to or greater than 1.6; 2 were equivocal and 2 were negative. Sera of 10 adults from a population with a clean health history were without CCV-neutralizing activity.

It is interesting to note that CCV can be rather easily isolated from the victims of an outbreak. The incubation time can be as short as 72 hr, and the epizootic may last but 7–10 days. Thereafter, for all practical purposes, it has not been possible to make an isolation. Field observations of CCV survivors have shown that the fish were unthrifty and did not fare as well as uninfected fish. That phenomenon was subsequently demonstrated experimentally. Although surviving fish fed well and lacked histological evidence of viral lesions, their weight averaged only 15% of that attained by the uninfected. The results are a tantalizing link in the hypothesis that the adult is the source of CCV.

The immune response of channel catfish to live CCV has been studied by several investigators, but the work of Heartwell (1975) was the most extensive. Such research typically involved fish that were older than 1 year, thus providing a host that would survive and would yield serum samples of adequate volume for study. Initial viral inoculation did not cause disease, but it did induce a humoral response that varied considerably from fish to fish but generally peaked at about 9 weeks at 22–28°C and then declined. A second injection evoked a weak and short-lived anamnestic reaction at about 10 days. Heat-killed virus apparently was not antigenic. Heartwell (1975), who also studied the viral-neutralizing component of the immune serum, characterized it as a macroglobulin with a molecular weight of $8–9 \times 10^5$ daltons, somewhat heat-labile, and subject to some reduction with 2-mercaptoethanol and 2 M urea. Complete reduction occurred with 6 M urea.

The first line of practical defense against CCV consists of avoidance—bypassing any source of stock that has a history of the disease. Second, if an epizootic occurs, the affected population can be destroyed and the ponds disinfected. Lability of CCV is rather marked, and in the absence of susceptible fish the virus is rapidly inactivated. Precautionary disinfection with 20–50 ppm chlorine is advised, to be followed by bioassay with susceptible fish in a holding cage.

Plumb and coworkers (Plumb, 1973a,b; Plumb and Gaines, 1975) have addressed problems associated with CCV. They found that reduction of temperature from 28 to 19°C markedly reduced losses when an epizootic occurs. He admitted, however, that under the common extensive aquaculture of catfish, temperature modulation may be difficult or impractical. Where CCV cannot be avoided, e.g., under conditions where well water is not available and a contaminated water source must be used, Plumb *et al.* (1975) found that several strains of catfish have significant degrees of resistance and that others are markedly susceptible. Under the test conditions employed, survival was only 30% in the Falcon strain, but nearly 90% in the Yazoo strain. In addition, hybrids of the Yazoo strain and strains with moderate resistance also had satisfactory levels of resistance.

Several workers have used antiviral drugs in *in vivo* or *in vitro* work with CCV, but practical chemical control measures have not been developed. As an example, Koment and Haines (1978), who compared herpes simplex and CCV, found that CCV required 20 times more phosphonacetic acid to inhibit its replication.

Attenuation of CCV has been achieved by passing CCV repeatedly in a cell line from the walking catfish. Early results indicate that protection against virulent virus is provided, but it is not known whether the attenuated strain is stable and will not revert to a virulent form.

E. Turbot Herpesvirus (*Herpesvirus scophthalmi*)

The turbot (*Scophthalmus maximus*) is a European species of marine flatfish of the family Bothidae (lefteye flounders). The turbot is highly regarded as a food fish, and considerable effort is being spent to rear the fish in commercial husbandry. When significant mortality occurred among cultured fry, a search for a possible cause revealed a herpesvirus in association with brachial and dermal polykaryocytes or giant cells (Fig. 31). The agent has not yet been isolated, nor have transmission studies been carried out. The size and shape of the agent leave little doubt about its grouping, and the original investigators have proposed the working name *Herpesvirus scophthalmi* (Buchanan *et al.*, 1978).

The agent came to light when heavy mortality (about 30% during fall and winter) occurred among fry being reared in heated effluent from a nuclear generating plant in Scotland. Search for an etiological agent failed to implicate bacteria or parasites. The fry were anorexic and lethargic and offered little resistance when netted. When they were in contact with tank bottoms, the head and tail were elevated abnormally. Externally and internally, the gross physical appearance of the fry was normal, but the polykaryocytes were abundantly evident in histological sections of skin and gills; other organs and tissues were without abnormalities (Richards and Buchanan, 1978).

Polykaryocytes apparently arose from the Malpighian layer, and commonly were 45–75 μm in diameter, though some were as large as 130 μm in greatest dimension (Fig. 32). The usual giant cell had only one

FIGURE 31. Differential interference phase-contrast micrograph of giant cells in skin of an infected turbot. Black dendritic forms are melanophores. Scale bar: 50 μm. From Buchanan *et al.* (1978). Reprinted with the permission of the British Veterinary Association.

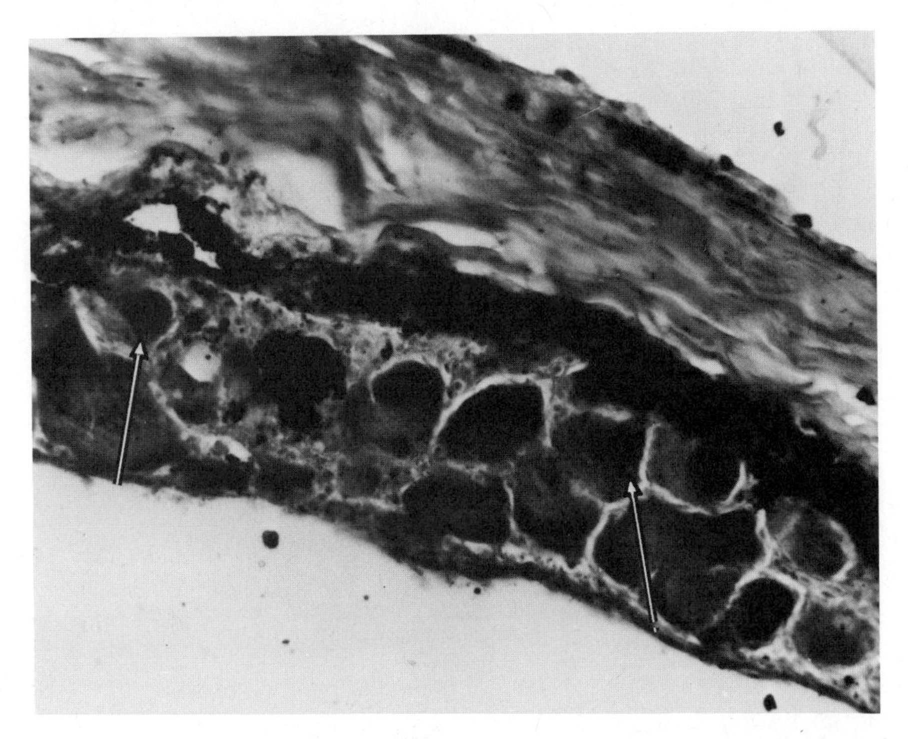

FIGURE 32. Section of skin from infected turbot showing prominent giant cells. Size and shape of nuclei (arrows) vary. Hematoxylin–eosin. ×200. Courtesy of J.S. Buchanan.

nucleus that occupied as much as 90% of the cell. A few cells had several nuclei, presumably destined to fuse. Giant-cell nuclei were often fenestrated and showed a basophilic Feulgen-positive, periodic acid–Schiff (PAS)-negative area. The nucleus also showed coarse Feulgen-positive granules. Cytoplasm was generally basophilic, but coarse PAS-negative and Feulgen-positive granules were present (Richards and Buchanan, 1978).

Heavy infection of gills resulted in hyperplasia and fusion of lamellae with attendant vascular stasis, fibrinous thrombosis, and reduction of functional respiratory exchange area (Fig. 33).

Buchanan and Madeley (1978) elegantly described and illustrated ultrastructural details of infected cells and *H. scophthalmi*. Naked capsids, some empty and some with a nucleoid, occur in the nucleus (Fig. 34), at times in paracrystalline array. Virion size is about 100 nm. Enveloped particles measure 200–220 nm and occur in the cytoplasm (Fig. 35). The envelope is trilaminar and shows an outer fringe of petal-like spikes about 18 nm long, much like those of a coronavirus (Fig. 36).

The occurrence of giant cells in the turbot is not peculiar to fish in

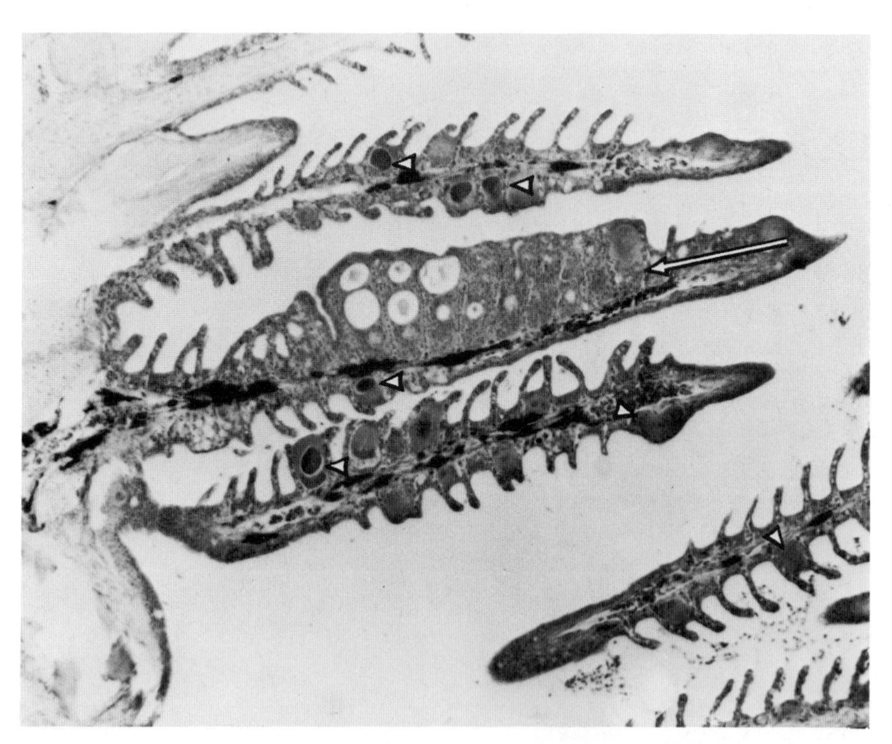

FIGURE 33. Gill section of infected turbot showing giant cells (points) and hyperplasia resulting in fusion of lamellae and reduction of respiratory surface (arrow). Hematoxylin–eosin. ×130. From Richards and Buchanan (1978). Reprinted with the permission of Blackwell Scientific Publications Ltd.

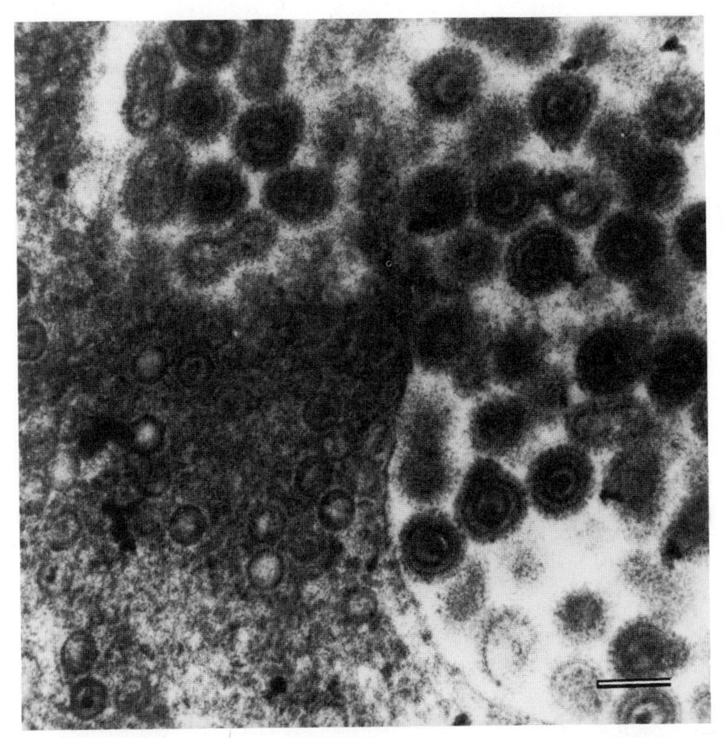

FIGURE 34. Thin section of infected turbot cell showing empty and full 100-nm capsids in the nucleus (left) and enveloped particles in the cytoplasm. Scale bar: 100 nm. From Buchanan *et al.* (1978). Reprinted with the permission of the British Veterinary Association.

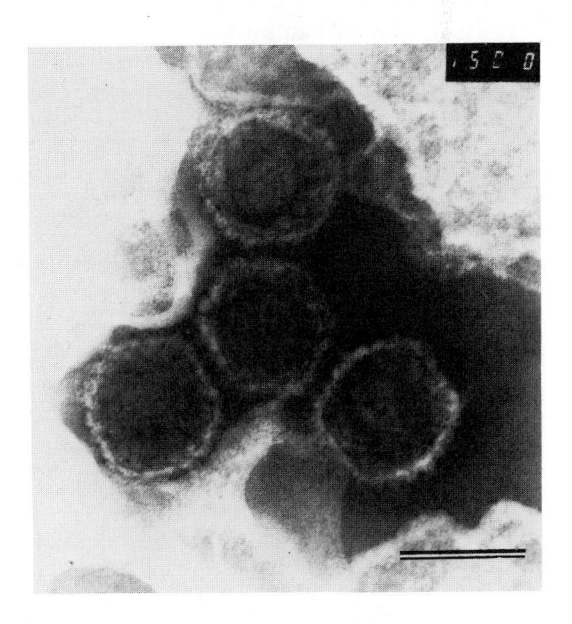

FIGURE 35. Negatively stained turbot herpesvirus particles. Scale bar: 100 nm. From Buchanan *et al.* (1978). Reprinted with the permission of the British Veterinary Association.

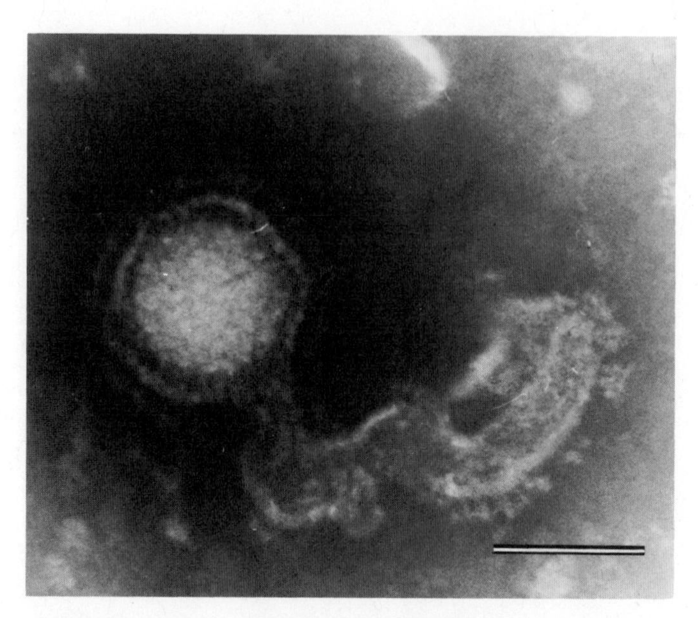

FIGURE 36. Negatively stained turbot herpesvirus particle showing fringe of 18-nm spikes. Scale bar: 100 nm. From Buchanan *et al.* (1978). Reprinted with the permission of the British Veterinary Association.

husbandry; they have also been found in otherwise normal fish taken directly from the wild. Also, hatchery fish normal in behavior and appearance were found with pathognomonic cells, although they were fewer than in clinically afflicted fish. Perhaps the virus plays a causal role in the mortality, and abnormal stress may also be involved. Typically, mortality has followed (within a matter of days) handling, transport, temperature fluctuations, or high chlorine levels.

F. Walleye Herpesvirus

The walleye (*Stizostedion vitreum*) is a percid highly regarded by anglers as choice table fare. Ecologically, walleyes are considered to be coolwater creatures and thus are found in fresh waters of the central and northern tiers of the United States and well into the Canadian Provinces.

Walleyes have long been known to harbor viruses. Walker (1969) noted that among walleyes of the New York region, at least three different viruses could be distinguished by electron microscopy—the large (200-nm) icosahedral cytoplasmic particles of lymphocystis disease, 100-nm intercellular virions associated with a dermal sarcoma, and abundant 80-nm virions budding from membranes of cells constituting a third and histologically distinctive condition known as epidermal hyperplasia. The particles associated with the two neoplasms have not yet been isolated.

Lymphocystis virus has been isolated from freshwater and marine fishes and presumably could be readily isolated from walleyes.

Kelly *et al.* (1980a) isolated a new herpeslike virus from a single adult walleye afflicted with epidermal hyperplasia. The source location was a lake in Saskatchewan. The lesions were present on fish taken in early spring, but regressed rapidly thereafter and were never observed later in the year (Kelly *et al.*, 1980b). Some specimens were tagged and recaptured a year later; accordingly, the authors concluded that the virus probably did not contribute significantly to postspawning mortality.

Details of histopathological findings have not been reported, but electron microscopy of thin-sectioned hyperplastic epidermis showed numerous incomplete virions present within cell nuclei. Size was not reported, but infectivity was said to pass through membranes with a mean pore diameter of 200 nm, but not 100 nm.

Data supporting the herpesvirus nature of the agent are inhibition of *in vitro* replication by IUdR and 5-bromodeoxyuridine (BUdR); inhibition by BUdR is reversed in the presence of excess thymidine. More to the point, replication was inhibited by 10^{-3} M phosphonacetate. The original isolation was made on the WO line of walleye ovary cells, but other cell lines of walleye origin, WC-1 and WC-2, were also susceptible. As would be expected for a herpesvirus, the walleye agent was not replicated in RTG-2 or CHSE-214 (salmonid), BB (ictalurid), or FHM (cyprinid) cell lines, which are collectively able to support most other isolated fish viruses.

The walleye herpesvirus is ether-labile, and *in vitro* it induces the formation of syncytia followed by lysis. Replication occurs at 4 and 15°C, but (consistent with the cool water preference of its host) not at 20°C. Plaquing is routinely carried out on the WC-1 cell line; at 15°C, plaques are produced in 10–13 days. Maximal titer is 10^5 PFU/ml.

Kelly *et al.* (1980b) noted that electron microscopy of infected cell cultures showed particles identical to the virus seen in the original lesion. The relationship of the virus to the induction of neoplasms is as yet undetermined, but the authors correctly speculated that the agent should be regarded as possibly being a cause of mortality in the young. Such a pattern of virulence for the young and association with skin neoplasms occurs with the *Oncorhynchus masou* virus (OMV) in Japan.

REFERENCES

Bowser, P.R., 1976, Fluorescent and antibody test for channel catfish virus, *Fish Health News* **5**:4.

Bowser, P.R., and Plumb, J.A., 1980, Channel catfish virus: Comparative replication and sensitivity of cell lines from channel catfish ovary and the brown bullhead, *J. Wildl. Dis.* **16**:451.

Buchanan, J.S., and Madeley, C.R., 1978, Studies on *Herpesvirus scophthalmi* infection of turbot *Scophthalmus maximus* (L.) Ultrastructural observations, *J. Fish Dis.* **1**:283.

Buchanan, J.S., Richards, R.H., Sommerville, C., and Madeley, C.R., 1978, A herpes-type virus from turbot (*Scophthalmus maximus* L.), *Vet. Rec.* **102**:527.

Chousterman, S., Lacasa, M., and Sheldrick, P., 1979, Physical map of the channel catfish virus genome: Location of sites for restriction endonucleases *Eco* RI, *Hind* III, *Hpa* I, and *Xba* I, *J. Virol.* **31**:73.

Clark, H.F., and Karzon, D.T., 1972, Iguana virus, a herpes-like virus isolated from cultured cells of a lizard, *Iguana iguana, Infect. Immun.* **5**:559.

Clark, H.F., and Lunger, P.D., 1981, Viruses of reptiles, in: *Diseases of Reptilia* (J.E. Cooper and D.F. Jackson, eds.), pp. 135–164, Academic Press, New York.

Cox, W.R., Rapley, W.A., and Barker, I.K., 1980, Herpesvirus-like infection in a painted turtle (*Chrysemys picta*), *J. Wildl. Dis.* **16**:445.

Fijan, N., 1968, Progress report on acute mortality of channel catfish caused by a virus, *Off. Int. Epizoot. Bull.* **69**:1167.

Frye, F.L., Oshiro, L.O., Dutra, F.R., and Carney, J.D., 1977, Herpesvirus-like infection in two Pacific pond turtles, *J. Am. Vet. Med. Assoc.* **171**:882.

Goodheart, C.R., and Plummer, G., 1975, The densities of herpesviral DNAs, *Prog. Med. Virol.* **19**:324.

Haines, H., 1978, A herpesvirus disease of green sea turtles in aquaculture, *Mar. Fish Rev. Paper* **1294**:33.

Haines, H., and Kleese, W.C., 1977, Effect of water temperature on a herpesvirus infection of sea turtles, *Infect. Immun.* **15**:756.

Heartwell, C.M., 1975, Immune response and antibody characterization of the channel catfish (*Ictalurus punctatus*) to a naturally pathogenic bacterium and virus, *U.S. Fish Wildl. Serv. Tech. Paper* **85,** 34 pp.

Kelly, R.K., Nielsen, O., and Yamamoto, T., 1980a, A new herpes-like virus (HLV) of fish (*Stizostedion vitreum vitreum*), *In Vitro* **16**:255.

Kelly, R.K., Nielsen, O., Campbell, J.S., and Yamamoto, T., 1980b, Current status of lymphocystis, dermal sarcoma/fibroma, and related diseases in walleye, *Stizostedion vitreum vitreum,* in western Canada, in: Proceedings of the Fourth Biennial Fish Health Section and Fish Disease Workshops, Seattle, Washington (mimeographed).

Kimura, T., Yoshimizu, M., Tanaka, M., and Sannohe, H., 1981a, Studies on a new virus (OMV) from *Oncorhynchus masou.* I. Characteristics and pathogenicity, *Fish Pathol.* **15**:143–147.

Kimura, T., Yoshimizu, M., and Tanaka, M., 1981b, Studies on a new virus (OMV) from *Oncorhynchus masou.* II. Oncogenic nature, *Fish Pathol.* **15**:149–153.

Kimura, T., Yoshimizu, M., and Tanaka, M., 1981c, Fish viruses: Tumor induction in *Oncorhynchus keta* by the herpesvirus, in: *Phyletic Approach to Cancer* (C.J. Dawe, J.C. Harshbarger, T. Sugimura, S. Takayama, and S. Kodo, eds.), pp. 59–68, Japan Sci. Soc. Press, Tokyo.

Koment, R.W., and Haines, H.G., 1977, A new reptilian herpesvirus isolated from the green sea turtle, *Chelonia mydas, Abstr. Annu. Meet. Am. Soc. Microbiol.* New Orleans, Louisiana, May 8–13, p. 347.

Koment, R.W., and Haines, H., 1978, Decreased antiviral effect of phosphonacetic acid on the poikilothermic herpesvirus of channel catfish disease, *Proc. Soc. Exp. Biol. Med.* **159**:21.

Lowenthal, W., 1907, Einschlussartige Zell und Kernveränderungen in der Karpfenpocke, *Z. Krebsforsch.* **5**:197.

Lunger, P.D., and Clark, H.F., 1978, Reptilia-related viruses, in: *Advances in Virus Research* (M.A. Lauffer, K. Maramorosch, F.B. Bang, and K.M. Smith, eds.), pp. 159–204, Academic Press, New York.

Major, R.D., McCraren, J.P., and Smith, C.E., 1975, Histopathological changes in channel catfish (*Ictalurus punctatus*) experimentally and naturally infected with channel catfish virus disease, *J. Fish. Res. Board Can.* **32**:563.

McCraren, J.P., 1972, Channel catfish virus disease (CCVD): A current review, *Proc. Annu. Conf. Western Assoc. Game Fish Comm.* **52**:528.

Monroe, J.H., Shibley, G.P., Schidlovsky, G., Nakai, T., Howatson, A.F., Wivel, N.W., and O'Conner, T.E., 1968, Action of snake venom on Rauscher virus, *J. Natl. Cancer Inst.* **40:**135.

Padgett, F., and Levine, A.S., 1966, Fine structure of the Rauscher leukemia virus as revealed by incubation in snake venom, *Virology* **30:**623.

Plumb, J.A., 1973a, Effects of temperature on mortality of fingerling channel catfish (*Ictalurus punctatus*) experimentally infected with channel catfish virus, *J. Fish. Res. Board Can.* **30:**568.

Plumb, J.A., 1973b, Neutralization of channel catfish virus by serum of channel catfish, *J. Wildl. Dis.* **9:**324.

Plumb, J.A., 1977, Channel catfish virus disease, *U.S. Fish Wildl. Serv. Fish Dis. Leaflet* **52,** 8 pp.

Plumb, J.A., and Gaines, J.L., Jr., 1975, Channel catfish virus disease, in: *Pathology of Fishes* (W.E. Ribelin and G. Migaki, eds.), pp. 287–302, University of Wisconsin Press, Madison.

Plumb, J.A., Gaines, J.L., Mora, E.C., and Bradley, G.G., 1974, Histopathology and electron microscopy of channel catfish virus in infected channel catfish, *Ictalurus punctatus* (Rafinesque), *J. Fish Biol.* **6:**661.

Plumb, J.A., Green, O.L., Smithermann, R.O., and Pardue, G.B., 1975, Channel catfish virus experiments with different strains of channel catfish, *Trans. Am. Fish. Soc.* **104:**140.

Raynaud, A., and Adrian, M., 1976, Lésions cutanées á structure papillomateuse associées á des virus chez le lézard vert (*Lacerta virdis* Laur.), *C. R. Acad. Sci. Ser. D* **283:**845.

Rebell, G., Rywlin, A., and Haines, H., 1975, A herpesvirus-type agent associated with skin lesions of green sea turtles in aquaculture, *Am. J. Vet. Res.* **36:** 1221.

Richards, R.H., and Buchanan, J.S., 1978, Studies on *Herpesvirus scophthalmi* infection of turbot *Scophthalmus maximus* (L.): Histopathological observations, *J. Fish Dis.* **1:**3.

Sano, T., 1976, Viral diseases of cultured fishes in Japan, *Fish Pathol.* **10:**221.

Schubert, G., 1964, Elektronenmikroskopische Untersuchungen zur Pockenkrankheit des Karpfens, *Z. Naturforsch.* **19:**675.

Schubert, G.H., 1966, The infective agent in carp pox, *Off. Int. Epizoot. Bull.* **65:**1011.

Simpson, C.F., Jacobson, E.R., and Gaskin, J.M., 1979, Herpesvirus-like infection of the venom gland of Siamese cobras, *J. Am. Vet. Med. Assoc.* **175:**941.

Sonstegard, R.A., and Sonstegard, K.S., 1978, Herpesvirus-associated epidermal hyperplasia in fish (carp), in: *Proceedings of the International Symposium on Oncogenesis and Herpesviruses III* (G. De-Thé, W. Henle, and F. Rapp, eds.), pp. 863–868, International Agency for Research on Cancer, Scientific Publication No. 24, Lyon, France.

Walker, R., 1969, Virus associated with epidermal hyperplasia in fish, *Natl. Cancer Inst. Monogr.* **31:**195.

Wolf, K., 1973, Herpesviruses of lower vertebrates, in: *The Herpesviruses* (A.S. Kaplan, ed.), pp. 495–520, Academic Press, New York.

Wolf, K., 1979, Comparative pathogenesis and virulence of poikilotherm vertebrate herpesviruses, in: *Proceedings of the 4th Munich Symposium on Microbiology on Mechanisms of Viral Pathogenesis and Virulence* (P.A. Bachmann, ed.), pp. 183–200, WHO Collaborating Centre for Collection and Evaluation of Data on Comparative Virology.

Wolf, K., and Darlington, R.W., 1971, Channel catfish virus: A new herpesvirus of ictalurid fish, *J. Virol.* **8:**525.

Wolf, K., and Mann, J.A., 1980, Poikilotherm vertebrate cell lines and viruses: A current listing for fishes, *In Vitro* **16:**179.

Wolf, K., and Smith,, C.E., 1981, *Herpesvirus salmonis:* Pathological changes in parenterally infected rainbow trout fry, *J. Fish Dis.*, 445–457.

Wolf, K., Herman, R.L., and Carlson, C.P., 1972, Fish viruses: Histopathologic changes associated with experimental channel catfish virus disease, *J. Fish. Res. Board Can.* **29:**149.

Wolf, K., Sano, T., and Kimura, T., 1975, Herpesvirus disease of salmonids, *U.S. Fish Wildl. Serv. Fish Dis. Leaflet* **44,** 8 pp.

Wolf, K., Darlington, R.W., Taylor, W.G., Quimby, M.C., and Nagabayashi, T., 1978, *Herpesvirus salmonis:* Characterization of a new pathogen of rainbow trout, *J. Virol.* **27**:659.

Zeigel, R.F., and Clark, H.F., 1972, Electron microscopy observations on a new herpes-type virus isolated from *Iguana iguana* and propagated in reptilian cells *in vitro, Infect. Immun.* **5**:570.

CHAPTER 7

Amphibian Herpesviruses

ALLAN GRANOFF

I. INTRODUCTION

> Frogs are strange creatures. One would describe them as peculiarly wary and timid, another as equally bold and imperturbable. All that is required in studying them is patience.
>
> THOREAU

Amphibian herpesviruses might never have been studied had Lucké (1934) not discovered that cells of the renal adenocarcinoma of the common leopard frog, *Rana pipiens*, frequently contain intranuclear acidophilic inclusion bodies similar to those found in herpes simplex infections and other viral diseases. On the strength of this observation, and results of subsequent experiments, Lucké concluded that the frog renal adenocarcinoma was of viral origin—an interpretation that was unequivocally substantiated more than three decades later when a herpesvirus was firmly established as the causative agent (reviewed in Naegele and Granoff, 1980). During experiments to isolate and cultivate the Lucké herpesvirus *in vitro*, Rafferty (1965) isolated a herpesvirus that was distinct from the Lucké tumor herpesvirus (Gravell *et al.*, 1968; Gravell, 1971). These remain the only two herpesviruses isolated from amphibia, and by comparison with other herpesviruses, little is known of their properties and interaction with host cells. This chapter will summarize the relevant available information on the two amphibian herpesviruses and, where possible, point out reasons for the paucity of information on them.

ALLAN GRANOFF • Division of Virology, St. Jude Children's Research Hospital, Memphis, Tennessee 38101.

II. LUCKÉ TUMOR

Since the frog renal adenocarcinoma (Lucké tumor) provided the impetus for studies on amphibian herpesviruses, a brief description of its general features is warranted. More detailed information can be found elsewhere (Rafferty, 1964; Lunger *et al.*, 1965; Zambernard and Mizell, 1965; Naegele and Granoff, 1980).

The Lucké tumor has a distinct geographic distribution, limited to the north central and northeastern United States and adjacent southern Canada (Lucké,1952; Lunger *et al.*, 1965; McKinnell, 1965). The frequency of tumor-bearing frogs taken directly from individual collection sites in nature has varied from less than 1% to as high as 9% (McKinnell and McKinnell, 1968); both sexes are equally susceptible to tumor formation (Lucké, 1952; McKinnell, 1965; McKinnell and McKinnell, 1968). When large adult frogs are kept in the laboratory, in either isolated or crowded conditions, at 25°C for about 8 months, the tumor incidence increases up to 50% (Rafferty and Rafferty, 1961; Rafferty, 1963). For some unknown reason, the *R. pipiens* population in regions where the tumor occurs has decreased drastically in recent years (Kemper, 1977). The resulting shortage of tumor-bearing frogs has seriously hampered research on this naturally occurring herpesvirus tumor.

Lucké tumors are usually well-differentiated adenocarcinomas composed of rather large basophilic cells; the growths are frequently cystic with pronounced papillary ingrowths. Occasionally, some tumors resemble adenomas rather than adenocarcinomas, with the component cells having a less atypical appearance and a more orderly arrangement. Tumor size ranges from small single or multicentric early tumors to large masses replacing most of the kidney. Several foci may develop simultaneously, and involvement of the kidneys is either unilateral or bilateral. The criteria by which a tumor is recognized as a malignant neoplasm, whether in man or other vertebrates, are met by the Lucké tumor: it invades and destroys its tissue of origin, and can metastasize.

An important characteristic of some Lucké tumors is the presence of acidophilic intranuclear inclusions reminiscent of virus infections. Lucké (1952) recognized a seasonal variation in the presence or absence of inclusions, and the relationship between temperature and inclusions (and virus) was eventually established experimentally (see Section III.C). Electron-microscopic examination of nuclear-inclusion-containing Lucké tumor cells invariably reveals the presence of herpesviruses (Fawcett, 1956; Lunger *et al.*, 1965; Zambernard and Mizell, 1965; Zambernard *et al.*, 1966).

The only successful transplants of the Lucké tumor have been to the anterior eye chamber (Lucké and Schlumberger, 1939; Schlumberger and Lucké, 1949). If tumor fragments or cell suspensions are implanted into other body sites, there is no appreciable local tumor growth, even when the immune system has been impaired by X-radiation, antilymphocyte serum, or immunosuppressive drugs (R.F. Naegele and A. Granoff, unpublished data).

III. LUCKÉ TUMOR HERPESVIRUS

A. Morphology and Structure

The Lucké herpesvirus (LHV) is similar in size and structure (Lunger, 1964; Toplin *et al.*, 1971) (Fig. 1B and C) to other herpesviruses of both higher vertebrates (e.g., humans) and lower vertebrates (e.g., fish). Figure 1A is an electron micrograph of a tumor cell containing numerous nuclear herpesvirus particles in various stages of development. The cells containing virus have degenerated, implying that cell survival is incompatible with virus replication. The developmental sequence of LHV in tumor cells, based on reconstruction from static images of virus seen by electron microscopy in thin sections of virus-containing tumor cells (Fig. 1A) (Lunger *et al.*, 1965), appears similar to that of other herpesviruses. The virus has not been grown *in vitro* in circumstances that would permit a temporal study of the kinetics and molecular and biological events of the replication cycle (see Section III.G).

B. Virus DNA

The DNA of the LHV is a linear double-stranded molecule with a base composition of 45–47% guanosine plus cytosine (G + C) (Wagner *et al.*, 1970; Gravell, 1971). We have determined the molecular weight of LHV DNA by contour-length measurements (Granoff and Naegele, 1980). DNA isolated from purified LHV obtained from inclusion-bearing Lucké tumors has an average molecular weight of 66×10^6 (range 62.5–73.4) (Table I and Fig. 2). For comparison, we also determined, by contour-length measurements, the molecular weights of the DNAs of a number of herpesviruses from both higher and lower vertebrates. From summary of the data (Table I), it can be seen that LHV DNA has a much lower molecular weight than the DNAs of other herpesviruses from lower vertebrates, as well as the DNA of human herpes simplex virus type 1 (HSV-1). The molecular weight of LHV DNA approximates the long (L) DNA segment of the unique base sequence present in other herpesvirus genomes (Honess and Watson, 1977). Although this correspondence in size may be fortuitous and of no biological import, it may represent the minimal but fundamental genetic information required for specification of a herpesvirus. A less provocative possibility is that the length of the L segment is shorter than that of other herpesviruses and that LHV may have a short (S) region as well, so that the total size of the LHV genome equals that of the L region of other members of this virus group. The lack

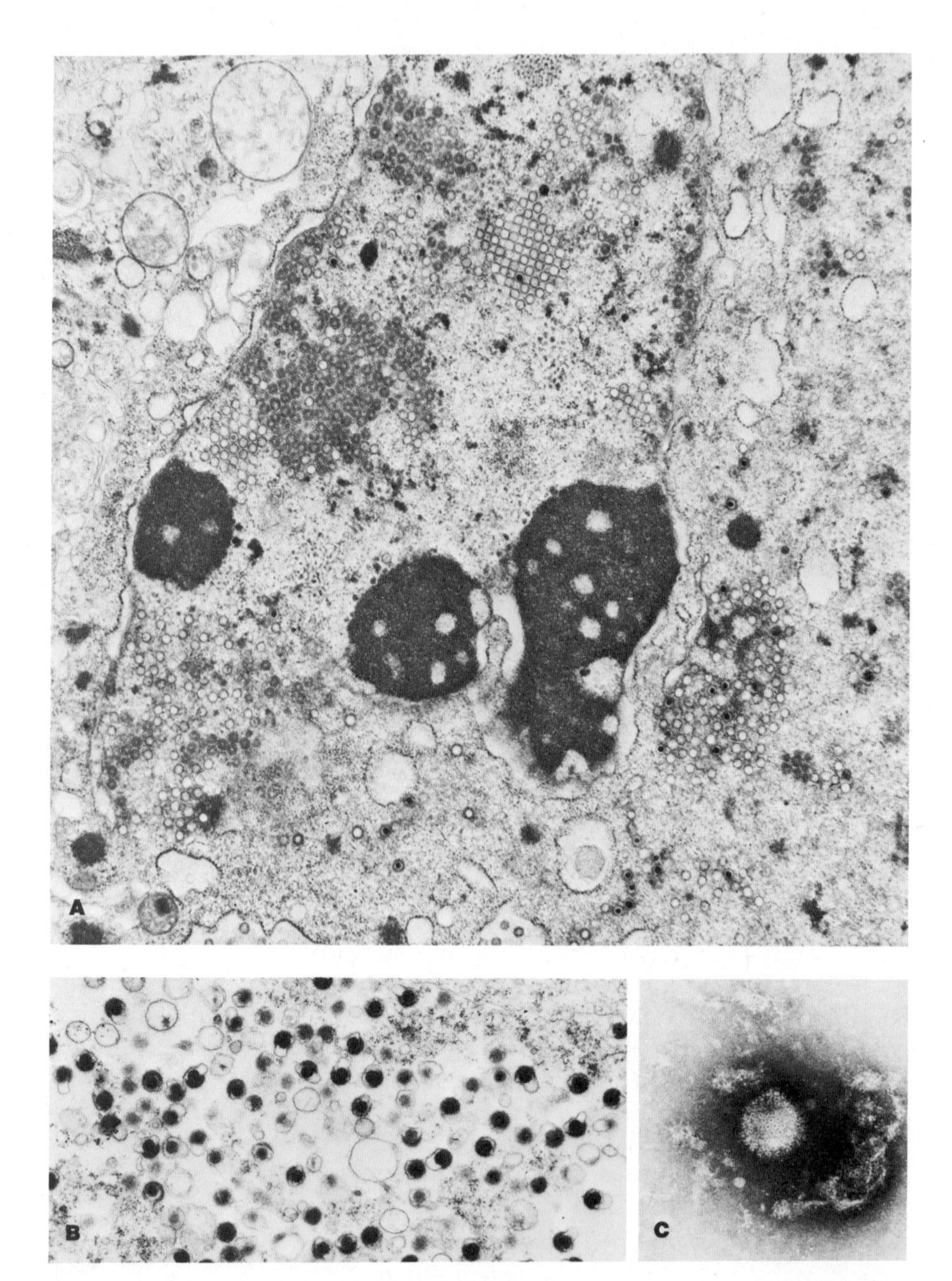

FIGURE 1. (A) Thin section of an inclusion-bearing Lucké tumor cell with typical herpesvirus particles in various stages of development in the nucleus. ×12,600. (B) Enveloped virions found extracellularly. ×15,600. (C) Negatively stained nonenveloped particle showing typical herpesvirus morphology. ×72,000. From Granoff (1972).

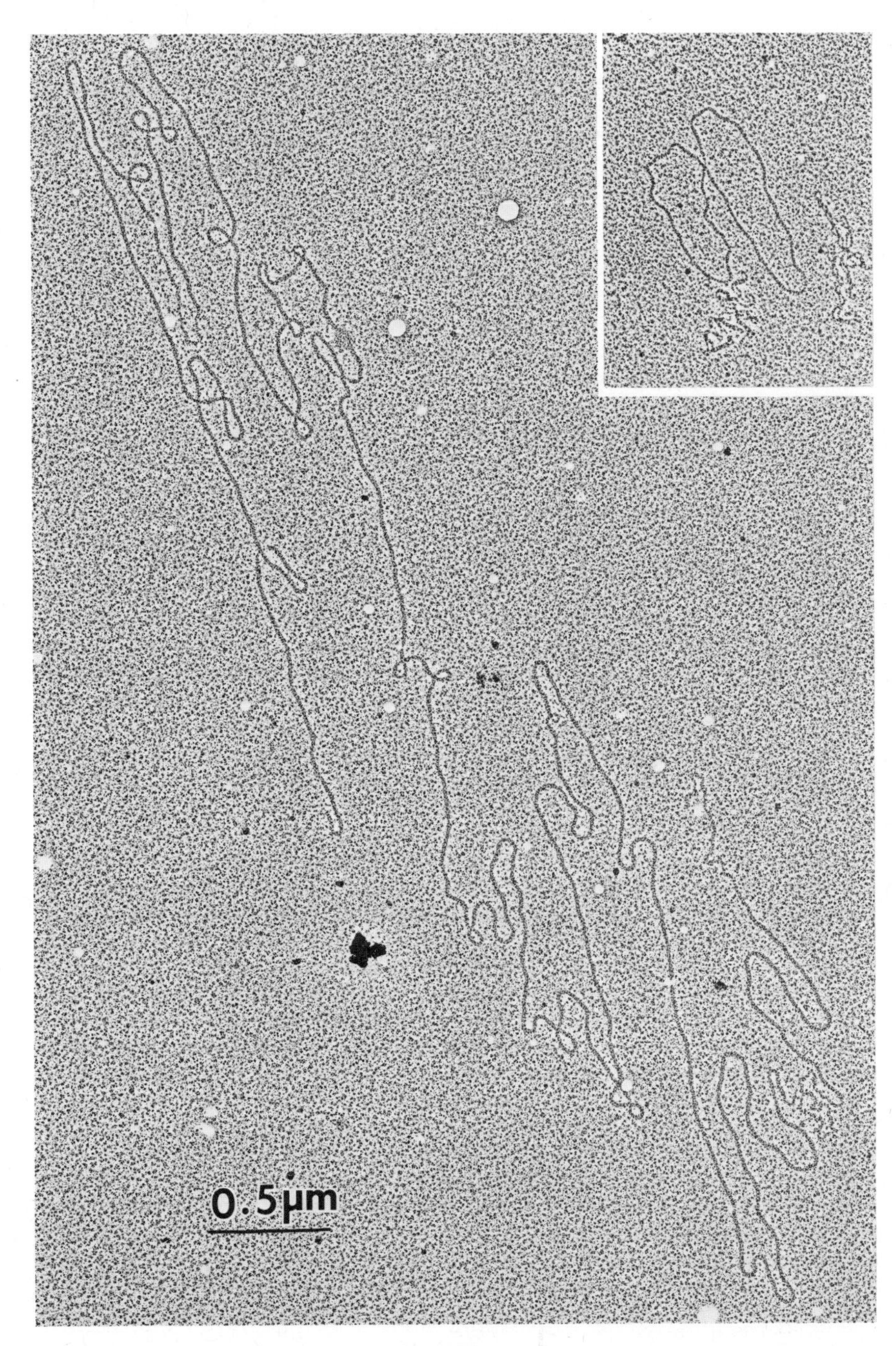

FIGURE 2. Electron micrograph of purified LHV DNA prepared by the Kleinschmidt technique. The molecule shown measures 10 PM_2 units or 64×10^6 daltons (1 PM_2 unit = 6.4×10^{-6} daltons). *Insert:* A PM_2 DNA molecule prepared and photographed under the same conditions. From Naegele and Granoff (1980).

TABLE I. Molecular Weights of Herpesvirus DNAs from a Human and from Lower Vertebrates[a]

Herpesvirus	Genome molecular weight ($\times 10^6$)	
	Range	Average
HSV-1	89.1–107.5	94.8
Channel catfish	66.6– 85.8	79.3
Trout	69.9– 83.8	78.4
Frog virus 4	73.9– 79.2	77.2
Lucké tumor	62.5– 73.4	65.5

[a] From Granoff and Naegele (1980). Molecular weights were determined by measurements from electron micrographs of contour lengths of each viral genome obtained from purified virions. PM_2 was used as a reference for obtaining the values given.

of virus-containing Lucké tumors as a source of virus and virus DNA has curtailed further experiments needed to establish a firm data base from which to draw important structural and evolutionary conclusions regarding the size of the LHV DNA genome.

C. Relationship of Temperature to Virus Replication and Expression of Virus Genes

Although Lucké (1952) recognized the seasonal variation of the presence of intranuclear inclusions, it was Roberts (1963) who first demonstrated clearly the temperature dependence of inclusion–body formation. She found that intranuclear inclusions were absent in Lucké tumor cells of frogs kept at 20–25°C, but were present in tumor cells when frogs were held at 5°C. In a temperature "shift-down" experiment, Rafferty (1965) demonstrated by sequential biopsies that intranuclear inclusions developed in previously inclusion-free tumors of frogs after 5–7 months when the animals were shifted from 20–25 to 4°C.

Mizell *et al.* (1968) and Skinner and Mizell (1972) transplanted the Lucké tumor to the anterior eye chamber of *R. pipiens* and other species, and then analyzed the relationship of temperature to the presence of intranuclear inclusions and herpesvirus. At various times after tumor fragments were transplanted into anterior eye chambers, frogs were placed at temperatures ranging from 4 to 26°C. A direct correlation was found between the presence of virus (and intranuclear inclusions) and the time and temperature of incubation. Replication of herpesvirus was induced by 3 months in 30% of the cells examined at temperatures below 11.5°C (optimal range, 7.5–9.5°C). The tumor cells remained virus-free at higher temperatures.

Further evidence for the effect of temperature on virus replication comes from the observation that urine from tumor-bearing frogs kept at 4°C frequently contains large amounts of herpesvirus, whereas urine from tumor-bearing frogs held at 25°C does not (Granoff and Darlington, 1969). Ascitic fluid from tumor-bearing frogs kept at low temperature may also contain herpesvirus (Naegele and Granoff, 1972).

The disappearance of virus particles with increases in temperature has likewise been demonstrated in the intact animal (Zambernard and Vatter, 1966).

Since each of the experiments described above ("shift-up" or "shift-down") involved intact animals, some part of the immune system or factors other than temperature might have determined the presence or absence of virus *in vivo*. Two experiments refute this possibility, demonstrating that temperature is in fact the controlling factor in virus replication. Morek and Tweedell (1969) reported the development of intranuclear inclusions in tumor cells of inclusion-free tumor fragments that had been explanted and held *in vitro* at 9°C. Moreover, Breidenbach *et al.* (1971) demonstrated that tissue-culture explants of virus-free tumors held *in vitro* at 7.5°C produced virus after 3 months. These results, indicating that the intact animal does not play a significant role in controlling the presence or absence of virus in tumor cells (e.g., by a humoral or cellular immune mechanism), were corroborated and extended by Naegele *et al.* (1974) and are discussed in more detail later in Section III.D.

It is abundantly clear that the replication of LHV is temperature-dependent both *in vivo* and *in vitro*; however, the mechanism of this dependence and its significance for virus replication and Lucké tumor induction remain unknown.

Although virus particles have never been demonstrated in tumor cells at the higher temperatures, some data indicate that these tumor cells do express LHV genetic information. Collard *et al.* (1973) detected LHV RNA in virus-free tumors, but could find no virus-specific RNA in normal frog kidney cells. Evidence for a gene product of viral RNA present in virus-free tumor cells was provided by Naegele and Granoff (1977). Antiserum prepared against LHV was tested by indirect immunofluorescence for the presence of LHV-associated antigens with both virus-containing and virus-free tumor cells. Cytoplasmic and nuclear intracellular fluorescence (Fig. 3, bottom) was detected only in acetone–methanol-fixed tumor cells that had been shown by electron microscopy to contain herpesvirus. There was good correlation between the number of positive cells and the number of cells containing virus. In sharp contrast, both virus-containing and virus-free tumor cells showed membrane fluorescence when viable, unfixed cells were tested (Fig. 3, top). Appropriate control tests with normal *R. pipiens* cells and with absorbed antisera established the specificity of the reaction.

The preceding results demonstrated that the LHV genome resident in virus-free tumor cells transcribes some information and that a trans-

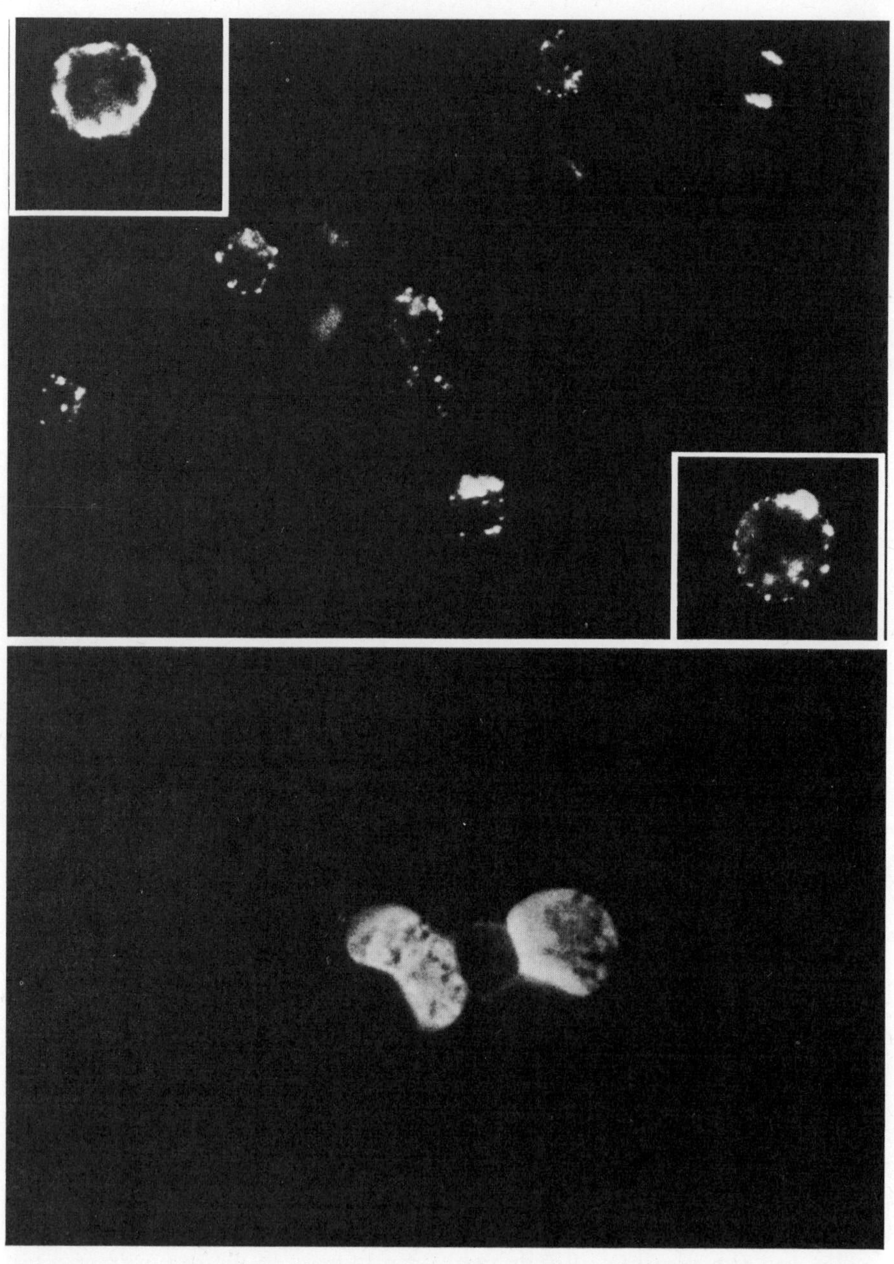

FIGURE 3. *Top:* Viable Lucké tumor cells from primary tissue culture demonstrating membrane fluorescence. ×240. (*Inserts:* ×390) *Bottom:* Fixed Lucké tumor cells demonstrating intracellular fluorescence. Many nonfluorescent cells are present in the field, but are not visible in the photograph. ×312. Both photographs were reproduced from color transparencies. Modified from Naegele and Granoff (1977).

lational product of the viral RNA can be demonstrated as a membrane-associated antigen in these cells.

D. Oncogenicity

Although Lucké's transmission experiments with cell-free preparations of tumors suggested a viral etiology (Lucké, 1938, 1952), subsequent tests for oncogenicity were generally inconclusive (Rafferty, 1964). This could be accounted for by the fact that investigators did not recognize the role of temperature in the presence or absence of virus.

A landmark advance in establishing the viral etiology of the Lucké tumor was made by Tweedell (1967), who demonstrated that as high as 90% of *R. pipiens* embryos inoculated with virus-containing cell fractions from Lucké tumors developed typical renal tumors as they reached metamorphosis. As demonstrated by electron microscopy, the morphology and structure of virus in the inoculum were compatible with herpesvirus. Extracts of normal adult frog kidneys or inclusion-free tumors failed to induce tumors.

Although the species of *Rana* in which the tumor naturally occurs is limited, Lucké tumors have been induced in embryos of *R. clamitans* and *R. palustris* and in hybrids between *R. palustris* and *R. pipiens* (Mulcare, 1969).

Confirmation of the results of Tweedell (1967) establishing LHV as the etiological agent of the Lucké tumor has come from several sources. LHV has been purified by rate zonal centrifugation of cytoplasmic fractions of virus-containing tumors, and preparations rich in enveloped herpesvirus are oncogenic when injected into developing frog embryos (Mizell *et al.*, 1969; Toplin *et al.*, 1971). Ascitic fluid from tumor-bearing frogs held at low temperature contains LHV, and when such ascitic fluid is injected into developing frog embryos, tumors develop (Naegele and Granoff, 1972).

Additional corroborative evidence for the role of LHV in Lucké tumor induction was provided by the experiments of Naegele *et al.* (1974), who attempted to fulfill the Koch–Henle postulates (Rivers, 1936) for identifying a causative disease agent. Figure 4 schematically illustrates the experiment, the results of which may be summarized as follows: LHV extracted from a naturally occurring Lucké tumor was injected into frog embryos, which were then raised at 20–22°C. After 3–8 months, 62% of these animals developed typical virus-free renal adenocarcinomas; tumors did not develop in sham-inoculated controls. Tissue explants of an induced tumor were placed at either 22 or 7.5°C and examined periodically by light and electron microscopy for the presence of intranuclear inclusions and herpesvirus. After 70 days at low temperature, 39% of the tumor cells contained virus, while explants maintained at the higher temperature were virus-free. A homogenate of the induced virus-con-

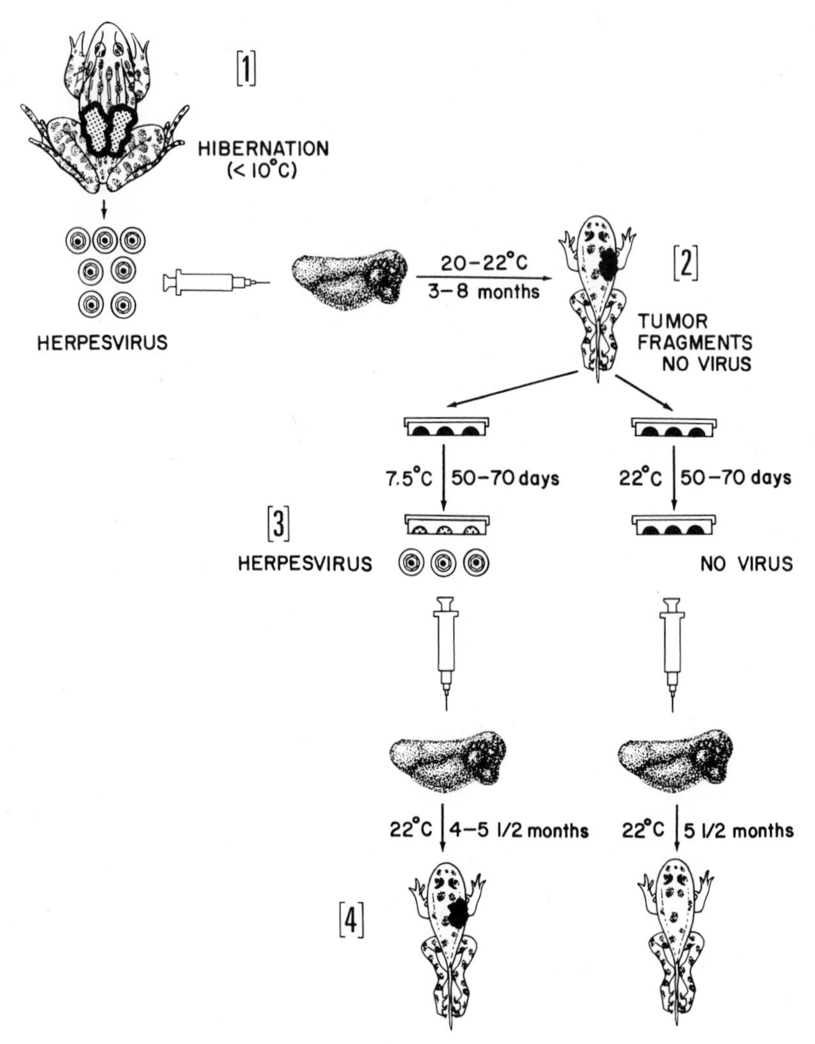

FIGURE 4. Schematic representation of the Lucké tumor virus transmission experiment fulfilling Koch–Henle postulates. [1] Association of the agent with the disease. [2] The agent must induce the same disease in a susceptible host. [3] The agent must be isolated from the induced disease. [4] The isolated agent must be identified as the same agent originally associated with the disease. From Naegele *et al.* (1974).

taining low-temperature tumor fragments was inoculated into frog embryos, and after 4–5 months, 65% of frog embryos developed tumors. This result indicated that the virus induced by low-temperature treatment had the same oncogenic property as the virus originally isolated from the naturally occurring Lucké tumor. Controls inoculated with homogenates of virus-free high-temperature tumor fragments did not develop tumors. With the exception of the "pure" culture requirement, the preceding experiment fulfills the Koch–Henle postulates and, together with all the available evidence, clearly supports the etiological role of LHV in Lucké tumor production.

From information on the relationship between temperature and LHV replication and tumor induction, it is clear that virus replication is not required for tumor formation; for example, virus-free tumors are induced in developing frog embryos inoculated and maintained at 22–25°C, even though virus replication takes place only at temperatures below 12°C. Since virus replication usually does not occur in every tumor cell at low temperature (Granoff, 1973), the tumor can be maintained in nature even during the yearly hibernation cycle of *R. pipiens*. Unfortunately, no information is available to explain why virus production is induced in only some tumor cells at low temperature. One explanation is that only a fraction of the LHV genome is present in certain tumor cells, whereas others contain the entire genome. In either situation, tumor cells would persist at elevated temperature, but in only those cells containing the complete genome would virus replication and subsequent cell death occur.

E. Effect of Physical and Chemical Agents on Lucké Herpesvirus Oncogenicity

Tumor-inducing activity of LHV is destroyed by ether (R.F. Naegele and A. Granoff, unpublished data), pronase, or lyophilization (Tweedell, 1972), but is retained after exposure to sonic vibration (Tweedell, 1972) and after more than 2 years at −70°C (R.F. Naegele and A. Granoff, unpublished data).

F. Natural Transmission

Reliable experimental data on the natural transmission of LHV or on factors that influence tumor formation in nature are lacking. Horizontal and nongenetic vertical transmission have been suggested as two modes of natural transmission. This subject has recently been reviewed (Naegele and Granoff, 1980) and will not be dealt with further here.

G. Cultivation *in Vitro*

Our attempts to grow tumor-inducing herpesvirus from Lucké tumors in a variety of cultured cells *in vitro* have been unsuccessful (Granoff, 1972). However, Tweedell and Wong (1974) presented evidence for the growth of LHV *in vitro* in a frog pronephros cell line. The cells were infected at 25°C with LHV obtained from a Lucké tumor. A cytopathic effect accompanied by the development of typical intranuclear inclusions occurred in about 5% of the cells and could be serially transferred at 25°C. Infected cells from the third passage contained typical herpesvirus particles, as demonstrated by electron microscopy, and produced tumors when inoculated into frog embryos. This is apparently the first report of successful cultivation of LHV *in vitro*. In a large number of tests, we have been unable to confirm this finding using the same cell line and numerous preparations of LHV (R.F. Naegele and A. Granoff, unpublished data). To date, this achievement has not been reproduced in other laboratories, nor have the data been significantly extended. It is worth noting that the virus replicated at 25°C in the pronephros monolayer cultures, in contrast to the low-temperature dependence of LHV *in vivo* and in fragment cultures *in vitro*. A reproducibly useful cell system for cultivation of LHV has not yet been developed.

IV. FROG VIRUS 4

A. Isolation

Rafferty (1965) isolated viruses from homogenates of Lucké tumors and from the pooled urine of tumor-bearing frogs and designated them frog viruses 4 through 8. These isolates multiplied to very low titers with an accompanying cytopathic effect (CPE) only in a frog embryo cell line derived from *R. sylvatica*. The CPE, which developed slowly, was characterized by elongation of cells and eventual detachment from the glass. Eosinophilic intranuclear inclusions containing DNA developed as a result of virus infection. Subsequently, two *R. pipiens* embryo cell lines (Freed *et al.*, 1969) were found to be susceptible to one of Rafferty's isolates, frog virus 4 (FV4) (Gravell *et al.*, 1968). This isolate had been obtained from the urine of a tumor-bearing frog, and as discussed below, it fulfills all the criteria of a herpesvirus. Unless indicated otherwise, all the published information available on FV4 since Rafferty's original report derives from the work of Gravell *et al.* (1968) and Gravell (1971).

B. Morphology and Structure

Electron–microscopic examination of infected *R. pipiens* embryo (RPE) cells and negative stains of cell-free virus have established the site of synthesis and the morphology and structure of FV4. The findings are consistent with those for other herpesviruses. Figure 5, a thin section of infected RPE cells, show typical herpesvirus particles in various stages of development; the insets illustrate extracellular enveloped and unenveloped virus from infected RPE culture fluid. The particle has cubic symmetry and contains 162 hollow capsomeres. Figure 5 may be compared with Fig. 1, which shows similar virus particles from a Lucké tumor

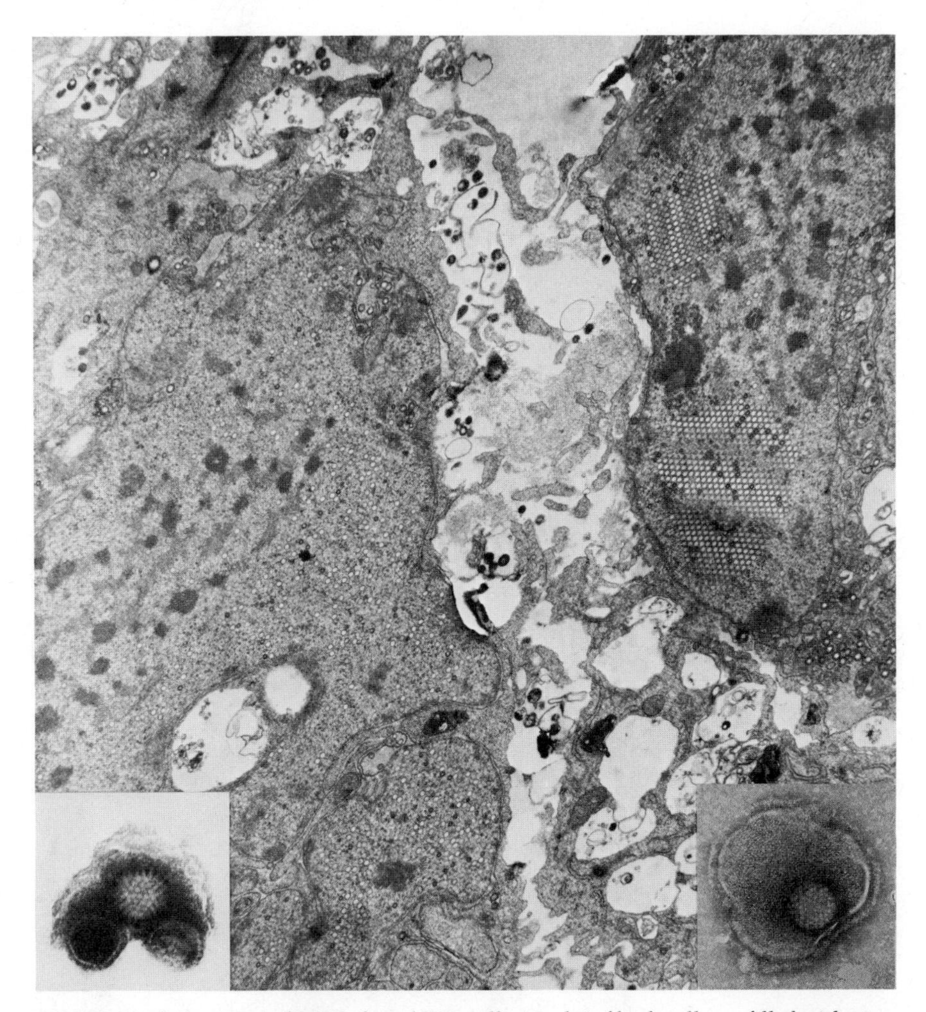

FIGURE 5. Thin section of FV4-infected RPE cells. Nuclei of both cells are filled with virus particles both single- and double-ringed. Virus in the nucleus of the cell at right is in crystalline array. Extracellular particles and particles in vacuoles at the cell periphery are enclosed in densely staining envelopes. ×8750. *Left insert:* Negative stain of unenveloped FV4 showing characteristic herpesvirus morphology. ×84,000. *Right insert:* Negative stain of enveloped FV4. ×77,000. From Granoff (1969).

cell. It is clear that FV4 and LHV are indistinguishable morphologically and in their nuclear site of synthesis.

C. Virus DNA

The DNA of FV4, extracted from purified virus, is a linear double-stranded molecule that by contour-length measurement has a molecular weight of about 77×10^6 daltons (Table I) (Granoff and Naegele, 1978).

This value is comparable to results obtained for the DNA of other herpesviruses from lower vertebrates, such as channel catfish and trout herpesvirus, and is significantly greater than that found for LHV (Table I). The base composition of FV4 differs appreciably from that of LHV (Gravell, 1971) and consists of 54–56% G+C. There appears to be no base sequence homology between FV4 and LHV as determined by DNA–DNA hybridization (Gravell, 1971).

D. Virus–Cell Interaction

A CPE characterized by elongation and contraction of infected cells followed by rounding, vacuolization, enlargement of nuclei, and polykaryon formation (Fig. 6) is observed at 25°C 10–21 days after infection of RPE cells with FV4. The cells typically have intranuclear inclusions that contain DNA. The virus can be plaque-assayed in RPE cells, and in an adult frog kidney cell line developed by Gravell (1971), 14–15 days are required for plaque formation. Because of low virus concentrations, it has not been possible to obtain a single-cycle growth curve, although the virus appears to have a rather long growth cycle (Fig. 7) compared to other herpesviruses. Virus yields in liquid media at 25°C are generally low,

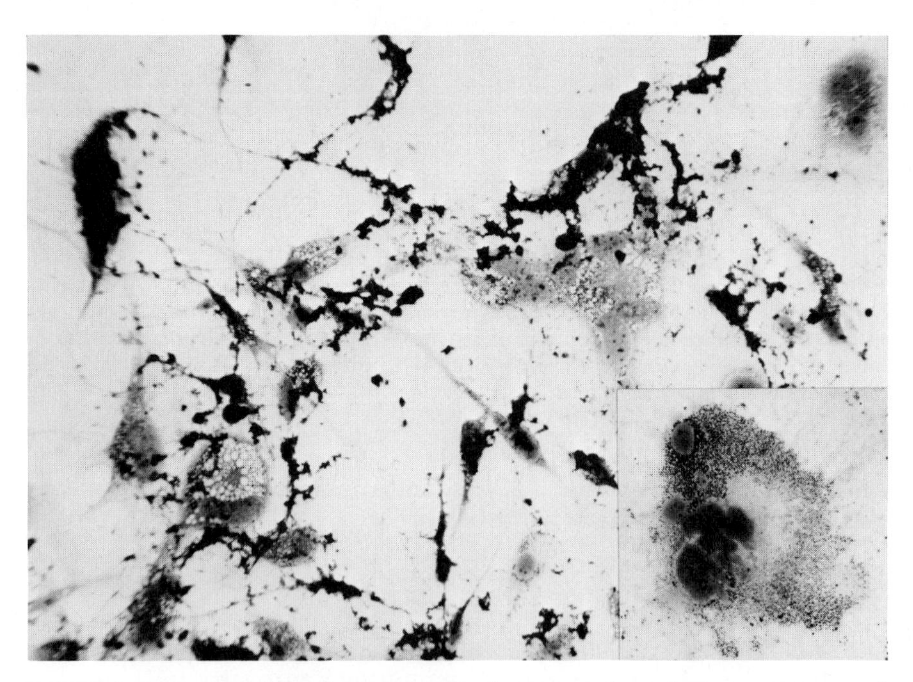

FIGURE 6. Cytopathic effect of FV4 on RPE cells 15 days after infection at 25°C stained with crystal violet. *Insert:* Polykaryon taken from another photomicrograph. ×175. From Gravell *et al.* (1968).

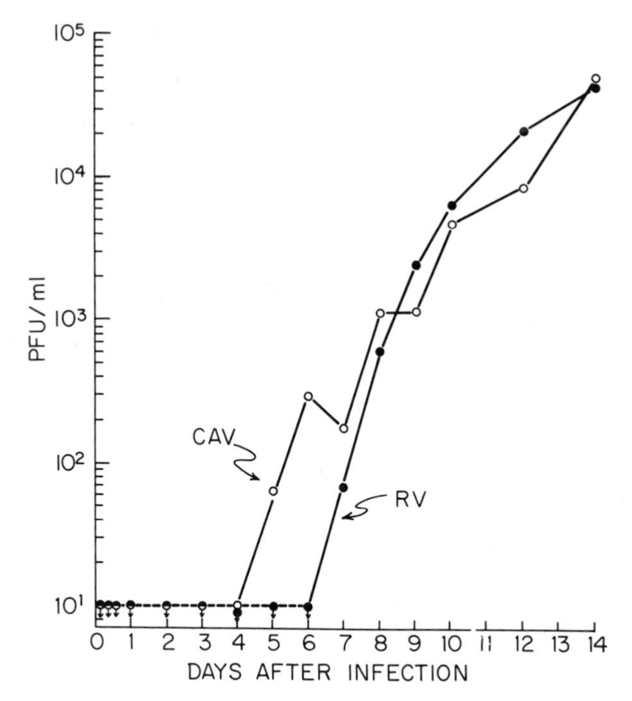

FIGURE 7. Growth curve at 25°C of FV4 in RPE cells. (CAV) Cell-associated virus; (RV) released virus. From Gravell *et al.* (1968).

ranging from 1×10^5 to 2×10^6 plaque-forming units (PFU)/ml. This represents an average of 1–10 PFU/cell. About equal amounts of virus remain cell-associated as are released (Fig. 7). Virus multiplication has also been observed at 10°C, but at a considerably slower rate than at 25°C.

FV4 has been tested for oncogenic activity in developing *R. pipiens* embryos and larvae (Granoff *et al.*, 1969). Animals injected with and surviving about 400 PFU of FV4 failed to develop tumors. Higher concentrations of FV4 killed embryos and produced vesicles, reminiscent of HSV infection, on the surface of the embryos 6 days after inoculation (R.F. Naegele, K.G. Murti, and A. Granoff, unpublished data). Embryos usually died around 13 days postinoculation, and FV4 was found in the cells of infected embryos by electron microscopy and could be cultured from diseased embryos.

E. Effect of Physical and Chemical Agents

Although sonic vibration releases cell-associated virus, only a brief treatment can be used (15 sec. maximum output, Raytheon 10kc sonic oscillator). Longer exposures result in loss of virus infectivity. Intact infected cells are not necessary for infectivity as indicated by infectious virus obtained from cells disrupted by ultrasonic treatment or freeze–thawing and from the results of the virus growth curve (Fig. 7). Exposure to ethyl ether or to pH 3 rapidly inactivates the virus.

V. CONCLUSIONS

Although much effort has been expended to cultivate the Lucké tumor herpesvirus, progress has been painfully slow. There is still no readily available and reproducible cell-culture system to provide adequate amounts of virus and to permit the various features of virus–cell interaction to be studied under controlled conditions. In addition, the *R. pipiens* population wherein the tumor occurs has been drastically reduced in recent years so that the source of tumor-bearing frogs is dwindling. The Lucké tumor and its related herpesvirus remain an unusually attractive virus–tumor system for study, but until the problems identified above are solved, progress will be slow. The preliminary information obtained from studies on FV4, an amphibian herpesvirus distinct from LHV, suggests that this virus would also be a useful model for the analysis of herpesvirus–cell interactions, for study of comparative virology, and for evolutionary evaluation of the origin of herpesviruses, generally.

ACKNOWLEDGMENTS. The preparation of this chapter and unpublished data from the author's laboratory were supported by Research Grant CA07055 and Cancer Center Support (CORE) Grant CA21765 from the National Cancer Institute, and by ALSAC. The author is indebted to Dr. K.G. Murti for contour-length measurements of herpesvirus DNA and the electron micrograph of LHV DNA.

REFERENCES

Breidenbach, G.P., Skinner, M.S., Wallace, J.H., and Mizell, M., 1971, *In-vitro* induction of a herpes-type virus in "summer-phase" Lucké tumor explants, *J. Virol.* **7**:679.

Collard, W., Thornton, H., Mizell, M., and Green, M., 1973, Virus-free adenocarcinoma of the frog (summer phase tumor) transcribes Lucké tumor herpesvirus-specific RNA, *Science* **181**:448.

Fawcett, D.W., 1956, Electron microscope observations of intracellular virus-like particles associated with the cells of the Lucké renal adenocarcinoma, *J. Biophys. Biochem. Cytol.* **2**:725.

Freed, J.J., Mezger-Freed, L., and Shatz, S.A., 1969, Characteristics of cell lines from haploid and diploid anuran embryos, in: *Recent Results in Cancer Research, Special Supplement* (M. Mizell, ed.), p. 101, Springer-Verlag, New York.

Granoff, A., 1969, Viruses of Amphibia, *Curr. Top. Microbiol. Immunol.* **50**:107.

Granoff, A., 1972, Herpesvirus and frog renal adenocarcinoma, *Fed. Proc. Fed. Am. Soc. Exp. Biol.* **31**:1626.

Granoff, A., 1973, The Lucké renal carcinoma of the frog, in: *The Herpesviruses* (A. Kaplan, ed.), p. 627, Academic Press, New York.

Granoff, A., and Darlington, R.W., 1969, Viruses and renal carcinoma of *Rana pipiens*. VIII. Electron microscopic evidence for the presence of herpesvirus in the urine of a Lucké tumor-bearing frog, *Virology* **38:**197.

Granoff, A., and Naegele, R.F., 1978, The Lucké tumor: A model for persistent virus infection and oncogenesis, in: *Persistent Viruses*, Vol. XI (J. Stevens, G. Todaro, and F.C. Fox, eds.), p. 15, ICN-UCLA Symposia on Molecular and Cellular Biology, Academic Press, New York.

Granoff, A., Gravell, M., and Darlington, R.W., 1969, Studies on the viral etiology of the renal adenocarcinoma of *Rana pipiens* (Lucké tumor), in: *Recent Results in Cancer Research* (M. Mizell, ed.), p. 279, Springer-Verlag, Berlin–Heidelberg–New York.

Gravell, M., 1971, Viruses and renal carcinoma of *Rana pipiens*. X. Comparison of herpes-type viruses associated with Lucké tumor-bearing frogs, *Virology* **43:**730.

Gravell, M., Granoff, A., and Darlington, R.W., 1968, Viruses and renal carcinoma of *Rana pipiens*. VII. Propagation of a herpes-type frog virus, *Virology* **36:**467.

Honess, R.W., and Watson, D.H., 1977, Unity and diversity in the herpesviruses, *J. Gen. Virol.* **37:**15.

Kemper, J.D., 1977, Leopard frog supply, *Science* **197:**106.

Lucké, B., 1934, A neoplastic disease of the kidney of the frog, *Rana pipiens*, *Am. J. Cancer* **20:**352.

Lucké, B., 1938, Carcinoma in the leopard frog: Its probable causation by a virus, *J. Exp. Med.* **68:**457.

Lucké, B., 1952, Kidney carcinoma in the leopard frog: A virus tumor, *Ann. N.Y. Acad. Sci.* **54:**1093.

Lucké, B., and Schlumberger, H., 1939, The manner of growth of frog carcinoma studied by direct microscopic examination of living intraocular transplants, *J. Exp. Med.* **70:**257.

Lunger, P.D., 1964, The isolation and morphology of the Lucké frog kidney tumor virus, *Virology* **24:**138.

Lunger, P.D., Darlington, R.W., and Granoff, A., 1965, Cell–virus relationships in the Lucké renal adenocarcinoma: An ultrastructure study, *Ann. N.Y. Acad. Sci.* **126:**289.

McKinnell, R.G., 1965, Incidence and histology of renal tumors of leopard frogs from the north central states, *Ann. N.Y. Acad. Sci.* **126:**85.

McKinnell, R.G., and McKinnell, B.K., 1968, Seasonal fluctuation of frog renal adenocarcinoma prevalence in natural populations, *Cancer Res.* **28:**440.

Mizell, M., Stackpole, C.W., and Halperen, S., 1968, Herpes-type virus recovery from "virus-free" frog kidney tumors, *Proc. Soc. Exp. Biol. Med.* **127:**808.

Mizell, M., Toplin, I., and Isaacs, J.J., 1969, Tumor induction in developing frog kidneys by a zonal centrifuge purified fraction of the frog herpes-type virus, *Science* **165:**1134.

Morek, D.M., and Tweedell, K.S., 1969, Interaction of normal and malignant tissues in organ culture, *Am. Zool.* **9:**1125.

Mulcare, D.J., 1969, Non-specific transmission of the Lucké tumor, in: *Recent Results in Cancer Research: Biology of Amphibian Tumors* (M. Mizell, ed.), p. 240, Springer-Verlag, New York.

Naegele, R.F., and Granoff, A., 1972, Viruses and renal carcinoma of *Rana pipiens*. XIII. Transmission of the Lucké tumor by herpesvirus-containing ascitic fluid from a tumor-bearing frog, *J. Natl. Cancer Inst.* **49:**299.

Naegele, R.F., and Granoff, A., 1977, Viruses and renal carcinoma of *Rana pipiens*. XV. The presence of virus-associated membrane antigen(s) on Lucké tumor cells, *Int. J. Cancer* **19:**414.

Naegele, R.F., and Granoff, A., 1980, The Lucké tumor and its herpesvirus, in: *Oncogenic Herpesviruses* (F. Rapp, ed.), p. 85, CRC Press, Boca Raton, Florida.

Naegele, R.F., Granoff, A., and Darlington, R.W., 1974, The presence of the Lucké herpes-virus genome in induced tadpole tumors and its oncogenicity: Koch–Henle postulates fulfilled, *Proc. Natl. Acad. Sci. U.S.A.* **71:**830.

Rafferty, K.A., Jr., 1963, Spontaneous kidney tumors in the frog: Rate of occurrence in isolated adults, *Science* **141:**720.

Rafferty, K.A., Jr., 1964, Kidney tumors of the leopard frog: A review, *Cancer Res.* **24**:169.

Rafferty, K.A., Jr., 1965, The cultivation of inclusion-associated viruses from Lucké tumor frogs, *Ann. N.Y. Acad. Sci.* **126**:3.

Rafferty, K.A., Jr., and Rafferty, N.S., 1961, High incidence of transmissible kidney tumors in uninoculated frogs maintained in a laboratory, *Science* **133**:702.

Rivers, T.M., 1936, Viruses and Koch's postulates, *J. Bacteriol.* **33**:1.

Roberts, M.E., 1963, Studies on the transmissibility and cytology of the renal carcinoma of *Rana pipiens, Cancer Res.* **23**:1709.

Schlumberger, H., and Lucké, B., 1949, Serial intraocular transplantation of frog carcinoma for 14 generations, *Cancer Res.* **9**:52.

Skinner, M.S., and Mizell, M., 1972, The effect of different temperatures on herpesvirus induction and replication in Lucké tumor explants, *Lab. Invest.* **26**:671.

Toplin, I., Mizell, M., Sottong, P., and Monroe, J., 1971, Zonal centrifuge applied to the purification of herpesvirus in the Lucké frog kidney tumor, *Appl. Microbiol.* **21**:132.

Tweedell, K.S., 1967, Induced oncogenesis in developing frog kidney cells, *Cancer Res.* **27**:2042.

Tweedell, K.S., 1972, Experimental alteration of the oncogenicity of frog tumor cell-viral fractions, *Proc. Soc. Exp. Biol. Med.* **140**:1246.

Tweedell, K., and Wong, W.Y., 1974, Frog kidney tumors induced by herpesvirus cultured in pronephric cells, *J. Natl. Cancer Inst.* **52**:621.

Wagner, E.K., Roizman, B., Savage, T., Spear, P.G., Mizell, M., Durr, F.E., and Sypowiez, D., 1970, Characterization of the DNA of herpesviruses associated with the Lucké adenocarcinoma of the frog and Burkitt lymphoma of man, *Virology* **42**:257.

Zambernard, J., and Mizell, M., 1965, Virus particles of the frog renal adenocarcinoma: Causative agent of passenger virus? I. Fine structure of primary tumors and subsequent intraocular transplants, *Ann. N.Y. Acad. Sci.* **126**:127.

Zambernard, J., and Vatter, A.E., 1966, The effect of temperature change upon inclusion-containing renal tumor cells of leopard frogs, *Cancer Res.* **26**:2148.

Zambernard, J., Vatter, A.E., and McKinnell, R.G., 1966, The fine structure of nuclear and cytoplasmic inclusions in primary renal tumors of mutant leopard frogs, *Cancer Res.* **26**:1688.

CHAPTER 8

B Virus (*Herpesvirus simiae*)

HANNS LUDWIG, G. PAULI, H. GELDERBLOM, G. DARAI, H.-G. KOCH, R.M. FLÜGEL, B. NORRILD, AND M.D. DANIEL

I. INTRODUCTION AND HISTORY

The early studies on the B virus (*Herpesvirus simiae*, herpesvirus B, simian herpesvirus, cercopithecine herpesvirus 1) are associated with the names of such pioneers in medical virology as Sabin and Wright (1934), Sabin (1934a–c), Burnet *et al.* (1939a,b), Melnick and Banker (1954), and Youngner (1956). In the 1930s, the virus was isolated and its cross-neutralization with herpes simplex virus (HSV) discovered. Some biological and morphological parameters were established in the 1950s. It was almost half a century, however, before the genetic material of the B virus could be characterized to allow for a clear differentiation from HSV and before the antigenic cross-reactivity with HSV and the bovine herpes mammillitis virus [bovid herpesvirus 2 (BHV-2)] could be pinpointed to major glycoprotein species (Sterz *et al.*, 1973–1974; Ludwig *et al.*, 1978; Norrild *et al.*, 1978b; Pauli *et al.*, 1981; and further details presented in this review).

The B virus is indigenous in primates of the *Macaca* species. The medical importance of the virus comes from its virulence in humans, in whom it causes almost invariably fatal ascending myelitis and encephalomyelitis. Infections in humans, the clinical symptoms and patho-

HANNS LUDWIG AND G. PAULI • Institut für Virologie, Freie Universität Berlin, West Berlin, Germany. H. GELDERBLOM • Robert Koch-Institut des Bundesgesundheitsamtes, West Berlin, Germany. G. DARAI AND H.-G. KOCH • Institut fur Medizinische Virologie, Universität Heidelberg, Heidelberg, Germany. B. NORRILD • Institute of Medical Microbiology, University of Copenhagen, Copenhagen, Denmark. M. D. DANIEL • New England Regional Primate Research Center, Harvard Medical School, Southborough, Massachusetts 01772.

genesis of the disease in both humans and monkeys, and the epidemiology of the infection in monkeys have been covered extensively in a number of reviews (Endo *et al.*, 1959; Davidson and Hummeler, 1960; Keeble, 1960; Hull, 1968, 1973; Kalter and Heberling, 1971; Daniel *et al.*, 1972; Weller and Pearson, 1972; España, 1974). However, a number of questions remain unanswered. Among these are the cause of the relatively low incidence of B virus infections in humans, particularly in light of the fact that the contact between monkeys and humans occurs frequently in some countries and therefore infection can be assumed, and whether latent B virus infections can occur in humans. In any event, the apparent relationship of B virus to HSV challenges scientific curiosity to search for the routes of these viruses in evolution.

The purpose of this review is to briefly summarize the available data on B virus infections and to present current progress in the study of the genetic and antigenic components of the B virion. For details on clinical–diagnostic problems and preventative measures, refer to the review of Hull (1973).

II. MORPHOGENESIS AND MORPHOLOGY

A. Virus–Cell Interaction

In vitro studies on B virus are hampered by the abundant evidence that accidentally acquired infections may be lethal. There are, however, a few fundamental reports covering major features of this virus, which was seen to grow on the chorioallantoic membrane (Burnet *et al.*, 1939a) and in tissue culture (Reissig and Melnick, 1955). The only growth curve reported (Fig. 1) shows that replication occurs in a one-step event with typical cell alterations, the appearance of Cowdry Type A intranuclear

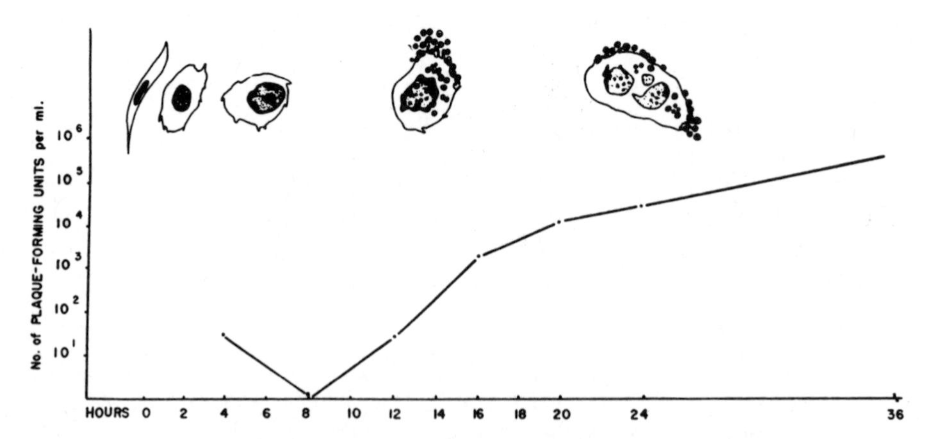

FIGURE 1. Herpes B virus infection of cultures of monkey kidney epithelium. Schematic diagrams to illustrate the correlation of morphological changes with the appearance of infectious virus in the culture fluid. The appearance of characteristic particles in the cells is coincident with the rise in virus titer. From Reissig and Melnick (1955).

inclusions, and complete lysis of cells. The duration of the reproductive cycle *in vitro* is comparable to those of the highly lytic HSV, pseudorabies (PsR) virus (Kaplan, 1973), and some bovine herpesviruses (Chapter 4).

The appearance of virus-induced cell alterations was correlated with the morphogenesis of the B virus in different cell systems (Reissig and Melnick, 1955; Bhutala and Mathews, 1963; España, 1973; Ruebner *et al.*, 1975). The prominent cytopathic changes in infected monolayers were studied using kinematography (Falke and Richter, 1961). Typical cell rounding or classic polykaryocytosis was found. Virus variants of such a different cytopathogenicity could even be isolated from the same stock solution (Falke, 1961) and were also found in another prototype virus stock (Benda and Cinátl, 1966). In our studies, cell fusion induced by B virus strain 130-65 (Daniel *et al.*, 1975) in rabbit kidney cells was detected at about 4 hr postinfection (p.i.), and complete fusion of the monolayer was reached at 10–12 hr p.i.—very similar to PsR-virus-induced fusion in the same cell system (Ludwig *et al.*, 1974). However, infected embryonic human oligodendrocyte and HeLa cell cultures became rounded, but did not fuse (see Fig. 7A). This suggests that the cell, too, plays a major role in determining of the fusion process. Similar conclusions have been reached from other virus cell systems (Compans and Klenk, 1979; Ruyechan *et al.*, 1979).

B. Morphogenesis

The sequence of virus morphogenesis, comprising the assembly of nucleocapsids in the cell nucleus, and the envelopment of the capsids at the inner lamellae of the nuclear membrane and at the cytoplasmic and plasma membranes (Fig. 2) follow in general the pattern reported for other herpesviruses (Morgan *et al.*, 1954; Nii *et al.*, 1968; Zee and Talens, 1972; Roizman and Furlong, 1974) (see also Chapter 4).

Several steps in the assembly of the nucleocapsid can be differentiated on the basis of a variety of more or less fully assembled cores. If the release from the nucleus into the cytoplasmic compartment is inhibited, paracrystalline nuclear arrays, densely packed with nucleocapsids, are found (Bhutala and Mathews, 1963; Ruebner *et al.*, 1975).

Beginning at 8–10 hr p.i., enveloped particles were seen inside the endoplasmic reticulum and adhering to the outer cell surface (Fig. 2.). Naked nucleocapsids are occasionally found free in the cytoplasm, and enveloped particles could be interpreted to be inside the cell nucleus. In BS-C-1 cells, virus progeny have been reported to adhere to thickened concave portions of the cell membrane and to be surrounded by extremely electron-dense material (Ruebner *et al.*, 1975).

In our studies, the formation and egress of subviral dense bodies with a diameter equal to or slightly smaller than that of a mature virus (Fig. 2) were observed and seemed to represent a peculiar feature seen also in other herpesvirus-infected cell systems (Gelderblom, unpublished).

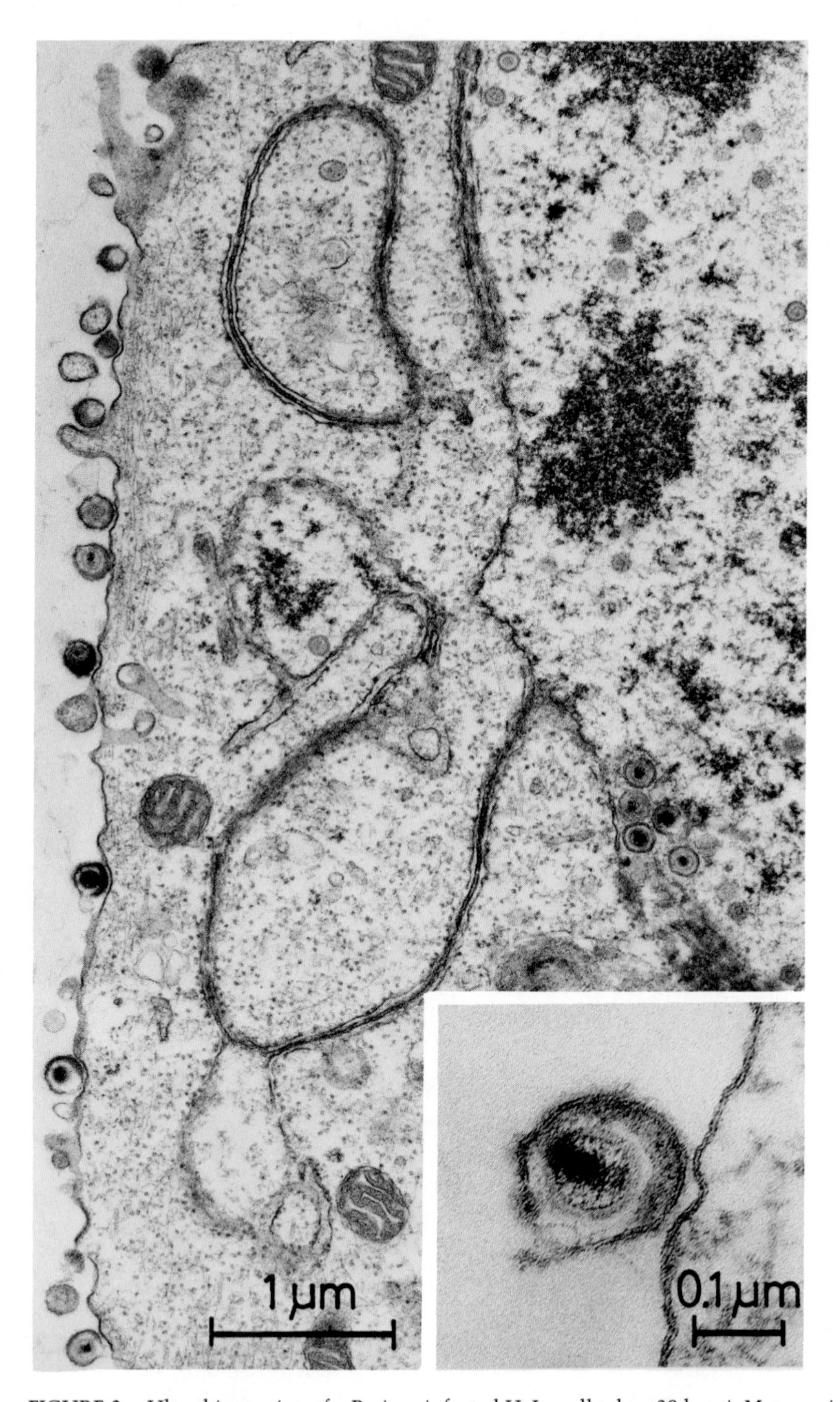

FIGURE 2. Ultrathin section of a B-virus-infected HeLa cell taken 30 hr p.i. Mature virus as well as empty envelope sacks adhere to the cell surface. Hypertrophic endoplasmic reticulum and perinuclear membrane systems are prominent in the cytoplasm. In the nucleus, herpesvirus nucleocapsids in different stages of assembly and envelopment are seen near

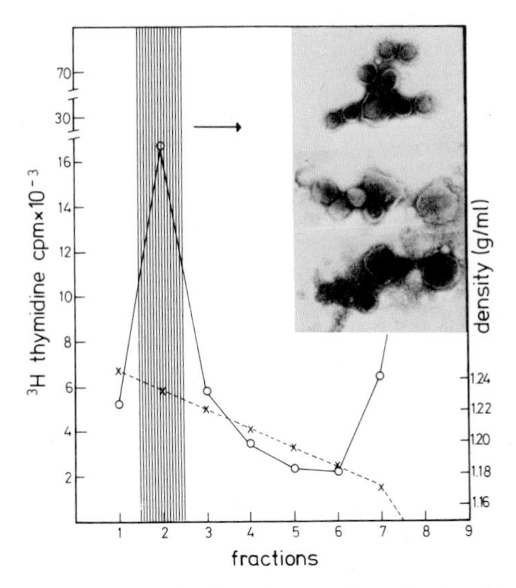

FIGURE 3. Purification of B virus by density centrifugation in a linear sucrose gradient. [^{3}H]Thymidine-labeled supernatant virus was purified as described earlier (Pauli and Ludwig, 1977). The fractions with highest radioactivity contained enveloped particles and nucleocapsids as shown (*inset*) by negative-staining electron microscopy with phosphotungstic acid (Brenner and Horne, 1959). Viral DNA extracted from this virion peak fraction, and banded by analytical ultracentrifugation, is shown in Fig. 4a.

C. Morphology

Although little information exists on the fine structure of the virion, there is no doubt that it follows the unified architectural principles recognized for members of the herpesvirus group (Wildy, 1973; Wildy *et al.*, 1960; Watson, 1973; Roizman and Spear, 1973).

From our studies using negative staining as well as ultrathin sections, the viral core measures 40 nm, the nucleocapsid 100 nm, and the enveloped particle 160–180 nm in diameter (Figs. 2 and 3, insets). These results are in agreement with those reported by Ruebner *et al.* (1975). The loosely fitting envelope is studded with surface projections and tends to disintegrate easily (see Fig. 2, inset). From differences in electron density, it can be deduced that virus particles may vary with respect to their content of DNA. The core structure of the B virion, as well as those of a variety of other herpesviruses investigated, show rotational symmetry, and in the case of the B virion, the core appears as an electron-dense

---

the nuclear membrane. ×25,000. *Inset:* Virion. ×125,000. The loosely fitting envelope is covered with surface projections. The space between the nucleocapsid and the membrane is filled with electron-dense material (tegument). The polygonal nucleocapsid exhibits an inner darkly stained toroidlike core structure; rotational symmetry is evident. For technical details, see Gelderblom *et al.* (1974).

toroid structure (Fig. 2, inset), similar to that reported by Roizman and Furlong (1974).

III. GENETIC MATERIAL

The present knowledge concerning the genetic material of the B virus is based on two strains of different origin. Isolate 130-65 was obtained from tissues of a rhesus monkey, which died of cerebral infarction (Daniel *et al.*, 1975). A second strain, used by Schneweis (1962) and by Falke (1961), had originally been obtained by R. Siegert, Marburg, and was kept frozen for 20 years. To ensure that the cytopathic agents were indeed herpesviruses, purification of the strains was performed on sucrose gradients. Labeled material from the supernatant fluids of infected tissue cultures banded at a density of approximately 1.24 g/ml and contained particles indistinguishable from herpesvirus capsids and virions by negative staining (Fig. 3).

A. Guanosine Plus Cytosine Content of the DNA

DNA extracted from the purified virus banded as a homogeneous peak when centrifuged to equilibrium in an analytical ultracentrifuge (Fig. 4a and b). Sheared DNA showed two major peaks in approximately the same density range (Fig. 4c). The banding pattern was significantly different from that of sheared *Herpesvirus saimiri* DNA (Fleckenstein *et al.*, 1975) and resembled more that of HSV-1 DNA (Ludwig *et al.*, 1972a). This kind of banding excludes the possibility that B virus DNA consists of long stretches of DNA sequences with low G+C content flanked by high-G+C DNA as in *Herpesvirus saimiri* genomes. From cocentrifugation of intact B virus DNA with different markers (Fig. 4A–D), a buoyant density of 1.734 g/ml was calculated. The G+C content of B virus DNA—75 moles %—is the highest among the known herpesvirus DNAs (Ludwig, 1972a; Roizman, 1978).

The melting point (T_m) of B virus DNA in 0.1 × SSC was obtained from comparisons with HSV DNA. The sigmoidal melting curve was consistent with a double-stranded DNA with a relatively uniform base distribution and yielded a T_m of 84°C. The T_m for HSV-1 DNA was 82°C, in good agreement with reported values (Graham *et al.*, 1972).

B. Measurement of the DNA Contour Length by Electron Microscopy

The molecular weight of B virus DNA was determined from its contour length using relaxed circular molecules of PM_2 DNA as internal length standard. Measurement of 47 molecules of B virus DNA yielded

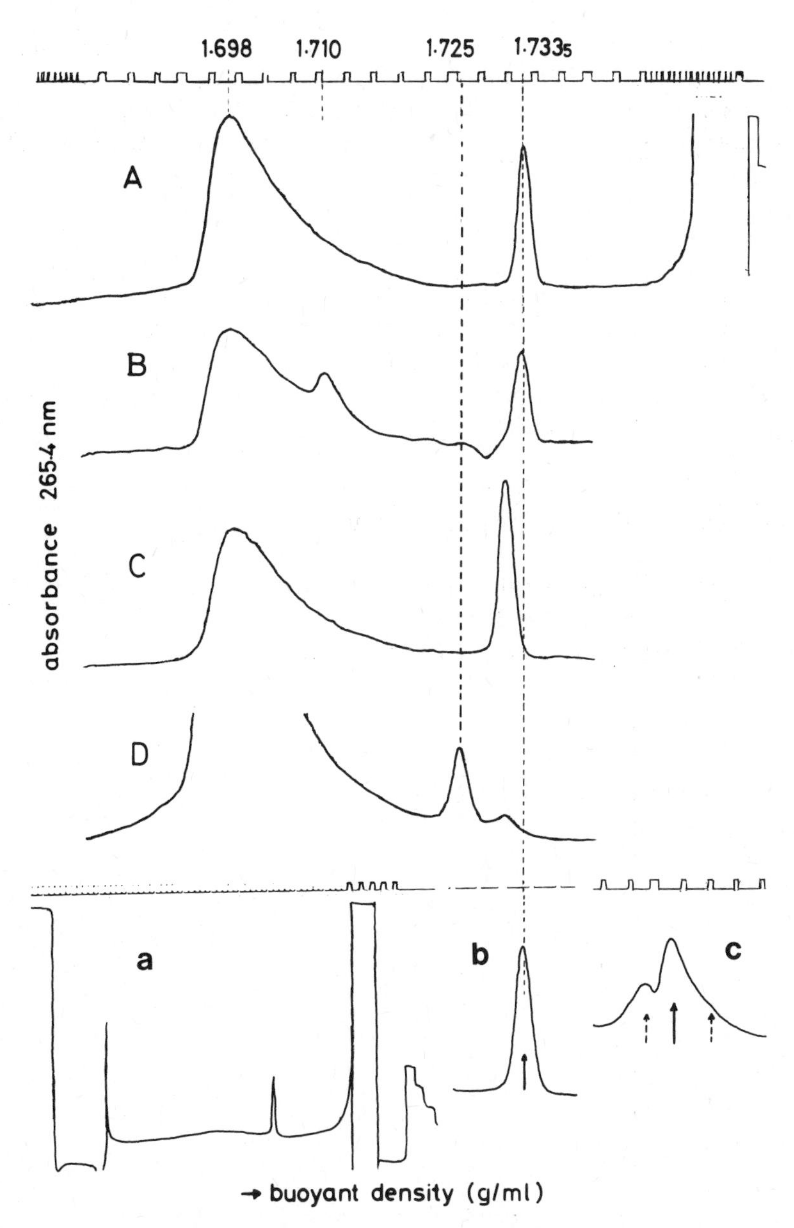

FIGURE 4. Photoelectric scanning of DNA preparations centrifuged to equilibrium in CsCl in a Beckman model E analytic ultracentrifuge (44,000 rpm, 20 hr, 25°C). The procedures followed those outlined in detail elsewhere (Ludwig *et al.*, 1972a,b). The DNA was obtained from the sources indicated. (A) B virus strain 135-65-infected rabbit kidney cells. (B) Virus-infected human oligodendroglia cells (kindly supplied by Y. Iwasaki). *Escherichia coli* DNA (ρ = 1.710 g/ml), 1.5 μg, was added as a marker. (C) Pseudorabies virus (strain DEK)-infected rabbit kidney cells (viral DNA ζ = 1.731 g/ml). (D) Rabbit kidney cells infected with an undiluted passage of HSV-1, strain XIII (HSV-1 DNA ρ = 1.725 g/ml; defective particle DNA ρ = 1.731 g/ml). (a) Purified B virus particles. (b) Same DNA as in (a) scanned with low chart speed. (c) DNA after shearing. Dotted arrows indicate DNA fragments of the intact DNA (solid arrow) with different buoyant densities.

a molecular weight of $107.63 \pm 8.1 \times 10^6$, assuming a molecular weight of 6.64×10^6 for PM_2 DNA (Stüber and Bujard, 1977) (Fig. 5). This value corresponds to 162.1 ± 12.3 kilo base pairs. A double-stranded B virus DNA molecule and reference PM_2 DNA molecules are shown in Fig. 6.

C. Infectivity of B Virus DNA

Using the calcium phosphate technique (Graham *et al.*, 1973), it was found that 0.1 μg B virus DNA induced 35–50 plaques on HeLa cells (1×10^6 cells per dish) 3 days after transfection. The morphology of these plaques could not be differentiated from that of plaques produced by the parent virus (Fig. 7A and B).

D. Restriction-Endonuclease Cleavage Patterns

The patterns and number of fragments produced by B virus DNA cleaved with restriction endonucleases *Bal*I, *Bam*HI, *Bst*EII, *Bgl*II, *Cla*I, and *Eco*RI are shown in Fig. 8 and summarized in Table I.

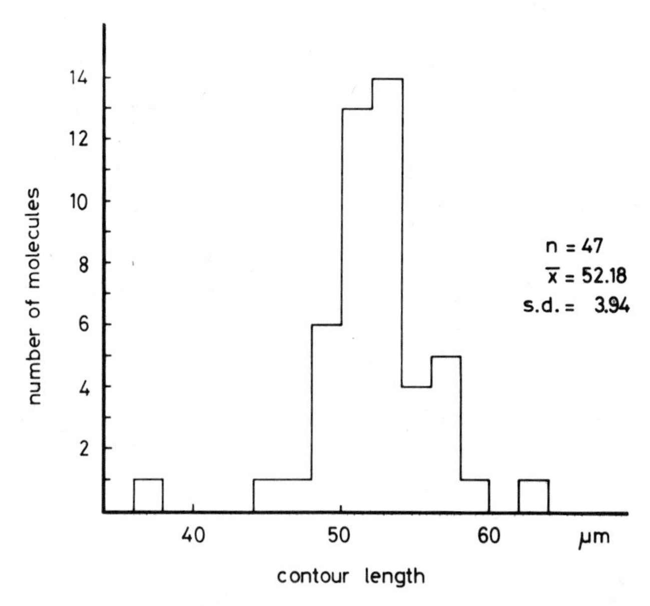

FIGURE 5. Length distribution of B virus DNA. Isolated DNA (3 μg/ml) was spread together with marker PM_2 DNA (4 μg/ml) from a cytochrome c (0.025% in 0.05 M Tris HCl buffer, pH 8.0) hyperphase onto a hypophase of 0.25 M ammonium acetate. The specimens were adsorbed onto 200-mesh pioloform-coated copper grids, dried in absolute ethanol, and rotary-shadowed under an angle of 7° with a layer of Pt/C. Electron micrographs containing both B virus DNA and marker DNA were taken at 2400 times magnification, 3 times enlarged, and evaluated using the Videoplan system (Kontron, Munich).

FIGURE 6. Electron micrograph of B virus DNA together with two PM_2 DNA molecules as internal length standard, prepared as detailed in the Fig. 5 caption. ×10,500.

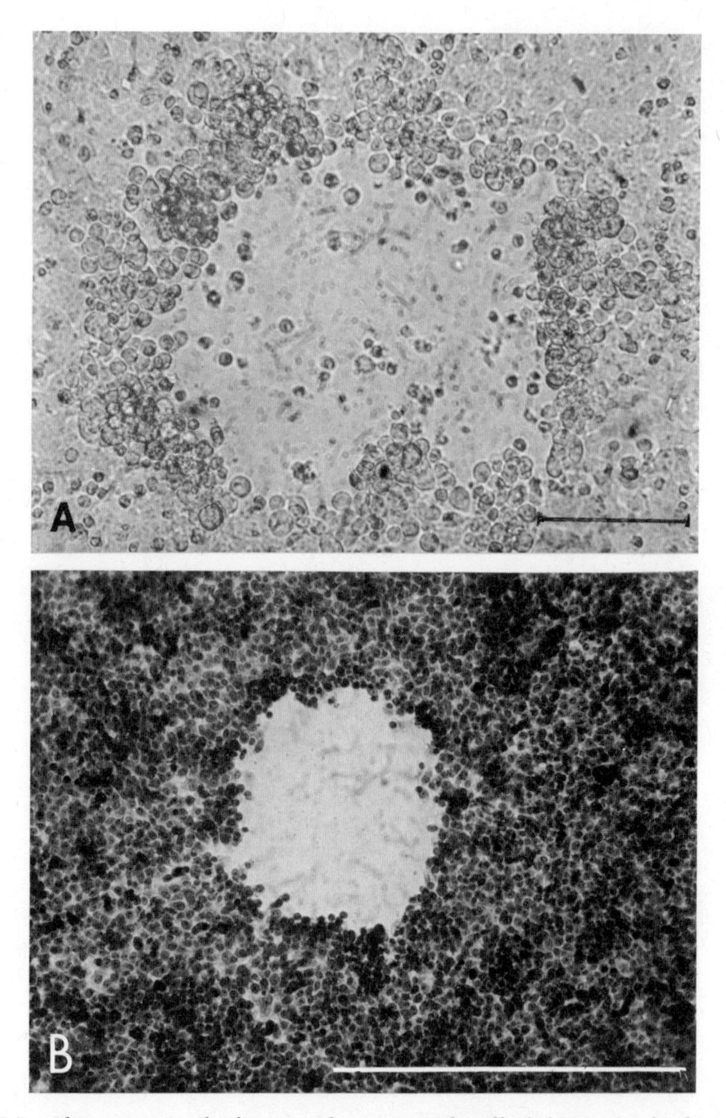

FIGURE 7. Photomicrograph of B virus plaques on Hela cells. (A) Unstained culture 3 days after infection. Scale bar: 0.1 mm. (B) Plaque morphology seen after transfection. The cells were fixed and stained with crystal violet. Scale bar: 0.5 mm.

The molecular weights of the DNA fragments obtained after digestion with restriction enzymes *Bgl*II and *Eco*RI were calculated using DNAs of bacteriophage lambda, HSV-1, human adenovirus type 2, and *Mycoplasma hyorhinis Bst*EII fragments A and B (molecular weights: 75 and 50×10^6) as markers (Darai *et al.*, 1982). The results are given in Table II. The terminal fragments of B virus DNA were determined by treatment with *Escherichia coli* exonuclease III and lambda exonuclease prior to restriction-enzyme cleavage. The results indicate that in both

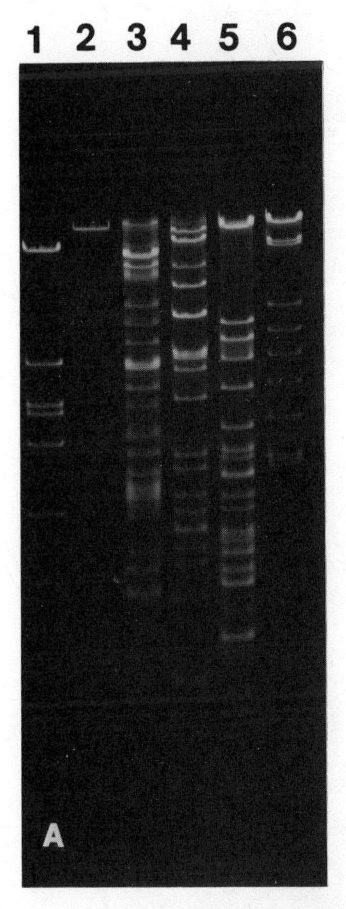

FIGURE 8. Agarose slab gel electrophoresis of B virus DNA and marker DNAs. The DNAs were extracted and prepared as follows: Virus pellets in TNE (pH 7.2) were lysed with sodium dodecyl sulfate (0.5% final concentration). Proteinase K (100 μg/ml) was added (incubation overnight, 37°C). Handling of the DNA followed known procedures (Ludwig *et al.*, 1972a; Sheldrick *et al.*, 1973; Hayward *et al.*, 1975b). Purified DNA had absorption ratios of 1.9–2.0 (A 260/A 280). (A) B virus DNA (1 μg) was digested with the restriction endonucleases *Bst*EII (3), *Bam*HI (4), and *Bal*I (5). The fragments were separated in 0.8% agarose slab gels (35 × 20 × 0.3 cm), run at 100 V, 16 hr, in electrophoresis buffer (4°C). The other lanes are: (1) lambda DNA, *Eco*RI; (2) lambda DNA, undigested; (6) tree shrew herpesvirus type 2 DNA, *Hind*III. (B) B virus DNA (0.5 μg) was digested with *Hind*III (2), *Eco*RI (3), and *Bgl*II (4). The other lanes are: (1) fragment A of *Mycoplasma hyorhinis* DNA, *Bst*EII (75 × 10^6 daltons); (5) tree shrew herpesvirus type 3 DNA, *Hind*III; (6) lambda DNA undigested and digested with *Eco*RI; (7) lambda DNA, *Eco*RI; (8) HSV-1 DNA, *Eco*RI; (9) HSV-2, *Eco*RI. The 0.6% agarose slab gels (35 × 20 × 0.3 cm) were run at 75 V, 4°C, for 45 hr (1–6) and for 20 hr (lanes 7–9). (C) PsR virus DNA (2.0 μg) was digested with *Bam*HI (1) and *Bst*EII (4). (2, 3) Lambda DNA digested with *Hpa*I or *Bst*EII; slab gel conditions as in (B).

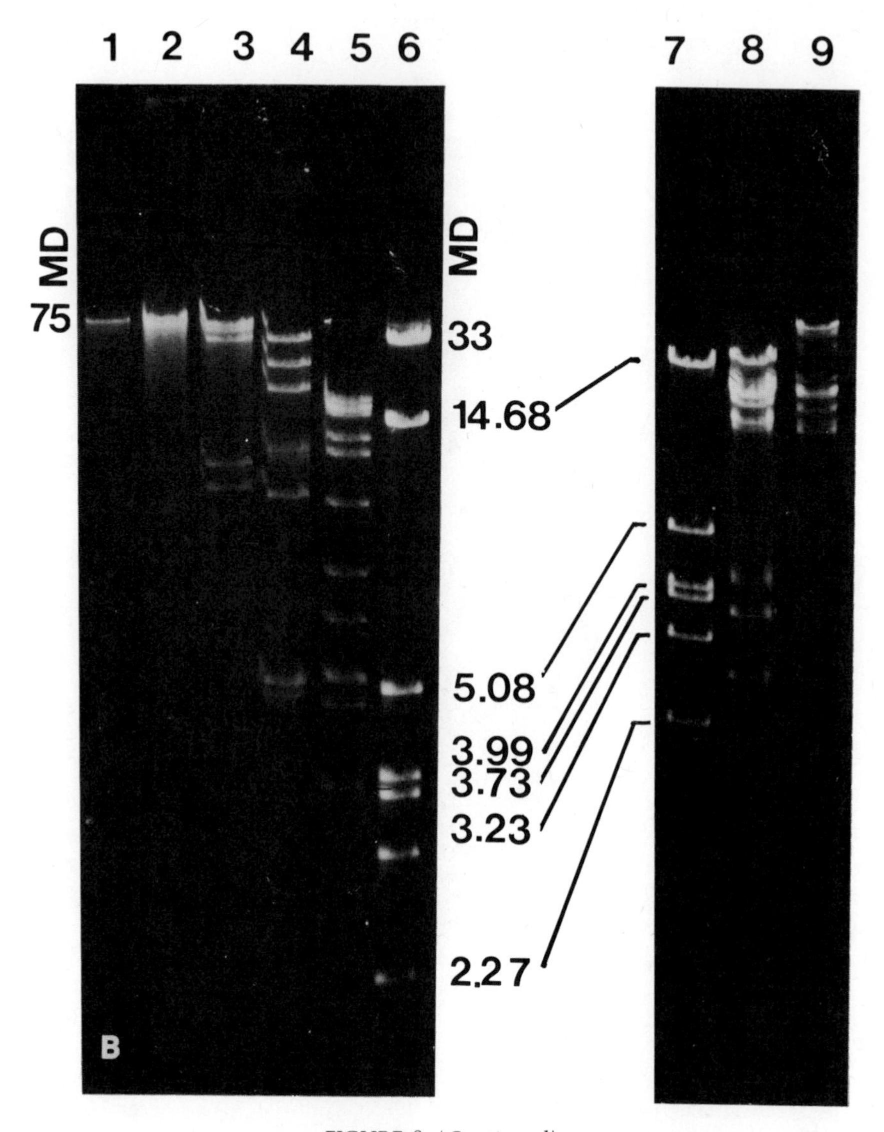

FIGURE 8. (*Continued*)

*Eco*RI and *Bgl*II digests, four of the DNA fragments were located at the termini of the B virus DNA. This result suggests that the B virus genome resembles the HSV genomes in that it forms several isomers (Hayward *et al.*, 1975b). Consistent with this conclusion, closer inspection of the DNA fragments revealed that submolar bands were present in the cleavage patterns obtained with each restriction enzyme (see representative pattern shown in Fig. 9).

Whereas the intact B virus DNA molecules have a molecular weight of 107×10^6, the sum of the molecular weights of the restriction-en-

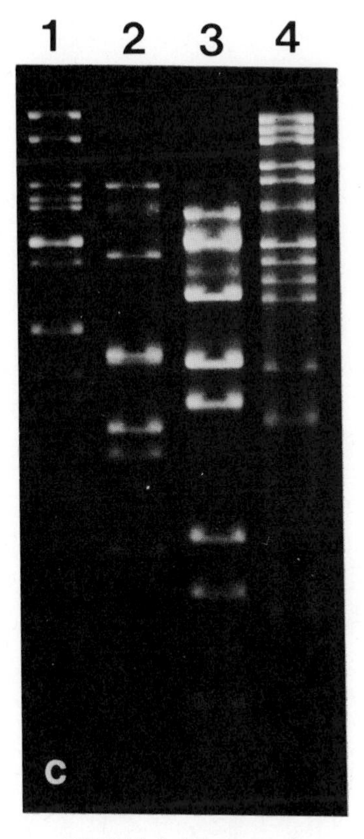

FIGURE 8. (*Continued*)

TABLE I. Analysis of DNA of B Virus Using Different Restriction Endonucleases and Electrophoresis on Agarose Gels

Restriction enzyme	Cleavage size	Apparent or minimum number of DNA bands[a]
*Bal*I	TGG/CCA	35
*Bam*HI	G/GATCC	30
*Bgl*II	A/GATCT	8
*Bst*EII	G/GTNAAC	30
*Cla*I	AT/CGAT	9
*Eco*RI	G/AATTC	5
Others[b]	—	>50

[a] Detectable at an agarose concentration of 0.5%; DNA fragments of less than 1×10^6 daltons are not included.

[b] The following enzymes produce more than 50 DNA fragments: *Alu*I, *Ava*I, *Ava*II, *Hae*II, *Hind*II, *Hpa*I, *Hpa*II, *Sac*I, *Xho*I, *Bgl*I, *Sac*II, *Sal*I, and *Sma*I.

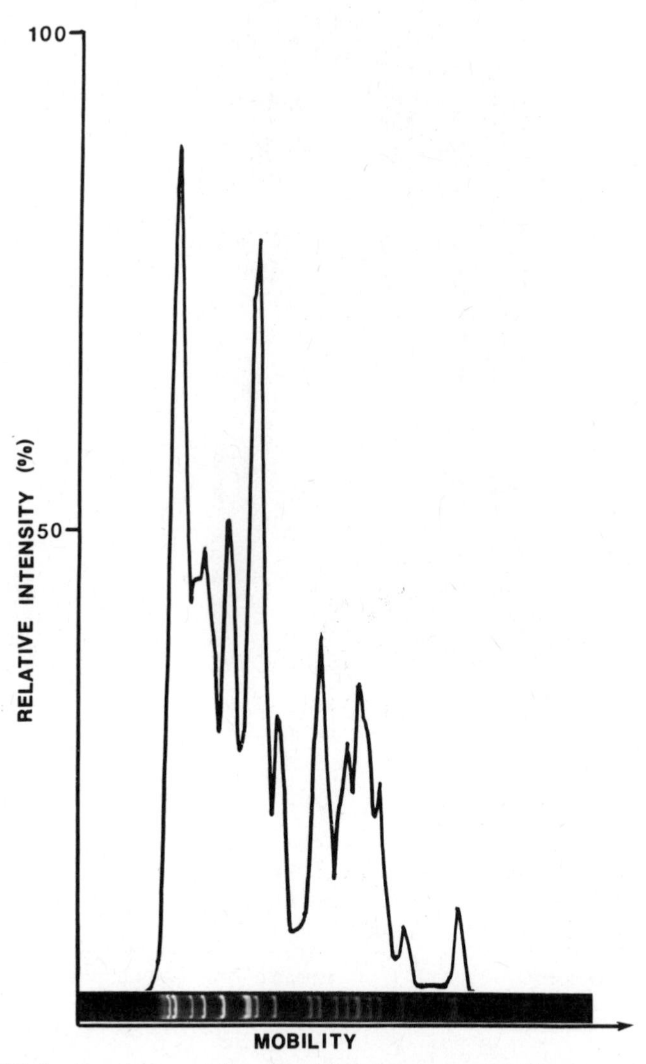

FIGURE 9. Densitogram of B virus DNA fragments after digestion with restriction endonuclease BAMHI. The different peak heights document different molar ratios of individual DNA fragments.

TABLE II. Nomenclature[a] and Molecular Weights of the *Eco*RI and *Bgl*II Fragments of B Virus DNA

Fragment	Molecular weight × 10^6	
	*Eco*RI	*Bgl*II
A	63	37
B	45	23
C	37	16.5
D	11.2	12.6
E	9.8	11.2
F	—	9.7
G	—	5.9
H	—	4.7

[a] The final fragment nomenclature will be established by measuring the molar ratios of each fragment of ^{32}P-labeled B virus DNA.

donuclease fragments was 166 × 10^6. This observation, similar to that made for HSV DNA (Hayward *et al.*, 1975a), is consistent with the hypothesis that the B virus genome forms several isomers (Hayward *et al.*, 1975b).

A comparison of the two B virus strains yielded identity in the DNA cleavage pattern with *Eco*RI, *Bam*HI, and *Cla*I restriction endonucleases. With the *Bgl*II and *Bst*EII enzymes, strain-specific differences could be detected (not shown). Since the B virus has been compared in a variety of biological parameters with PsR virus (Sabin, 1934a; Hurst, 1936; Burnet *et al.*, 1939a,b; Plummer, 1964; Watson *et al.*, 1967), it should be emphasized that no similarity exists in the endonuclease cleavage patterns of the DNAs of these viruses (Ludwig *et al.*, 1982; Herrmann, 1981) (see Fig. 8C).

Nothing is known about the synthesis of B virus DNA during the replication cycle. The few biological similarities of this virus to HSV or PsR virus, however, suggest that the same mechanisms for initiation and replication of viral DNA, as well as host-cell DNA shutdown, are at work in these highly lytic herpesviruses (Ben-Porat and Kaplan, 1973; Ludwig and Rott, 1975; Roizman, 1978).

IV. ANTIGENS AND PROTEINS

Research in the 1930s led to the early suggestion that the B virus and HSV were different entities. This view has withstood the test of time.

Although an initial isolate by Gay and Holden (1933) was claimed to be identical to HSV, and the serological data collected over the years from infected monkeys and humans could have reinforced this conclusion, recent studies on B virus DNA summarized in Section III have removed remaining uncertainties.

Serological reactions that detect more closely the immune response against B virus antigens are dealt with in Sections IV and V. They cover recent studies on the structural proteins and on components that participate in homologous and heterologous reactions.

A. Proteins

On electrophoresis under denaturing conditions in sodium dodecyl sulfate–polyacrylamide gels (SDS-PAGE) (Fig. 10), the purified B virus yields approximately 23 major polypeptides ranging in molecular weight

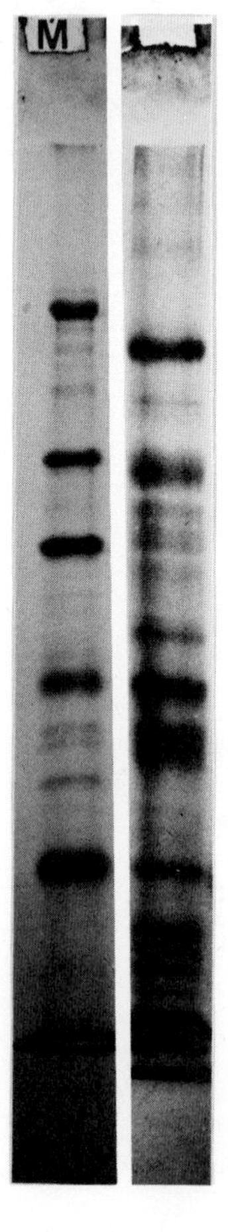

FIGURE 10. Polypeptide pattern of solubilized B virus proteins after electrophoresis on a 10% polyacrylamide slab gel. (M) Marker proteins. Monolayers of HeLa cell cultures were infected with B virus at a multiplicity of 1.0 and incubated at 37°C. B virus was isolated 36–48 hr after infection and purified (Allen and Randall, 1979). B virus particles were inactivated by treatment with 5% formaldehyde. PAGE was performed in a vertical electrophoresis cell using the discontinuous system of Lämmli (1970). A 3% stacking gel was performed above the resolving gel at a concentration of 10%. Diallyltartratdiamide was used for cross-linking as described by Heine *et al.* (1974). Virions purified by flotation in a CsCl gradient were freed from CsCl by dialysis and denatured by heating at 100°C in the presence of SDS (2%), 2-mercaptoethanol (5%), glycerol (10%), and bromophenol blue (0.05%). Volumes of B virus containing about 100 μg viral protein were applied to each well. Electrophoresis was carried out at a constant current of 30 mA for 14 hr. The gel was subsequently stained with Coomassie brilliant blue R. The following proteins (Biorad) were used as markers (M): myosin (200 K), β-galactosidase (116 K), phosphorylase B (94 K), bovine serum albumin (68 K), ovalbumin (43 K), carbonic anhydrase (30 K), soybean trypsin inhibitor (21 K), and lysozyme (14 K).

from 18,000 to 220,000. The polypeptides in the molecular weight range of 150,000, 125,000, and 40,000–60,000 appear to be among the most abundant. Immunoprecipitation tests with convalescent rhesus monkey serum (see Fig. 12B) suggest that B virus may specify four to six glycoproteins ranging in apparent molecular weight from 50,000 to 150,000. The properties of these glycoproteins are not yet known.

B. Antigens

The B virus antigens contained in infected-cell lysates have been investigated by a variety of techniques including immunoelectrophoresis and indirect immunoprecipitation, followed by electrophoresis of the denatured precipitates in polyacrylamide gels (Pauli and Ludwig, 1977; Norrild *et al.*, 1978b).

1. Homologous Reactions

Crossed immunoelectrophoresis of nonionic-detergent-solubilized infected cells with rhesus monkey convalescent serum yielded at least four distinct B-virus-specific antigens. Of these, one formed a major precipitation arc (Fig. 11A). Further studies (data not shown) revealed that these antigens were in the molecular-weight range of 50,000–150,000. As noted above, this serum reacted with four to six glycoproteins of which those with apparent molecular weights of 100,000 and 125,000 were present in greatest abundance (Fig. 12B).

2. Antigenic Cross-Reactivity

B virus or B-virus-infected cells have been reported to show cross-reactivity with a variety of other herpesviruses. The relevant data are summarized in Table III.

The identification and characterization of common antigens were reported by our group and are described in more detail below. HSV-1, HSV-2, and BHV-2 (bovine mammillitis virus) antisera predominantly recognize one of the major glycoproteins of B virus that has an apparent molecular weight of 125,000 (Figs. 11 and 12). Previous reports have shown that the peak I glycoprotein group and, more specifically, HSV antigen Ag-11, corresponding to glycoprotein gA/B of HSV, share common antigenic sites with the B virus and with BHV-2 (Pauli and Ludwig, 1977; Ludwig *et al.*, 1978; Norrild *et al.*, 1978b). More sensitive studies using crossed immunoelectrophoresis combined with the intermediate-gel technique revealed that a second HSV antigen complex, designated as Ag-8 and corresponding to glycoprotein gD, is also involved in this cross-precipitation, although the majority of the antibodies react with Ag-11 (Fig. 13A and B). A quantitative analysis of the common antigenic sites

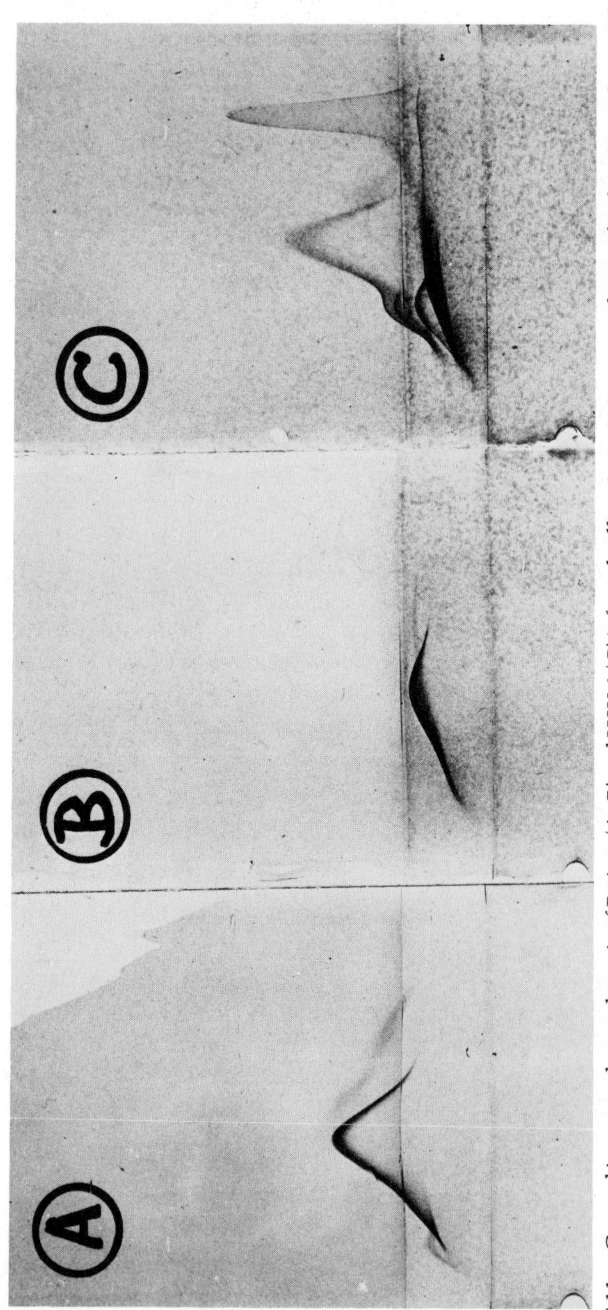

FIGURE 11. Crossed immunoelectrophoresis of B virus (A, B) and HSV-1 (C) infected cell antigens in B-virus-polyspecific rhesus monkey antibodies [(A) second-dimension gel: 20/ul/cm^2) and HSV specific human antibodies (B, C; second dimension gel: 15μl/cm^2). The plates were stained with Coomassie brilliant blue. For further details, see Ludwig *et al.* (1978) and Norrild *et al.* (1978b). Note: The human immune globulin pool recognizes a major and a minor B-virus-specific antigen.

TABLE III. Cross-Reactions between the B Virus and Other Herpesviruses[a,c]

Herpes-virus	Natural host	Antigenic cross-reactivity detected in the different test systems using B-virus-specific antigen or antibodies[b]							
		CF	ELISA	ID	FA	IE+PAGE	Ippt+PAGE	NT	Protection
B virus	Rhesus monkeys	++++[7,8]	++++[6]	++++[23]	++++[1]	++++[16]	++++[16]	++++	++++[9]
SA8	Vervets, papio monkeys		++[6]					+++[9,14,22]	
HSV-1	Humans	+++[7,8]	++[6]	++[23]	+[1]	++[12,13,15]	++[16]	+++[2–5,8,9,11,12,16–23]	++[9,18]
HSV-2	Humans	+++[7]	++[6]			++[12,13,15]	++[16]	+++[11,19]	
VZV	Humans							–[16]	
BHV-1	Bovines							–[16]	
BHV-2	Bovines					++[12,13,15]		–[16]	
BHV-6	Goats							–[5]	
PRV	Swine			+[23]		+[16]	+[16]	–[2,16,17,23]	+/±[9,10,18]
EHV-1	Horses							–[16]	
EHV-3	Horses							–[16]	
FHV	Cats							–[16]	

[a] This table updates previous ones by Hull (1968) and Wildy (1973) and gives a comprehensive view of cross-reactions of other herpesviruses with the B virus, though crossreactivities among other herpesviruses are not shown.

[b] (CF) Complement fixation; (ELISA) enzyme-linked immunosorbent assay; (ID– immunodiffusion; (FA) fluorescent antibody; (IE+PAGE) immunoelectrophoresis with PAGE; (Ippt+PAGE) indirect immunoprecipitation with PAGE; (NT) neutralization test. It should be mentioned that the original NT, by Sabin (1934a) and Hurst (1936) were done in animals. Burnet *et al.* (1939b) used focus reduction in the egg chorioallantoic membrane, and then all subsequent tests were done in tissue cultures. (+ to ++++) Estimated amount of cross-reaction, here omitting the observation of socalled "one-way reactions." These "one-way reactions" are clarified to some extent by Fig. 18, which shows whether antigenic components are expressed and accessible, or buried in the membrane. (–) Negative reaction. Where no symbol is given, there has been no test.

[c] The table is based on the references given below and does not claim to be exhaustive for all the cross-neutralization tests. Protection experiments showed complete protection (++++), partial protection (+) and modified disease (±) depending on the type of animal and ways of immunization. 1. Benda et al. (1966); 2. Burnet et al. (1939b); 3. Cabasso et al. (1967); 4. Daniel et al. (1975); 5. Engels et al. (1981); 6. Eichberg et al. (1980); 7. Falke (1964); 8. Gary and Palmer (1977); 9. Hull (1968; 1973); 10. Hurst (1936); 11. Hutt et al. (1981); 12. Ludwig et al. (1978); 13. Ludwig (Chapter 4); 14. Malherbe and Harwin (1963); 15. Norrild et al. (1978b); 16. Pauli and Ludwig (unpubl.); 17. Plummer (1964); 18. Sabin (1934a); 19. Schneweis (1962); 20. Ueda et al. (1968); 21. Van Hoosier and Melnick, (1961); 22. Vizoso (1974); 23. Watson et al. (1967).

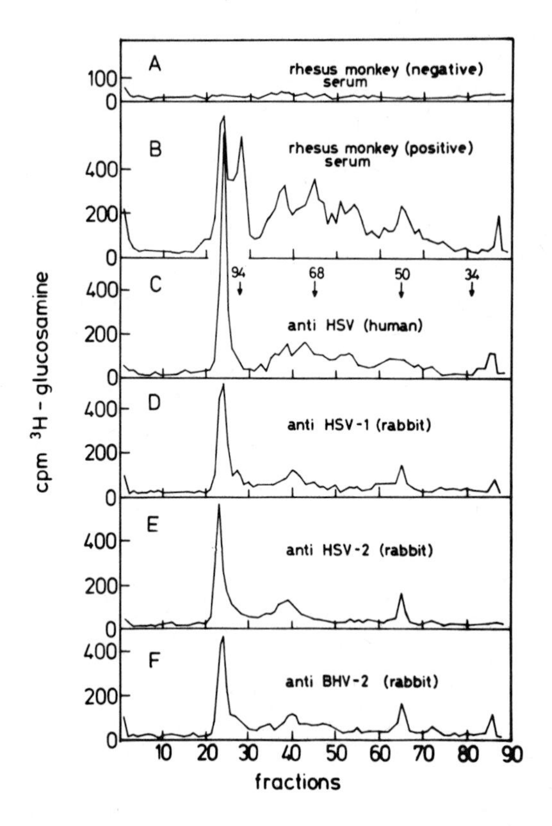

FIGURE 12. Characterization of the major B-virus-specific immunogenic components in homologous and heterologous reactions. The labeled B-virus-specific infected-cell antigens (proteins) were immunoprecipitated by the indirect technique, using the antisera noted in the figure and appropriate amounts of anti-IgG antibodies. Characterization of the proteins was done by PAGE. For details, see Pauli and Ludwig (1977). The arrows indicate the molecular-weight markers labeled with [^{14}C]amino acids [Semliki Forest virus proteins (Kaluza and Pauli, 1980)] run in the same gels.

by electrophoresis of immune precipitates in SDS–polyacrylamide gels is shown in Fig. 14A–C.

The preabsorption technique successfully used to localize HSV antigens (Pauli and Ludwig, 1977) permits a further insight into common antigenic determinants located on the B virus and HSV antigens. As demonstrated in Fig. 15A–F, monoprecipitine antibodies against Ag-11 (glycoprotein gA/B) of HSV recognize the antigen in both viruses. These antibodies can be absorbed out by the HSV antigen, but not by the B virus antigen. This finding indicates that Ag-11 of HSV and the corresponding antigen of the B virus share only a few antigenic determinant sites.

The major antigen(s), or, to be more specific, the common antigenic sites on the major antigen(s) of the B virus, HSV, and BHV-2, are schematically drawn in Fig. 18 (see Section V.C). The location, biochemical properties, and possible function of the antigen cross-reacting between HSV and BHV-2 are dealt with extensively in Chapter 4.

3. Discrimination of B Virus and HSV Infection

The antigenic relationship between the monkey B virus and the human HSV makes it difficult to differentiate between them. Numerous

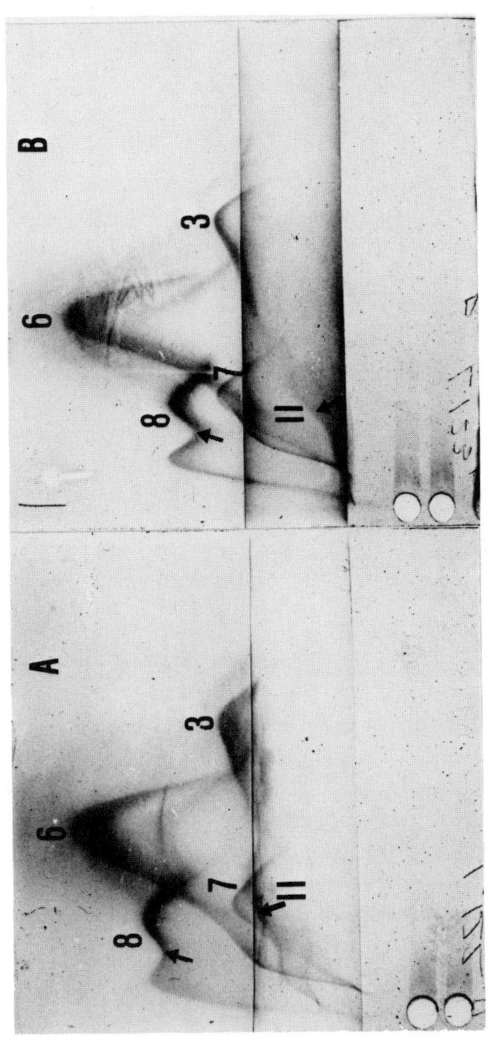

FIGURE 13. Crossed immunoelectrophoresis with intermediate gels. Coomassie brilliant blue staining. HSV-1-infected-cell antigens were separated by electrophoresis and run into rabbit polyspecific antibodies directed against HSV-1 (15 $\mu l/cm^2$). The intermediate gels contained 20 $\mu l/cm^2$ each of normal monkey (A) and B-virus-specific monkey (B) antibodies. For further details, see Ludwig *et al.* (1978) and Norrild *et al.* (1978b). Note: Ag-11 and Ag-8 (arrows) are reduced in height, which means they are cross-reacting.

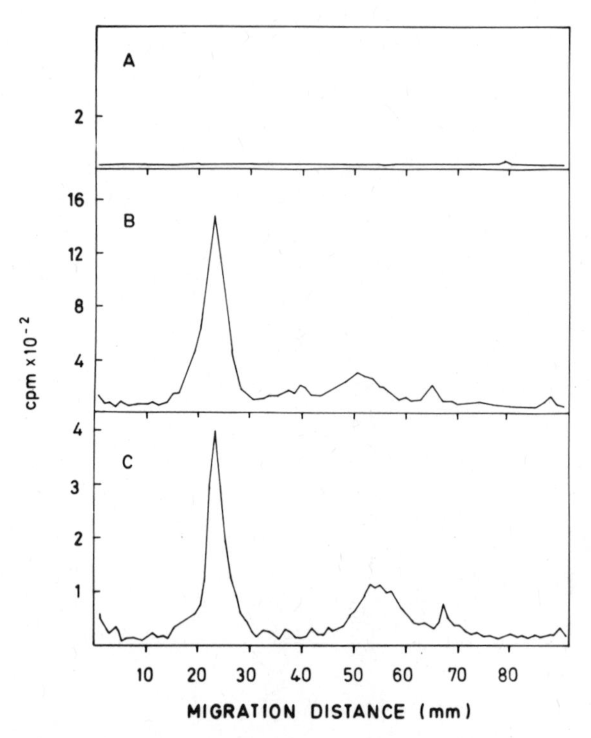

FIGURE 14. Analysis of the major immunogenic components of HSV-1, cross-reacting with B virus. For technical details, see the Fig. 12 caption. The following sera were employed in the indirect immunoprecipitation test: (A) normal serum from SPF-rhesus monkeys; (B) pool serum from humans with a high antibody titer against HSV; (C) pool serum from rhesus monkeys with a high antibody titer against B virus.

tests involving different kinds of antibodies have been reported (see also Section V). It is feasible, however, to differentiate between these viruses by the preabsorption technique outlined earlier in the text. Specifically, anti-HSV sera of human or animal origin were pretreated with unlabeled HSV antigen. After such a procedure, no antibodies remained to react with radioactively labeled HSV antigen. When, however, the B virus antigen was used for pretreatment, only antibodies directed against common determinants were eliminated, leaving the majority of antibodies directed against HSV available for the precipitation reaction (Pauli *et al.*, 1981).

We have used this technique to identify HSV- and B-virus-specific antibodies in the serum of a patient clinically diagnosed to have survived a B virus infection (Bryan *et al.*, 1975). The serum was kindly supplied by Dr. España. From the report in the literature and our own serological data (Table IV), it became clear that sera collected after onset of the disease neutralized HSV as well as the B virus. In a series of immunoprecipitation experiments, in which either the HSV or the B virus antigen was used for pretreatment, it could be demonstrated that specific anti-

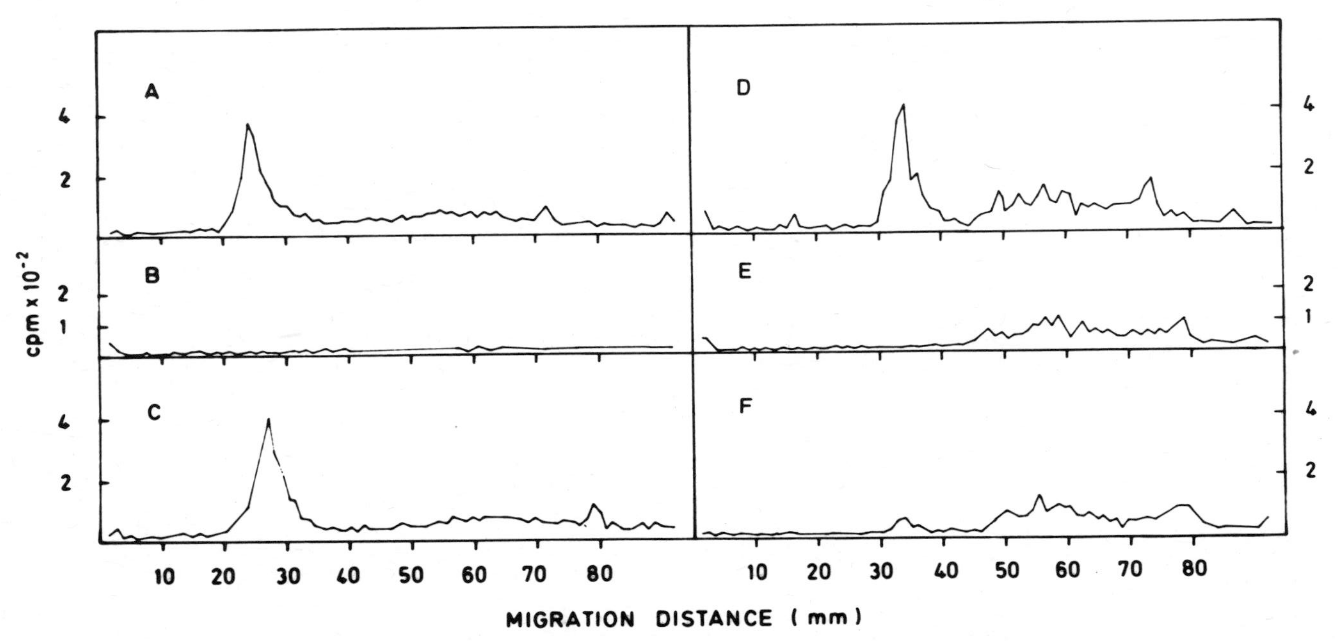

FIGURE 15. Indirect immunoprecipitation and PAGE analysis of HSV-1- specific or B-virus-specific antigens reacted with polyspecific rabbit anti-Ag-11 serum. The antiserum was pretreated with lysates of normal rabbit kidney (NRK) cells (A, D), HSV-1-infected NRK cells (B, E), B-virus-infected NRK cells (C, F) prior to addition of [^{3}H]glucosamine-labeled HSV-1 (A–C) or B-virus-infected-cell antigens (D–F). For further technical details, see Pauli and Ludwig (1977).

TABLE IV. Common Antigenic Determinants on B Virus, HSV-1, and BHV-2 as Detected by Neutralization Tests

Designation[a]	Origin	Kind of serum	Directed against:	NT titers tested with:[b] B virus	HSV-1	BHV-2	Reaction in immunoprecipitation[c] B virus	HSV-1	BHV-2
Roger	Rhesus monkey	SPF serum pool	—	<1:4	<1:4	<1:4	–	–	–
März	Rhesus monkey	Individual serum	B virus	1:68	1:35	<1:4	+	+	+
RM 1-5	Rhesus monkey	Five individual sera	—	<1:4	<1:4	<1:4	–	–	–
B renset	Rhesus monkey	Serum pool (IgG)	B virus	1:133	1:260	n.t.	+	+	+
D pool	Rhesus monkey	Serum pool	B virus	1:500	1:600	n.t.	+	+	+
L. preill	Human	—	—	<1:4	1:23	<1:4	–	–	–
L. ill	Human	—	—	1:360	1:2400	n.t.	+	+	+
L. convalescent	Human	—	—	1:440	1:6000	n.t.	+	+	+
Sh A	Sheep	Hyperimmune	HSV-1	1:20	1:1200	1:25	n.t.	+	+
HSV-1 ref.	Rabbit	Hyperimmune	HSV-1	1:58	1:640	1:25	+	+	+
HSV-2 ref.	Rabbit	Hyperimmune	HSV-2	1:410	1:720	n.t.	+	+	+
M0	Rabbit	Preimmune	—	<1:4	<1:4	<1:4	–	–	–
M1	Rabbit	Hyperimmune	HSV-1	1:110	1:1820	n.t.	+	+	+
M4	Rabbit	Hyperimmune	HSV-1	1:85	1:1500	n.t.	+	+	+

M8	Rabbit	Hyperimmune	HSV-1	1:190	1:240	1:4	+	+	+
M10	Rabbit	Hyperimmune	HSV-1	1:110	1:1500	n.t.	+	+	+
Sh B	Sheep	Hyperimmune	BHV-2	<1:4	1:50	1:200	+	+	+
AS-4	Bovine	Hyperimmune	BHV-2	<1:4	1:12	1:120	n.t.	+	+
R 12/0	Rabbit	Preimmune	—	<1:4	<1:4	<1:4	−	−	n.t.
R 11/3	Rabbit	Hyperimmune	Band II	1:16	1:810	n.t.	+	+	n.t.
R 11/3B	Rabbit	Hyperimmune	Band II	<1:4	1:170	n.t.	+	+	n.t.
Anti-Ag-11 (1977)	Rabbit	Hyperimmune	Ag-11	1:21	1:280	<1:4	+	+	+
Anti-Ag-11 (1981)	Rabbit	Hyperimmune	Ag-11	1:8	1:52	<1:4	+	+	+
Anti-Ag-8	Rabbit	Hyperimmune	Ag-8	1:75	1:478	<1:4	+	+	−
Anti-Ag-6	Rabbit	Hyperimmune	Ag-6	<1:4	1:71	<1:4	−	+	−
Anti-VZV	Guinea pig	Hyperimmune	VZV	<1:4	<1:4	<1:4	−	−	−

[a] The human serum came from a man who was supposed to be infected with B virus; the serum was kindly provided by Dr. Carlos España (see Bryan *et al.*, 1975). The sheep sera were prepared by hyperimmunization of the animals with virus-infected homologous skin cells grown in the preimune sera. The rabbit sera M0–M10 came from an animal that had been hyperimmunized with lysates of infected cells orginating from one testicle of this animal, the cells of which had been grown in the animal's preimmune serum. Sera were taken after the first (M1), second (M4), third (M8), and fourth (M10) booster. The anti-band II sera were kindly provided by Dr. Douglas Watson. The anti-Ag-11, Ag-8, and Ag-6 sera were those reported earlier (Vestergaard and Norrild, 1978); two different anti-Ag-11 sera prepared in 1977 and 1981 were used. The anti-VZV serum was kindly provided by Dr. Meyers.
[b] The titers indicate 50% plaque reduction obtained with inactivated sera. (n.t.) Not tested.
[c] Judgment of positive immunoprecipitation reaction is based on results obtained with two techniques (Pauli and Ludwig, 1977; Norrild *et al.*, 1978b). (n.t.) Not tested.

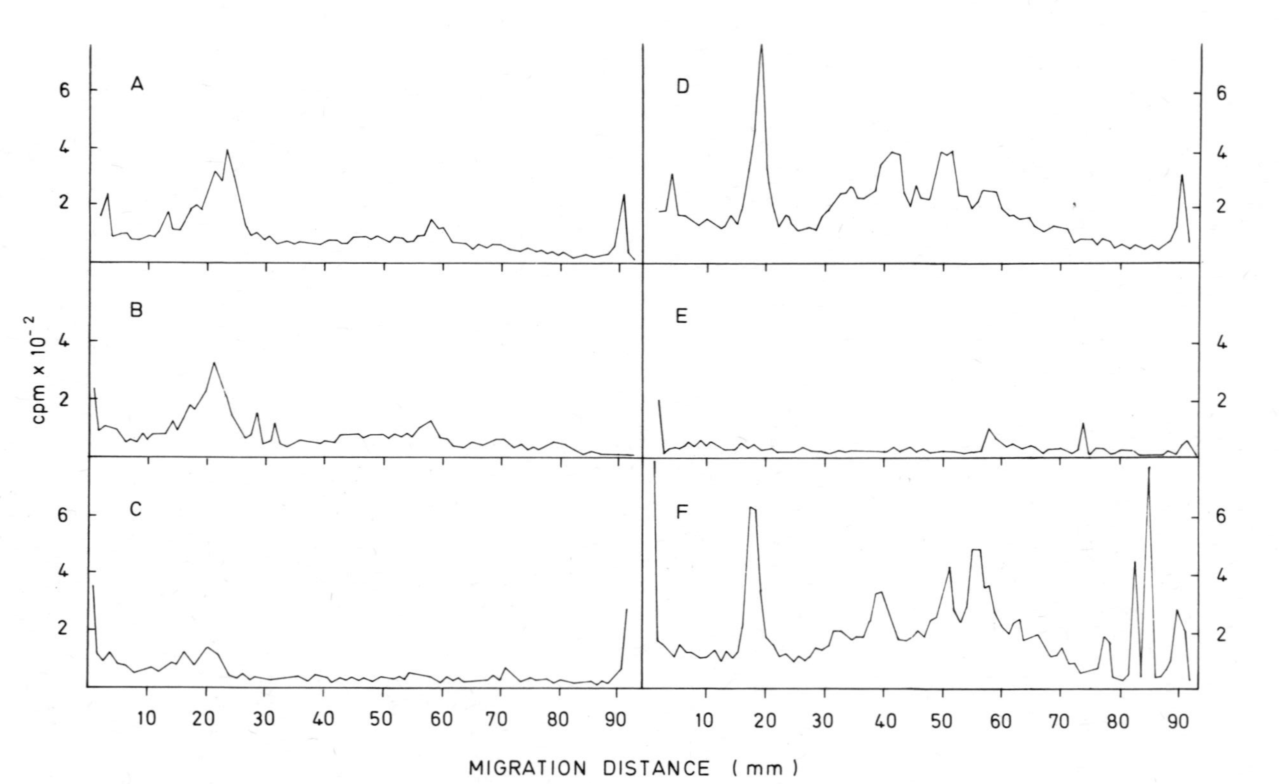

FIGURE 16. Analysis of HSV-1 and B-virus-specific antibodies in a convalescent serum of a B-virus-infected man. The serum was preincubated either with lysates of unlabeled normal rabbit kidney (NRK) (A, D), B-virus-infected NRK (B, E), or HSV-1-infected NRK (C, F) cells before addition of labeled HSV-1 (A–C) or B virus (D–F) antigens. For further details, see Figs. 12 and 15 and Pauli and Ludwig (1977). The presence of HSV-1- and B-virus-specific antibodies in this serum is demonstrated by the total elimination of these antibodies after preincubation with the homologous antigens (C, E).

bodies directed against both viruses were present in this serum (Fig. 16A–F).

To date, this experimental procedure is the only one that readily and successfully discriminates between antibodies directed against the B virus and HSV.

V. SEROLOGY

The majority of investigations dealing with B virus infections use conventional serological techniques to elucidate diagnostic or epidemiological problems. Most of the investigations include—or are even based solely on—cross-reactions with HSV.

We have tried to explain the antigenic cross-reactivity of the two viruses in Section IV. In this section, we will briefly summarize the serological problems arising from the cross-reactivity of B virus with HSV and present some new aspects of the differentiation of B and herpes simplex viruses.

A. B Virus and Other Herpesvirus Infections: Serological Diagnosis

Complement-fixation (CF) tests (Falke, 1964; Gary and Palmer, 1977) do not unequivocally differentiate antibodies induced during B virus and HSV infections. The highest titers were found in the homologous system. Most of the human sera showed higher titers against HSV, whereas monkey sera usually yielded higher CF titers against B virus. On the other hand, some sera reacted equally well with both viral antigens.

An enzyme immunoassay used to detect antibodies against HSV-1, HSV-2, SA 8, and B virus was less sensitive than neutralization and could not solve problems related to cross-reactivity between these viruses (Eichberg *et al.*, 1980).

A review of the published reports (see Table III) and our data (see Tables IV and V) concerned with neutralization of one or the other of the viruses and with attempts to differentiate them indicates that human sera or experimentally raised antisera against HSV in general possess higher neutralizing–antibody titers against HSV than against B virus. Surprisingly, sera derived from rhesus monkeys in most cases also neutralized HSV better than B virus, although antibodies seemed to originate from B virus infections.

The results obtained during our studies with neutralization tests and immunoprecipitation tests (see also Section IV) essentially confirm this conclusion. In these studies, antibodies exclusively directed against HSV or B virus were not detectable (Table IV) in human or monkey sera.

The reports in the literature and our experience make it clear that as yet no simple serological test system exists to separate clearly the two primate viruses (Table IV). Neutralization—as well as CF tests—gives a hint to suggest infection by one or the other. It is even more difficult to identify the etiological agent of disease when both viruses infect the same host. This can be seen from rhesus monkeys, which may be infected with both viruses. Our neutralization data on a human serum, which definitely has antibodies derived from a B virus infection, point in the same direction.

1. Serological Identification of B Virus

A possible solution to the problem of differentiating infections by B and herpes simplex viruses emerges from the feasibility of separating the viruses on the basis of their genetic material (see Section III). Another alternative not involving the preabsorption technique (see Section IV) is described here and employs neutralization tests (Table IV). The procedure is based on the observation that both HSV-1 anti-Ag-11 (glycoprotein gA/B) and anti-HSV-1 Ag-8 (glycoprotein gD) sera neutralize HSV and the B virus. The anti-Ag-8 serum seems to inactivate the B virus especially well. However, the anti-HSV-1 Ag-6 (glycoprotein gC) serum, which is HSV-specific (Vestergaard and Norrild, 1978; Norrild *et al.*, 1979; Spear, 1980), does not neutralize the infectivity of the B virion. Furthermore, polyvalent antisera against BHV-2, which precipitate B virus antigen, do not neutralize the B virus. They serve as a further tool to distinguish the B virus from herpes simplex viruses.

B. Antibody-Dependent Cellular Cytoxicity Tests

Little or no information exists on the cellular immune response following B virus infections. In an effort to extend the studies concerning exposure of HSV-specific antigens on the cell surface (Pauli and Ludwig, 1977; Norrild *et al.*, 1978a), antibody-dependent cellular cytotoxicity (ADCC) tests were performed to detect cross-reacting determinants on infected cells. As shown in Table V, the major findings are that B-virus- and BHV-2-specific antibodies recognize determinants on HSV-infected cells. Most likely, HSV glycoprotein gD (Ag-8) and glycoprotein gA/B (Ag-11) determinants are involved in the reaction with B virus antibodies and HSV glycoprotein gA/B determinants in the interaction with BHV-2 antibodies. This conclusion would be consistent with the results of the neutralization studies. Because of the potential hazard, it was not possible to perform the experiments on B-virus-infected cells.

TABLE V. Detection of Common Cell-Surface Antigens among B Virus, HSV-1, and HSV-2 in ADCC Tests and Comparison of the Serum Neutralizing Capacity

			Neutralization titer against[b]:			Specific ^{51}Cr release (%) of vero cells infected with[c]:	
Serum source	Kind of serum[a]	Directed against:	HSV-1	B virus	BHV-2	HSV-1	HSV-2
Rhesus monkey	Convalescent (pool)	B virus	1:500	1:600	<1:4	12.5	0.6
Rhesus monkey	SPF (pool)		<1:4	<1:4	<1:4	<0.1	<0.1
Sheep	Hyperimmune syngeneic	HSV-1	1:1200	<1:10	1:25	11.9	4.7
Sheep	Hyperimmune syngeneic	BHV-2	1:50	<1:4	1:200	1.8	0.8
Rabbit	Hyperimmune	BHV-2	1:80	<1:4	1:260	17.8	18.6
Rabbit	Hyperimmune	HSV-1	1:800	1:300	n.t.	22.0	10.7
Rabbit	Hyperimmune	HSV-2	1:1000	1:450	n.t.	13.8	3.0

[a] The preparation of antisera is described in previous reports (Pauli and Ludwig, 1977; Daniel *et al.*, 1975; Sterz *et al.*, 1973–1974); preparation of syngeneic sheep sera was performed by injection of infected homologous cells grown in preimmune serum of the same animal.
[b] Neutralization titers reflected 50% plaque reduction using inactivated sera. (n.t.) Not tested.
[c] The chromium-release assay was performed as detailed by Norrild, *et al.* (1979).

C. Considerations on the Function of Common Antigenic Determinants in B Virus, Herpes Simplex Virus, and Bovid Herpesvirus 2

A model showing the location and distribution of these common antigenic determinants on the glycoprotein species of these viruses is shown in Fig. 17. The major immunogenic components of HSV-1 are contained in the peak I (gA/B) and peak II (gD) glycoproteins (Pauli and Ludwig, 1977). The results of neutralization, immunoprecipitation, and absorption tests with virus, intact infected cells, or lysates from infected cells indicate that the antigenic determinants shared by B virus and HSV-1 and present on gA/B are different from those shared by HSV-1 and BHV-2. This conclusion is consistent with the observation that none of the BHV-2-specific antisera neutralize the B virus. On the other hand, a common antigen with approximately the same molecular weight can be precipitated from all three viruses. We could predict, therefore, that the antigenic sites shared by B virus and BHV-2 are not exposed on the surface of the virus or of the infected cell. Neutralization of the B virus by antibodies to glycoprotein gD (Ag-8, peak II) indicates that these common antigenic sites are exposed on the surface of the virion. This conclusion is consistent with the presence of these determinants also on the surface of HSV-1-infected cells, first demonstrated by Pauli and Ludwig (1977).

In a previous report, we showed that HSV-1 and HSV-2 were both

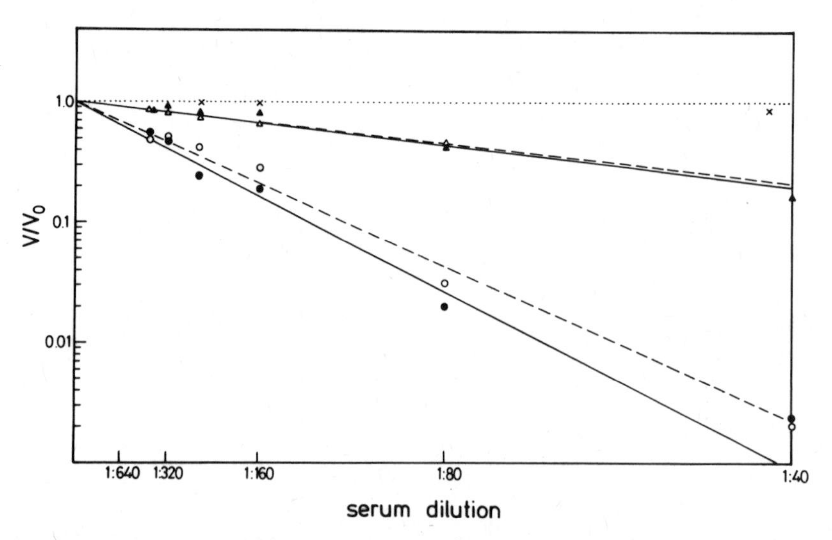

FIGURE 17. Neutralization tests of HSV-1 and HSV-2 (strains KOS and 196, respectively) with different B-virus-specific sera: HSV-1 or HSV-2 with a serum pool from specific pathogen-free rhesus monkeys (× ..... ×). HSV-1 (△– – –△) or HSV-2 (▲——▲) with an individual rhesus monkey antiserum, and (○– – –○, ●——●) with a pool serum of latently infected rhesus monkeys from a large animal colony. (V/V_0) Fractional reduction in plaque counts.

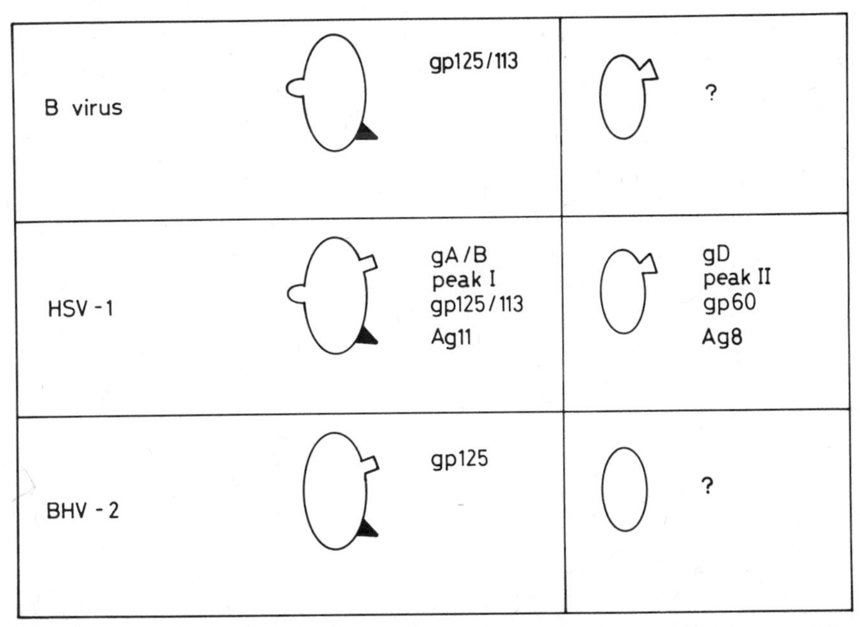

FIGURE 18. Our present view of the cross-reacting determinants on the antigens of three herpesviruses indigenous to different mammalian species. The open symbols on the antigens indicate that these determinants are available for antibodies on the surface of the infected cells and on the virus. The closed symbols mean that the sites are recognized by antibodies only after solubilization with detergents. The model is based on published work (Sterz *et al.*, 1973–74; Pauli and Ludwig, 1977; Killington *et al.*, 1977; Ludwig *et al.*, 1978; Norrild *et al.*, 1978a,b; Gelderblom *et al.*, 1980; Pauli *et al.*, 1981; Lewis *et al.*, 1982; D. Watson, personal communication) (see also Chapter 4) and data presented in this chapter.

neutralized to the same extent by anti-BHV-2 and anti-B virus antibodies (Fig. 18), (Ludwig *et al.*, 1978). This finding could be interpreted to suggest that neutralization of the two types of HSV with heterologous antisera occurs at the same antigenic surface component available on the glycoprotein gA/gB in HSV. The nonneutralizability of the B virus by anti-BHV-2 sera, however, reinforces our proposal that neutralization of HSV (types 1 and 2) involving glycoprotein gA/B by antibodies directed against B virus or against BHV-2 occurs with different determinants on this antigen complex, as shown in Fig. 17. There is no evidence that gD of HSV-1 shares antigenic determinants with a corresponding glycoprotein of BHV-2.

VI. PATHOGENESIS AND PATHOLOGY

The first isolate of the B virus came from a human patient (Sabin and Wright, 1934; Gay and Holden, 1933). Only some years later was the virus also obtained from rhesus monkey kidney cultures used for poliovirus vaccine production (Wood and Shimada, 1954) and from the brain tissue of a monkey (Melnick and Banker, 1954). It is now accepted that the B virus is indigenous in Asian monkeys of the *Macaca* species. From B virus infections, a characteristic phenomenon applicable to a variety of other herpesviruses became apparent. Infections that are subclinical or latent in one species can be fatal in another host, when crossing the species barrier (Hsiung, 1969; Holmes *et al.*, 1966; Hunt and Meléndez, 1969; Hurst, 1934, 1936; Katzin *et al.*, 1967). Burnet *et al.* (1939b) reasoned that the reservoir host and the virus must have been associated for a long period and have therefore developed a *status vivendi*. In most cases, the incidence of infection in the natural host is very high. An overt disease, however, is only rarely seen and is usually limited to young or adolescent populations.

A. Disease in Monkeys

Although B virus isolates were first recovered from the brain necropsy of a man, it was immediately recognized that rhesus monkeys might be carriers. Subsequent experiments suggested the presence of antibodies to the virus in these animals (Sabin, 1934a,b). It was, however, 25 years before evidence of a B-virus-specific clinical syndrome in naturally infected monkeys was discovered (Keeble *et al.*, 1958).

Clinically, B virus infections in rhesus monkeys are characterized by lingual or labial vesicles or both that can give rise to ulcers. The lesions are found most frequently on the dorsal surface of the tongue and on the mucocutaneous junctions of the lip, but also occur on the skin or conjunctiva (Fig. 19A–C).

Usually, no signs of the disease can be observed, and without critical examination of the mouth, the infection may remain undetected. Examination of 14,400 rhesus monkeys revealed that 332 monkeys (2.3%) had lip or tongue lesions. The animals had been kept in groups of 60 per cage. There were some indications that separation in smaller groups reduced the incidence of the disease (Keeble, 1960).

Histological examination of the lesions showed that the degeneration is characterized by ballooning of cells, necrosis of epithelial cells, and the presence of intranuclear inclusion bodies. Besides these effects, multinucleated epithelial cells containing intranuclear inclusions can be observed. Only a scant inflammatory reaction to the ulcer is usually present, but this varies with secondary invading bacteria. In the rhesus monkey, visceral lesions may also occur during the course of clinical disease (Sabin and Hurst, 1935). In several diseased animals, foci of necroses and cells with intranuclear inclusion bodies were found in the liver.

Specific lesions such as neuronal necrosis and gliosis were also seen in the central nervous system (CNS). Minimal perivascular cuffing with lymphocytes was associated with the lesions. Glial cells and neurons

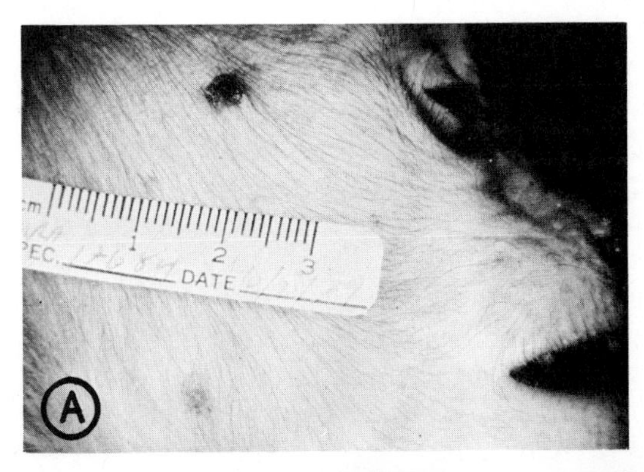

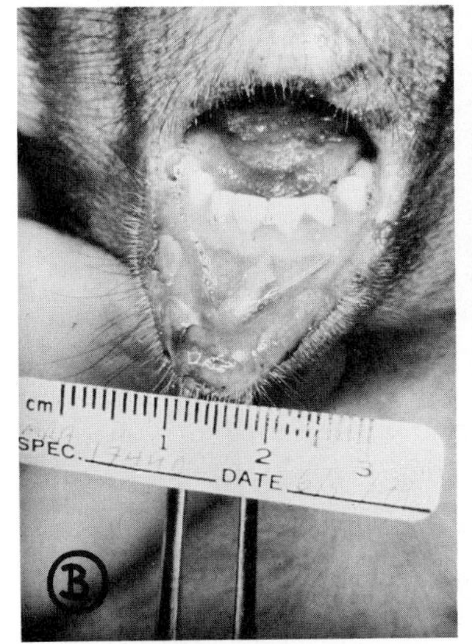

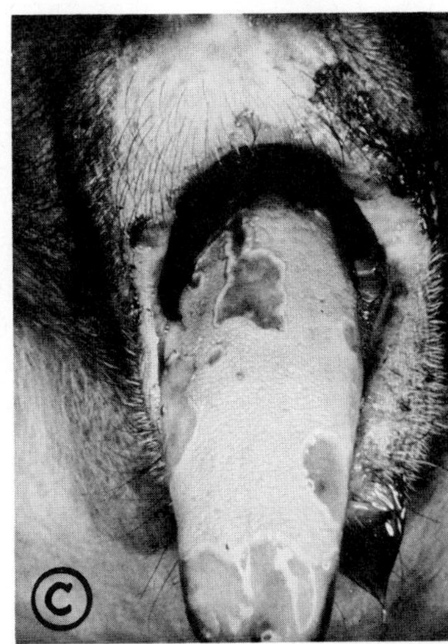

FIGURE 19. Clinical expression of B virus infection in the rhesus monkey (*Macaca mulatta*). (A) Skin ulcers; (B) labial and lingual ulcers with gingivostomatitis; (C) glossitis with multiple ulcerative plaques. Courtesy of Dr. J. Storz.

contained intranuclear inclusion bodies. Lesions are most common in the different brain nuclei and in the nerve tract of the descending branch of the trigeminal nerve, between the roots of the facial and auditory nerves, and in the trigeminal and facial nerves (Hunt and Meléndez, 1969; Keeble *et al.*, 1958).

B. Disease in Man

Interest in the B virus has been maintained because of the severity of the disease in humans. Although only 25 cases are recorded, all but 5 resulted in fatalities (Hull, 1973; Bryan *et al.*, 1975; Roth and Purcell,

1977). Infection may occur through monkey bites or scratches or by contamination of wounds with infected saliva. The nature of the virus transmission could not be clarified in every case and remained obscure especially when infection followed contact with tissue cultures of rhesus monkey kidney cells. The generalized disease can develop within a week, but an exact incubation period has not always been documented (Davidson and Hummeler, 1960; Hartley, 1966; Fierer *et al.*, 1973).

First signs and symptoms include local erythema and induration at the point of inoculation. These are usually followed by fever, abdominal cramps, muscular pains, headaches, lymphangitis, and lymphadenitis. In most cases, encephalitis and encephalomyelitis occur and lead to coma and death in 3–21 days. The histopathological changes are comparable to those of fatal systemic HSV infections in infants, with encephalitis, myelitis, and focal necrosis in the liver, spleen, lymph nodes and adrenals.

C. Latency

B virus infections in monkeys are in many aspects similar to infections caused by HSV in that the primary infection is almost invariably followed by a lifelong latency, with the possibility of virus reactivation from the sensory ganglia (Boulter, 1975; Baringer and Swoveland, 1973). The latent B virus can be activated to produce the overt disease, and the infectious virus may be shed following a variety of stimuli. Recovery of the B virus has been confirmed by examination of oral secretions collected from seropositive monkeys that had no detectable lesions (Hunt and Meléndez, 1969).

Studies on the contamination of poliomyelitis virus vaccines grown in rhesus monkey cells also suggested that apparently healthy monkeys harbor the B virus in a latent form. Six isolates were recovered from 650 different pools of rhesus monkey kidney cells (Wood and Shimada, 1954). The original isolation of the virus from the CNS of a rhesus monkey that had been inoculated during studies on polio vaccine (Melnick and Banker, 1954) is consistent with this conclusion. The search for the latent virus in the trigeminal ganglia of seropositive rhesus monkeys was successful in about 40% of the animals (Boulter and Grant, 1977).

Under controlled conditions, however, it was not possible to induce the secretion of infectious virus by immunosuppressive drugs (Boulter and Grant, 1977). On the other hand, toxicological studies with 60 rhesus and cynomolgus monkeys revealed that reactivation of the virus could be triggered. These animals had been found to be negative for B virus antibodies before the start of the experiment. Seven days after an acute toxicity study, one cynomolgus developed herpetic lesions and a positive B virus antibody titer. During the next 20 months, 20–28% of the animals converted to seropositivity (Gralla *et al.*, 1966). A similar case of drug-mediated induction was observed by another group (McClure, 1970).

These observations indicate that monkeys can carry the B virus without showing any clinical signs or detectable antibody titers (Hull, 1973).

1. Latency in Man

One possible case of B virus latency in man was described in detail by Fierer *et al.* (1973) and Hull (1973). The patient had no recent exposure to monkeys, and direct contact dated back 10 years before he became ill. The disease started with pain followed by headaches, ataxia, and fever, accompanied by a vesicular rash in the region of the ophthalmic branch of the left trigeminal nerve. After intense medical treatment, the patient survived. He had shown all the neurological signs and symptoms typical of B virus infection. The medical history and the encephalomyelitis, connected with the peculiar clinical sign of an ophthalmic zoster, all hinted toward the activation of a latent B virus. A virus isolated from scrapings taken of a zoster-like rash was identified as the B virus on the basis of cross-neutralization tests.

2. Latency in Rabbits

Rabbits are very sensitive to B virus infections by any route of inoculation, and the ensuing disease is almost uniformly lethal (Hartley, 1966; Sabin, 1934a). Vizoso (1975a,b) demonstrated that the B virus enters latency in rabbits protected by an immune response to HSV. The B virus could be recovered from the dorsal root ganglia for up to 2 years after challenge with the virus. This discovery led the author to speculate on the likelihood of B virus latency in humans. Since many people have antibodies to HSV, B virus infections in humans could occur without showing clinical disease. This possibility of latent infection under the protection of HSV immune response should always be kept in mind, especially with individuals who have intensive contact with B-virus-infected animals.

VII. EPIDEMIOLOGY

On the basis of clinical disease, virus isolation, and the presence of serum neutralizing antibodies, it has been established that several species of the genus *Macaca* act as a reservoir host for the B virus. These are *Macaca mulatta, M. fascicularis, M. fuscata, M. arctoides, M. cyclopis,* and *M. radiata* (Endo *et al.*, 1960; Hunt and Meléndez, 1969; Keeble *et al.*, 1958; Keeble, 1960; Valerio, 1971; Hartley, 1964; España, 1974). Careful examination after capture from the wild revealed clinical symptoms of herpetic infections in 2.3% of the rhesus (*M. mulatta*) monkeys (Keeble, 1960) and 3% of cynomolgus (*M. irus*) monkeys (Hartley, 1964). There seems to be a seasonal increase of natural disease during the monsoon and postmonsoon period (October to February).

Serological investigations of freshly captured animals indicated that the age-dependent distribution of antibody-positive animals follows comparable distribution of anti-HSV antibody in humans. Young animals had a rather low incidence of anti-B virus antibodies (approximately 10% of the animals) (Christopher *et al.*, 1971; Orcutt *et al.*, 1976), whereas young adults had detectable anti-B virus antibody titers in about 36% of those examined. Among adults, the seropositive animals constituted 70–80% of total. The age distribution of positive animals seems to depend on the population density in a given area. On Santiago Island, where the population density is high, the monkeys obviously acquire antibodies at a relatively early age (Shah and Morrison, 1969). In a recent report that includes a survey on the antibodies present in monkeys from South-East Asia, about 60% of the newly captured animals had antibodies directed against HSV (CF test), probably reflecting the percentage of animals with anti-B virus titers. Monkeys originating from Indonesia had the highest incidence of positive CF titers (77%) followed by the animals from Malaysia (64%) and from the Philippines (28%). Unfortunately, no age distribution of the tested animals is given (Suzuki *et al.*, 1981). In captive juvenile rhesus monkeys from India, the rate of animals that have antibodies to the B virus is higher than that of wild animals of a comparable age (Burnet *et al.*, 1939b; Keeble *et al.*, 1958; Pille, 1960). In 13–18% of such animals, antibodies directed against B virus were detected.

In this context, it should be mentioned that only 4.2% of animals that were tested soon after delivery had antibodies, but after 60 days, the number had already increased to 22%. Comparable results were found in other investigations (Hutt *et al.*, 1981) concerning the prevalence of seropositive monkeys used in research laboratories. A serological survey of laboratory rhesus monkeys from 1 to 7 years old revealed the presence of neutralizing antibodies to the B virus in only 1 of 39 animals. The monkeys were housed in individual cages for a number of years among rhesus monkeys obtained from India, of which many had B virus antibodies (Di Giacomo and Shah, 1972). The results described by these authors show that in a well-established rearing colony, the spread of B virus infection to individually kept animals is extremely low and it is therefore possible to rear B-virus-free colonies.

From the data available, it can be seen that B virus infections epidemiologically resemble HSV infections in man: The percentage of individuals with antibodies increases with age, most of the adults having antibodies. The infection is very common but usually subclinical. The virus is transmitted only by direct contact, and the overcrowding of monkey populations is associated with a higher prevalence of antibodies.

Because of the hazard for people in close contact with infected animals or monkey tissues, a number of attempts have been made to prevent B virus infection in such individuals. One of the proposals was immunization with the inactivated virus (Hull, 1973). Most probably because of the small number of infected persons over the last few years, this idea has not been pursued further. Because of the infectivity of B virus DNA and the ability of the virus to remain latent, such vaccines are not without risk. Vaccination against B virus infection—for example, with subunit vaccines—must still remain in an experimental stage, as long as the antigen makeup of even the closely related HSV is not fully understood.

Another approach was based on the storage of high-titer antibody preparations for treatment during the infection (MacLeod *et al.*, 1960; Boulter and Grant, 1977). However, the effectiveness of such hyperimmune sera in treatment of herpetic infections is questionable. Recently, promising experiments with acyclovir in B-virus-infected rabbits suggest that at least the replication of the virus might also be blocked by such antiviral drugs in humans (Boulter *et al.*, 1980).

VIII. CONCLUSIONS

Overwhelming evidence indicates that the B virus is very hazardous for humans. The infection is almost uniformly fatal once the patient exhibits overt clinical signs of the disease. This is, however, quite in contrast with a very similar herpesvirus—the pseudorabies (PsR) virus—which is fatal in a variety of other mammals, but which does not cause serious illness in humans. The only known protection against the B virus in humans appears to stem from the close link between the virus and HSV.

Consequently, the research on this virus is closely related and parallels studies on HSV, and the discovery by Sabin (1934a) that antigenic cross-reactivity exists has been the basis of much of the subsequent research. However, this also makes differentiation difficult. The genetic differences among the B virus, HSV, and PsR virus, discussed in this review, are certainly of major importance. The B virus is unique in many features. It has the highest G+C content of any known herpesvirus (75 moles %); it replicates relatively rapidly and has a broad host range *in vitro* and *in vivo*, including humans. This again differentiates it from the PsR virus—the porcine herpesvirus—against which humans are refractory.

The unique and strong cross-reactivity with HSV predicts homology of B virus and HSV-1 DNAs, particularly, between 0.3 and 0.4 and between 0.9 and 0.95 on the HSV physical map, which contains the sequences coding glycoproteins gA/B and glycoprotein gD, respectively (Ruyechan *et al.*, 1979). It will be of interest to know whether the genetic homology that must exist between the B virus and HSV is greater than that between BHV-2 and HSV, or even than that between HSV-1 and HSV-2 (Sterz *et al.*, 1973–1974; Ludwig *et al.*, 1972a; Ludwig, 1972b).

The antigenic cross-reactivity between the B virus and HSV, which also extends to the bovine herpesvirus, BHV-2, can now be focused more specifically to the gA/B, in the case of all three, and to the gD in the case of the two primate herpesviruses. In the latter case, cross-neutralization seems to involve the two major antigenic components of HSV. Glycoprotein peak I (gA/B, Ag-11), which, from our experience, is the major immunogenic component of HSV, may well in addition be involved in cross-protection between the B virus and HSV. The exposure of the common site on the infected cell surface lends further support to this hypothesis (Pauli and Ludwig, 1977).

The interplay of the cellular and humoral immune response, recognizing the two glycoprotein regions where the common sites are exposed or covered, certainly contributes to the strength of this cross-reactivity. The existence of common determinants between the two viruses on both the major immunogenic components, which have shown minimal divergence in the course of evolution, suggests that these antigenic sites play a significant functional role. A similar situation is known to exist between HSV and BHV-2 (Sterz *et al.*, 1973—1974; Gelderblom *et al.*, 1980) (see also Chapter 4), but the relationship between the B virus and HSV is probably much closer (see Fig. 17). The view on unity and diversity of herpesviruses has also been extensively discussed by Honess and Watson (1977).

Despite all the research, the B virus remains highly hazardous to humans. This hazard should reflect the frequency of contact between the carriers of the disease, the *Macaca* monkeys, and humans. On the other hand, it is of great significance that there have been only 25 recorded cases of the disease in man. This paradox cannot yet be fully explained. However, it is clear that the relationship between the B virus and HSV is of extreme importance here. Possibly man's natural defenses against HSV provide him with immunity to the B virus. The extension of this immunity would appear to be the most promising way of reducing the incidence of diseases resulting from B virus infection.

ACKNOWLEDGMENTS. This review was made possible only by the unselfish cooperation of several laboratories: The Department of Microbiology, New England Primate Research Center, Harvard Medical School, Boston (M.D. Daniel); Institute of Medical Microbiology, University of Copenhagen (B. Norrild); Institut für Virusforschung, Deutsches Krebsforschungszentrum, Heidelberg (R.M. Flügel); Institut für Medizinische Virologie, Universität Heidelberg (G. Darai and H.-G. Koch); Abteilung Virologie, Robert Koch-Institut des Bundesgesundheitsamtes, Berlin (H. Gelderblom); Institut für Virologie, Freie Universität Berlin, Nordufer 20, 1 Berlin 65 (G. Pauli and H. Ludwig). Stimulating discussions with colleagues of the Institut für Virologie, Giessen, where some of the early experiments were done, the excellent technical assistance of Tine Leiskau, editorial help from Jan G. Lewis and Nick Hunter, and the continuous support given by Meike and Viviana L. were all of invaluable help. Financial aid by grants from the Deutsche Forschungsgemeinschaft, the Bundesministerium für Forschung und Technologie, and the Danish Cancer Society, as well as the chancellor of the Freie Universität Berlin, D. Borrmann, is gratefully acknowledged.

REFERENCES

Allen, G.P., and Randall, C.C., 1979, Proteins of equine herpes virus type 3: Polypeptides of the purified virion, *Virology* **92:**252.

Baringer, J.R., and Swoveland, P., 1973, Recovery of herpes simplex virus from human trigeminal ganglions, *N. Engl. J. Med.* **288:**648.

Benda, R., and Cinátl, 1966, Isolation of two plaque variants from the prototype strain of B virus (*Herpesvirus simiae*), *Acta Virol.* **10:**178.

Benda, R., Prochazka, O., Cerva, L., Rohn, F., Dubanska, H., and Hronovsky, V., 1966, Demonstration of B virus (*herpesvirus simiae*) by the direct fluorescent antibody technique, *Acta Virol.* **10:**149.

Ben-Porat, T., and Kaplan, A.S., 1973, Replication–biochemical aspects, in: *The Herpesviruses* (A.S. Kaplan, ed.), pp. 164–220, Academic Press, New York, San Francisco, London.

Bhutala, B.A., and Mathews, J., 1963, Studies on *Herpesvirus simiae* (B virus): Intracellular development, *Cornell Vet.* **53:**494.

Boulter, E.A., 1975, The isolation of monkey B virus (*Herpesvirus simiae*) from the trigeminal ganglia of a healthy seropositive rhesus monkey, *J. Biol. Stand.* **3:**279.

Boulter, E.A., and Grant, D.P., 1977, Latent infection of monkeys with B virus and prophylactic studies in a rabbit model of this disease, *J. Antimicrobiol. Chemother.* **3:**107.

Boulter, E.A., Thornton, B., Bauer, D.J., and Bye, A., 1980, Successful treatment of experimental B virus (Herpesvirus simiae) infection with acyclovir, *Br. Med. J.* **280:**681.

Brenner, S., and Horne, R.W., 1959, A negative staining method for high resolution electron microscopy of viruses, *Biochim. Biophys. Acta* **34:**103.

Bryan, B.L., España, C.D., Emmons, R.W., Vijayan, N., and Hoeprich, P.D., 1975, Recovery from encephalomyelitis caused by *Herpesvirus simiae, Arch. Intern. Med.* **135:**868.

Burnet, F.M., Lush, D., and Jackson, A.V., 1939a, The propagation of herpes B and pseudorabies viruses on the chorioallantois, *Aust. J. Exp. Biol. Med. Sci.* **17:**35.

Burnet, F.M., Lush, D., and Jackson, A.V., 1939b, The relationship of herpes and B-virus; immunological and epidemiological considerations, *Aust. J. Exp. Biol. Med. Sci.* **17:**41.

Cabasso, V.J., Chappel, W.A., and Avampato, J.E., 1967, Correlation of B virus and herpes simplex virus antibodies in human sera, *J. Lab. Clin. Med.* **70:**170.

Christopher, S., John, T.J., and Feldman, R.A., 1971, Virus infections of bonnet monkeys, *Amer. J. Epidem.* **94:**608.

Compans, R.W., and Klenk, H.-D., 1979, Viral membranes, in: *Comprehensive Virology,* Vol. 13 (H. Fraenkel-Conrat and R.R. Wagner, eds.), pp. 293–407, Plenum Press, New York.

Daniel, M.D., Meléndez, L.V., Hunt, R.D., and Trum, B.F., 1972, The herpesvirus group, Pathology of simian primates, Part II, (R.N.T.-W-Fiennes, ed.) pp. 592–611, Karger, Basel.

Daniel, M.D., Garcia, F.G., Meléndez, L.V., Hunt, R.D., O'Connor, J., and Silva, D., 1975, Multiple *Herpesvirus simiae* isolation from a rhesus monkey which died of cerebral infarction, *Lab. Anim. Sci.* **25:**303.

Darai, G., Zöller, L., and Matz, B., 1982, Analysis of mycoplasma hyorhinis genome by restriction endonucleases and by electron microscopy, *J. Bacteriol.* **150:**788.

Davidson, W.L., and Hummeler, K., 1960, B-virus infection in man, *Ann. N. Y. Acad. Sci.* **85:**970.

Di Giacomo, R.F., and Shah, K.V., 1972, Virtual absence of infection with *Herpesvirus simiae* in colony-reared rhesus monkeys (*Macaca mulatta*), with a literature review on antibody prevalence in natural and laboratory rhesus populations, *Lab. Anim. Sci.* **22:**61.

Eichberg, J.W., Heberling, R.L., Gueijardo, J.E., and Kalter, S.S., 1980, Detection of primate herpesvirus antibodies including *Herpesvirus simiae* by enzyme immunoassay, *Dev. Biol. Stand.* **45:**61.

Endo, M., Kamimura, T., Aoyama, Y., Hyashida, T., Kinjo, T., Ono, Y., Kotera, S., Suzuki, K., Tajima, Y., and Ando, K., 1959, Etude du virus B au Japon. I. Recherche des anticorps neutralisant le virus B chez les singes d'origine japonaise et les singes étrangers importés au Japon, *Japan J. Exp. Med.* **30:**227.

Endo, M., Kamimura, T., Kusano, N., Kawai, K., Aoyama, Y., Tajima, Y., Suzuki, K., and Kotera, S., 1960, Etude du virus B au Japon. II. Le premier isolement du virus B au Japon, *Japan J. Exp. Med.* **30:**385.

Engels, M., Darai, G., Gelderblom, H., and Ludwig, H., 1981, Properties of the goat herpesvirus, International Workshop on Herpesviruses, Bologna, Italy, July 27–31, 1981, p. 20.

España, C., 1973, *Herpesvirus simiae* infection in *Macaca radiata, Am. J. Phys. Anthropol.* **38:**447.

España, C., 1974, Viral epizootics in captive nonhuman primates, *Am. Lab. Anim. Sci.* **24:**167.

Falke, D., 1961, Isolation of two variants with different cytopathic properties from a strain of herpes B virus, *Virology* **14:**492.

Falke, D., 1964, Über die serologischen Beziehungen zwischen B- und Herpes Simplex Virus in der Komplementbindungsreaktion, *Z. Hyg. Infektionkr.* **150:**185.

Falke, D., and Richter, J.E., 1961, Mikrokinematographische Studien über die Enstehung von Riesenzellen durch Herpes B Virus in Zellkulturen. I. Mitt.: Vorgänge an den Zellgrenzen und Granulabewegungen. II. Mitt.: Morphologisches Verhalten und Bewegungen der Kerne, *Arch. Gesamte Virusforsch.* **11:**71; **11:**86.

Fierer, J., Bazeley, P., and Braude, A.I., 1973, Herpes B virus encephalomyelitis presenting as ophthalmic zoster, *Ann. Intern. Med.* **79:**225.

Fleckenstein, B., Bornkamm, G.W., and Ludwig, H., 1975, Repetitive sequences in complete and defective genomes of *Herpesvirus saimiri, J. Virol.* **15:**398.

Gary, W.G., and Palmer, E.L., 1977, Comparative complement fixation and serum neutralization antibody titers to herpes simplex virus type 1 and *Herpesvirus simiae* in *Macaca mulatta* and humans, *J. Clin. Microbiol.* **5:**465.

Gay, F.P., and Holden, M., 1933, The herpes encephalitis problem II, *J. Infect. Dis.* **53:**287.

Gelderblom, H., Ogura, H., and Bauer, H., 1974, On the occurrence of oncornavirus-like particles in HeLa cells, *Cytobiologie* **8:**339.

Gelderblom, H., Pauli, G., Schäuble, H., and Ludwig, H., 1980, Detection of a cross-reacting antigen between herpes simplex virus type 1 and bovine herpes mammillitis virus by immunoelectron microscopy, 7th European Congress on Electron Microscopy, Den Haag, *Electron Microscopy* **2:**482.

Graham, B.J., Ludwig, H., Bronson, D.L., Benyesh-Melnick, M., and Biswal, N., 1972, Physicochemical properties of the DNA of herpes viruses, *Biochim. Biophys. Acta* **259:**13.

Graham, F.L., Velduisen, G., and Wilkie, N.M., 1973, Infectious herpesvirus DNA, *Nature (London) New Biol.* **245:**265.

Gralla, E.J., Ciecura, S.J., and Delahunt, C.S., 1966, Extended B virus antibody determinants in a closed monkey colony, *Lab. Anim. Care* **16:**510.

Hartley, E.G., 1964, Naturally-occurring "B" virus infection in cynomolgus monkeys, *Vet. Rec.* **76:**555.

Hartley, E.G., 1966, "B" virus disease in monkey and man, *Br. Vet. J.* **122:**46.

Hayward, G.S., Frenkel, N., and Roizman, B., 1975a, Anatomy of herpes simplex virus DNA: Strain differences and heterogeneity in the location of restriction endonuclease cleavage sites, *Proc. Natl. Acad. Sci. U.S.A.* **72:**1768.

Hayward, G.S., Jacob, R.J., Wadsworth, S.C., and Roizman, B., 1975b, Anatomy of herpes simplex virus DNA: Evidence for four populations of molecules that differ in the relative orientations of their long and short components, *Proc. Natl. Acad. Sci. U.S.A.* **72**:4243.

Heine, I.W., Honess, R.W., Cassai, E., and Roizman, B., 1974, Proteins specified by herpes simplex virus. XII. The virion polypeptides of type 1 strains, *J. Virol.* **14**:640.

Herrmann, S., 1981, Differentiation of pseudorabies virus strains, International Workshop on Herpesviruses, Bologna, Italy, July 27–31, 1981, p. 21.

Holmes, A.W., Devine, J.A., Nowakowski, E., and Deinhardt, F., 1966, The epidemiology of a herpes virus infection of new world monkeys, *J. Immunol.* **90**:668.

Honess, R.W., and Watson, D.H., 1977, Unity and diversity in the herpesviruses, *J. Gen. Virol.* **37**:15.

Hsiung, G.D., 1969, Recent advances in the study of simian viruses, *Ann. N. Y. Acad. Sci.* **162**:483.

Hull, R.N., 1968, The simian viruses, in: *Virology Monographs Vol. 2*, (S. Gard, C. Hallauer, K.F. Meyer, eds.), Springer-Verlag, New York.

Hull, R.N., 1973, The simian herpesviruses, in: *The Herpesviruses* (A.S. Kaplan, ed.), pp. 389–426, Academic Press, New York, San Francisco, and London.

Hunt, R.D., and Meléndez, L.V., 1969, Herpes virus infections of non-human primates: A review, *Lab. Anim. Care* **19**:21.

Hurst, E.W., 1934, Studies on pseudorabies (infectious bulbar paralysis, mad itch). II. Routes of infection in the rabbit, with remarks on the relation of the virus to other viruses affecting the nervous system, *J. Exp. Med.* **59**:729.

Hurst, E.W., 1936, Studies on pseudorabies (infectious bulbar paralysis, mad itch). III. The disease in the rhesus monkey, *Macaca mulatta, J. Exp. Med.* **63**:449.

Hutt, R., Guajardo, J.E., and Kalter, S.S., 1981, Detection of antibodies to *Herpesvirus homonis* in nonhuman primates, *Lab. Anim. Sci.* **31**:184.

Kalter, S.S., and Heberling, R.L., 1971, Comparative virology of primates, *Bacteriol. Rev.* **35**:310.

Kaluza, G., and Pauli, G., 1980, The influence of intramolecular disulfide bonds on the structure and function of Semliki Forest virus membrane glycoproteins, *Virology* **102**:300.

Kaplan, A.S., (ed.), 1973, *The Herpesviruses*, Academic Press, New York, San Francisco, and London.

Katzin, D.S., Connor, J.D., Wilson, L.A., and Sexton, R.S., 1967, Experimental herpes simplex infection in the owl monkey, *Proc. Soc. Exp. Biol. Med.* **125**:391.

Keeble, S.A., 1960, B. virus infection in monkeys, *Ann. N. Y. Acad. Sci.* **85**:960.

Keeble, S.A., Christofinis, G.J., and Wood, W., 1958, Natural virus-B infection in rhesus monkeys, *J. Pathol. Bacteriol.* **76**:189.

Killington, R.A., Yeo, J., Honess, R.W., Watson, D.H., Duncan, B.E., Halliburton, I.W., and Mumford, J., 1977, Comparative analysis of the proteins and antigens of five herpesviruses, *J. Gen. Virol.* **37**:297.

Lämmli, U.K., 1970, Cleavage of structural proteins during the assembly of the head of bacteriophage T_4, *Nature (London)* **227**:680.

Lewis, J.G., Kucera, L.S., Eberle, R., and Courtney, R.J., 1982, Detection of Herpes Simplex Virus type 2 glycoproteins expressed in virus-transformed rat cells, *J. Virol.* **42**:275.

Ludwig, H., 1972a, Untersuchungen am genetischen Material von Herpesviren. I. Biophysikalisch-chemische Charakterisierung von Herpesvirus-Desoxyribonukleinsäuren, *Med. Microbiol. Immunol.* **157**:186.

Ludwig, H., 1972b, Untersuchungen am genetischen Material von Herpesviren. II. Genetische Verwandtschaft verschieder Herpesviren, *Med. Microbiol. Immunol.* **157**:212.

Ludwig, H., and Rott, R., 1975, Effect of 2-deoxy-D-glucose on herpesvirus-induced inhibition of cellular DNA synthesis, *J. Virol.* **16**:217.

Ludwig, H.O., Biswal, N., and Benyesh-Melnick, M., 1972a, Studies on the relatedness of herpesviruses through DNA–DNA hybridization, *Virology* **49**:95.

Ludwig, H., Haines, H.G., Biswal, N., and Benyesh-Melnick, M., 1972b, The characterization of varicella–zoster virus DNA, *J. Gen. Virol.* **14**:111.

Ludwig, H., Becht, H., and Rott, R., 1974, Inhibition of herpes virus-induced cell fusion by Concanavalin A, antisera, and 2-deoxy-D-glucose, *J. Virol.* **14:**307.

Ludwig, H., Pauli, G., Norrild, B., Vestergaard, B.F., and Daniel, M.D., 1978, Immunological characterization of a common antigen present in herpes simplex virus, bovine mammillitis virus and *Herpesvirus simiae* (B virus), in: *Oncogenesis and Herpesviruses III/1* (G. de Thé, W. Henle, and F. Rapp, eds.), p. 235, IARC Sci. Publ. No. 24, Lyon.

Ludwig, H., Heppner, B., and Herrmann, S., 1982, The genomes of different field isolates of Aujeszky's Disease, in: *Aujeszky's Disease,* (G. Wittmann and S.A. Hall, eds.), pp. 15–20, Martinus Nijhoff Publishers, The Hague, Boston, London.

MacLeod, D.R.E., Shimada, F.T., and Walcroft, M.J., 1960, Experimental immunization against B virus, *Ann. N. Y. Acad. Sci.* **85:**980.

Malherbe, H., and Harwin, R., 1963, The cytopathic effects of vervet monkey viruses, *S. Afr. Med. J.* **37:**407.

McClure, H.M., 1970, Disseminated herpesvirus infection in a rhesus monkey, Primate Zoonoses Surveillance Report No. 1, Jan.–Feb. 1970.

Melnick, J.L., and Banker, D.C., 1954, Isolation of B-virus (herpes group) from the central nervous system of a rhesus monkey, *J. Exp. Med.* **100:**181.

Morgan, C., Ellison, S.A., Rose, H.M., and Moore, D.H., 1954, Structure and development of viruses as observed in the electron microscope. I. Herpes simplex virus, *J. Exp. Med.* **100:**195.

Nii, S., Morgan, M., and Rose, H.M., 1968, Electron microscopy of herpes simplex virus. II. Sequence of development, *J. Virol.* **2:**517.

Norrild, B., Bjerrum, O.J., Ludwig, H., and Vestergaard, B.F., 1978a, Analysis of herpes simplex virus type 1 antigens exposed on the surface of infected tissue culture cells, *Virology* **87:**307.

Norrild, B., Ludwig, H., and Rott, R., 1978b, Identification of a common antigen of herpes simplex virus, bovine herpes mammillitis virus, and B virus, *J. Virol.* **26:**712.

Norrild, B., Shore, S.L., and Nahmias, A.J., 1979, Herpes simplex virus glycoproteins: Participation of individual herpes simplex virus type 1 glycoprotein antigens in immunocytolysis and their correlation with previously identified glycopolypeptides, *J. Virol.* **32:**741.

Orcutt, R.P., Pucak, G.J., Foster, H.L., Kilcourse, J.T., and Ferrell, T., 1976, Multiple testing for the detection of B virus antibody in specially handled rhesus monkeys after capture from virgin trapping ground, *Lab. Anim. Sci.* **26:**70.

Pauli, G., and Ludwig, H., 1977, Immunoprecipitation of herpes simplex virus type 1 antigen with different antisera and human cerebrospinal fluids, *Arch. Virol.* **53:**139.

Pauli, G., Ludwig, H., Norrild, B., Daniel, M.D., and Darai, G., 1981, Differentiation of *Herpesvirus simiae* and herpes simplex virus, Fifth International Congress of Virology, Strasbourg, France, August 2–7, 1981, p. 317.

Pille, E.R., 1960, Virus-B infection in monkeys, *Probl. Virol.* (*London*) **6:**542.

Plummer, G., 1964, Serological comparison of the herpes viruses, *Br. J. Exp. Pathol.* **45:**135.

Reissig, M., and Melnick, J.L., 1955, The cellular changes produced in tissue cultures by herpes B virus correlated with concurrent multiplication of the virus, *J. Exp. Med.* **101:**341.

Roizman, B., 1978, The herpesviruses, in: *The Molecular Biology of Animal Viruses*, Vol. 2 (D.P. Nayak, ed.), pp. 769–848, Marcel Dekker, New York and Basel.

Roizman, B., and Furlong, D., 1974, The replication of herpesviruses, in: *Comprehensive Virology*, Vol. 3 (H. Fraenkel-Conrat and R. R. Wagner, eds.), pp. 229–403 Plenum Press, New York.

Roizman, B., and Spear, P., 1973, Herpes viruses, in: *Ultrastructure of Animal Viruses and Bacteriophages: An Atlas* (A.J. Dalton and F. Haguenau, eds.), p. 83, Academic Press, New York and London.

Roth, A.M., and Purcell, Th.W., 1977, Ocular findings associated with encephalomyelitis caused by herpesvirus simiae, *Amer. J. Ophthalmol.* **84:**345.

Ruebner, B.H., Devereux, D., Rorvik, M., España, C. and Brown, J.F., 1975, Ultrastructure of *Herpesvirussimiae* (herpes B virus), *Exp.Mol.Pathol.***22:**317.

Ruyechan, W.T., Morse, L.S., Knipe, D.M., and Roizman, B., 1979, Molecular genetics of herpes simplex virus. II. Mapping of the major viral glycoproteins and of the genetic loci specifying the social behavior of infected cells, *J. Virol.* **29:**677.

Sabin, A.B., 1934a, Studies on the B virus. I. The immunological identity of a virus isolated from a human case of ascending myelitis associated with visceral necrosis, *Br. J. Exp. Pathol.* **15:**248.

Sabin, A.B., 1934b, Studies on the B virus. II. Properties of the virus and pathogenesis of the experimental disease in rabbits, *Br. J. Exp. Pathol.* **15:**268.

Sabin, A.B., 1934c, Studies on the B virus. III. The experimental disease in macacus rhesus monkeys, *Br. J. Exp. Pathol.* **15:**321.

Sabin, A.B., and Hurst, W.E., 1935, Studies on the B virus. IV. Histopathology of the experimental disease in rhesus monkeys and rabbits, *Br. J. Exp. Pathol.* **16:**133.

Sabin, A.B., and Wright, A.M., 1934, Acute ascending myelitis following a monkey bite, with the isolation of a virus capable of reproducing disease, *J. Exp. Med.* **59:**115.

Schneweis, K.E., 1962, Die serologischen Beziehungen der Typen 1 und 2 von *Herpesvirus hominis* zu *Herpesvirus simiae, Z. Immunitaetsforsch.* **124:**335.

Shah, K.V., and Morrison, J.A., 1969, Comparison of three rhesus groups for antibody patterns to some viruses: Absence of active simian virus 40 transmission in the free-ranging rhesus of Cayo Santiago, *J. Epidemiol.* **89:**308.

Sheldrick, P., Laithier, M., Lando, D., and Ryhiner, L., 1973, Infectious DNA from herpes simplex virus: Infectivity of double-stranded and single-stranded molecules, *Proc. Natl. Acad. Sci. U.S.A.* **70:**3621.

Spear, P.G., 1980, Composition and organization of herpesvirus virions and properties of some of the structural proteins, in: *Oncogenic Herpesviruses*, Vol. 1 (F. Rapp, ed.), pp. 53–84, CRC Press, Boca Raton, Florida.

Sterz, H., Ludwig, H., and Rott, R., 1973–1974, Immunologic and genetic relationship between herpes simplex virus and bovine herpes mammillitis virus, *Intervirology* **2:**1.

Stüber, D., and Bujard, H., 1977, Electron microscopy of DNA: Determination of absolute molecular weights and linear density, *Mol.Gen.Genet.* **154:**299.

Suzuki, M., Sasagawa, A., Inayoshi, T., Nakamura, F., and Honjo, S., 1981, Serological survey for SV5, measles and herpes simplex infections in newly imported cynomolgus monkeys, *J. Med. Sci. Biol.* **34:**69–80.

Ueda, Y., Tagaya, I., and Shiroki, K., 1968, Immunological relationship between herpes simplex virus and B virus, *Arch. Gesamte Virusforsch.* **24:**231.

Valerio, D.A., 1971, Colony management as applied to disease control with mention of some viral diseases, *Lab. Anim. Sci.* **21:**1011.

Van Hoosier, G., and Melnick, J.L., 1961, Neutralizing antibodies in human sera to *Herpesvirus simiae* (B virus), *Tex. Rep. Biol. Med.* **19:**376.

Vestergaard, B.F., and Norrild, B., 1978, Crossed immunoelectrophoretic analysis and viral neutralization activity of five monospecific antisera against five different herpes simplex virus glycoproteins, in: *Oncogenesis and Herpesviruses III/1* (G. de Thé, W. Henle, and F. Rapp, eds.), pp. 225–234, IARC Sci. Publ. No. 24, Lyon.

Vizoso, A.D., 1974, Heterogeneity in *Herpes simiae* (B virus) and some antigenic relationship in the herpes group, *Br. J. Exp. Pathol.* **55:**471.

Vizoso, A.D., 1975a, Recovery of *Herpes simiae* (B virus) from both primary and latent infections in rhesus monkeys, *Br. J. Exp. Pathol.* **56:**485.

Vizoso, A.D., 1975d, Latency of *Herpes simiae* (B virus) in rabbits, *Br. J. Exp. Pathol.* **56:**489.

Watson, D.H., Wildy, P., Harvey, B.A.M., and Shedden, W.H.I., 1967, Serological relationship among viruses of the herpes group, *J. Gen. Virol.* **1:**139.

Weller, T.H., and Pearson, R., 1972, Herpes-like simian viruses: Retrospective and prospective considerations, *J. Natl. Cancer Inst.* **49:**209.

Wildy, P., 1973, Herpes: History and classification, in: *The Herpesviruses* (A.S. Kaplan, ed.), pp. 1–25, Academic Press, New York.

Wildy, P., Russel, W.C., and Horne, R.W., 1960, The morphology of herpesvirus, *Virology* **12**:204.

Wood, W., and Shimada, F.T., 1954, Isolation of strains of virus B from tissue cultures of cynomolgus and rhesus kidney, *Can. J. Public Health* **45**:509.

Youngner, J.S., 1956, Virus adsorption and plaque formation in monolayer cultures of trypsin-dispersed monkey kidney, *J. Immunol.* **76**:288.

Zee, Y.C., and Talens, L., 1972, Electron microscopic studies on the development of infectious bovine rhinotracheitis virus in bovine kidney cells, *J. Gen. Virol.* **17**:333.

Index